GEOMETRY AND TRIGONOMETRY

Geometry, Plane Trigonometry, Natural Trigonometric Functions, Logarithmic Trigonometric Functions

MAVEN BOOKS

GEOMETRY AND TRIGONOMETRY

Geometry, Plane Trigonometry, Natural Trigonometric Functions,
Logarithmic Trigonometric Functions

INTERNATIONAL
CORRESPONDENCE SCHOOLS

MAVEN BOOKS

Chennai   Trichy   Tirunelveli   New Delhi

MAVEN BOOKS

An Imprint of **MJP Publishers**

ISBN 978-93-88191-21-0   **MAVEN Books**

All rights reserved   No. 44, Nallathambi Street,
Printed and bound in India   Triplicane, Chennai 600 005

MJP   595   © Publishers, 2018

Publisher :   **C. Janarthanan**

Publisher's Note

The legacy of a country is in its varied cultural heritage, historical literature, developments in the field of economy and science. The top nations in the world are competing in the field of science, economy and literature. This vast legacy has to be conserved and documented so that it can be bestowed to the future generation. The knowledge of this legacy is slowly getting perished in the present generation due to lack of documentation.

Keeping this in mind, the concern with retrospective acquiring of rare books has been accented recently by the burgeoning reprint industry. MAVEN Books is gratified to retrieve the rare collections with a view to bring back those books that were landmarks in their time.

In this effort, a series of rare books would be republished under the banner, "MAVEN Books". The books in the reprint series have been carefully selected for their contemporary usefulness as well as their historical importance within the intellectual. We reconstruct the book with slight enhancements made for better presentation, without affecting the contents of the original edition.

Most of the works selected for republishing covers a huge range of subjects, from history to anthropology. We believe this reprint edition will be a service to the numerous researchers and practitioners active in this fascinating field. We allow readers to experience the wonder of peering into a scholarly work of the highest order and seminal significance.

MAVEN Books

PREFACE

The volumes of the International Library of Technology are made up of Instruction Papers, or Sections, comprising the various courses of instruction for students of the International Correspondence Schools. The original manuscripts are prepared by persons thoroughly qualified both technically and by experience to write with authority, and in many cases they are regularly employed elsewhere in practical work as experts. The manuscripts are then carefully edited to make them suitable for correspondence instruction. The Instruction Papers are written clearly and in the simplest language possible, so as to make them readily understood by all students. Necessary technical expressions are clearly explained when introduced.

The great majority of our students wish to prepare themselves for advancement in their vocations or to qualify for more congenial occupations. Usually they are employed and able to devote only a few hours a day to study. Therefore every effort must be made to give them practical and accurate information in clear and concise form and to make this information include all of the essentials but none of the non-essentials. To make the text clear, illustrations are used freely. These illustrations are especially made by our own Illustrating Department in order to adapt them fully to the requirements of the text.

In the table of contents that immediately follows are given the titles of the Sections included in this volume, and under each title are listed the main topics discussed.

INTERNATIONAL TEXTBOOK COMPANY

B

CONTENTS

CONTENTS

TRIGONOMETRIC TABLES—(At Back of Book)

Natural Trigonometric Functions.
Logarithmic Trigonometric Functions.

LOGARITHMIC TABLES—(At Back of Book)

GEOMETRY
(PART 1)

Serial 778AEdition 1

PRELIMINARY DEFINITIONS

NOTE.—The study of Geometry is a process of systematic and orderly reasoning rather than a matter of memory. The student is advised to study the principles and propositions stated until he understands them thoroughly and sees their relation one to another, and, when a proposition is accompanied by an explanation in small type, to read over the explanation carefully one or more times, until he clearly understands the matter, following out the references to the figure when a figure is given. If he will do this he will find Geometry to be of great benefit and assistance to him in his subsequent studies. But he is not required to commit to memory the explanations or any part of the text except a few of the more important principles and propositions, such as those to which the Examination Questions relate.

1. Every material body possesses two general properties without regard to any other condition, namely: **form, or shape**, which is due to the relative positions of its parts; and **magnitude, or size**, which is due to the distance of its parts from one another.

The form and magnitude of a body can be described by the relative positions of *points*, *lines*, and *surfaces*.

2. A **point** has position without magnitude. A dot is commonly used to represent a point; but a dot, no matter how small, has length, breadth, and thickness, while a theoretical point has position only.

3. A **line** is the path of a point in motion; it has one dimension—length. Thus, if a point is moved from the position A, Fig. 1, to the position B, its path, or trace, is the line $A B$.

$$A \text{———————} B$$

FIG. 1

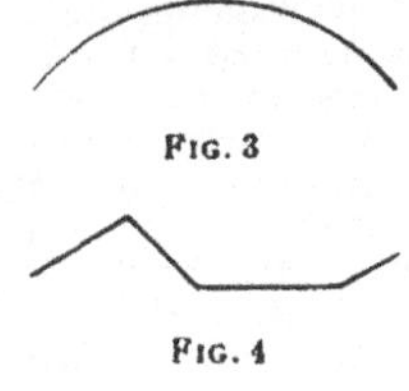

FIG. 2

4. **A straight line, or right line,** Fig. 2, is a line that does not change its direction.

5. The **distance** between two points is the length of the straight line joining them.

FIG. 3

6. A **curved line,** Fig. 3, is a line that changes its direction at every point.

FIG. 4

7. **A broken line,** Fig. 4, is a line that changes its direction at only certain points. It is made up wholly of different straight lines.

The word *line*, when not qualified by any other word, is understood to mean a straight line.

8. A **surface** is the path of a line when moved in a direction other than its length. Thus, if a line is moved from the position AB, Fig. 5, to the position CD, the line describes the surface $ABDC$.

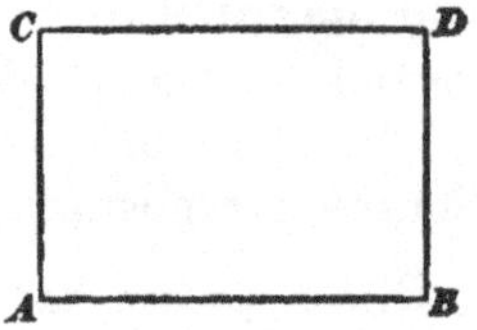

FIG. 5

9. **A flat surface, plane surface,** or simply a **plane,** is a surface such that a straight line between any two of its points lies wholly in the surface. If a straightedge is laid on a plane surface in any direction, every point of the straightedge will touch the surface.

10. A **figure** is any combination of points and lines. A figure that lies entirely in one plane is a **plane figure.**

In referring to a figure, a point is designated by a letter placed conveniently near it; thus, in Fig. 1, the left end of the line is referred to as the point A. The entire line is referred to as "the line AB," the letters A and B designating two points, usually the ends of the line. If a line is broken or curved, as many points are named as are considered necessary to designate the line.

11. **Geometry** is that branch of mathematics that treats of the construction and properties of figures.

12. To **produce** a line is to prolong it or to increase its length. A straight line can be prolonged or produced to any extent in either direction. Thus, in Fig. 6, the straight line AB is produced to the points C and D.

Fig. 6

13. To **bisect** any given magnitude is to divide it into two equal parts. Thus, the

Fig. 7

straight line AB, Fig. 7, is bisected at the point C if AC is equal to CB. When a given magnitude is bisected, each of the parts into which it is divided is one-half the given magnitude.

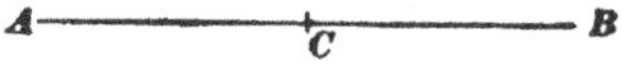

STRAIGHT-LINE FIGURES

ANGLES AND PERPENDICULARS

14. An **angle**, Fig. 8, is the opening between two straight lines that meet in a point. The two straight lines are the **sides**, and the point where the lines meet is the

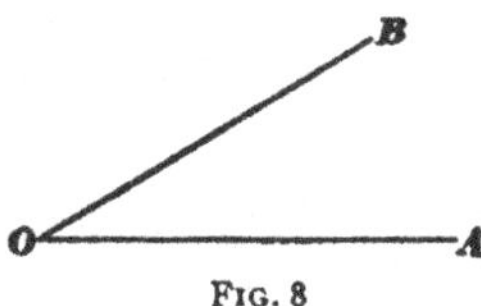

Fig. 8

vertex, of the angle. Thus, in Fig. 8, the straight lines OA and OB form an angle at the point O; the lines OA and OB are the sides of this angle, and the point O is its vertex.

An angle is usually referred to by naming a letter on each of its sides and a third letter at the vertex, the letter at the vertex being placed between the other two. Thus, the angle in Fig. 8 is called angle AOB or angle BOA.

An angle may also be designated by a letter placed between its sides near the vertex. Thus, the two angles XCY and YCZ, Fig. 9, may be referred to as the angles A and B, respectively.

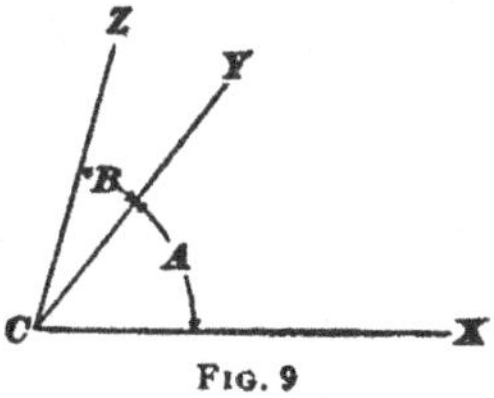

Fig. 9

An isolated angle, that is, an angle whose vertex is not the vertex of any other angle, may be designated by naming the letter at its vertex. For example, the angle in Fig. 8 may be called the angle O.

15. Two angles, as *A* and *B*, Fig. 9, having the same vertex and a common side *C Y*, are called **adjacent angles.**

16. Two angles are equal when one can be placed on the other so that they will coincide. Thus, in Fig. 10, the angles *A O B* and *A′ O′ B′* are equal, because *A′ O′ B′* can be superimposed on *A O B*, so that with *O′* upon *O* and *A′ O′* along *A O, B′ O′* will take the direction of *B O* and coincide with it.

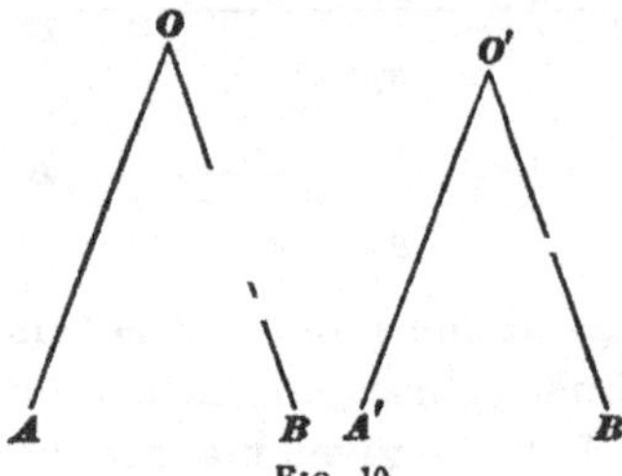

Fig. 10

17. Any angle may be thought of as being formed, or **generated,** by a line turning about the vertex as a pivot, from the position of one side to the position of the other. Thus, the angle *A O B*, Fig. 8, may be conceived as generated by a line turning about *O* from the position *O A* to the position *O B*. The size of the angle does not depend on the length of the sides, which are supposed to be of indefinite length, but on the opening between the sides; or, what is the same thing, on the amount of turning necessary to bring one side to the position of the other.

18. If a straight line, as *A B*, Fig. 11, meets another straight line, as *C D*, so as to make with it two equal adjacent angles, each of these angles is a **right angle,** and the first line is said to be perpendicular to the second. The point where the first line meets the second is called the foot of the perpendicular. It is evident that all right angles are equal.

19. A **horizontal line** is a line parallel to the horizon, or to the surface of still water.

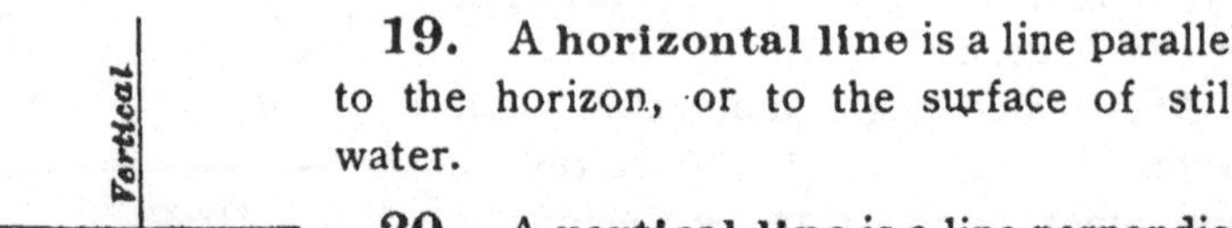

Fig. 12

20. A **vertical line** is a line perpendicular to a horizontal line, and having, therefore, the direction of a plumb-line. See Fig. 12.

21. An **oblique angle** is any angle that is not a right angle. An **acute angle** is an oblique angle that is less than a right angle. An **obtuse angle** is an oblique angle that is greater than a right angle. In Fig. 13, BOC and

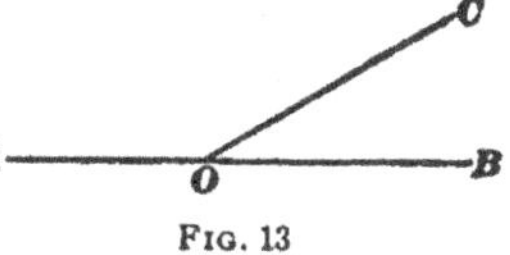

FIG. 13

AOC are oblique angles, BOC being an acute angle, and AOC an obtuse angle.

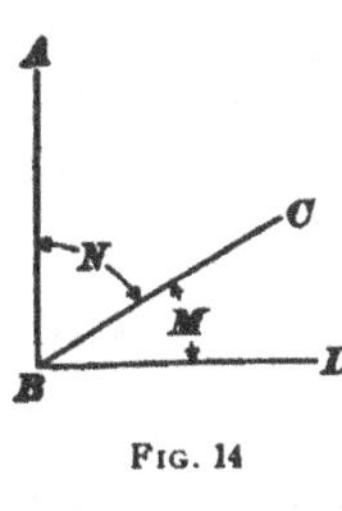

FIG. 14

22. Two angles are said to be **complementary** when their sum is equal to one right angle. Each of two complementary angles is called the **complement** of the other. Thus, in Fig. 14, in which AB is perpendicular to BD, the angles M and N are complementary, their sum being equal to the right angle ABD.

23. Two angles are said to be **supplementary** when their sum is equal to two right angles. Each of two supplementary angles is called the **supplement** of the other. In Fig. 15, AOD and DOB are supplementary angles, their sum being evidently equal to the sum of the two right angles POB and POA.

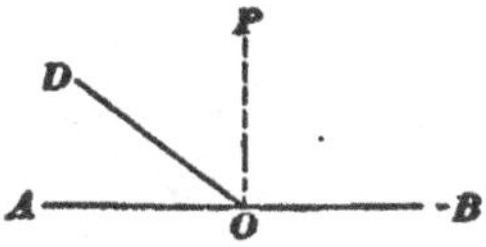

It will be seen from this illustration that two adjacent angles whose non-common sides are in the same straight line are always supplementary. Conversely, if two adjacent angles are supplementary, their non-common sides are in the same straight line.

FIG. 15

24. At a given point in a straight line, one perpendicular to the line and only one can be drawn.

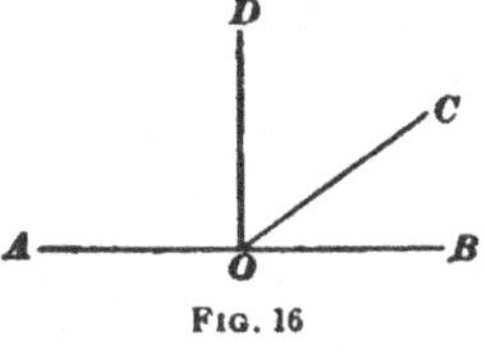

FIG. 16

Let O, Fig. 16, be the given point in the line OB. Suppose that with the point O fixed, the line OC starts from the position OB and revolves about O. In any position, as OC, it makes two angles with the line AB; one AOC, the other BOC. As OC revolves from the position OB to the position OA, the angle BOC will continually increase, and the angle *A O C* will continually decrease. There will therefore be one position, as *O D*, where the two angles are equal, and there can evidently be but one such position.

25. The sum of all the angles formed on the same side of a straight line about the same point in the line is equal to two right angles.

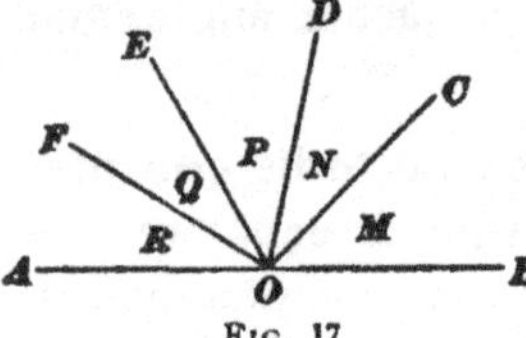

FIG. 17

In Fig. 17, the sum of the three angles *M*, *N*, and *P* is evidently equal to the angle *B O E*, and the sum of the angles *Q* and *R* is equal to the angle *E O A*. But, by Art. **23**, *B O E + E O A* is equal to two right angles. Hence, $M + N + P + Q + R =$ two right angles.

26. The sum of all the angles formed in the same plane about one point is equal to four right angles. Thus, in Fig. 18, $M + N + P + Q + R + S + T + U =$ four right angles.

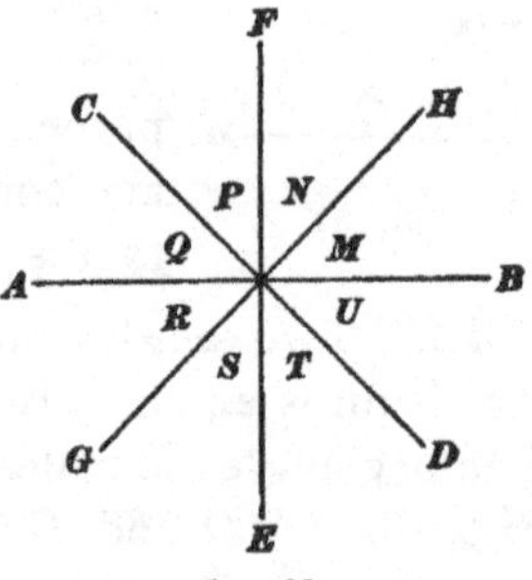

FIG. 18

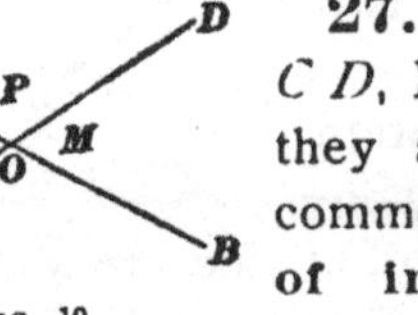

FIG. 19

27. When two lines, as *A B* and *C D*, Fig. 19, cut or cross each other, they are said to **intersect**. Their common point *O* is called their **point of intersection**, or simply their **intersection**.

28. Two intersecting straight lines determine four angles having a common vertex. Any one of these angles and the angle on the opposite side of both lines, as the angles *M* and *N*, Fig. 19, are called **vertical angles** with respect to each other. Vertical angles may also be defined as those having a common vertex and in which the sides of the one are the prolongations of the sides of the other.

Since *M* and *N* are each the supplement of *P*, they are equal to each other. Any angle is equal to its vertical angle.

29. If two straight lines intersect and one of the angles is a right angle, the other three angles are right angles, and the lines are perpendicular to each other.

30. Two oblique lines drawn from the same point in a perpendicular to a line, and cutting off on that line equal distances from the foot of the perpendicular, are equal.

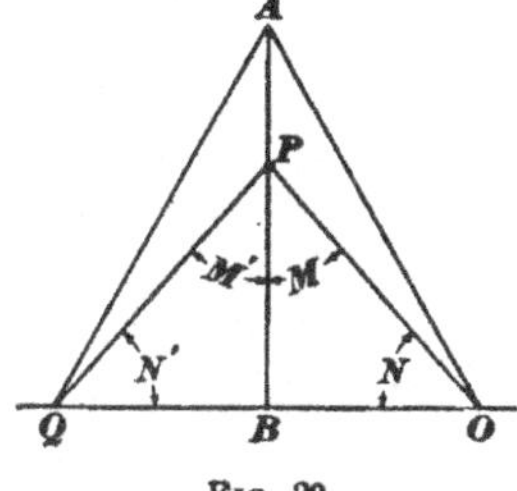

Fig. 20

Let PO and PQ, Fig. 20, be two oblique lines drawn from the point P in the perpendicular AB, and let BO and BQ be equal. Then, by turning the right side of the figure about AB, it will coincide with the left side; O will fall on Q, and PO will coincide with PQ. Hence, PO is equal to PQ.

31. Every point in the perpendicular at the middle point of a straight line is equally distant from the ends of the line. Thus, in Fig. 20, P, which may be any point in the perpendicular AB at the middle point B of OQ is equally distant from Q and O.

32. Two equal oblique lines drawn from the same point in the perpendicular to a straight line make equal angles with the straight line and with the perpendicular.

Since, when PBO, Fig. 20, is brought to coincide with PBQ, PO coincides with PQ and BO with BQ, the angle M = angle M', and angle N = angle N'.

33. A line that divides an angle into two equal angles is called the bisector of that angle. In Fig. 20, PB is the bisector of OPQ, since $M = M'$.

34. Two points, each of which is equally distant from the two extremities of a line, determine a perpendicular bisecting the line. Thus, in Fig. 20, A and P are two points equally distant from Q and O and determine the perpendicular bisecting the line OQ.

EXAMPLES FOR PRACTICE

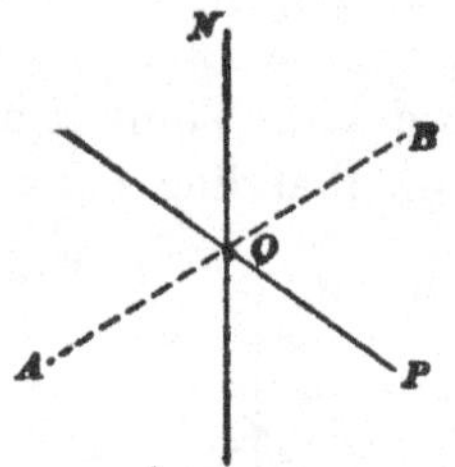

FIG. 21

1. Show that the bisectors of two vertical angles are in the same straight line.

SUGGESTION.—In Fig. 21, show that the sum of the angles on one side of the bisector *A B* of the angle *N O P* is equal to the sum of the angles on the other side.

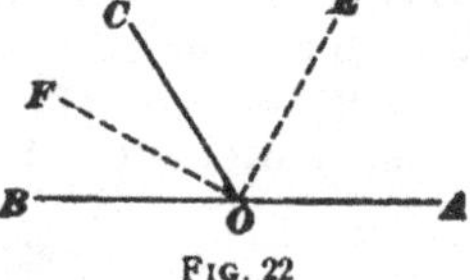

FIG. 22

2. Show that the bisectors of two supplementary adjacent angles are perpendicular to each other.

SUGGESTION.—In Fig. 22, show that the angle *E O F* is one-half of two right angles.

PARALLELS

35. **Parallel lines,** Fig. 23, are straight lines that lie in the same plane and never meet, however far they are produced. Any two parallel lines have the same direction and are everywhere equally distant from each other.

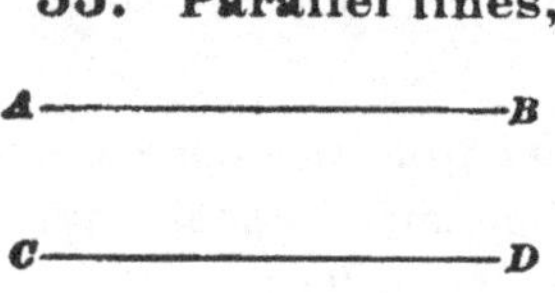

FIG. 23

36. When two parallel lines, as *P Q* and *R S*, Fig. 24, are cut by a third line, as *X Y*, the cutting line *X Y* is called a **secant line** or a **transversal.**

The eight angles thus formed are named as follows: The angles *a*, *A*, *d*, and *D* are **exterior angles.** The angles *b*, *B*, *c*, and *C* are **interior angles.** The pairs of angles *a* and *d* or *A* and *D* are **alternate-exterior angles.** The pairs of angles *b* and *c* or *B* and *C* are **alternate-interior angles.** The pairs of angles *a* and *c*, *A* and *C*, *b* and *d*, or *B* and *D* are **exterior-interior** or **corresponding angles.**

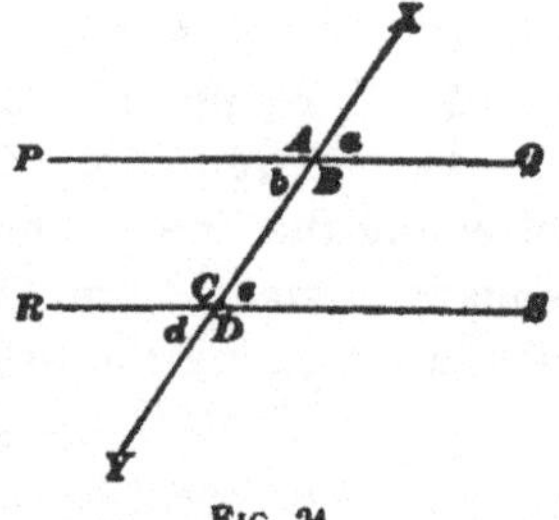

FIG. 24

37. When two parallel lines are cut by a transversal, the alternate-interior angles are equal.

Let CD and EF, Fig. 25, be the parallel lines and AB the transversal. The angles M and M' have their sides GD and HF parallel and AG and GH in the same line; hence, the turning in changing from the direction HF to the direction HG is equal to the turning in changing from the direction GD to the direction GA. That is, angle AGD, or M, is equal to the angle GHF, or M', Art. **17.** But angle M is equal to angle N, Art. **28**; therefore, angle N is equal to angle M'. In like manner, it can be shown that the angle DGH is equal to the angle GHE.

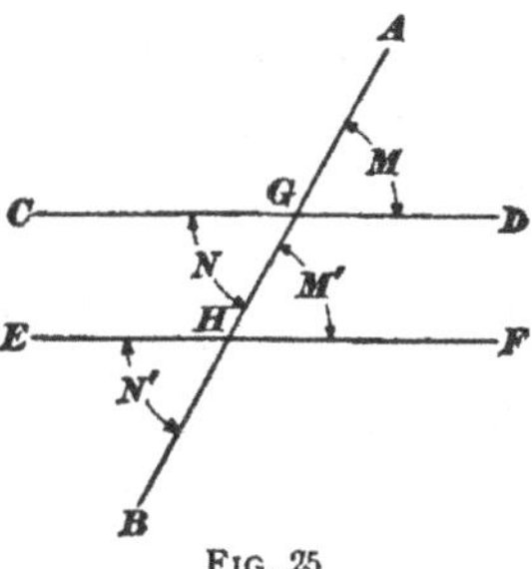

FIG. 25

38. It follows from the preceding article that the alternate-exterior angles are equal; also, the exterior-interior angles. Thus, in Fig. 24, we have $a = d$, $A = D$; $B = D$, $b = d$; $B = C$, $b = c$.

39. In Fig. 24, the angle a and the angle A are supplementary adjacent angles, and their sum is, therefore, equal to two right angles. From this, and from the principle stated in the preceding article, it follows that any angle in Fig. 24 marked by a capital letter and any angle marked by a small letter are together equal to two right angles.

The principles stated in this and in the two preceding articles may be summed up as follows: When two parallel lines are cut by an oblique transversal, the four obtuse angles are equal to one another; the four acute angles are equal to one another; and any of the obtuse angles is the supplement of any of the acute angles.

40. If a straight line is perpendicular to one of two parallel lines, it is perpendicular to the other also.

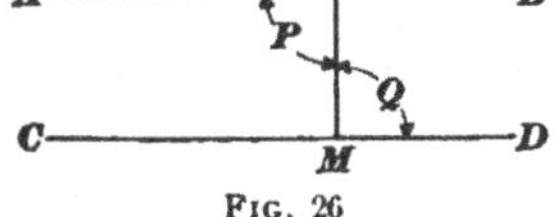

FIG. 26

In Fig. 26, AB and CD are parallel, and LM is drawn perpendicular to AB. Then, since the alternate-interior angles P and Q are equal, and since P is a right angle, Q must be a right angle also; that is, LM is perpendicular to CD.

I L T 36F—2

41. The distance between two parallel lines is the length intercepted by the two parallels on any line perpendicular to them. Thus, LM, Fig. 26, is the distance between AB and CD.

42. If two straight lines AB and CD, Fig. 27, are cut by a third straight line EF so that the exterior-interior angles M and N are equal, the two straight lines are parallel.

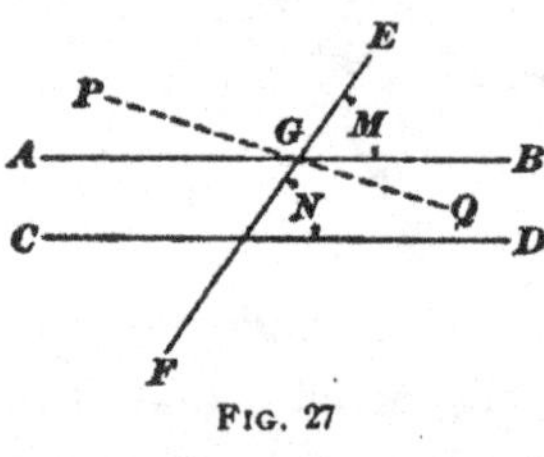

FIG. 27

If AB were not parallel to CD, we might draw through G a line PQ that was parallel to CD. But then the exterior-interior angles N and EGQ would be equal (Art. 38), which is obviously inconsistent with the supposition that N is equal to M.

43. If two lines, as AB and CD, Fig. 28, are parallel to a third line, as EF, they are parallel to each other.

Draw a transversal GH. Then, since AB is parallel to EF, the alternate-interior angles M and N are equal; and, since CD is parallel to EF, the alternate-interior angles P and N are equal. We have, therefore, $N = M$, $N = P$, and, consequently, $M = P$. As M and P are exterior-interior angles, it follows, from Art. 42, that AB and CD are parallel.

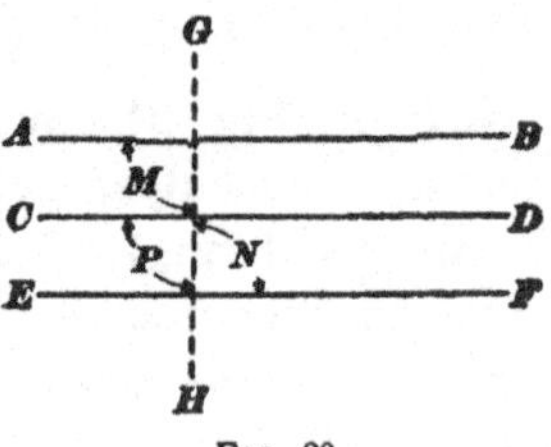

FIG. 28

44. Two angles whose sides are respectively parallel and lie in the same or opposite directions from their vertexes are equal.

In Fig. 29 (*a*), BA and ED are parallel and extend in the same direction; also, BC and EF are parallel and extend in the same direction from the vertexes. Let O be the point of intersection of the sides BC and ED produced. Then, since BC and EF are parallel, the exterior-interior angles E and M are equal; and, since BA and EG are parallel, the exterior-interior angles B and M are equal. Therefore, the angles B and E, being each equal to M, are equal to each other.

In Fig. 29 (*b*), BA and EF are parallel and extend in opposite directions; also, BC and ED are parallel and extend in opposite directions from the vertexes. Producing FE and DE, we have. by the preceding case, $B = D'EF'$. As DEF and $D'EF'$ are vertical angles, they are equal, and, therefore, B, which is equal to $D'E F'$, is also equal to $D E F$.

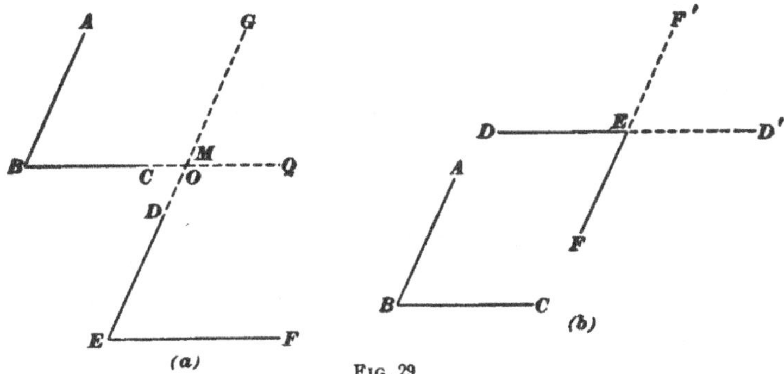

Fig. 29

45. If one side of an angle is parallel to one side of another angle, the two extending in the same direction from the vertexes, and if the other sides of the two angles are also parallel, but extend in opposite directions from the vertexes, the two angles are supplementary.

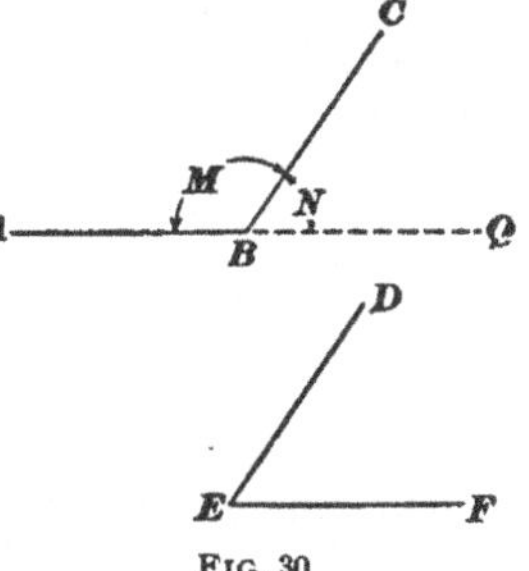

Fig. 30

In Fig. 30, $B C$ and $E D$ are parallel and extend in the same direction, while $B A$ and $E F$ are parallel and extend in opposite directions from the vertexes. Producing $A B$, we have, by Art. **44**, $N = E$. Now, $M + N =$ two right angles; therefore, $M + E =$ two right angles.

46. Two angles that have their sides perpendicular, each to each, are either equal or supplementary; they are equal if both are acute or both obtuse; and supplementary if one is acute and the other obtuse.

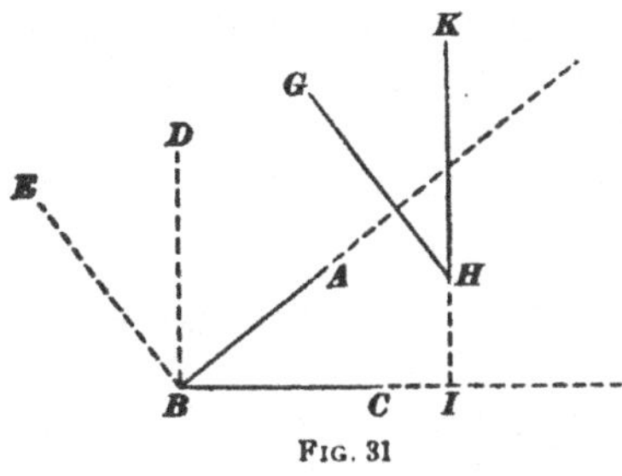

Fig. 31

In Fig. 31, let $G H$ be perpendicular to $A B$, and $K H$ perpendicular to $B C$. Draw $B D$ parallel to $K H$, and $B E$ parallel to $G H$. Then, by Art. **44**, $D B E$ is equal to $K H G$. Since $E B A$ and $D B C$ are right angles, by taking $D B A$

from each of them EBD is seen to be equal to ABC. Hence, the acute angle ABC is equal to the acute angle KHG. Also, when one angle is the acute angle ABC and the other is the obtuse angle GHI, since GHI is the supplement of KHG, it must be the supplement of ABC.

POLYGONS

DEFINITIONS

47. A **polygon** is a portion of a plane bounded by straight lines. The boundary lines are the **sides** of the polygon. The angles formed by the sides are the **angles** of the polygon. The vertexes of the angles of the polygon are the **vertexes** of the polygon. The broken line that bounds it, or the whole distance around it, is the **perimeter** of the polygon. Thus, $ABCDE$, Fig. 32, is a polygon; the sides of this polygon are AB, BC, CD, DE, and EA; its angles are ABC, BCD, CDE, DEA, and EAB; and its vertexes are A, B, C, D, and E.

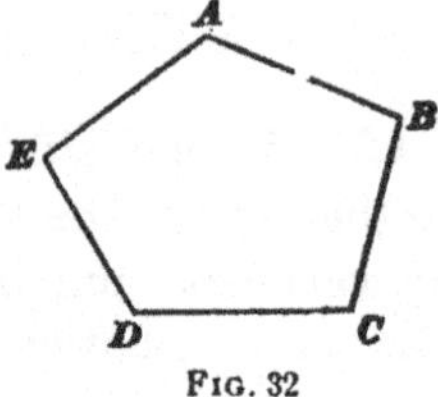
FIG. 32

48. The number of vertexes of a polygon is the same as the number of sides.

49. The least number of sides that a polygon can have is three, since two straight lines cannot enclose space.

50. Polygons are classified in various manners. One of these classifications is based on the number of sides. A polygon of three sides is a **triangle**; a polygon of four sides, a **quadrilateral**; a polygon of five sides, a **pentagon**; a polygon of six sides, a **hexagon**; a polygon of seven sides, a **heptagon**; a polygon of eight sides, an **octagon**; a polygon of nine sides, a **nonagon**; a polygon of ten sides, a **decagon**; a polygon of twelve sides, a **dodecagon**.

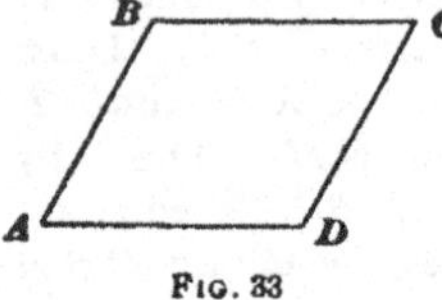
FIG. 33

51. An **equilateral polygon** is a polygon whose sides are all equal. Thus, in Fig. 33, $AB = BC = CD = DA$; hence, $ABCD$ is an equilateral polygon.

52. An **equiangular polygon** is
a polygon whose angles are all equal.
Thus, in Fig. 34, angle A = angle B
= angle D = angle C; hence, $ABDC$
is an equiangular polygon.

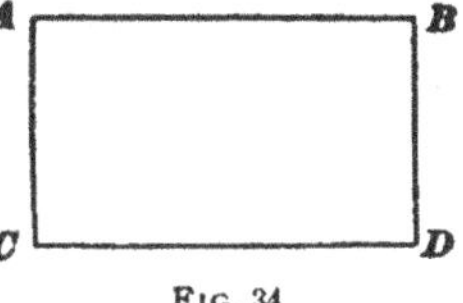
Fig. 34

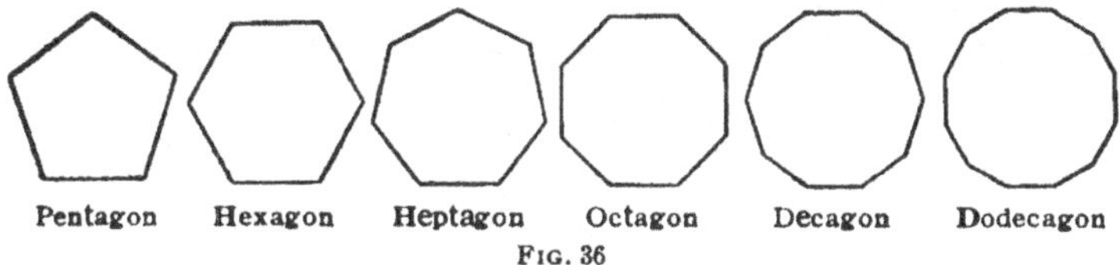

53. A **regular polygon** is a polygon
in which all the sides and all the angles are
equal. Thus, in Fig. 35, $AB = BD = DC$
$= CA$; and angle A = angle B = angle D
= angle C; hence, $ABDC$ is a regular
polygon. Some regular polygons are shown
in Fig. 36.

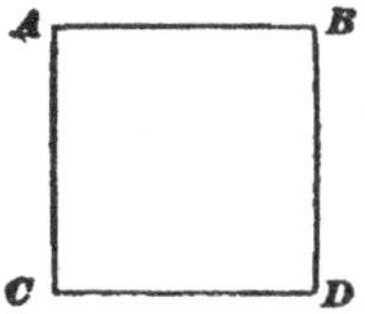

Fig. 35

Pentagon   Hexagon   Heptagon   Octagon   Decagon   Dodecagon

Fig. 36

54. A **reentrant angle** of a polygon is an angle whose sides if produced
through the vertex will enter the surface
bounded by the perimeter of the polygon. Thus, BCD, Fig. 37, is a reentrant angle.

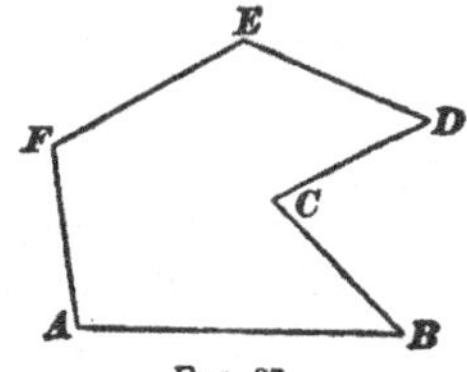
Fig. 37

TRIANGLES

55. Triangles are classified with regard to their sides
into *scalene*, *isosceles*, and *equilateral* triangles.

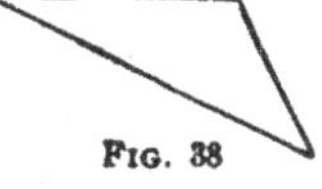
Fig. 38

56. A **scalene triangle**, Fig. 38, is a
triangle that has no two of its sides equal.

57. An **isosceles triangle**, Fig. 39, is a
triangle that has two of its sides equal.

Fig. 39

58. An **equilateral triangle,** Fig. 40, is a triangle that has its three sides equal. An equilateral triangle is a

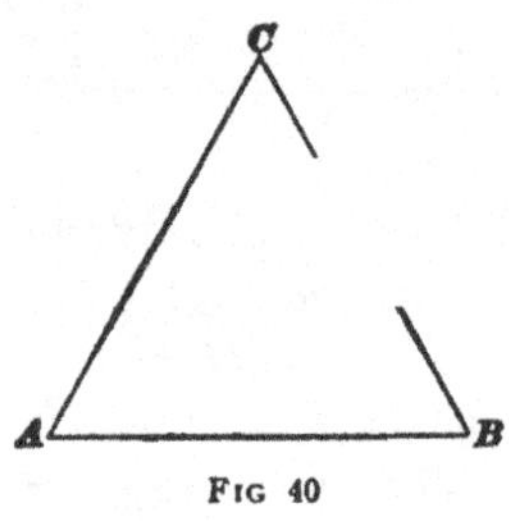

FIG 40

particular kind of isosceles triangle. Thus, the triangle ABC, Fig. 40, may be regarded as an isosceles triangle whose equal sides are AB and AC, as an isosceles triangle whose equal sides are BA and BC, or as an isosceles triangle whose equal sides are CA and CB. All the statements made with regard to isosceles triangles are, therefore, true of equilateral triangles.

59. Triangles are classified with regard to their angles into *right-angled*, *obtuse-angled*, and *acute-angled triangles*. See Fig. 41.

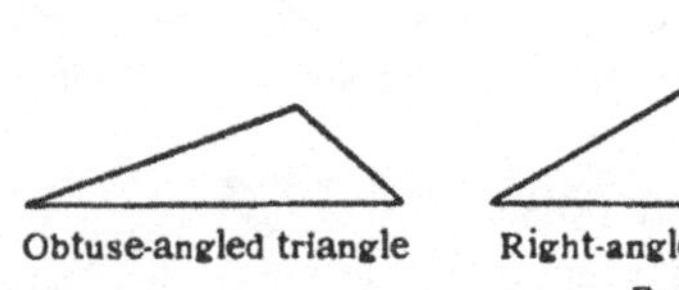
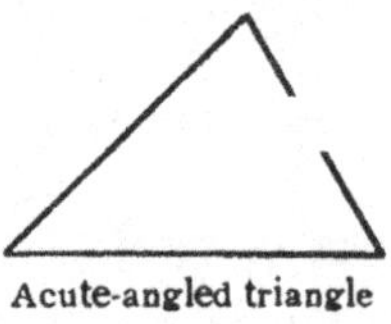

Obtuse-angled triangle　　　Right-angled triangle　　　Acute-angled triangle

FIG. 41

60. A **right-angled triangle,** or a **right triangle,** is a triangle having a right angle. The **hypotenuse** of a right triangle is the side opposite the right angle. The **legs** of a right triangle are the sides that include the right angle.

61. An **obtuse-angled triangle** is a triangle having an obtuse angle.

62. An **acute-angled triangle** is a triangle all the angles of which are acute.

63. An **oblique triangle** is a triangle that has no right angle. The class oblique triangles includes all obtuse-angled and acute-angled triangles.

64. An **equiangular triangle** is a triangle whose three angles are equal.

65. The **base** of a triangle is the side on which the triangle is supposed to stand. In a scalene triangle. any side may be considered as the base.   In an isosceles triangle, the unequal side is usually, though not necessarily, taken as the base.

The angle opposite the base of a triangle is sometimes called the **vertical angle** of the triangle.   In Figs. 42 and 43, $A\,C$ is the base.

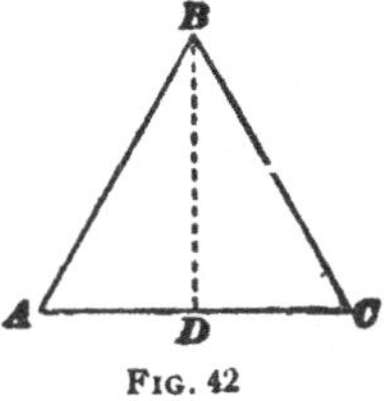

FIG. 42

66.   The **altitude** of a triangle is the length of a line drawn from the vertex of the angle opposite the base perpendicular to the base.   Thus, in Figs. 42 and 43, the length of $B\,D$ is the altitude.

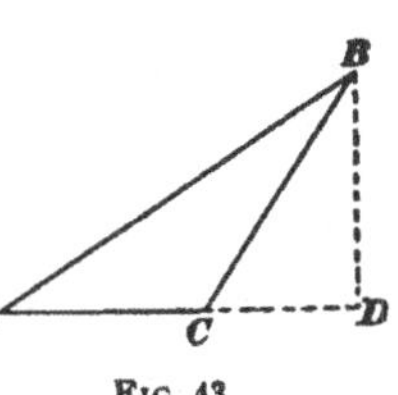

FIG. 43

67.   An **exterior angle** of a triangle is an angle formed by a side and the prolongation of another side.   Thus, in Figs. 43 and 44, the angle $B\,C\,D$, formed by the side $B\,C$ and the prolongation of the side $A\,C$, is an exterior angle of the triangle $A\,B\,C$. The angle $B\,C\,A$ is adjacent to the exterior angle $B\,C\,D$.   The angles A and B are **opposite-interior** angles to the angle $B\,C\,D$.

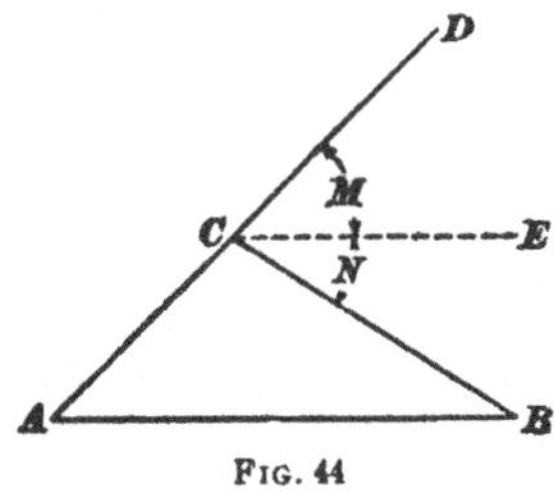

FIG. 44

68.   In any triangle, an exterior angle is equal to the sum of the opposite-interior angles.

Let $D\,C\,B$, Fig. 44, be an exterior angle of the triangle $A\,B\,C$. Draw $C\,E$ through C parallel to $A\,B$. Then, the angles M and A, being exterior-interior angles, are equal. Also, N and B, being alternate-interior angles, are equal.   Hence, angle M plus angle N, that is, the exterior angle $D\,C\,B$, is equal to angle A plus angle B, or the sum of the opposite-interior angles.

69.   The sum of the interior angles of a triangle is equal to two right angles.

In Fig. 44, the angles $B\,C\,D$ and $B\,C\,A$, being supplementary adjacent angles, are together equal to two right angles.   But, by the preceding article, the angle $B\,C\,D$ is equal to the sum of the angles A and B.   Hence, the sum of the three interior angles A, B, and $B\,C\,A$ is equal to two right angles.

70. The following important propositions are immediate consequences of that stated in Art. **69:**

1. If two angles of a triangle are known, or if their sum is known, the third angle can be found by subtracting their sum from two right angles.

2. If two angles of a triangle are equal, respectively, to two angles of another triangle, the third angle of the first-mentioned triangle is equal to the third angle of the other triangle.

3. A triangle can have but one right angle, or one obtuse angle.

4. In any right triangle, the two acute angles are complementary.

5. Each angle of an equiangular triangle is equal to one-third of two right angles, or two-thirds of one right angle.

6. From a point without a line, only one perpendicular to the line can be drawn.

EXAMPLES FOR PRACTICE

1. If one acute angle of a right triangle is one-third of a right angle, what is the value of the other?   Ans. Two-thirds of a right angle

2. If one angle of a triangle is one-half of a right angle, and another is five-sixths of a right angle, what is the third angle?

Ans. Two-thirds of a right angle

3. The exterior angle of a triangle is $1\frac{2}{3}$ right angles, and one of the opposite-interior angles is one-fourth of a right angle; what are the other angles of the triangle?

Ans. $\begin{cases} \text{Other opposite-interior angle} = \frac{23}{20} = 1.15 \text{ right angles} \\ \text{Angle adjacent to exterior angle} = \text{three-fifths of a right angle} \end{cases}$

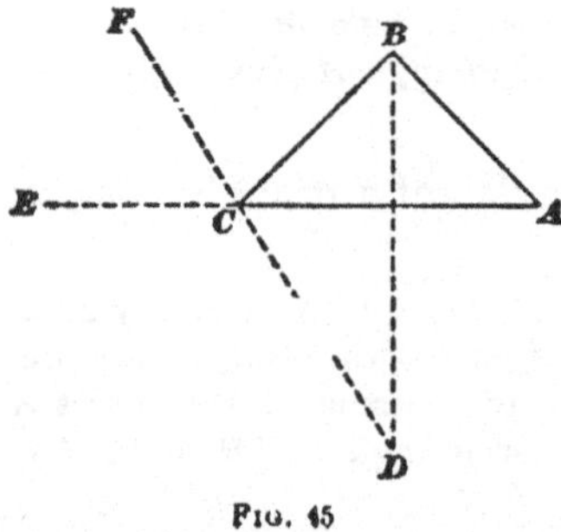

Fig. 45

4. Show that in the triangle ABC, Fig. 45, the bisector of the right angle ABC forms with the bisector of the exterior angle at C an angle that is equal to one-half of the angle A.

Suggestion.—Let BD be the bisector of ABC and FD the bisector of BCE. Then BCF is equal to CBD plus CDB, or CDB is equal to BCF minus CBD. Also, ECB is equal to CBA plus A, or A is equal to ECB minus CBA. Furthermore, ECB is equal to twice BCF and CBA is equal to twice CBD.

5. One angle of a triangle is one-half of a right angle: (*a*) What are the remaining two angles, if one is twice as large as the other? (*b*) What kind of triangle is this?

$$\text{Ans.}\begin{cases}(a)\ \text{One-half of a right angle and one right angle}\\(b)\ \text{An isosceles right triangle}\end{cases}$$

---

71. Two plane figures are **equal** when one can be placed on the other so that they will coincide in all their parts.

Thus, the triangles *A B C* and *A′ B′ C′*, Fig. 46, are equal, because if *A′ B′ C′* is imagined to be lifted off the paper, moved over and placed on *A B C*, the sides *A′ B′*, *B′ C′*, and *C′ A′* can be made to coincide

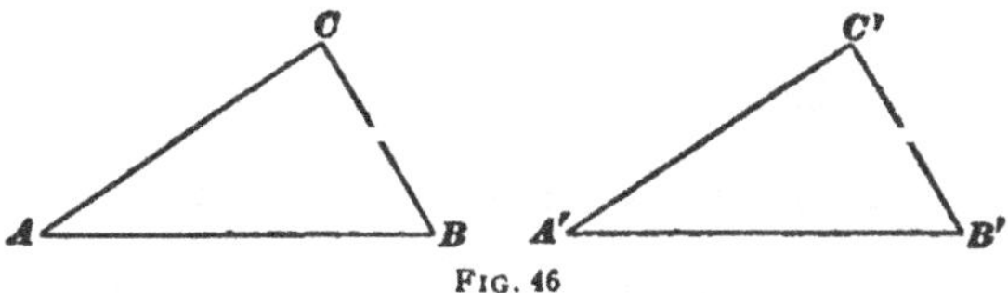

FIG. 46

with *A B*, *B C*, and *C A*, respectively, and the angles *A′*, *B′*, and *C′* to coincide with the angles *A*, *B*, and *C*. It is evident, from the figure, that if the vertexes of the two triangles coincide, the triangles will coincide throughout, and are, therefore, equal.

The polygons *A B C D E* and *A′ B′ C′ D′ E′*, Fig. 47, are equal, because *A′ B′ C′ D′ E′* can be imagined to be lifted, turned over, and placed on *A B C D E* so as to make the two polygons coincide in all their parts.

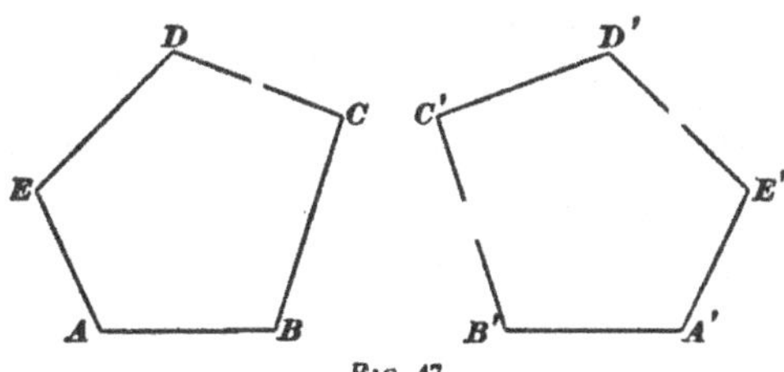

FIG. 47

72. Two triangles are equal when a side and two adjacent angles of one are equal to a side and two adjacent angles of the other.

Let *A′ B′*, Fig. 48, equal *A B*, the angle *A′* equal the angle *A*, and the angle *B′* equal the angle *B*. Now, if *A′ B′ C′* is placed on *A B C* so that *A′ B′* coincides with its equal *A B*, with *A′* on *A* and *B′* on *B*, *A′ C′* will take the direction *A C*; since the angle *A′* is equal to the angle *A*, and as *B′* is equal to *B*, *B′ C′* will take the direction *B C*.

Now, the point C will fall somewhere on the line $A\,C$, and also some-where on the line $B\,C$, and since two lines can intersect in only one point, C must fall at the intersection of $A\,C$ and $B\,C$, or at C. Hence, the vertexes of the triangles coincide and the triangles are equal.

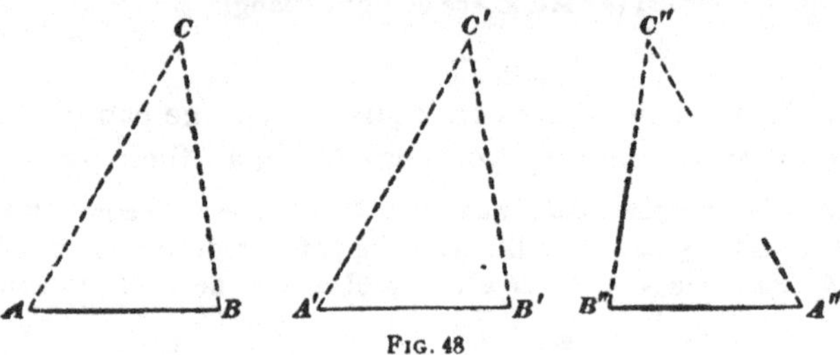

Fig. 48

The same reasoning applies to the triangles $A\,B\,C$ and $A''\,B''\,C''$, in which $A\,B = A''B''$, and $A = A''$, $B = B''$; but the triangle $A''\,B''\,C''$ must be imagined to be lifted and turned over before it can be placed on $A\,B\,C$.

73.   The following important principles are consequences of the preceding proposition:

1.   Two triangles are equal when one side and any two angles of one are equal, respectively, to one side and the two similarly situated angles of the other.

2.   Two right triangles are equal when one side and one acute angle of the one are equal, respectively, to one side and the similarly situated acute angle of the other.

74.   Two triangles are equal when two sides and the included angle of one are equal to two sides and the included angle of the other.

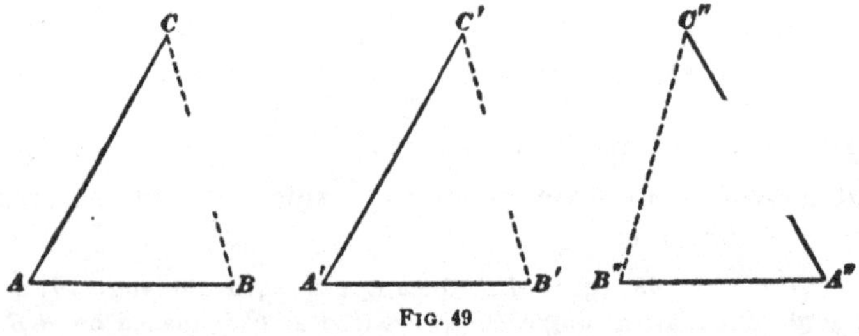

Fig. 49

In Fig. 49, $A\,B = A'\,B' = A''\,B''$, $A\,C = A'\,C' = A''\,C''$, and $A = A' = A''$. If $A'\,B'\,C'$ is placed on $A\,B\,C$, so that A' will coincide with A, and $A'\,B'$ with $A\,B$, the rest of the triangles will evidently coincide; for since $A' = A$, $A' C'$ will take the direction $A C$, and since $A' C' = A C$, C' will coincide with C. The same reasoning applies to $A'' B'' C''$, after the latter triangle has been turned over.

75. Two triangles are equal when the three sides of the one are equal, respectively, to the three sides of the other.

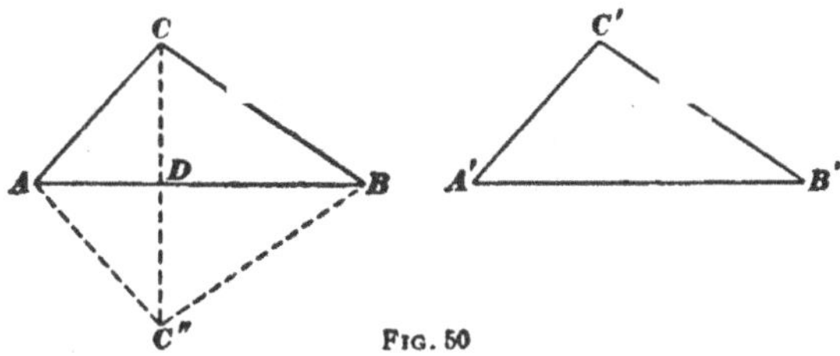

Fig. 50

In Fig. 50, let $A' B'$, $B' C'$, and $C' A'$ be equal, respectively, to $A B$, $B C$, and $C A$. Place $A' B' C'$ in the position $A B C''$, with its longest side $A' B'$ coinciding with $A B$, and C'' on the opposite side of $A B$ from C; then join C and C''. Now, $A C$ is equal to $A C''$, and $B C$ is equal to $B C''$; hence, A and B determine a perpendicular to $C C''$ at its mid-point (Art. **34**). Then, by Art. **32**, the angle $C A D$ is equal to the angle $C'' A D$, or to $C' A' B'$, and by Art. **74** the triangles $C A B$ and $C' A' B'$ are equal.

76. In an isosceles triangle, the angles opposite the equal sides are equal.

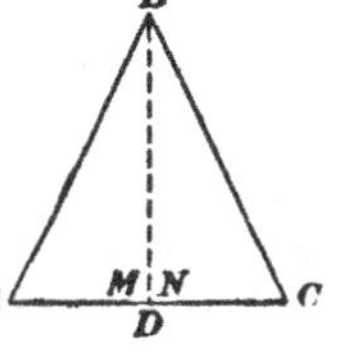

Fig. 51

Let $A B C$, Fig. 51, be an isosceles triangle in which $A B = B C$. Draw the bisector $B D$ of the angle B. Then, by Art. **74**, the triangles $A B D$ and $C B D$ are equal. Therefore, $A = C$.

77. The equality of the triangles $A B D$ and $C B D$, Fig. 51, gives $A D = D C$, and angle $M =$ angle $N =$ one right angle (since $M + N =$ two right angles). Hence,

1. The bisector of the vertical angle of an isosceles triangle bisects the base and is perpendicular to it.

2. Conversely, the perpendicular bisecting the base of an isosceles triangle passes through the vertex of the opposite angle and bisects that angle.

3. Also, the perpendicular drawn from the vertical angle of an isosceles triangle to the base, bisects both the base and the vertical angle.

78. If two angles of a triangle are equal, the sides opposite these two angles are equal, and the triangle is therefore isosceles.

In Fig. 51, let $A = C$. Draw BD perpendicular to AC. The right triangles BDC and BDA have the common side BD, and acute angle $A = C$. Therefore (Art. **73**), they are equal, and their hypotenuses BA and BC are equal.

79. It follows from Art. **76** that an equilateral triangle is, also equiangular, and from the preceding article that an equiangular triangle is also equilateral.

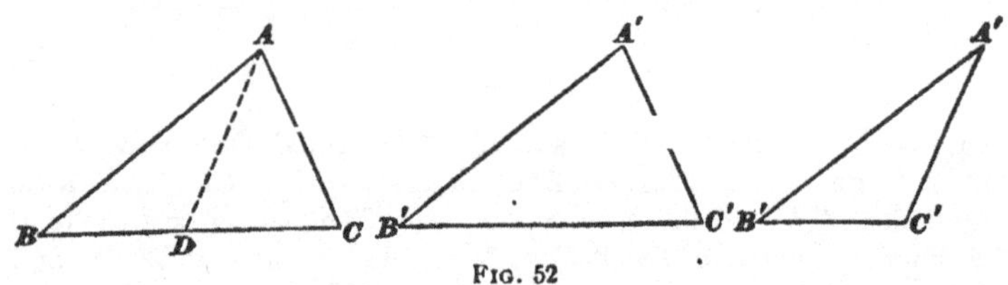

FIG. 52

80. If two sides of a triangle are equal, respectively, to two sides of another triangle, and the angle opposite one of these two sides in the first triangle is equal to the corresponding angle in the second triangle, the angles opposite the other two equal sides are either equal or supplementary.

In Fig. 52, let $A'C' = AC$, $A'B' = AB$, and the angle $B' = B$. Place $A'B'C'$ on ABC so that $A'B'$ coincides with AB. Then since $B' = B$, $B'C'$ will take the direction BC, and since $A'C'$ joins $B'C'$, C' must fall on BC, at either C or D. If C' falls at C, the triangles are equal and the angle $C' = C$; but if C' falls at D, ADB is the angle C', and ADB, the supplement of ADC, is the supplement of C, since, by Art. **76**, $ADC = C$.

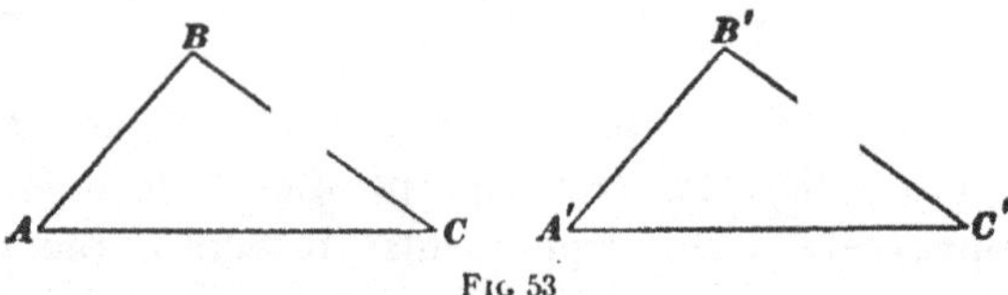

FIG. 53

81. If two triangles have two sides of the one equal to two sides of the other, and the angles opposite one pair of the equal sides are right angles or equal obtuse angles, the triangles are equal.

Since a triangle can have but one right or one obtuse angle, when the angles B and B', Fig. 53, are obtuse, the angle C' cannot be the supplement of C, hence C' must equal C.

82. Of two sides of a triangle, that is greater which is opposite the greater angle.

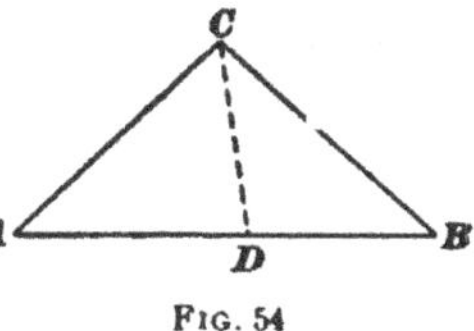

FIG. 54

In the triangle ABC, Fig. 54, let the angle C be greater than the angle B. Draw CD, making with CB an angle BCD equal to the angle B. Then BCD is an isosceles triangle, and $CD = DB$. Therefore, $AD + DB$, or AB, is the same as $AD + DC$, which is evidently greater than AC.

83. Of two angles of a triangle, that is greater which is opposite the greater side.

Let A and B be two angles of a triangle, a the side opposite A, and b the side opposite B. Suppose that a is greater than b. If A were equal to B, the triangle would be isosceles, and $a = b$. If B were greater than A, then by the preceding article, b would be greater than a. Therefore, since B cannot be equal to or greater than A, it must be less, or A must be greater than B.

84. If from a point O, Fig. 55, without a line AB, a perpendicular OP to the line is drawn, and also two oblique lines OL and OL', the oblique line OL, whose foot L is farther from the foot P of the perpendicular, is the greater of the two oblique lines.

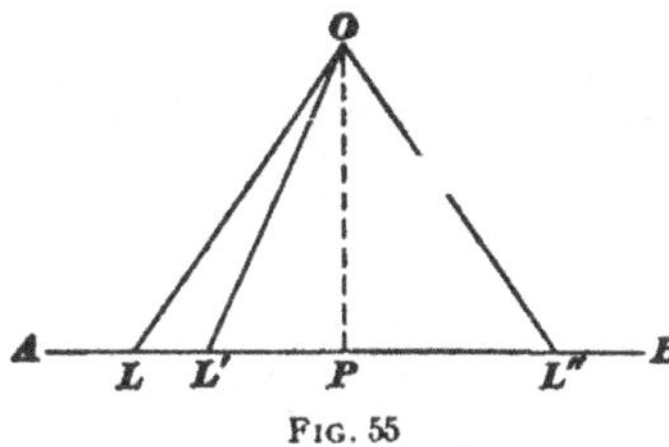

FIG. 55

Suppose the two oblique lines OL and OL' to be on the same side of the perpendicular. Since $OL'P$ is a right triangle and OPL' the right angle, the angle $OL'P$ is acute; also, the angle $OL'L$ is obtuse, since it is the supplement of $OL'P$. As the triangle OLL' can have but one obtuse angle, $OL'L$ is greater than OLP, and, therefore (Art. **82**), OL is greater than OL'. If OL lies on the opposite side of the perpendicular from OL', as in the position OL'', and if $PL'' = PL$, which is greater than PL', then, by Art. **30**, $OL'' = OL$, which is greater than OL'.

85. If the hypotenuse, as AB, Fig. 56, and one leg, as BC, of a right triangle are equal, respectively, to the

hypotenuse and one leg of another right triangle, as $A'B'C'$,
the two triangles are equal.

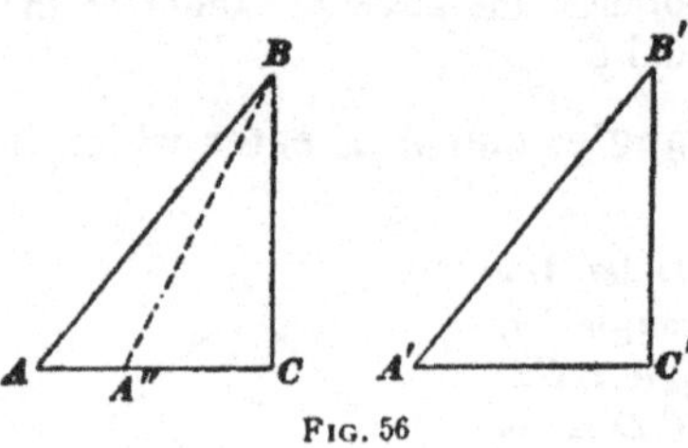

FIG. 56

Place $A'B'C'$ on ABC, so that $B'C'$ will coincide with its equal BC. Since $B'C'A'$ is a right angle, $C'A'$ will take the direction CA; and, since $B'A' = BA$, A' must fall on A; for if it fell to the right of A, as at A'', the hypotenuse BA'', or $B'A'$, would be less than BA (Art. 84); and, if A' fell on the left of A, the hypotenuse $B'A'$ would be greater than BA.

EXAMPLES FOR PRACTICE

1. Show that, if two intersecting lines, as AB and DC, Fig. 57, bisect each other, the lines AC and DB are parallel.

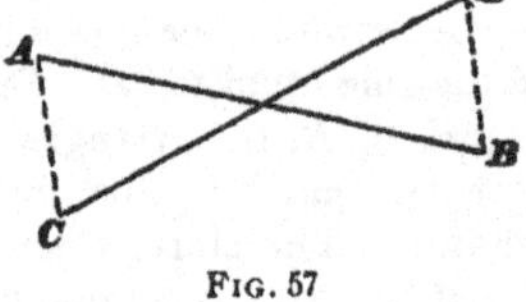

FIG. 57

2. If the value of the unequal or vertical angle of an isosceles triangle is two-fifths of a right angle, what is the value of each of the base angles?   Ans. Four-fifths of a right angle

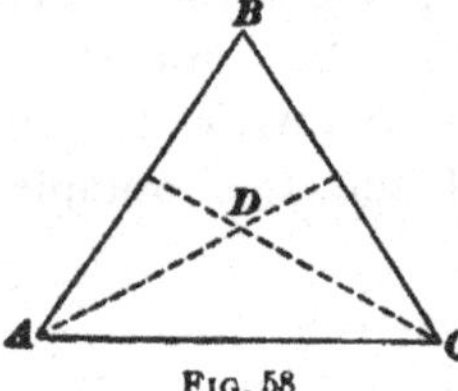

FIG. 58

3. Show that the bisectors of the base angles of an isosceles triangle form with the base an isosceles triangle; or that, ADC, Fig. 58, is an isosceles triangle.

4. Show that the length of the inaccessible line AB, Fig. 59, can be found by measuring AO and BO, then making $OD = OB$ and $OC = OA$, and finally measuring CD.

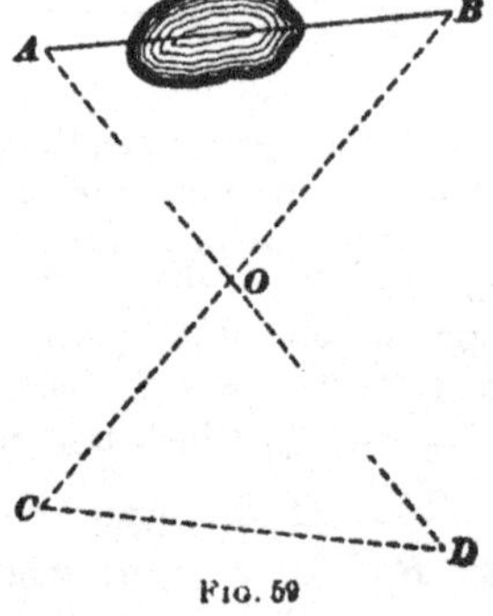

FIG. 59

QUADRILATERALS

86.   There are three kinds of quadrilaterals: the *parallelogram*, the *trapezoid*, and the *trapezium*.

87.   A **parallelogram** is a quadrilateral whose opposite sides are parallel.   There are four kinds of parallelograms: the *rectangle*, the *square*, the *rhomboid*, and the *rhombus*.

~ **88.**   A **rectangle**, Fig. 60, is a parallelogram whose angles are all right angles.

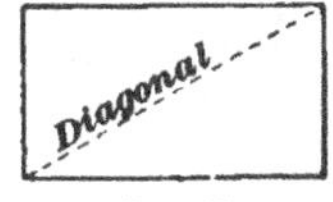

Fig. 60

Fig. 61

89.   A **square**, Fig. 61, is a rectangle whose sides are equal.

90.   A **rhomboid**. Fig. 62, is a quadrilateral whose opposite sides are parallel, and whose angles are not right angles.

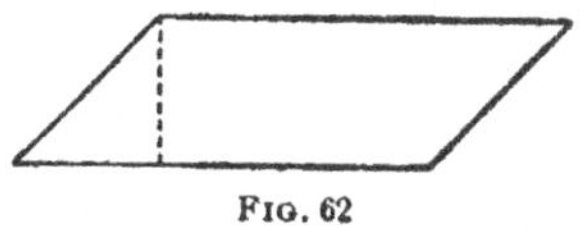
Fig. 62

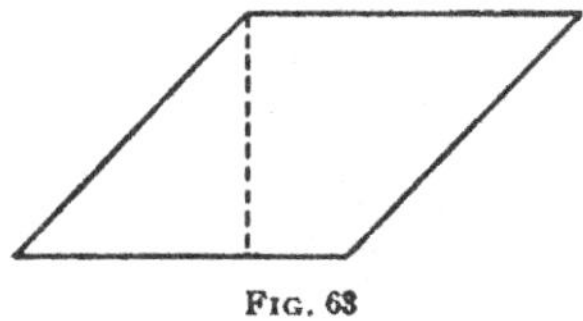
Fig. 63

91.   A **rhombus**, Fig. 63, is a rhomboid having equal sides.

92.   A **trapezoid**, Fig. 64, is a quadrilateral that has only two of its sides parallel.

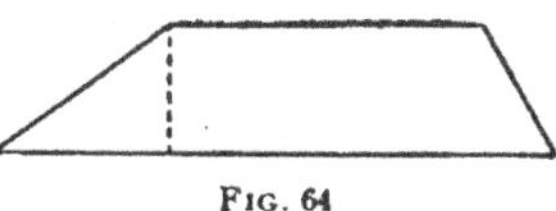
Fig. 64

93.   A **trapezium**, Fig. 65, is a quadrilateral having no two sides parallel.

94.   The **altitude** of a parallelogram, or of a trapezoid, is the length of the perpendicular distance between the

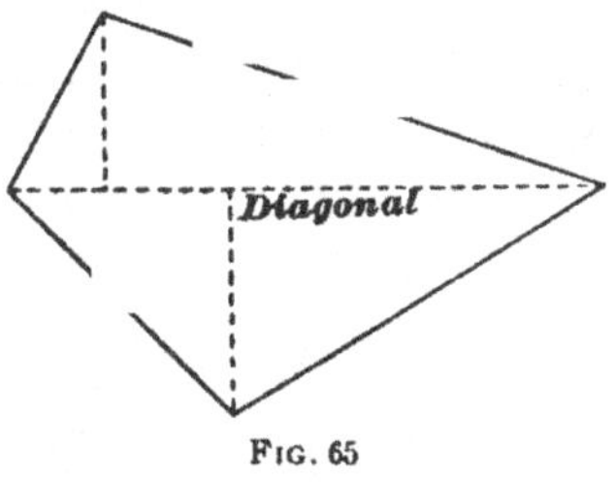

Fig. 65

parallel sides.   See dotted line in Figs. 62, 63, and 64.

95. A **diagonal** of a quadrilateral is a straight line drawn from the vertex of any angle of the quadrilateral to the vertex of the angle opposite. A diagonal divides a quadrilateral into two triangles. See Figs. 60 and 65.

96. In a parallelogram, as $ABCD$, Fig. 66, the opposite sides and opposite angles are equal; that is, $AB = DC, AD = BC$, angle A = angle C, angle B = angle D.

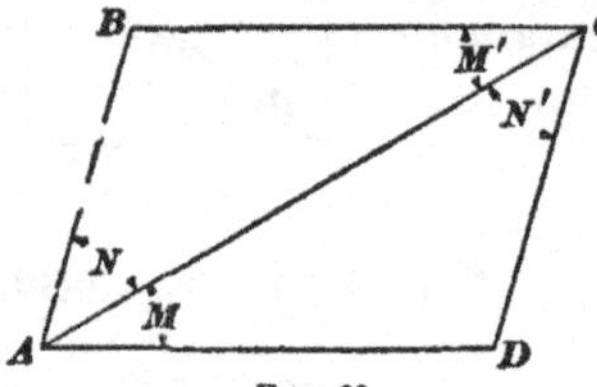

Fig. 66

Draw the diagonal AC. Then, angle M = angle M', and $N = N'$ (Art. **37**). The triangles ADC and ABC, having the common side AC and the adjacent angles M and N' equal, respectively, to M' and N, are equal (Art. **72**). Therefore, $AD = BC$, $AB = DC$, and angle B = angle D. Also, since $M = M'$ and $N = N'$, it follows that $M + N$, or BAD, is equal to $M' + N'$, or BCD.

97. The diagonal of a parallelogram divides the parallelogram into two equal triangles.

98. Parallel lines intercepted between parallel lines are equal. Thus, if the parallels AB and CD, Fig. 67, are cut by the parallels EF, GH, IJ, KL, we have, from Art. **96,** $MN = OP = QR = ST$.

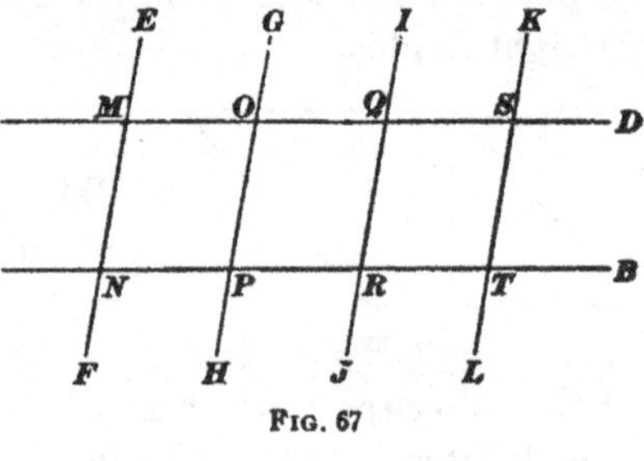

Fig. 67

99. The diagonals of a parallelogram, as AC and BD, Fig. 68, bisect each other; that is, denoting by O the point of intersection of the diagonals, $OA = OC$ and $OB = OD$.

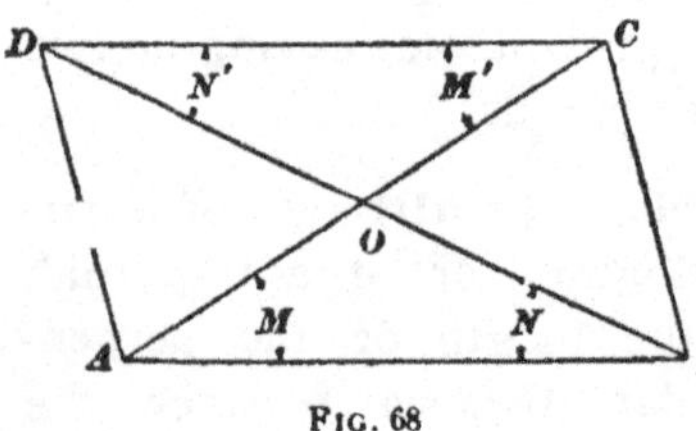

Fig. 68

In the triangles AOB and DOC, $AB = DC$ (Art. **96**), $M = M'$ and $N = N'$ (Art. **37**). Therefore, the triangles are equal (Art. **72**) and $OA = OC$. $OB = OD$.

EXAMPLES FOR PRACTICE

1. Show that if the diagonals of a quadrilateral bisect each other the figure is a parallelogram.

SUGGESTION.—In Fig. 68, assume that $OA = OC$, $OB = OD$. Then show that triangle BOC = triangle AOD, and triangle AOB = triangle DOC.

2. Show that the diagonals of a rectangle are equal.

SUGGESTION.—Show that in any rectangle $ABCD$ the triangle ABC = triangle ABD.

3. Show that if the opposite sides of a quadrilateral are equal, the figure is a parallelogram.

SUGGESTION.—Draw the diagonal. Then, by Art. 75, the triangles formed are equal.

4. Show that if two sides of a quadrilateral are equal and parallel, the figure is a parallelogram.

5. Show that if one angle of a parallelogram is a right angle, the parallelogram is a rectangle.

ADDITIONAL PROPERTIES OF TRIANGLES

100. The bisectors of the three angles of a triangle meet in a point.

In the triangle ABC, Fig. 69, draw the bisectors of the angles A and B and let them meet at O. Join C and O. and draw the perpendiculars from O to the sides of the triangle. Then, in the right triangles BOF and BOE, BO is common and the angle OBF = angle OBE. Hence, by Art. 73, these triangles are equal. Therefore, $OF = OE$. In a similar manner it can be shown that $OD = OE$. Therefore, $OD = OF$. The right triangles OFC and ODC, having $OD = OF$ and OC common, are equal (Art. 85). Hence, angle OCF = angle OCD; that is, OC, which meets the bisectors AO and BO in O, is the bisector of the angle C.

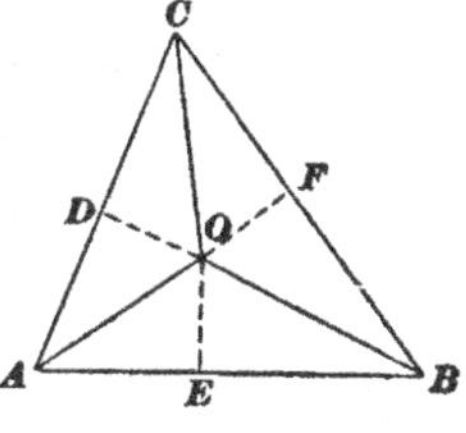

FIG. 69

101. Any point in the bisector of an angle is equally distant from the sides of the angle. For it has just been shown that, in Fig. 69, $OF = OE$.

102. The perpendiculars erected at the middle points of the three sides of a triangle meet in a point equally distant from the vertexes of the triangle.

1 L T 36F—3

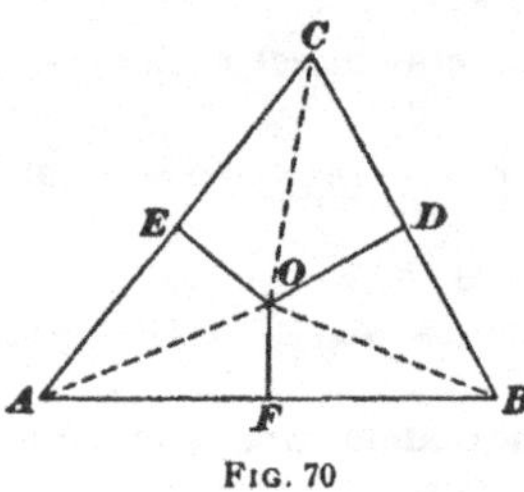

FIG. 70

In Fig. 70, draw the perpendiculars to CB and AC at their mid-points D and E, and let O be the point in which these perpendiculars meet. Now, O, being in OD, is equally distant from C and B (Art. **31**), that is, $OB = OC$: and being in OE, is equally distant from A and C; that is, $OA = OC$. From these two equalities it follows that $OB = OA$. Therefore, the perpendicular to AB at its middle point F passes through O. (Art. **77**).

103. If several parallel lines intercept equal distances on one transversal, they intercept equal distances on any other transversal.

In Fig. 71, let the parallels AE, BG, CI, DK intercept the equal distances AB, BC, and CD on the transversal PQ, and let RS be any other transversal. Draw EF, GH, IJ, parallel to AB. Then, by Art. **98**, $EF = AB$, $GH = BC$, $IJ = CD$. Hence, $EF = GH = IJ$. In the triangles EFG, GHI, and IJK, angle E = angle G = angle I (Art. **38**), and angle F = angle H = angle J (Art. **44**).

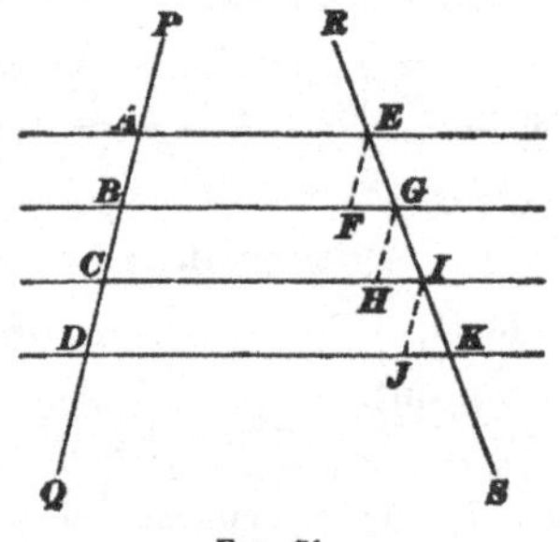

FIG. 71

Hence, by Art. **72**, these triangles are equal, and, therefore, $EG = GI = IK$.

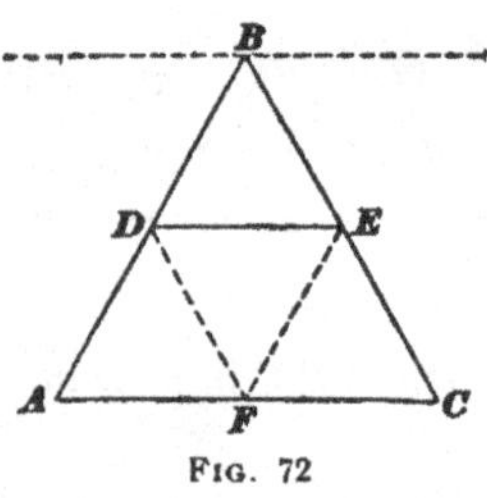

FIG. 72

104. A line parallel to one of the sides of a triangle and bisecting one of the other sides, bisects the third side also.

In Fig. 72, let DE bisect AB and be parallel to AC. Draw a line through B parallel to AC. Then since the three parallels intercept equal parts on AB, they intercept equal parts on BC; that is, $BE = EC$.

105. A line joining the middle points of two sides of a triangle is parallel to the third side and equal to one-half of that third side.

In Fig. 72, let DE join D and E, the middle points of AB and BC. The first part of this proposition follows at once from the preceding article. Let F be the middle point of AC, and draw FE. This line is parallel to AB, and, therefore, $ADEF$ is a parallelogram. Consequently (Art. 96), $DE = AF = \frac{1}{2}AC$.

106. The lines joining the middle points of the three sides of a triangle divide it into four equal triangles.

The diagonal DF, Fig. 72, divides the parallelogram $ADEF$ into two equal triangles AFD and DFE. Likewise, the diagonal EF divides $DECF$ into two equal triangles DFE and EFC; and the diagonal DE divides the parallelogram $BDFE$ into the two equal triangles DFE and BDE. Hence, triangle AFD = triangle DFE = triangle EFC = triangle BDE.

107. Any of the parallelograms $ADEF$, $FCED$, $DBEF$, Fig. 72, is equal to one-half the given triangle, since it contains two of the four equal triangles into which the given triangle is divided.

108. A line, as EF, Fig. 73, parallel to the bases AB and DC of a trapezoid and passing through the middle point E of one of the non-parallel sides, passes through the middle point of the other non-parallel side and is equal to one-half the sum of the parallel sides or bases.

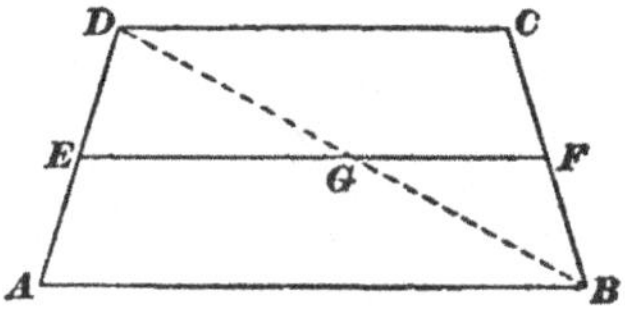

Since the parallels $AB, EF,$ and DC intercept equal parts on AD, they intercept equal parts on BC (Art. **103**); that is, $BF = FC$.

Draw BD, meeting EF in G. Then, by Art. **105**, in the triangle DCB, FG is one-half CD. Also, in the triangle ADB, GE is one-half BA. Hence, $FG + GE$, or FE, $= \frac{1}{2}(CD + BA)$.

FIG. 73

109. The medians of a triangle are the lines drawn from the vertexes to the middle points of the opposite sides.

110. The medians of a triangle meet in a point whose distance from any vertex is two-thirds the length of the median from that vertex.

In Fig. 74, AD, BE, CF are the median lines of the triangle ABC; they meet at O, and $AO = \frac{2}{3}AD$, $BO = \frac{2}{3}BE$ and $CO = \frac{2}{3}CF$.

Let AD and CF meet at O. Join I and H, the mid-points of CO and

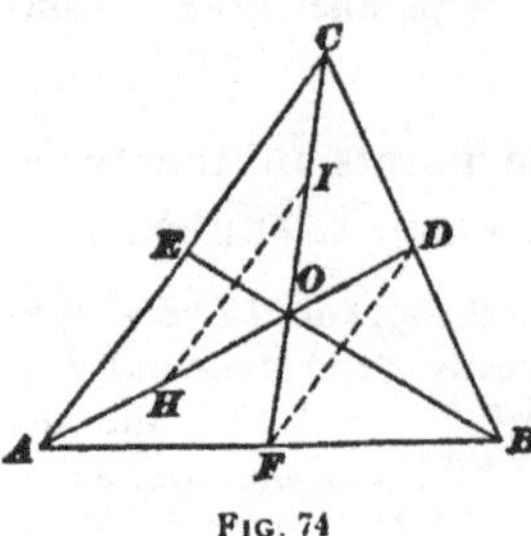

FIG. 74

AO, respectively; also join D and F. Then, in the triangle AOC, IH is parallel to AC and equal to one-half AC (Art. **105**). Also, in triangle ABC, DF is parallel to AC and equal to one-half AC. Hence, IH and DF are equal and parallel. It follows that the triangles DOF and HOI are equal, and that, therefore, $HO = OD$. But, by construction, $AH = HO$. Hence, $AH = HO = OD$, whence, $AO = \frac{2}{3}AD$. Similarly $CO = \frac{2}{3}CF$. That is, one median cuts off on the other median two-thirds of the distance from the vertex to the opposite side.

POLYGONS IN GENERAL

111. Two polygons are equal when they can be divided into the same number of triangles equal each to each and

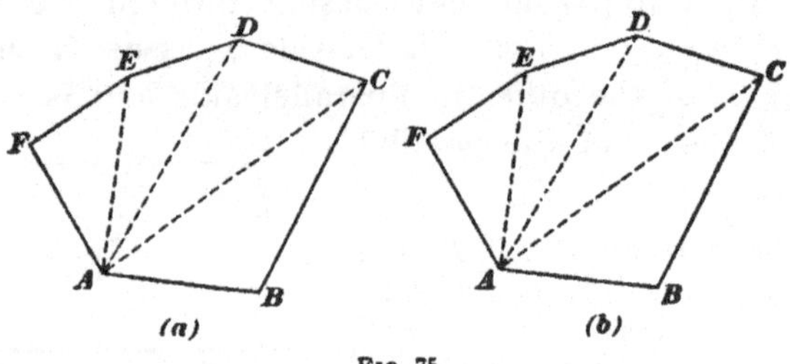

(a)   (b)

FIG. 75

similarly placed. Thus, the polygons shown at (*a*) and (*b*) in Fig. 75 are composed of the same number of triangles equal each to each and similarly placed, and it is evident that one polygon can be placed on the other so that they will coincide throughout; hence, they are equal.

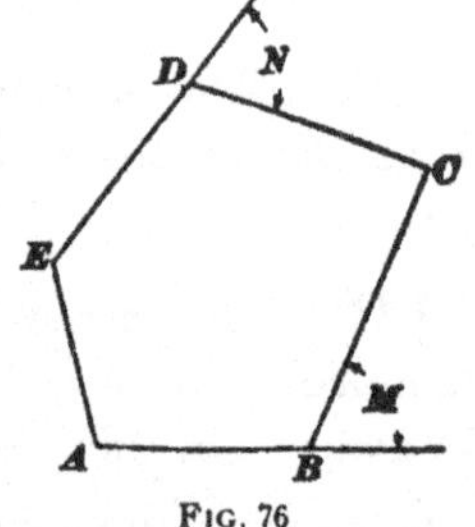

FIG. 76

112. An **exterior angle** of a polygon is an angle formed by any side and the prolongation of an adjacent side. In Fig. 76, the angles M and N are exterior angles of the polygon $ABCDE$.

113. A **diagonal** of a polygon is any line joining two vertexes not adjacent to the same side of the polygon. Thus, in Fig. 75, AC, AD, and AE are diagonals of the polygon $ABCDEF$.

114. The sum of the interior angles of any polygon is equal to two right angles multiplied by a number that is two less than the number of sides of the polygon.

Let (*a*), Fig. 75, be any polygon. Draw the diagonals from one vertex and thus divide the polygon into triangles. It is seen that the first triangle ABC and the last triangle AFE, each contains two sides of the polygon, while each of the other triangles contains but one side of the polygon. Thus, the number of triangles formed is two less than the number of the sides of the polygon. Hence (Art. **69**), the sum of the angles of the triangles, or of the polygon, is two right angles multiplied by a number that is two less than the number of sides of the polygon.

115. Let n = number of sides of a polygon;

S = sum of interior angles of the polygon, expressed in right angles.

Then, $$S = 2\,(n - 2) = 2n - 4$$

If $n = 4$, then $S = 2 \times 4 - 4 = 4$ right angles; that is, the sum of the angles of a quadrilateral is equal to four right angles.

Example 1.—What is the value of one of the interior angles of an equiangular hexagon?

Solution.—The number of sides of a hexagon is six; hence, applying the formula, $S = 2 \times (6 - 2) = 8$ right angles, that is, the sum of the interior angles of a hexagon is equal to eight right angles. Since the hexagon is equiangular, one of the angles is equal to one-sixth of eight right angles, or $1\frac{1}{3}$ right angles. Ans.

Example 2.—If one of the interior angles of an equiangular polygon is equal to $1\frac{3}{7}$ right angles, what is the name of the polygon?

Solution.—If one of the interior angles is equal to $1\frac{3}{7}$ or $\frac{10}{7}$ right angles, their sum S is equal to $\frac{10}{7} \times n = \frac{10n}{7}$. But from the formula, $S = 2n - 4$. Therefore, $\frac{10n}{7} = 2n - 4$; whence, $n = 7$. A polygon of seven sides is a heptagon; therefore, the polygon is a heptagon. Ans.

EXAMPLES FOR PRACTICE

1. Show that if two angles of a quadrilateral are supplementary the other two angles are supplementary.

2. In a triangle ABC, the angle C is twice the angle B. Show that the line that bisects the angle C meets the line AB at a point D so that $CD = BD$.

SUGGESTION.—Half the angle C = angle B. Then in the triangle CDB, angle BCD = angle CBD.

3. What is the value of one of the interior angles of an equiangular octagon?   Ans. $1\frac{1}{2}$ right angles

4. (*a*) What is the value of one of the interior angles of an equiangular quadrilateral? (*b*) What kind of quadrilateral is it?

Ans. $\begin{cases} (a)\ \text{One right angle} \\ (b)\ \text{Rectangle} \end{cases}$

5. If one of the interior angles of an equiangular polygon is equal to $1\frac{5}{9}$ right angles, what is the name of the polygon?   Ans. Nonagon

---

THE CIRCLE

DEFINITIONS AND GENERAL PROPERTIES

116. A **circle**, Fig. 77, is a plane figure bounded by a curved line every point of which is equally distant from a point within called the **center**.

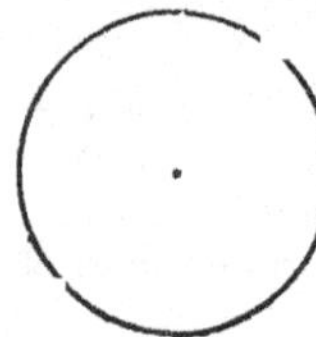

FIG. 77

117. The **circumference** of a circle is the line that bounds the circle. The term *circle* is often used in the sense of *circumference*.

118. The **diameter** of a circle is a straight line drawn through the center and terminated at both ends by the circumference. Thus, AE, Fig. 78, is a diameter of the circle whose center is O.

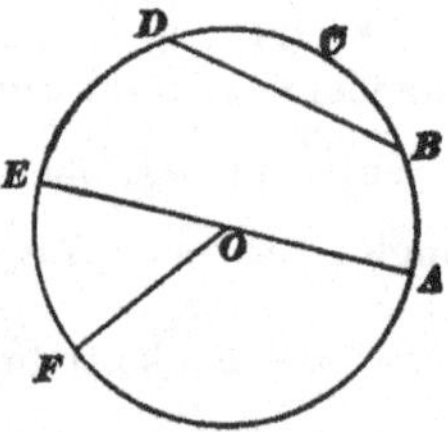

FIG. 78

119. The **radius** of a circle is any straight line drawn from the center to the circumference. The plural of *radius* is *radii*. Thus, OA, OE, and OF, Fig. 78, are radii of the circle whose center is O.

120. The distance from the center to the circumference is, by the definition of a circle, the same for all points in the same circle; hence, all radii are equal.

121. When any two radii, as OA and OE, Fig. 78, are in the same straight line, they form a diameter. Hence, the length of the diameter is twice the length of the radius.

122. An **arc** of a circle is any part of its circumference, as DCB, Fig. 78.

123. An arc equal to one-half the circumference is a **semi-circumference;** and an arc equal to one-fourth the circumference is a **quadrant.**

124. A **chord** is a straight line, as BD, Fig. 78, joining any two points in a circumference, or it is a line joining the extremities of an arc.

125. The longest chord that can be drawn in a circle is a chord that passes through the center and is, therefore, a diameter.

126. An arc of a circle is said to be **subtended** by its chord. Thus, the arc BCD, Fig. 78, is subtended by the chord BD.

Every chord in a circle subtends two arcs. Thus, BD subtends both the arcs BCD and $BAFED$.

When an arc and its chord are spoken of, the arc less than a semi-circumference is meant, unless the contrary is stated. The shorter arc is usually referred to by naming the letters at its extremities; thus, the arc BCD is called the arc BD.

127. A **segment** of a circle is a part of the circle enclosed by an arc and its chord. In Fig. 78, the part of the circle between the chord BD and the arc BD is a segment.

A segment equal to one-half the circle is a **semicircle.**

128. A **sector** of a circle is the space included between an arc and the two radii drawn to the extremities of the arc. In Fig. 78, the space included between the arc FE and the radii OF and OE is a sector.

129. Two circles are equal when the radius or diameter of one is equal to the radius or diameter of the other.

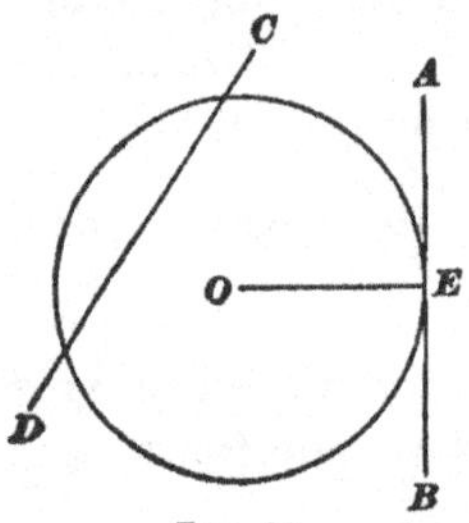

FIG. 79

130. A **tangent** to a circle is a line that touches the circumference in only one point. In Fig. 79, AB is tangent to the circle whose center is O.

The point E at which the tangent touches the circumference is the **point of contact,** or **point of tangency.**

131. Two circles are tangent when they touch each other in one point only, as in Fig. 80. When two circles are tangent, they are tangent to the same straight line at the point of tangency.

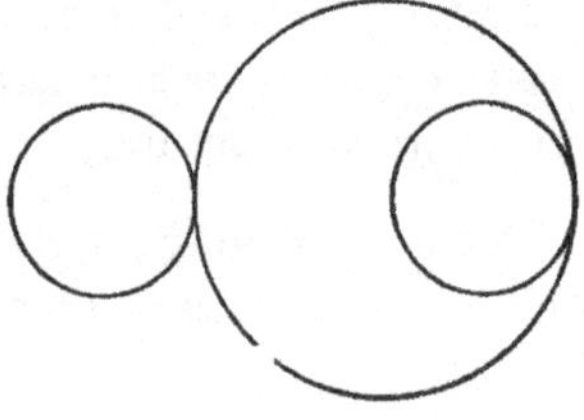

FIG. 80

132. A **secant,** as the term is used in geometry, is a line that intersects the circumference of a circle in two points. In Fig. 79, CD is a secant to the circle whose center is O.

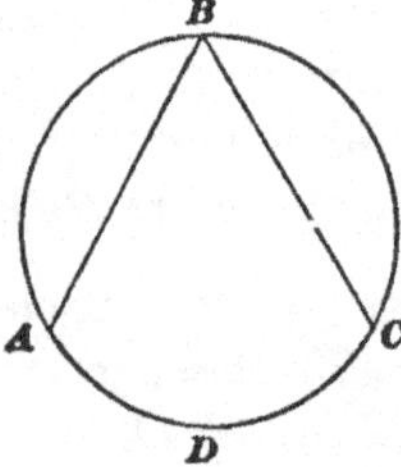

FIG. 81

133. An **inscribed angle** is an angle whose vertex lies on the circumference of a circle, and whose sides are chords. In Fig. 81, ABC is an inscribed angle.

134. **A central angle,** or an **angle at the center,** is an angle whose vertex is at the center of a circle and whose sides are radii. Thus, in **Fig. 82,** AOB is a central angle.

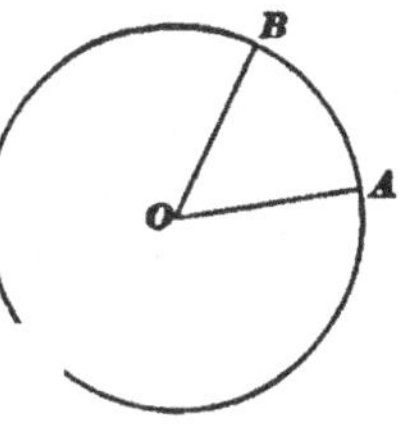

FIG. 82

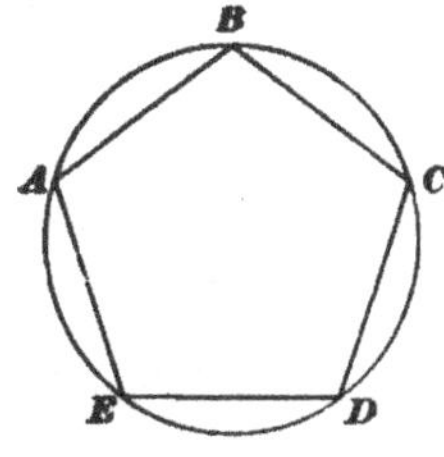

FIG. 83

135. An **inscribed polygon** is a polygon each of whose vertexes lies on the circumference of a circle, as in Fig. 83. The circle is said to be **circumscribed** about the polygon.

136. An **inscribed circle** is a circle whose circumference touches but does not intersect each of the sides of a polygon, as in Fig. 84. The polygon is said to be **circumscribed** about the circle.

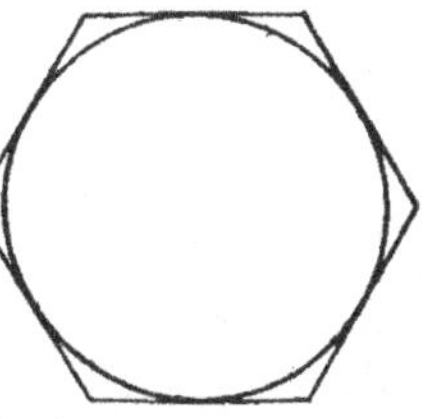

FIG. 84

FIG. 85

137. **Concentric circles** are circles having the same center. See Fig. 85.

138. Every diameter of a circle bisects the circle and its circumference. Thus, in Fig. 86, both the arc and the portion of the circle on one side of the diameter AB are equal, respectively, to the arc and the portion of the circle on the other side.

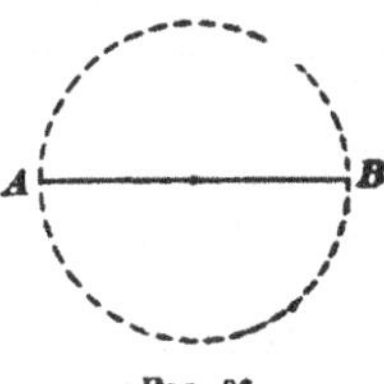

FIG. 86

139. In the same circle, or equal circles, equal angles at the center intercept equal arcs on the circumference.

Let O and O', Fig. 87, be equal circles, and $A O B$ and $A' O' B'$ equal angles. Place the circle O' on O so that the point O'

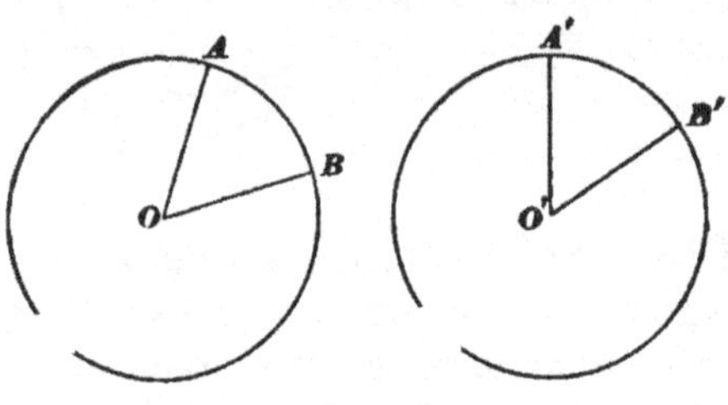

coincides with O and the line $O' B'$ takes the direction $O B$. Then, since $O B$ and $O' B'$ are equal, being radii of equal circles, B' will fall on B, and, since the angle O' is equal to the angle O, the line $O' A'$ will take the direction of $O A$, and, being equal to $O A$, its extremity A'

Fig. 87

will fall on A. Hence, the arcs $A B$ and $A' B'$ will coincide and are equal.

140. In the same circle, or equal circles, equal arcs are intercepted by equal angles at the center.

Let O and O', Fig. 87, be equal circles, and $A B$ and $A' B'$ equal arcs. Place the circle O' on the circle O, with the points O' and A' on O and A, respectively. Then, since the arc $A' B'$ is equal to the arc $A B$, B' will fall on B. Then the angle O' is equal to the angle O, as the vertex and the sides of the angles coincide.

141. In the same circle or equal circles, equal chords subtend equal arcs.

Let $A B$ and $C D$, Fig. 88, be equal chords. Draw the radii $A O, B O, C O$, and $D O$, joining A, B, C and D to O. Then the triangles $A O B$ and $C O D$, having three sides of one equal to three sides of the other, are equal. Hence, the angle $A O B$ is equal to the angle $C O D$, and, therefore (Art. **139**), the arc $A B$ is equal to the arc $C D$.

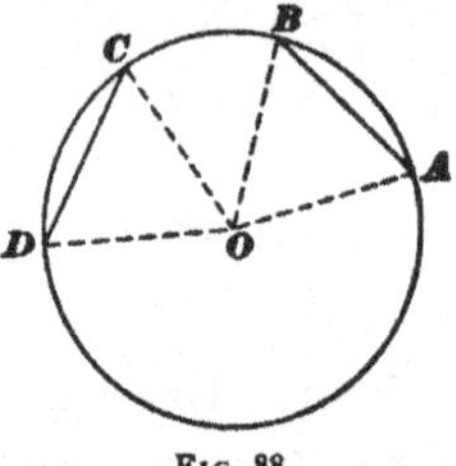

Fig. 88

142. In the same circle, or equal circles, equal arcs are subtended by equal chords.

143. A perpendicular from the center of a circle to a chord bisects the chord and the arc subtended by it.

Let OM, Fig. 89, be drawn from O perpendicular to the chord AB. Join O to A and B. The triangle AOB is isosceles, since the two sides OA and OB are radii of the same circle. Therefore (Art. **77**), $AM = MB$. Also, $AOM = MOB$ (Art. **77**); therefore (Art. **139**), arc $AC =$ arc CB.

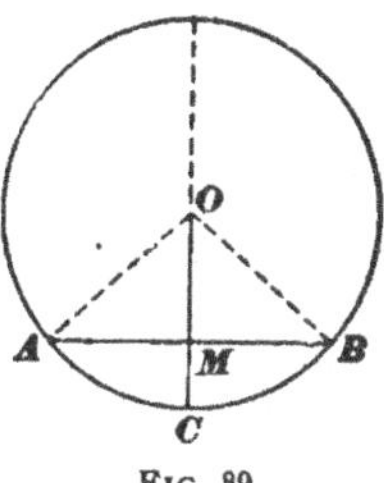

FIG. 89

144. The perpendicular erected at the middle of a chord passes through the center of the circle and bisects the arc subtended by the chord.

145. Through any three points not in a straight line a circumference can be passed.

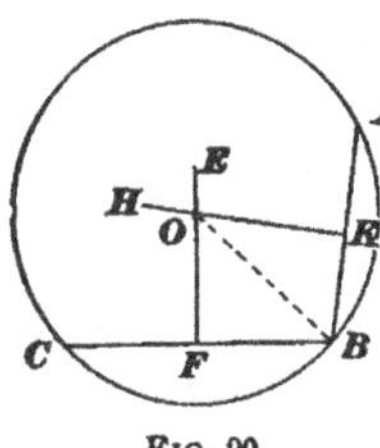

FIG. 90

Let A, B, and C, Fig. 90, be any three points. Draw AB and BC. At the middle point of AB draw KH perpendicular to AB; at the middle point of CB draw FE perpendicular to BC and meeting KH at O. As O is a point in the perpendiculars at the middle points of AB and BC, it is equally distant from A, B, and C. Therefore, a circle with O as center and OB as radius will pass through A, B, and C.

146. A straight line perpendicular to a radius at its extremity is tangent to the circle.

Let AB, Fig. 91, be perpendicular to OH at its extremity H. As OH is perpendicular to AB it is shorter than any other line, as OM, drawn from O to AB. Hence, M is without the circle, and any point in AB other than H is without the circle. Therefore, AB touches the circle in only the point H, and is, consequently, tangent to the circle.

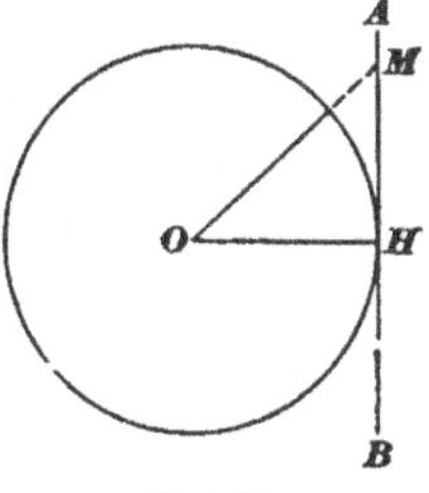

FIG. 91

147. A perpendicular to a tangent at the point of tangency passes through the center of the circle.

148. A tangent to a circle is perpendicular to the radius drawn to the point of tangency.

149. If two circles intersect, the line joining their centers bisects at right angles the line joining the points where the circles intersect.

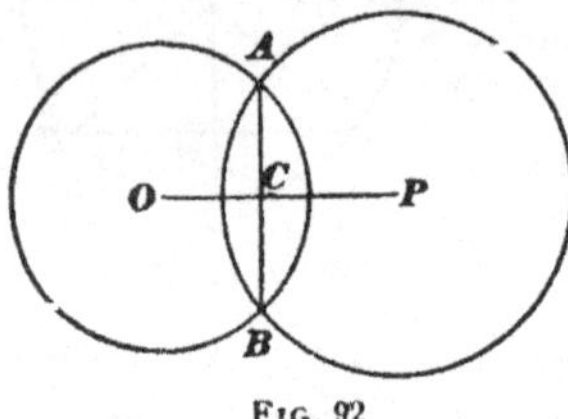

FIG. 92

Let the two circles whose centers are O and P, Fig. 92, intersect at A and B. The point P, being the center of a circle, is equally distant from A and B, points on the circumference. Similarly, O is equally distant from A and B. Hence, by Art. **34**, O and P determine the perpendicular bisecting AB.

150. The two tangents from a point to a circle are equal.

Let PA and PB, Fig. 93, be tangents from P to the circle whose center is O. Draw OA, OP, OB. Then the triangles POB and POA are right triangles (Art. **148**). In these triangles, PO is common and OA is equal to OB. Hence, the triangles are equal, and $PA = PB$.

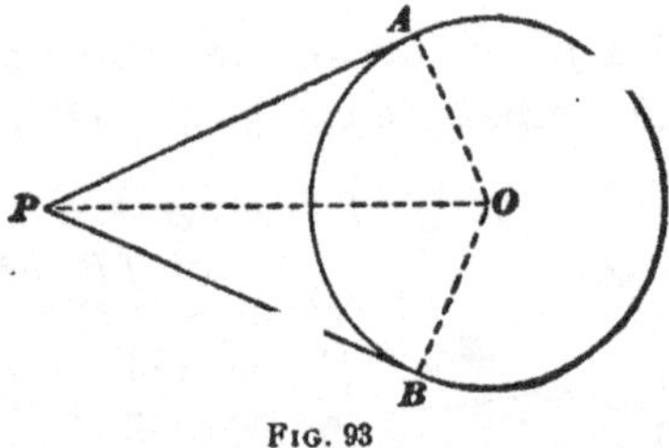

FIG. 93

151. The line joining an external point to the center of a circle bisects the angle made by the two tangents drawn from the point to the circle. Thus, the angle OPA, Fig. 93, is equal to the angle OPB.

EXAMPLES FOR PRACTICE

1. Show that the line joining the intersection of two tangents to the center of the circle bisects the chord joining the points of tangency.

2. Show that the bisector of the angle between two tangents passes through the center of the circle.

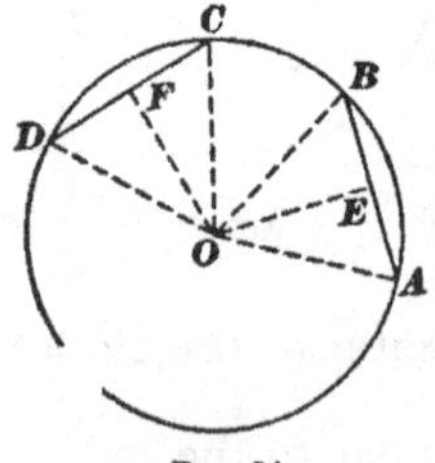

FIG. 94

3. Show that in the same circle, or equal circles, equal chords are equally distant from the center.

SUGGESTION.—Draw OE and OF, Fig. 94, perpendicular to the equal chords AB and CD. Then what is true of the triangles AEO and DOF?

4. Show that the tangents to a circle at the extremities of a diameter are parallel.

5. Show that in any circle a chord parallel to a tangent is bisected by the diameter drawn to the point of contact.

MEASUREMENT OF ANGLES

152. The ratio of one quantity to another of the same kind is the number of times that the first contains the second. When both quantities are represented by numbers, their ratio is the same as the quotient obtained by dividing one of the numbers by the other.

153. In the same circle, or equal circles, two central angles have the same ratio as their intercepted arcs; that is, in Fig. 95, angle AOB : angle $COD =$ arc AB : arc CD.

Suppose the arc AB to be three-fifths of the arc CD. Divide AB into three equal parts, and CD into five equal parts, as shown, and join the points of division with the center.

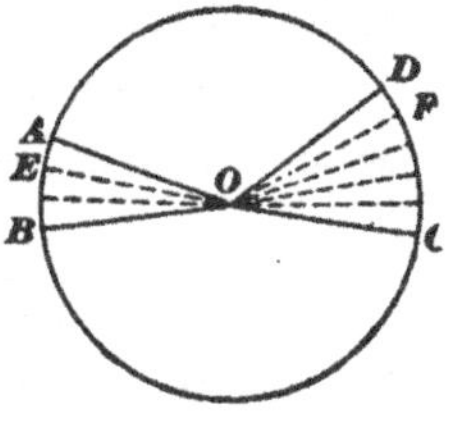
FIG. 95

Since $AB : CD = 3 : 5$, or $\dfrac{AB}{CD} = \dfrac{3}{5}$, it follows that one-third of AB is one-fifth of CD; that is, arc $AE =$ arc DF, and, therefore, angle $AOE =$ angle DOF. We have, therefore, angle $AOB = 3 \times$ angle AOE, angle $COD = 5 \times$ angle $DOF = 5 \times$ angle AOE; whence,

$$\frac{\text{angle } AOB}{\text{angle } COD} = \frac{3 \times \text{angle } AOE}{5 \times \text{angle } AOE} = \frac{3}{5} = \frac{\text{arc } AB}{\text{arc } CD}.$$

154. Since the angle at the center and its intercepted arc increase and decrease in the same ratio, it is said that *an angle at the center is measured by its intercepted arc.*

155. The whole circumference of a circle is divided into 360 equal parts, called **degrees**. A degree is divided into 60 equal parts, called **minutes**; and a minute is divided into 60 equal parts, called **seconds**. Degrees, minutes, and seconds of arc are used as units for measuring circular arcs. Since the circumference of every circle contains 360 degrees, the length of a degree differs in different circles. Thus, if AOB, Fig. 96, is an angle of 1°,

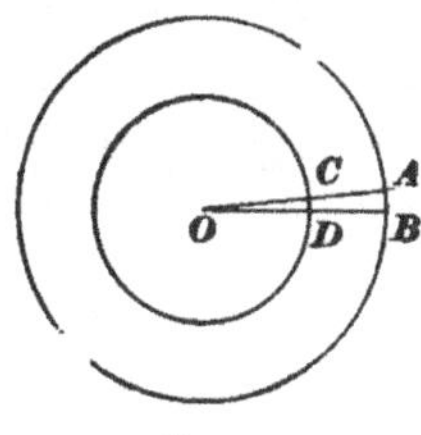
FIG. 96

AB is an arc of 1° in the larger circle and CD is also and arc of 1° in the smaller concentric circle. A degree of the earth's equator is a little more than 69 miles long; and a degree of the circumference of a circle whose diameter is 360 inches is 3.1416 inches long.

Degrees, minutes, and seconds are indicated by °, ′, ″. Thus, 25° 3′ 10″ means 25 degrees, 3 minutes, and 10 seconds.

Since a right angle intercepts one quarter of a circumference, the number of degrees measuring it is $360 \div 4 = 90°$. The number of degrees measuring an angle equal to one-half of a right angle is $90° \div 2 = 45°$.

Usually, the magnitude of an angle is expressed by stating the number of degrees that it subtends. Thus, a right angle is referred to as an angle of 90°; one-third of a right angle, as an angle of 30°, etc.

156. An inscribed angle is measured by one-half the intercepted arc. Thus, in Fig. 97, the angle $A B C$ is measured by one-half the arc $A D C$.

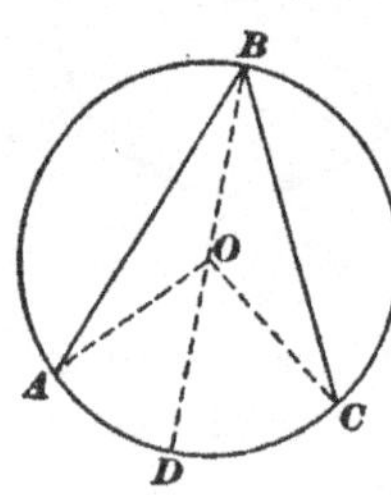

FIG. 97

Draw the diameter $B O D$ and the radii $O C$ and $O A$. The angle $C O D$, the exterior angle of the triangle $O B C$, is equal to the angle $O B C$ plus the angle $O C B$. But the angle $O C B$ is equal to the angle $O B C$, as they are opposite the equal sides of an isosceles triangle. Hence, the angle $C O D$, which is measured by the arc $C D$, is equal to $2 \times O B C$. Therefore, $O B C$ is measured by one-half the arc $C D$. Similarly, the angle $O B A$ is measured by one-half the arc $A D$. Therefore, the angle $A B C$ is measured by one-half the arc $A D$ plus one-half the arc $D C$; that is, by one-half the arc $A C$.

157. In the same circle, or equal circles, equal arcs are intercepted by equal inscribed angles.

158. All angles inscribed in the same segment are equal.

159. Any angle inscribed in a semicircle is a right angle.

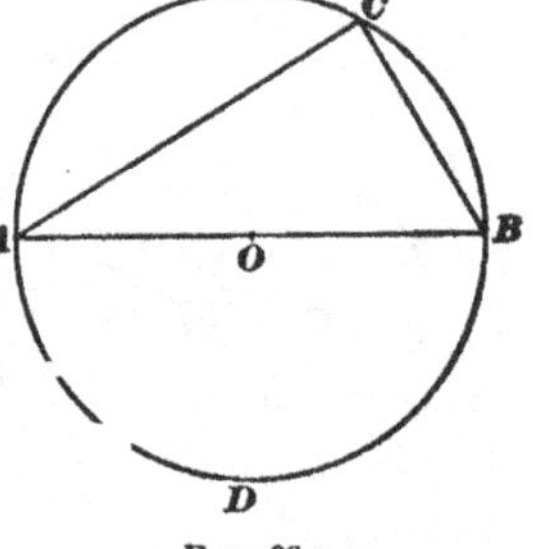

FIG. 98

The angle $A C B$, Fig. 98, is measured by one-half the arc $A D B$, which is a semi-circumference. As a semi-circumference contains 180°, the angle $A C B$ is measured by one-half of 180°, or 90°, and is, therefore, a right angle.

160. The vertexes of all the angles of a given magnitude whose sides pass through two fixed points, lie on a circle that passes through the two fixed points and any one of the vertexes.

In Fig. 99, let APB be an angle of the given magnitude and A and B the fixed points. Through A, B, and P, pass a circle. Now, any angle, as AP_1B or AP_2B, whose sides pass through A and B and whose vertex lies on the arc APB is (Art. 158) equal to the given angle APB.

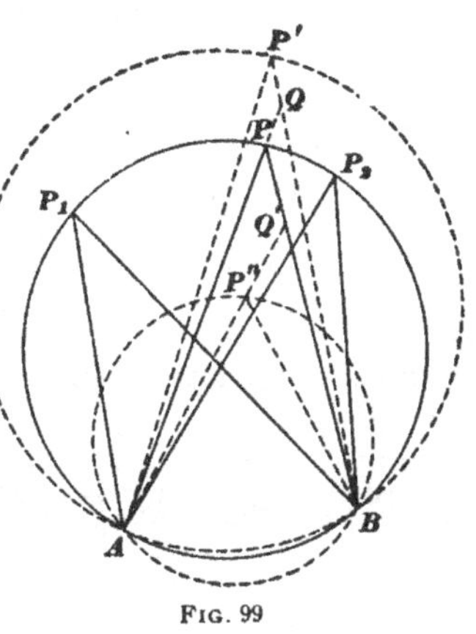

FIG. 99

Again, any angle, as $AP'B$, whose sides pass through A and B and whose vertex lies without the arc APB is less than the angle APB. For if AP is produced to meet BP' at Q, the angle APB being an exterior angle of the triangle BPQ, is equal to $PQB + PBQ$ and is therefore greater than PQB, and, as PQB is greater than $AP'B$ (since $PQB = AP'B + QAP'$), it follows that APB is greater than $AP'B$.

In like manner it can be shown that any angle, as $AP''B$, whose sides pass through A and B and whose vertex lies within the arc APB, is greater than the given angle APB.

161. An angle formed by a tangent, as TM, Fig. 100, and a chord, as TP, is measured by one-half the intercepted arc TEP.

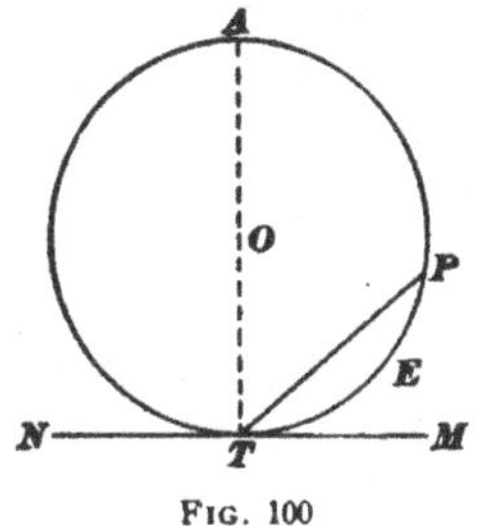

FIG. 100

Draw the diameter TOA. Then MTA is a right angle and is, therefore, measured by one-half the semi-circumference $TEPA$. The angle PTA is measured by one-half the arc PA. Hence, the angle MTP, equal to MTA minus PTA, is measured by one-half the difference between the semi-circumference and PA; that is, by one-half the arc TEP.

EXAMPLES FOR PRACTICE

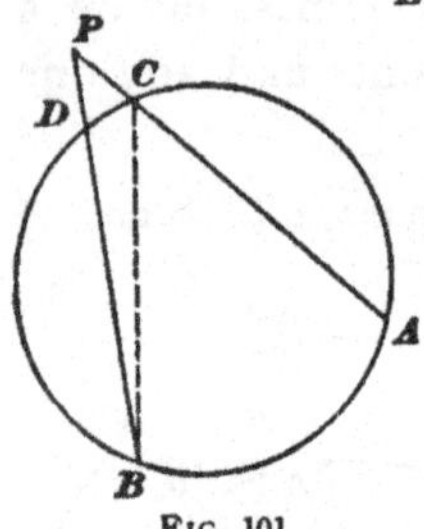

FIG. 101

1. Prove that the angle BPA, Fig. 101, formed by two secants intersecting without the circumference is measured by one-half the difference of the intercepted arcs AB and CD; that is, by $\frac{1}{2}(AB - CD)$.

SUGGESTION.—Join C and B. Then angle BCA is an exterior angle of the triangle BCP, and angle BPC is equal to angle ACB minus angle DBC.

2. Show that the angle APB, Fig. 102, formed by two tangents PT and PT' is measured by one-half the difference of the intercepted arcs TQT' and TRT'.

SUGGESTION.—Join T and T'. Then ATT' is an exterior angle of triangle $TT'P$, while ATT' and $PT'T$ are angles formed by a tangent and a chord.

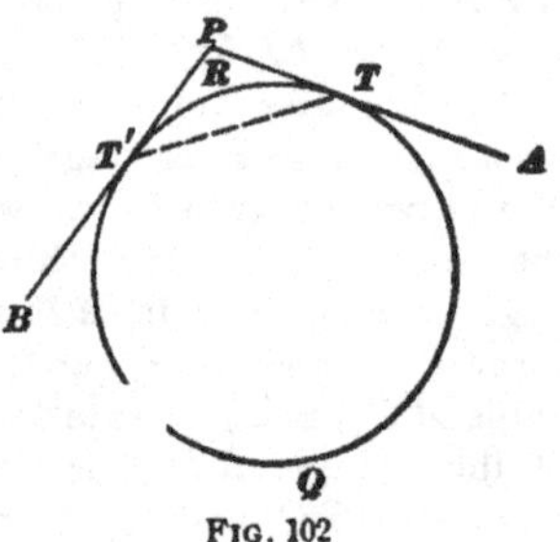

FIG. 102

162. The **angle of intersection** of two tangents is the angle formed by one tangent with the prolongation of the other tangent. Thus, the angle APT', Fig. 103, is the angle of intersection of the two tangents TP and PT'.

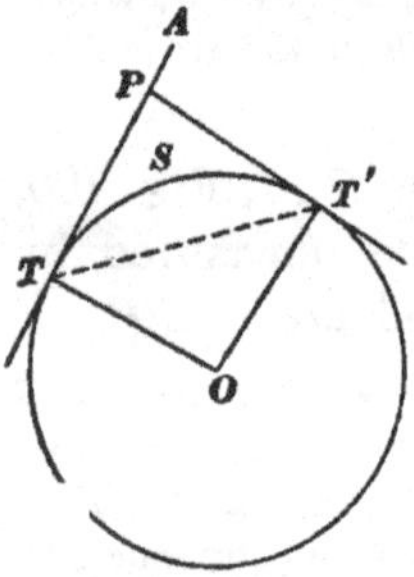

163. The angle of intersection of two tangents is equal to the central angle whose sides pass through the points of tangency.

FIG. 103

In Fig. 103, join TT'. Then, the angle APT' is equal to the sum of the equal angles PTT' and $PT'T$. But each of these angles is made by a tangent and a chord and is, therefore, measured by one-half of the arc TST'. Hence, the angle APT' is measured by the arc TST'. The central angle O is also measured by this arc; therefore, the angle O is equal to the angle APT'.

164. The opposite angles of an inscribed quadrilateral are supplementary; that is, their sum is equal to two right angles or 180°.

In Fig. 104, the angle B is measured by one-half the arc ADC, and the opposite angle D is measured by one-half the arc ABC. The sum of the arcs ADC and ABC is a circumference, or $360°$. Hence, the sum of the angles ADC and ABC is measured by one-half of $360°$, or $180°$.

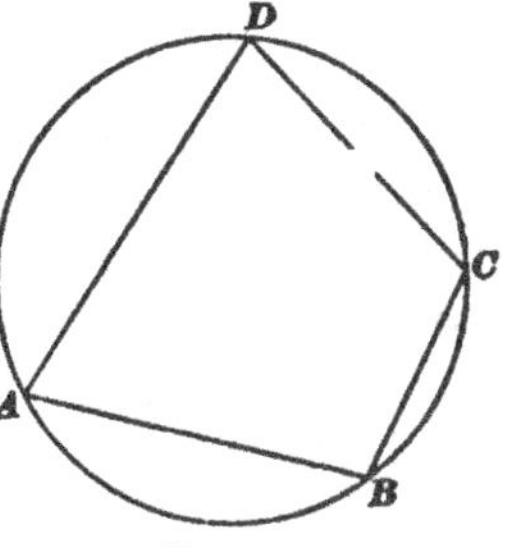

Fig. 104

165. If the opposite angles of a quadrilateral are supplementary, the quadrilateral can be inscribed in a circle.

EXAMPLE 1.—What is the number of degrees in each angle of an equilateral triangle?

SOLUTION.—The sum of the three angles of the triangle is two right angles, or $180°$. Since the three angles are equal, each angle is one-third of $180°$, or $\dfrac{180°}{3} = 60°$. Ans.

EXAMPLE 2.—The unequal angle of an isosceles triangle is $75°\ 32'\ 10''$; what is the magnitude of each of the equal angles?

SOLUTION.—Since the sum of the three angles is $180°$, the sum of the two equal angles is $180°$ minus the other angle, or $180° - 75°\ 32'\ 10'' = 104°\ 27'\ 50''$, and each of them is one-half of this sum, or $(104°\ 27'\ 50'') \div 2 = 52°\ 13'\ 55''$. Ans.

EXAMPLE 3.—The exterior angle of a triangle is $124°\ 3'\ 40''$, and one of the opposite-interior angles is $60°$; find the other two angles of the triangle.

SOLUTION.—Let the given exterior angle be denoted by A, the given interior angle by B, the other opposite-interior angle by C, and the third angle of the triangle by A'. (Let the student draw the triangle and mark these angles.) Then, $A = B + C$; whence, $C = A - B = 124°\ 3'\ 40'' - 60° = 64°\ 3'\ 40''$. Ans. Also, $A + A' = 180°$; whence, $A' = 180° - A = 180° - 124°\ 3'\ 40'' = 55°\ 56'\ 20''$. Ans.

EXAMPLES FOR PRACTICE

1. Show that the only parallelogram that can be inscribed in a circle is a rectangle.

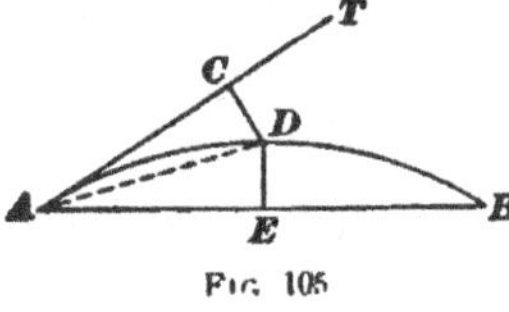

Fig. 105

2. Show that if from a point A, Fig. 105, on the arc of a circle a chord AB and a tangent AT are drawn, the perpendiculars DC and DE drawn to them from the middle point D of the subtended arc are equal.

I L T 36F—4

3. The angle of intersection of two tangents is 100°; find the number of degrees in each angle formed by the tangents and the chord through the points of contact. Ans. 50°

4. One of the acute angles of a right triangle is 50°; what is the magnitude of the other acute angle? Ans. 40°

5. Each of the equal angles of an isosceles triangle is 45°; show that the triangle is right-angled.

6. Two angles of a triangle are 37° 41′ 38″ and 86° 51′ 2″; what is the value of the other angle? Ans. 55° 27′ 22″

GEOMETRY

PROPORTION

DEFINITIONS AND GENERAL PRINCIPLES

1. A **proportion** is an equality of ratios or of fractions. Thus, the fractions $\frac{4}{5}$ and $\frac{8}{10}$, being equal, form a proportion. In general, if $\frac{a}{b}$ is equal to $\frac{c}{d}$, these two ratios or fractions form a proportion, which may be written in any of the following forms $\frac{a}{b} = \frac{c}{d}$, $a : b = c : d$, $a : b :: c : d$. When written in either of the last two forms, the proportion is read a is to b as c is to d.

2. **Properties of Proportions.**—The first and the fourth term of a proportion are called the **extremes**; the second and the third, the **means**. Thus, in the proportion $a : b = c : d$, the extremes are a and d, and the means, b and c.

3. If any four quantities are in proportion, the product of the extremes is equal to the product of the means. This principle follows at once from the definition of a proportion, as will be explained presently. If a, b, c, and d, are in proportion, then, by the definition,

$$\frac{a}{b} = \frac{c}{d} \qquad (1)$$

This equation may be treated the same as any other algebraic equation. Both members of the equation may be multiplied or divided by the same quantity, or the same quantity may be added to or subtracted from both members, and the proportion may thus be changed to a great number of forms without destroying the equality of the ratios. Different names are applied to these changes, some of the most common of which are given in the following articles.

In order to show that the product of the means is equal to the product of the extremes, multiply both members of equation (1) by bd to clear of fractions; the equation then becomes

$$a\,d = b\,c \qquad (2)$$

4. It is evident that if two fractions are equal, their reciprocals are also equal. If $\dfrac{a}{b} = \dfrac{c}{d}$, then $\dfrac{b}{a} = \dfrac{d}{c}$; that is, if $a : b = c : d$, we have also $b : a = d : c$.

Taking the reciprocal of a fraction is called **inverting** the fraction. The operation of inverting the two fractions of a proportion is called **inversion.**

5. If both members of equation (2), Art. **3**, are divided by cd, there results

$$\frac{a}{c} = \frac{b}{d}, \text{ or } a : c = b : d$$

Or, if both members of equation (2) are divided by ba, the result is

$$\frac{d}{b} = \frac{c}{a}, \text{ or } d : b = c : a$$

Therefore, either the means or the extremes of a proportion can be interchanged. This operation is called **alternation.**

6. If 1 is added to each member of equation (1), Art. **3**, the equation becomes

$$\frac{a}{b} + 1 = \frac{c}{d} + 1$$

Reducing each member to an improper fraction,

$$\frac{a + b}{b} = \frac{c + d}{d}, \text{ or } a + b : b = c + d : d \qquad (1)$$

In a similar manner it can be shown that

$$a + b : a = c + d : c \qquad (2)$$

The proportions (1) and (2) are said to be derived from the original proportion by **composition.**

7. If 1 is subtracted from each member of equation (1), Art. **3**, the equation becomes

$$\frac{a}{b} - 1 = \frac{c}{d} - 1$$

Reducing each member to an improper fraction,

$$\frac{a-b}{b} = \frac{c-d}{d}, \text{ or } a - b : b = c - d : d \qquad (1)$$

In a similar manner it can be shown that

$$a - b : a = c - d : c \qquad (2)$$

The proportions (1) and (2) are said to be derived from the original proportion by **division.**

LINES DIVIDED PROPORTIONALLY

8. Two straight lines are **divided proportionally** when the corresponding segments or parts are in proportion; or when the ratio of the two segments of one is the same as the ratio of the two segments of the other. Thus, the lines AB

FIG. 1

and CD, Fig. 1, are divided proportionally in the points E and F if $AE : EB = CF : FD$.

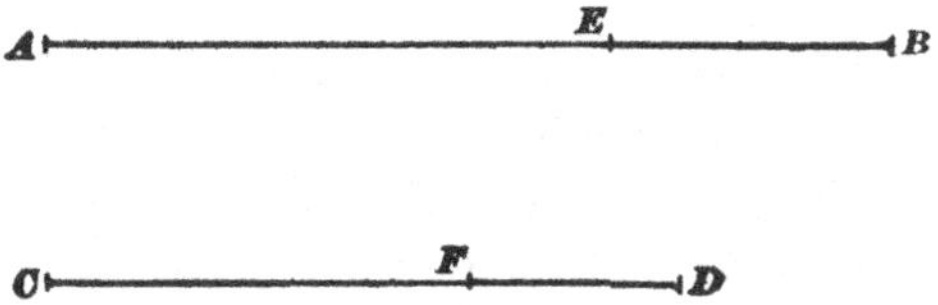

9. A line parallel to one of the sides of a triangle divides the other two sides proportionally. Thus, in Fig. 2, where DE is parallel to BC, $AD : DB = AE : EC$.

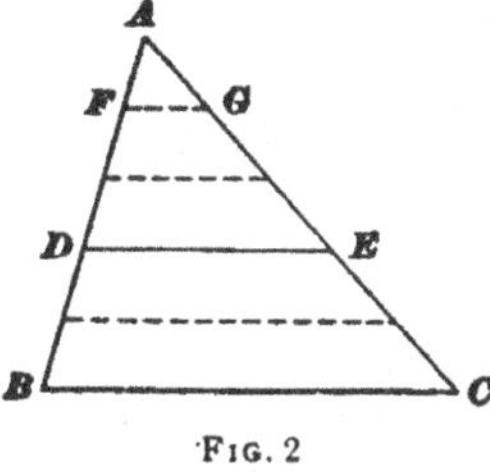

FIG. 2

Suppose that the ratio of AD to DB is as 3 to 2; that is, let $\dfrac{AD}{DB} = \dfrac{3}{2}$. Divide AB into five equal parts, and through the points of division draw lines parallel to BC. These lines will intercept equal distances on AC (see *Geometry*, Part 1).

As the ratio of AD to DB is that of 3 to 2, AD will contain three, and DB will contain two, of the equal parts into which AB is divided.

Also, AE will contain three and EC two of the equal parts into which AC is divided; so that $AE = 3 \times AG$, and $EC = 2 \times AG$; whence

$$\frac{AE}{EC} = \frac{3 \times AG}{2 \times AG} = \frac{3}{2} = \frac{AD}{DB}$$

or,
$$AE : EC = AD : DB$$

10. Any two sides of a triangle are to each other as the segments into which they are divided by any line parallel to the third side. Thus, in Fig. 2, $AB : AC = AD : AE = DB : EC$.

From the preceding article, we have
$$AD : DB = AE : EC$$
whence (Art. **6**),
$$AD + DB : DB = AE + EC : EC$$
that is,
$$AB : DB = AC : EC$$
and, interchanging the means (Art. **5**),
$$AB : AC = DB : EC$$

In the same manner it may be shown that
$$AB : AC = AD : AE$$

11. If a line divides two sides of a triangle proportionally, it is parallel to the third side. Thus, if DE, Fig. 3, divides AB and AC so that $AD : DB = AE : EC$, then DE is parallel to BC.

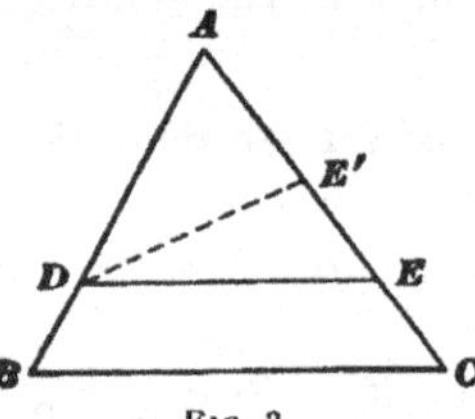

FIG. 3

If DE were not parallel to BC, a line DE' could be drawn through D parallel to BC. Then, by Art. **9**, we should have $\dfrac{AD}{DB} = \dfrac{AE'}{E'C}$; whence, since we have assumed that $\dfrac{AD}{DB} = \dfrac{AE}{EC}$,

$$\frac{AE}{EC} = \frac{AE'}{E'C}$$

By interchanging the means of this proportion, we obtain

$$\frac{AE}{AE'} = \frac{EC}{E'C}$$

This equality is evidently absurd, since AE is greater than AE', whereas EC is less than $E'C$. Therefore, no other line than DE can pass through D and be parallel to BC.

EXAMPLE 1.—Find the length of the line AB, Fig. 4, of which the end B is inaccessible.

SOLUTION.—There are several ways of solving this problem in practice. The one illustrated in the figure is as follows: Any convenient distance AC is measured and the angle C observed with a transit or compass. From C, a distance CE is measured, and at E an angle AED equal to C is turned off. The point D where the line of sight ED meets AB is marked, and the distances AD and AE are measured. Then, since AED equals C, the lines ED and CB are parallel (see *Geometry*, Part 1) and, therefore (Art. 9),

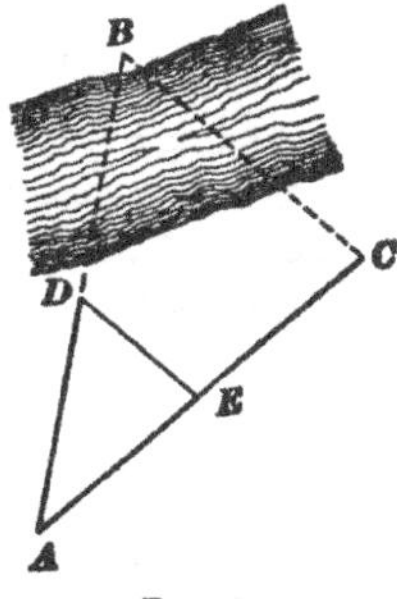

FIG. 4

$$AB : AC = AD : AE$$

whence (Art. 3),

$$AB \times AE = AC \times AD$$

and, dividing by AE,

$$AB = \frac{AC \times AD}{AE}. \quad \text{Ans.}$$

EXAMPLE 2.—Divide a line AB, Fig. 1, of given length into two parts AE and EB whose ratio shall be the same as that of two given numbers m and n; that is, so that $AE : EB = m : n$.

SOLUTION.—Since $AE : EB = m : n$, we must have (Art. 6),

$$\frac{AE + EB}{EB} = \frac{m + n}{n}, \text{ or, } \frac{AB}{EB} = \frac{m + n}{n},$$

whence, solving for EB,

$$EB = \frac{n \times AB}{m + n}. \quad \text{Ans.}$$

AE can be found in a similar manner, or by subtracting the value of EB from AB.

EXAMPLES FOR PRACTICE

1. If the measured distances in Fig. 4 are AC = 100 feet, AE = 45.2 feet, AD = 48.36 feet, what are the distances AB and DB?

Ans. $\begin{cases} AB = 106.99 \text{ ft.} \\ DB = 58.63 \text{ ft.} \end{cases}$

2. If, in Fig. 3, AD = 75 feet, DB = 16.25 feet, and AC = 80 feet, find AE and EC.

Ans. $\begin{cases} AE = 65.75 \text{ ft.} \\ EC = 14.25 \text{ ft.} \end{cases}$

3. If AB, Fig. 1, is equal to 125 feet, find the distances AE and EB so that the line will be divided at E in the ratio of 5 to 2.

Ans. $\begin{cases} AE = 89.286 \text{ ft.} \\ EB = 35.714 \text{ ft.} \end{cases}$

12. If two lines, as AB and CD, Fig. 5, are cut by **any** number of parallel lines, as EM, GN, IO, etc., the corresponding intercepts are proportional; that is, $EG : GI = MN : NO;\ GI : IK = NO : OP$, or, by interchanging the means, $EG : MN = GI : NO = IK : OP$, etc.

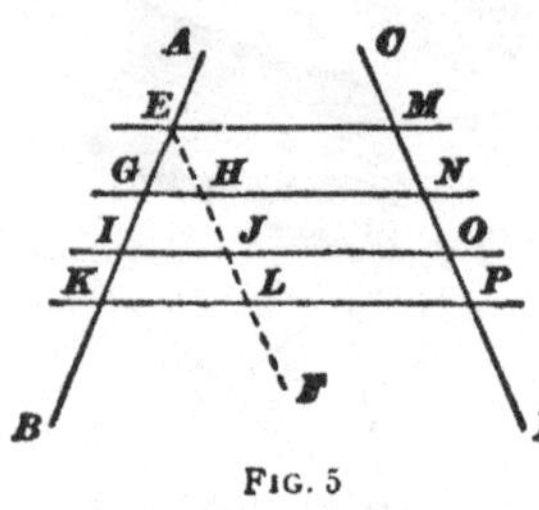

FIG. 5

Through E, draw EF parallel to CD. Then, $EH = MN$, $HJ = NO$, $JL = OP$. (See *Geometry*, Part 1.) Also, by Arts. 9 and **10**,

$$\frac{EK}{EL} = \frac{EG}{EH} = \frac{GI}{HJ} = \frac{IK}{JL}$$

that is,

$$\frac{EK}{MP} = \frac{EG}{MN} = \frac{GI}{NO} = \frac{IK}{OP}$$

or, $EK : MP = EG : MN = GI : NO = IK : OP$

13. In any triangle ABC, Fig. 6, the bisector BD of an angle divides the side opposite proportionally to the including sides; that is, $AB : BC = AD : DC$.

Draw CE parallel to BD and meeting AB produced in E. Then, in the triangle AEC, by Art. 9,

$$AB : BE = AD : DC \qquad (1)$$

The angles DBC and M, being alternate-interior angles, are equal; that is, $M = \frac{1}{2}B$. The angles DBA and E, being exterior-interior angles, are equal; that is, $E = \frac{1}{2}B$. Therefore, $E = M$, and $BE = BC$. (See *Geometry*, Part 1.)

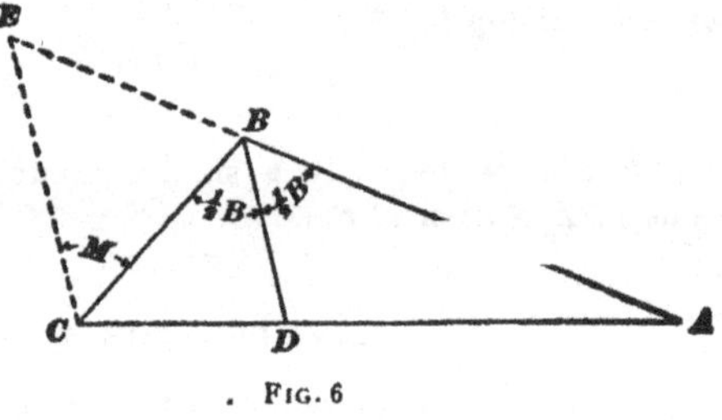

FIG. 6

Substituting, in equation (1), BC for its equal BE,

$$AB : BC = AD : DC$$

POLYGONS

SIMILAR POLYGONS

SIMILAR TRIANGLES

14. Similar polygons are those whose corresponding angles are equal and whose corresponding sides are proportional.

In order that two polygons may be similar, it is manifestly necessary that each angle of the one shall be equal to the corresponding angle of the other. But this is not sufficient; the corresponding sides must be proportional. For example,

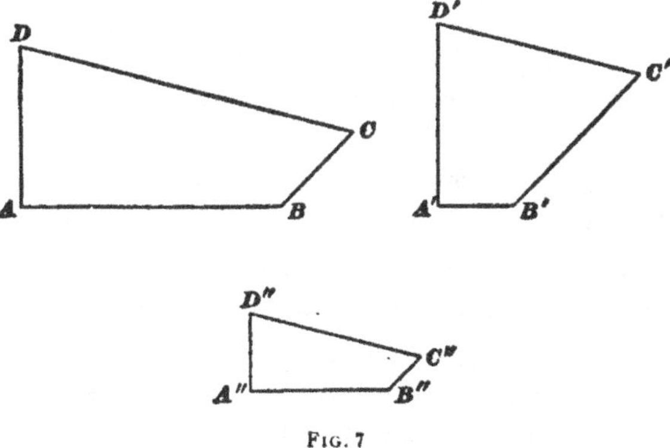

Fig. 7

the quadrilaterals $ABCD$ and $A'B'C'D'$, Fig. 7, have their corresponding angles equal, but they are not similar, because their corresponding sides are not proportional. The quadrilaterals $ABCD$ and $A''B''C''D''$ have their corresponding angles equal and their corresponding sides proportional, and are, therefore, similar.

The corresponding sides of similar polygons are called homologous sides.

15. Two triangles are similar when the angles of one are equal to the angles of the other.

In Fig. 8, let the angles of the triangle $A B C$ be equal, respectively, to those of the triangle $A' B' C'$. Place the triangle $A' B' C'$ upon $A B C$, so that the angle A' will coincide with its equal A. Then B' will fall along $A B$ and C' along $A C$, as at B'' and C'', respectively, and $B' C'$ will take the position $B'' C''$. The angle B'', which is equal to B', is equal to B, and the angle C'', which is equal to C', is equal to C; hence, $B'' C''$ is parallel

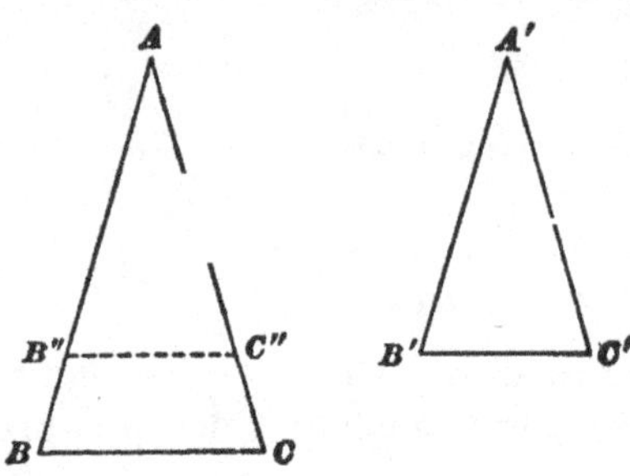

FIG. 8

to $B C$ (see *Geometry*, Part 1). Then, by Art. **10,**

$$A B : A B'' = A C : A C''$$

Substituting $A' B'$ and $A' C'$ for their respective equals $A B''$ and $A C''$,

$$A B : A' B' = A C : A' C'$$

In like manner, it can be proved that

$$A B : A' B' = B C : B' C'$$

Therefore, the triangles, having their angles equal and their corresponding sides proportional, are similar.

16. Two triangles are similar when two angles of the one are equal respectively to two angles of the other.

17. Two right triangles are similar when an acute angle of one is equal to an acute angle of the other.

18. A triangle is similar to any triangle formed by a line parallel to one of its sides and the segments it intercepts on the other two sides or the other two sides prolonged.

19. Two triangles are similar when the three sides of one are either parallel or perpendicular to the three sides of the other.

20. Two triangles are similar when their corresponding sides are proportional.

In Fig. 9, $A B : A' B' = A C : A' C = B C : B' C$

On AB, lay off AD equal to $A'B'$; on AC, lay off AE equal to $A'C$, and join DE. Then, since $AB:AD = AC:AE$, DE is parallel to BC. Hence, by Art. **18**, triangles ABC and ADE are similar, and, consequently, triangles ABC and $A'B'C'$ are similar if it can be shown that $DE = B'C'$. Now,

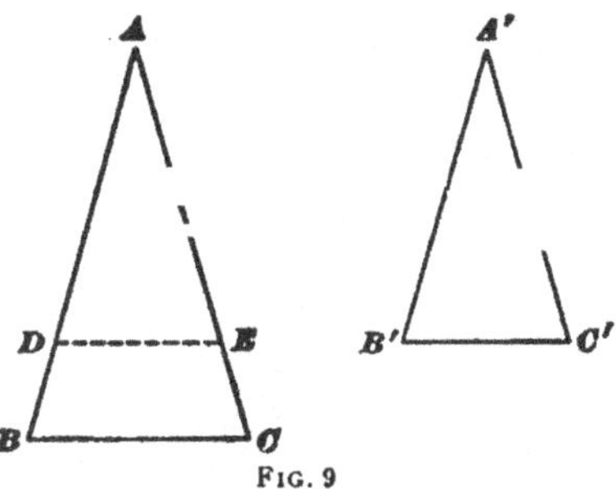

Fig. 9

$$AB:AD = BC:DE, \text{ or } AB:A'B' = BC:DE$$

But, $$AB:A'B' = BC:B'C$$

The last two proportions are the same, term for term, excepting the last term; hence, DE is equal to $B'C'$, and the triangles ADE and $A'B'C'$ are equal. Therefore, the triangles ABC and $A'B'C'$ are similar.

21. Two triangles are similar when an angle of the one is equal to an angle of the other and the including sides are proportional.

22. In two similar triangles, corresponding altitudes have the same ratio as any two corresponding sides.

Let CD and $C'D'$, Fig. 10, be the corresponding altitudes of the two similar triangles ABC and $A'B'C'$. The right triangles ACD and $A'C'D'$, having angle A equal to the angle A', are similar; hence,

$$CD:C'D' = AC:A'C'$$

But, since the triangles ABC and $A'B'C'$ are similar,

$$AC:A'C' = AB:A'B'$$
$$= BC:B'C'$$

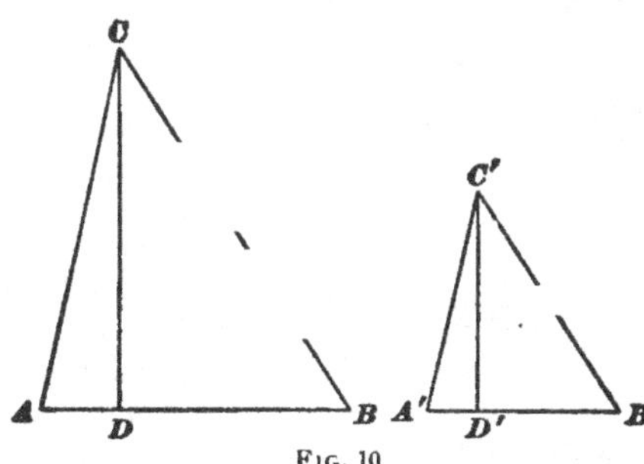

Fig. 10

Therefore,

$$CD:C'D' = AC:A'C' = AB:A'B' = BC:B'C'$$

23. As stated in Art. **14**, two polygons are similar when their corresponding angles are equal and their corresponding sides are proportional. It has now been shown that, in triangles, either of these conditions includes the other. This could have been expected from the fact that either the three angles or the three sides of a triangle fix its shape. This is not true of a polygon of more than three sides, as the angles can be changed without altering the sides, or the proportions of the sides can be changed without altering the angles.

EXAMPLE 1.—In the triangles ABC and $A'B'C'$, Fig. 11, angle A = angle A', angle B = angle B', and angle C = angle C', and the

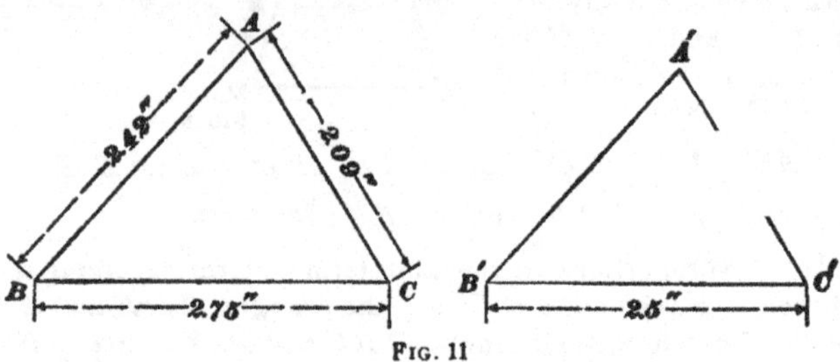

FIG. 11

sides BC, CA, AB, and $B'C'$ have the dimensions that are marked on them; find the lengths of the sides $C'A'$ and $A'B'$.

SOLUTION.—Since the two triangles are equiangular, they are similar, and hence the value of $A'B'$ and that of $A'C'$ are conveniently found as follows:

$$2.75 : 2.42 = 2.5 : A'B'$$

whence
$$A'B' = \frac{2.42 \times 2.5}{2.75} = 2.2 \text{ in.} \quad \textbf{Ans.}$$

$$2.75 : 2.09 = 2.5 : A'C'$$

whence
$$A'C' = \frac{2.09 \times 2.5}{2.75} = 1.9 \text{ in.} \quad \textbf{Ans.}$$

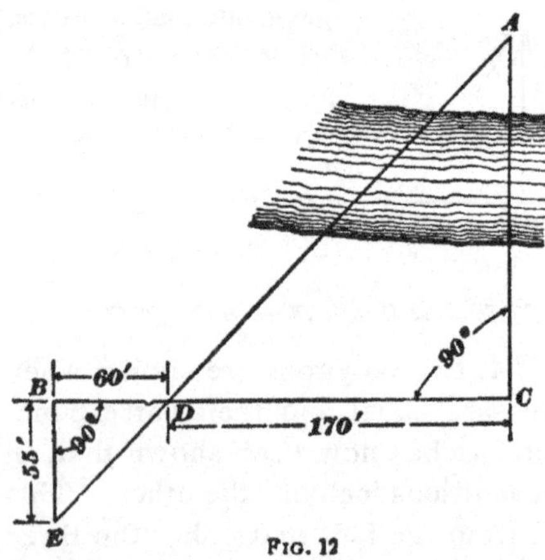

FIG. 12

EXAMPLE 2.—In Fig. 12, CD is perpendicular to AC and is 170 feet long; DB is 60 feet; and BE, perpendicular to BD, is 55 feet long; find the distance AC.

SOLUTION.—The right triangles ACD and DBE have the angles ADC and BDE equal; hence, they are similar, and

$$AC : CD = BE : BD,$$
or $\quad AC : 170 = 55 : 60$

whence,

$$A C = \frac{170 \times 55}{60} = 155.83 \text{ ft. Ans.}$$

EXAMPLE 3.—It is required to cut from a triangular plate $A\,B\,C$, Fig. 13, having the dimensions shown, a trapezoidal plate $B\,D\,E\,C$ whose upper base $D\,E$ shall be 6 inches; find the distances $A\,D$ and $A\,E$ that must be cut off.

SOLUTION.—The similar triangles $A\,D\,E$ and $A\,B\,C$ give,

$$\frac{A\,D}{A\,B} = \frac{D\,E}{B\,C}; \quad A\,D = \frac{A\,B \times D\,E}{B\,C} = \frac{12.5 \times 6}{10} = 7.5 \text{ in. Ans.}$$

$$\frac{A\,E}{A\,C} = \frac{D\,E}{B\,C}; \quad A\,E = \frac{A\,C \times D\,E}{B\,C} = \frac{15 \times 6}{10} = 9 \text{ in. Ans.}$$

EXAMPLE 4.—In order to measure the height $R\,H$ of a pier $A_1\,P\,Q\,R$, Fig. 14, whose base and top are, respectively, 22 feet and 12 feet square and whose sides all have the same inclination, a transit was set at a point O distant 250 feet from the side A_1 of the pier; that is, so that $O\,M_1 = 250$ feet. $A_1\,B_1$ was a rod on which the horizontal line of sight $O\,M_1$ intercepted a distance $A_1\,M_1 = 4.5$ feet. The same rod was

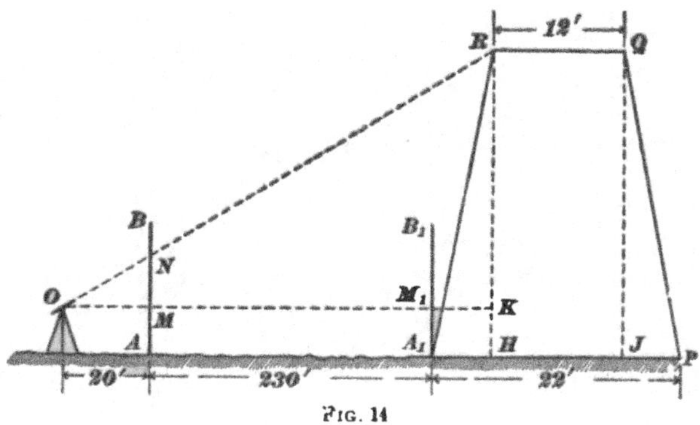

FIG. 14

held at a distance $O\,M = 20$ feet from the instrument, and the height $A\,M$ above the ground noted. Then the telescope of the transit was directed to the top R of the pier, and, with the rod still held at A, the height $A\,N$ was read on the rod. By subtracting $A\,M$ from $A\,N$, the distance $M\,N$ intercepted between the lines $O\,M_1$ and $O\,R$ was found to be 6 feet. What was the height $R\,H$?

SOLUTION.—The inclination of $A_1 R$ and that of PQ being equal we have, $A_1 H = JP$, and $HJ = RQ = 12$ ft. Now,

$$A_1 P = A_1 H + HJ + JP = 2 A_1 H + 12$$

whence, $$A_1 H = \frac{A_1 P - 12}{2} = \frac{22 - 12}{2} = 5 \text{ ft.}$$

The similar triangles OMN and OKR give

$$\frac{OM}{MN} = \frac{OK}{KR}; \quad KR = \frac{OK \times MN}{OM}$$

$$= \frac{(OM_1 + M_1 K) \times MN}{OM} = \frac{(OM_1 + A_1 H) \times MN}{OM}$$

$$= \frac{(250 + 5) \times 6}{20} = \frac{255 \times 6}{20} = 76.5 \text{ ft.}$$

Finally,

$$RH = RK + KH = RK + M_1 A_1 = 76.5 + 4.5 = 81 \text{ ft.} \quad \textbf{Ans.}$$

EXAMPLES FOR PRACTICE

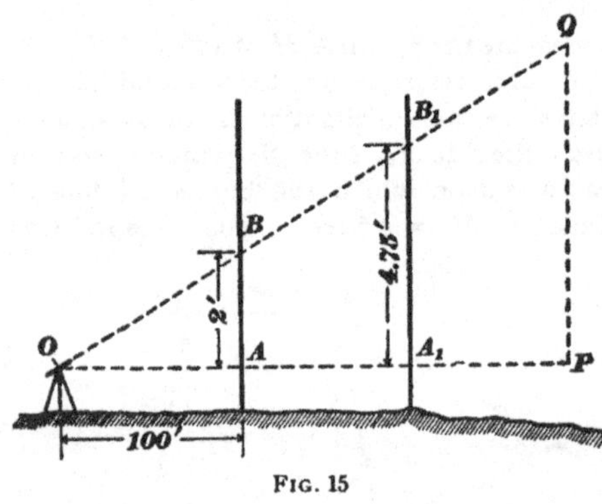

FIG. 15

1. In Fig. 15, the lines of sight OP and OQ of a transit intercept on a rod distances, $AB = 2$ feet and $A_1 B_1 = 4.75$ feet; if the distance OA is 100 feet, what is the distance OA_1?   Ans. 237.5 ft.

2. In order to find the stress in the member BD, Fig. 16, by the method of moments, it is necessary to find the distance DO from D to

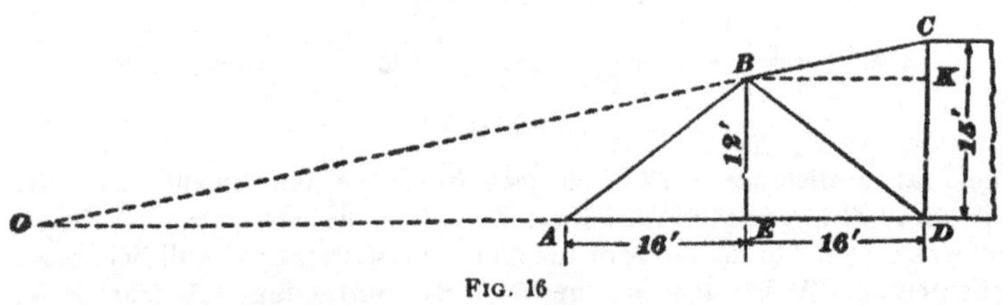

FIG. 16

the point of intersection O of DA and CB, both produced; the dimensions being as shown, what is that distance?   **Ans.** $DO = 80$ ft.

2. *A B C D*, Fig. 17, is a trapezoid whose non-parallel sides produced meet at *O*; the line *M N* is parallel
to the bases *A D* and *B C*; the dimensions
of *A D*, *B C*, *B M*, and *M A* being as
shown, find *O B* and *M N*.

Ans. $\begin{cases} O B = 315 \text{ ft.} \\ M N = 85.714 \text{ ft.} \end{cases}$

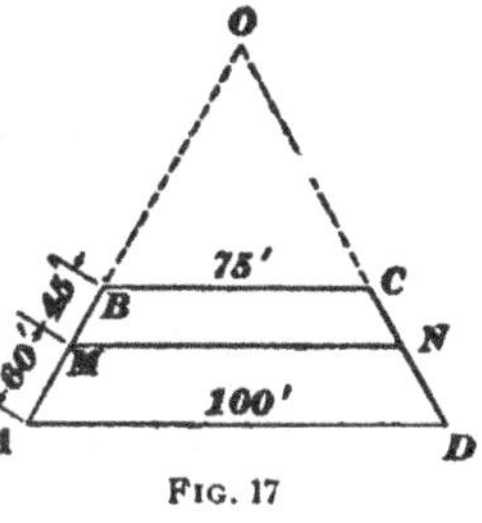

FIG. 17

4. In a triangle *A B C*, side *A B*
= 32 feet, *B C* = 34 feet, and *A C* = 48 feet;
if side *A' B'* of a similar triangle *A' B' C'* is
72 feet long, what are the lengths of the
other two sides?

Ans. $\begin{cases} A' C' = 108 \text{ ft.} \\ B' C' = 76.5 \text{ ft.} \end{cases}$

5. The base of a right triangle is 24 inches, and its altitude 72 inches;
at what distance from the top is the triangle 16 inches wide?

Ans. 48 in.

IMPORTANT CONSEQUENCES OF THE THEORY OF SIMILAR TRIANGLES

24. When the first of three quantities is to the second
as the second is to the third, the three quantities are in
continued proportion; the second is a **mean proportional** between the first and third; and the third is a **third
proportional** to the first and second. Thus, if $a : b = b : c$,
the three quantities a, b, and c are in continued proportion;
b is a mean proportional between a and c; and c is a third
proportional to a and b.

25. In a right triangle, as *A B C*, Fig. 18, the perpendicular *C D* drawn from the vertex of the right angle
to the hypotenuse, divides the triangle into two triangles
A C D and *C D B* that are similar to the whole triangle and
to each other.

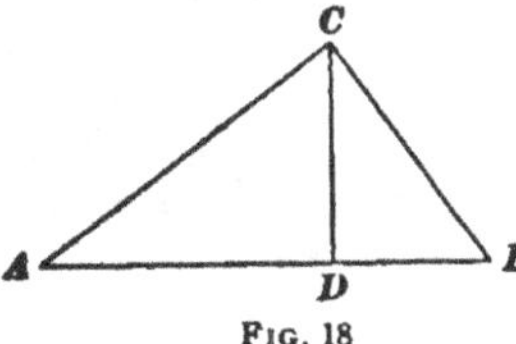

FIG. 18

The right triangles *A B C* and *A C D*
are similar, by Art. **17**, as the angle *A*
is common. Also, the triangles *A B C*
and *C B D*, having angle *B* in common,
are similar. Again, the triangles *A C D*
and *C B D*, being each similar to *A B C*,
are similar to each other.

26. In a right triangle, the perpendicular to the hypotenuse from the vertex of the right angle is a mean

proportional between the two parts or segments into which it divides the hypotenuse; that is, Fig. 18, $AD:CD = CD:DB$.

As the triangles BCD and ABC are similar, and the angle B is common, the angle BCD must equal the angle A, and similarly the angle ACD must equal the angle B. The triangles ACD and BCD are similar, hence the sides opposite equal angles are in proportion; that is,

$$\frac{AD \text{ (side opposite } ACD)}{CD \text{ (side opposite } B)} = \frac{CD \text{ (side opposite } A)}{DB \text{ (side opposite } BCD)}$$

Or, $$AD:CD = CD:DB$$

27. The side AC, Fig. 18, is a mean proportional between the whole hypotenuse and the segment AD on the same side of CD as the side AC; that is, $AB:AC = AC:AD$. Similarly, $AB:BC = BC:BD$.

The triangles ABC and ACD are similar, hence the sides opposite equal angles are proportional; that is,

$$\frac{AB \text{ (opposite right angle)}}{AC \text{ (opposite right angle)}} = \frac{AC \text{ (opposite } B)}{AD \text{ (opposite } ACD)}$$

Or, $$AB:AC = AC:AD$$

EXAMPLE 1.—In the right triangle ABC, Fig. 19, find the length of the perpendicular CD.

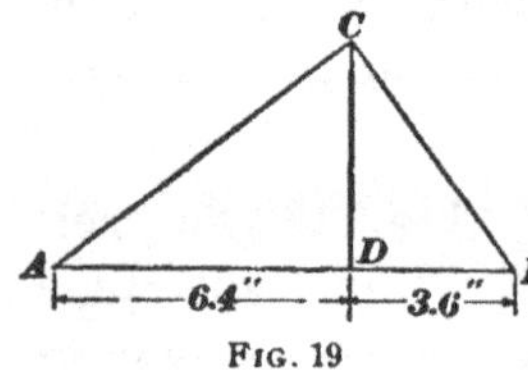
FIG. 19

SOLUTION.—The perpendicular is a mean proportional between the parts AD and DB into which it divides the hypotenuse; therefore,

$$6.4 : CD = CD : 3.6$$

whence, $$\overline{CD}^2 = 6.4 \times 3.6$$

and $CD = \sqrt{6.4 \times 3.6} = 4.8$ in. Ans.

EXAMPLE 2.—Find the length of the sides of the right triangle ABC, Fig. 20, in which CD is the perpendicular from the vertex of the right angle to the hypotenuse.

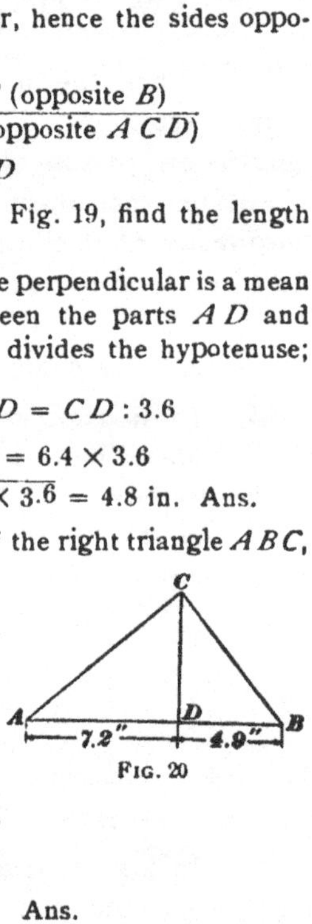

SOLUTION.—The hypotenuse is 7.2 in. + 4.9 in. = 12.1 in. The side CB is a mean proportional between the hypotenuse AB and the part DB; therefore,

$$12.1 : CB = CB : 4.9$$

$$\overline{CB}^2 = 12.1 \times 4.9$$

$$CB = \sqrt{12.1 \times 4.9} = 7.7 \text{ in. Ans.}$$

FIG. 20

The leg AC is a mean proportional between AB and AD; that is,

$$AB:AC = AC:AD$$
$$AC = \sqrt{AB \times AD}$$
$$= \sqrt{12.1 \times 7.2} = 9.34 \text{ in. } \textbf{Ans.}$$

28. Since an angle inscribed in a semicircle is a right angle, it follows from Arts. **26** and **27**, that:

(a) A perpendicular CD, Fig. 21, drawn from any point on the circumference of a circle to a diameter AB, is a mean proportional between the segments into which it divides the diameter; that is,

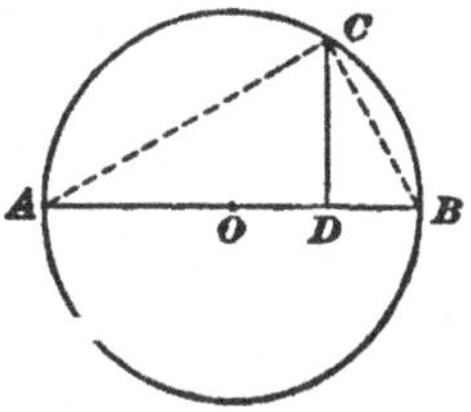

$$AD:CD = CD:DB$$

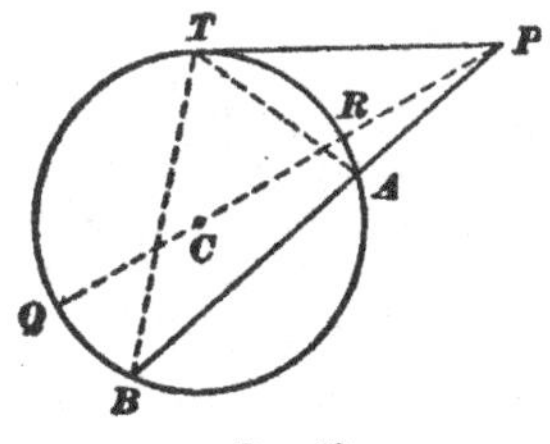

Fig. 21

(b) A chord CA drawn from a point in a circumference to the end of a diameter is a mean proportional between the whole diameter and the adjacent segment AD; that is,

$$AB:AC = AC:AD$$

29. If from a point without a circle, a tangent and a secant are drawn, the tangent is a mean proportional between the whole secant and the exterior segment; that is, in Fig. 22, $PB:PT = PT:PA$.

In the triangles BPT and APT, the angle P is common. The angle B, an inscribed angle, and the angle PTA, an angle formed by a tangent and a chord, are equal, since each is measured by one-half the same arc AT. Hence, the triangles are similar by Art. **16**, and

Fig. 22

angles are similar by Art. **16**, and

$$\frac{PB \text{ (opposite angle } PTB)}{PT \text{ (opposite angle } PAT)} = \frac{PT \text{ (opposite angle } B)}{PA \text{ (opposite angle } PTA)}$$
$$\overline{PT}^2 = PB \times PA$$
$$PT = \sqrt{PB \times PA}$$

30. If from a point without a circle any two secants are drawn, the product of one secant and its external segment is equal to the product of the other secant and its external segment.

I L T 36F—5

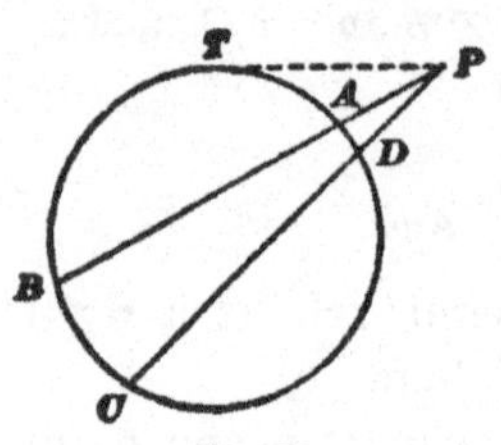

Fig. 23

In Fig. 23, PB and PC are secants. Draw the tangent PT. Then, from Art. **29**,

$$\overline{PT}^2 = PA \times PB$$

and $$\overline{PT}^2 = PC \times PD$$

hence, $$PA \times PB = PC \times PD$$

31. If any two chords be drawn through a point within a circle, the product of the segments of one is equal to the product of the segments of the other.

In Fig. 24, the angles D and B, being measured by one-half the arc AC, are equal. The angles BPC and DPA, being vertical angles, are equal. Hence, by Art. **16**, the triangles CBP and ADP are similar. Therefore,

$$\frac{AP}{CP} = \frac{PD}{PB}$$

and $$AP \times PB = CP \times PD$$

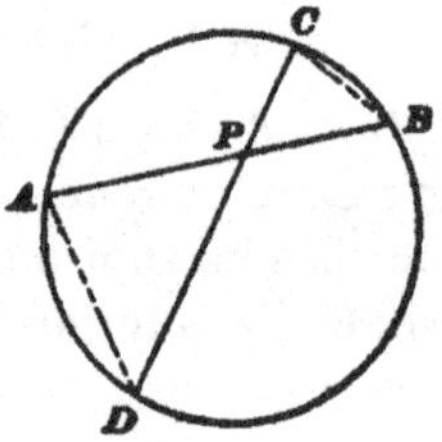

Fig. 24

EXAMPLES FOR PRACTICE

1. The perpendicular from the vertex of the right angle of a right triangle divides the hypotenuse into parts of 23.04 inches and 1.96 inches. Find: (*a*) the length of the perpendicular; (*b*) the length of the two sides of the triangle. Ans. $\begin{cases}(a)\ 6.72 \text{ in.}\\(b)\ 24 \text{ in. and } 7 \text{ in.}\end{cases}$

2. If, in Fig. 22, the distance CP of the point P from the center of the circle is 65 feet, and the radius CR is 25 feet, what is the length of the tangent PT? Ans. 60 ft.

3. The chord of the arc of a segment is 14 inches long and the height of the segment is 2 inches; what is the radius? Ans. $13\frac{1}{4}$ in.

OTHER SIMILAR POLYGONS

32. Two polygons are similar when they are composed of the same number of triangles similar each to each and similarly placed.

Thus, in Fig. 25, the polygons $ABCDE$ and $A'B'C'D'E'$ are composed of the same number of similar triangles similarly placed.

Since the triangle $A E D$ is similar to the triangle $A' E' D'$, angle E = angle E' and angle $A D E$ = angle $A' D' E'$. Also, in the similar triangles $A D C$ and $A' D' C'$, angle $A D C$ = angle $A' D' C'$. Hence, the sum of the angles $A D E$ and $A D C$, or the angle $E D C$, is equal to the sum of the angles $A' D' E'$ and $A' D' C'$, or the angle $E' D' C'$. In like manner, angle $D C B$ = angle $D' C' B'$,

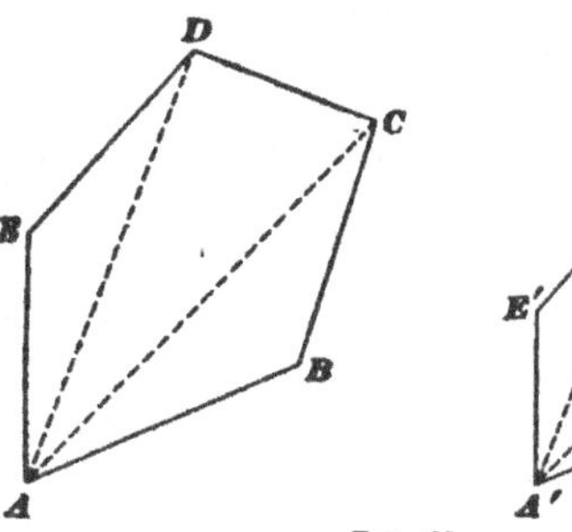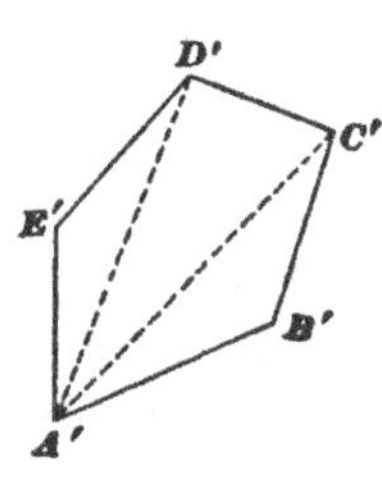

Fig. 25

angle B = angle B', and angle $B A E$ = angle $B' A' E'$. Since the triangles are similar,

$$E D : E' D' = A D : A' D' \text{ and } A D : A' D' = D C : D' C'$$

hence,
$$E D : E' D' = D C : D' C'$$

In like manner,

$$D C : D' C' = C B : C' B' = B A : B' A' = A E : A' E'$$

Therefore, as the angles of the one polygon are equal to the corresponding angles of the other and the sides of the one polygon are proportional to the sides of the other, the polygons are similar.

33. Two similar polygons can be divided into the same number of similar triangles similarly placed.

34. The perimeters of two similar polygons are in the same ratio as any two homologous sides.

In Fig. 25, let P be the perimeter of the polygon $A B C D E$, and P' the perimeter of the polygon $A' B' C' D' E'$. Since the polygons are similar

$$\frac{A E}{A' E'} = \frac{E D}{E' D'} = \frac{D C}{D' C'} = \frac{C B}{C' B'} = \frac{B A}{B' A'} \qquad (1)$$

Let each of these equal ratios be denoted by R; that is, let

$$\frac{A E}{A' E'} = R, \frac{E D}{E' D'} = R, \frac{D C}{D' C'} = R, \frac{C B}{C' B'} = R, \frac{B A}{B' A'} = R.$$

From these equations we obtain,

$$A E = R \times A' E', E D = R \times E' D', D C = R \times D' C',$$
$$C B = R \times C' B', B A = R \times B' A'$$

Adding the sides of these equalities,

$$A E + E D + D C + C B + B A$$
$$= R \times A' E' + R \times E' D' + R \times D' C' + R \times C' B' + R \times B' A'$$
$$= R (A' E' + E' D' + D' C' + C' B' + B' A')$$

whence
$$\frac{A E+E D+D C+C B+B A}{A' E'+E' D'+D' C'+C' B'+B' A'} = R$$

But
$$R = \frac{A E}{A' E'} = \frac{E D}{E' D'} = \frac{D C}{D' C'}, \text{ etc.};$$

therefore,
$$\frac{P}{P'} = \frac{A E}{A' E'} = \frac{E D}{E' D'} = \frac{D C}{D' C'}, \text{ etc.}$$

35. Equation (1) of the preceding article is a series of equal ratios, of which the numerators are the antecedents and the denominators the consequents. The general truth was shown in that article, that in a series of equal ratios the sum of the antecedents is to the sum of the consequents as any antecedent is to its consequent.

AREAS OF POLYGONS

36. Definitions.—The area of a surface is the superficial space included within its boundary lines. Area is expressed by the ratio of the surface to a surface of fixed value chosen as a unit and called the **unit of area.**

37. A square whose side is equal in length to the unit of length is usually taken as the unit of area, and its area is called the **square unit.** For example, if the unit of length is 1 inch, the unit of area, or square inch, is the square whose sides measure 1 inch, and the area of any surface is expressed by the number of square inches that the surface contains. If the unit of length were 1 foot, the unit of area would measure 1 foot on each side, and the area of the surface would be expressed in square feet. Square inch and square foot are abbreviated to sq. in. and sq. ft., respectively, and are often indicated by the symbols $\square''$ and $\square'$.

38. Two surfaces are **equivalent** when their areas are equal.

39. Comparison of the Areas of Two Rectangles. The areas of two rectangles $A B C D$ and $A' B' C' D'$, Fig. 26, having equal altitudes are to each other as their bases; that is, area $A B C D$: area $A' B' C' D' = A B : A' B'$.

Suppose that $A' B'$ is four-fifths of $A B$, or that $A B : A' B' = 5 : 4$. Divide $A B$ into five equal parts $A E$, $E F$, etc., and $A' B'$ into four equal parts $A'E'$, $E'F'$, etc. It is evident that $A'E' = AE$, for, since AB is to $A'B'$ in the ratio of 5 to 4, any quantity, as AE, that is contained five times in AB must be contained four times in $A'B'$. Through the points of division E, F, E', F', etc., draw perpendiculars

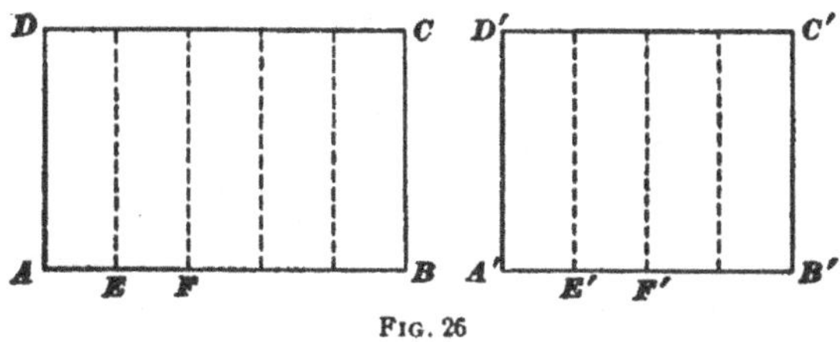

Fig. 26

to AB and $A'B'$. Each large rectangle is thus divided into small rectangles, all of which are equal. As $ABCD$ contains five, and $A'B'C'D'$ contains four, of the small rectangles, the ratio of the two large rectangles is that of 5 to 4, which is also the ratio of their bases.

40. Since any of the sides of a rectangle can be considered as the base, it follows that the area of two rectangles having equal bases are to each other as their altitudes.

41. The areas of any two rectangles are to each other as the products of their bases by their altitudes.

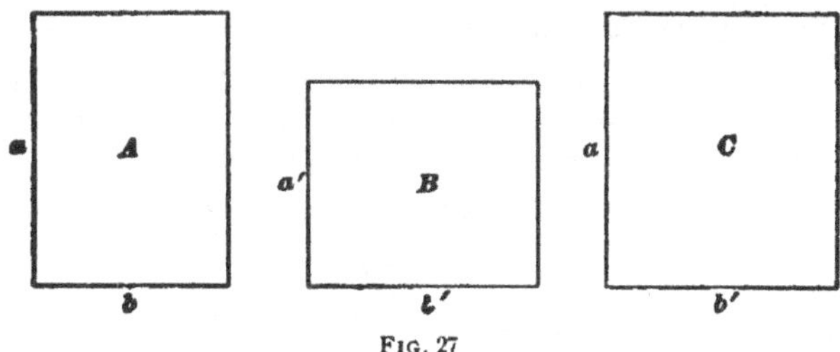

Fig. 27

Let A and B, Fig. 27, be two rectangles whose altitudes are a and a' and whose bases are b and b', respectively. Construct a rectangle C with an altitude a and a base b'. Then, by Arts. **39** and **40**,

$$A : C = b : b' \qquad (1)$$

and
$$C : B = a : a' \qquad (2)$$

Multiplying equation (1) by equation (2),

$$AC : BC = ab : a'b' \qquad (3)$$

Dividing the terms of the first member of equation (3) by C,

$$A : B = ab : a'b'$$

42. Area of a Rectangle.—The area of a rectangle is equal to its base multiplied by its altitude; that is, in Fig. 28,

$$A = b h.$$

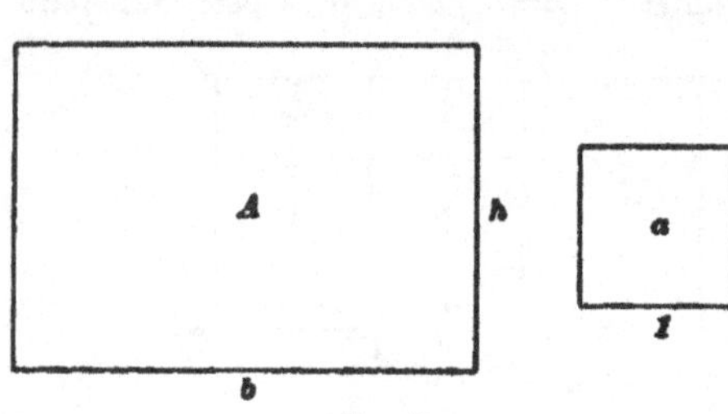

FIG. 28

Construct a unit square a. Then (Art. 41),

$$A : a = h \times b : 1 \times 1;$$

or,

$$\frac{A}{a} = \frac{h \times b}{1 \times 1}$$

But a is a unit square, and its area is therefore equal to 1; hence,

$$A = b h$$

43. Area of a Triangle.—The area of a right triangle is equal to one-half the product of the two legs of the triangle; that is, in Fig. 29, area $A B C = \frac{1}{2} a b$.

For the triangle $A B C$ is one-half the rectangle $A B C D$ and the area of the latter is $a b$.

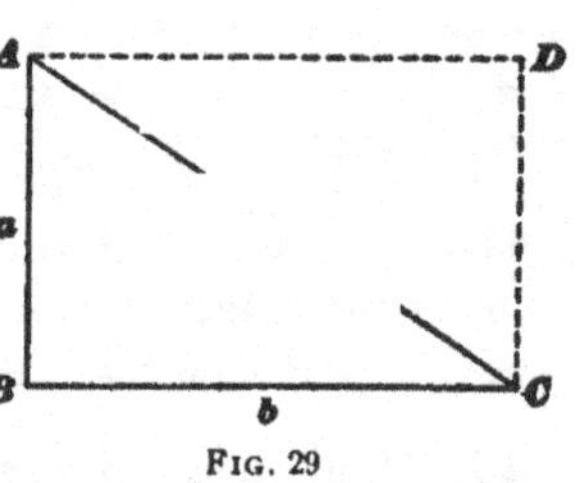

FIG. 29

44. The area of any triangle is equal to one-half the product of its base and altitude.

In Fig. 30 (a), let $A C$ be the base and $B H$ the altitude of the triangle $A B C$. The area $A B C$ is equal to the sum of the right triangles $A H B$ and $C H B$, which, by the last article, is

$$\tfrac{1}{2} B H \times A H + \tfrac{1}{2} B H \times H C = \tfrac{1}{2} B H \times (A H + H C) = \tfrac{1}{2} B H \times A C$$

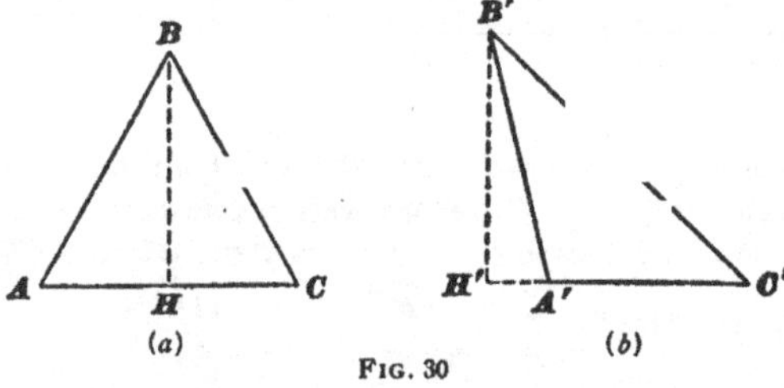

FIG. 30

In Fig. 30 (b), the area $A' B' C'$ is the difference between the areas of the right triangles $B' H' C'$ and $B' H' A'$; that is,

$$\tfrac{1}{2} B' H' \times H' C' - \tfrac{1}{2} B' H' \times H' A' = \tfrac{1}{2} B' H' \times (H' C' - H' A')$$
$$= \tfrac{1}{2} B' H' \times A' C'$$

Let b be the base, h the altitude, and A the area of any triangle; then,

$$A = \tfrac{1}{2} b h$$

45. Two triangles having the same base are to each other as their altitudes, and two triangles having the same altitude are to each other as their bases.

46. Two triangles having the same base and the same altitude are equivalent.

It should be borne in mind that any side of a triangle can be taken as the base, the altitude being the perpendicular to that side from the opposite vertex.

47. To find the area of a triangle from the lengths of its three sides, apply the following:

Rule.—*From half the sum of the three sides subtract each side separately; multiply together the half sum and the three remainders and extract the square root of the product.*

Let a, b, and c be the three sides of a triangle, and A the area; let

$$s = \tfrac{1}{2}(a + b + c)$$

Then
$$A = \sqrt{s(s - a)(s - b)(s - c)}$$

The geometrical proof of this rule is very laborious, and will not be given here. A proof will be found in *Trigonometry*.

Example.—What is the area of a triangle having two sides 19.8 feet long, and one side 28 feet long?

Solution.—It is immaterial which side is called a, b, or c.
$s = \dfrac{a + b + c}{2} = \dfrac{28 + 19.8 + 19.8}{2} = 33.8$; taking b and c as the short sides, $s - a = 33.8 - 28 = 5.8$, and $s - b$ and $s - c$ are each $33.8 - 19.8 = 14$. Then, applying the formula
$A = \sqrt{s(s - a)(s - b)(s - c)} = \sqrt{33.8 \times 5.8 \times 14 \times 14} = 196$ sq. ft., nearly.
Ans.

48. A triangle equivalent to any given polygon may be constructed as follows:

Let $ABCDEF$, Fig. 31, be the given polygon. Produce any of the sides, as AF, in both directions, as indicated by XY. This line will be referred to as the base. Starting from one of the ends of AF, as A, draw a diagonal AC forming a triangle with AB and BC. Draw BB_1 parallel to CA, meeting the base at B_1, and join C to B_1.

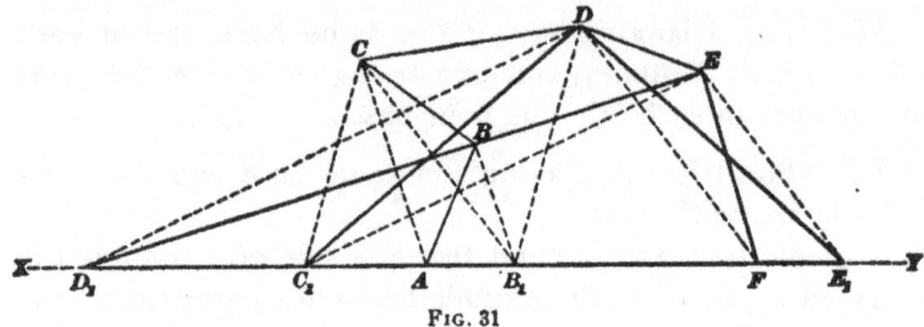

Fig. 31

The polygon B_1CDEF has one side less than the given polygon, and is equivalent to it. For

$$ABCDEF = B_1BCDEF + \text{triangle } B_1BA$$
$$B_1CDEF = B_1BCDEF + \text{triangle } B_1BC$$

The two triangles B_1BA and B_1BC are equivalent, for they have the common base B_1B, and their altitudes, being each equal to the distance between the parallels AC and B_1B, are equal. Proceeding with the polygon B_1CDEF as with the original polygon, draw the diagonal B_1D, forming a triangle with B_1C and CD. Draw CC_1 parallel to DB_1, and join D and C_1. It can be shown as before that the polygon C_1DEF is equivalent to B_1CDEF, and, therefore, to the original polygon. Finally, draw the diagonal C_1E, and DD_1 parallel to it, meeting the base at D_1. Then will the triangle D_1EF be the required triangle equivalent to the given polygon.

In practice, it is more convenient, as well as more accurate, to reduce about one-half of the polygon on one side of A and the rest on the other side of F. Thus, having reduced the polygon to the quadrilateral C_1DEF, the diagonal FD is drawn from F; EE_1 is drawn through E parallel to DF, and E_1 joined to D. This gives C_1DE_1 as the required triangle.

49. Area of a Parallelogram.—The area of a parallelogram is equal to its base multiplied by its altitude; that is, in Fig. 32, area $ABCD = AD \times MN$.

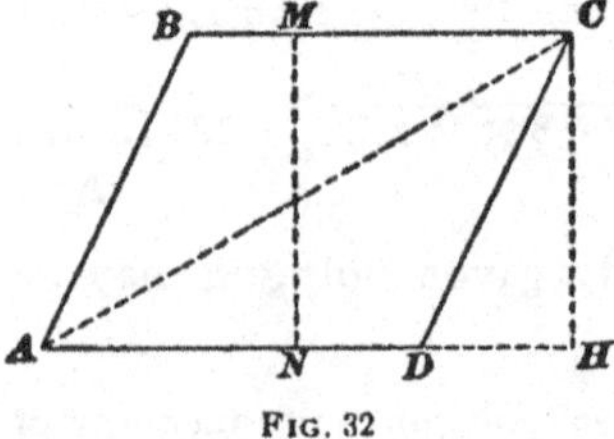

Fig. 32

For $ABCD$ is equal to the sum of the equal triangles ABC and ADC, or to twice either of them, as ADC; that is, $ABCD = 2 \times \frac{1}{2} AD \times CH = AD \times CH = AD \times MN$.

EXAMPLE 1.—What is the cost of paving a street 1,800 feet long and 36 feet wide with asphalt, the price being \$2 per square yard?

SOLUTION.—The surface to be covered is a rectangle whose sides are 36 ft. and 1,800 ft., or 12 yd. and 600 yd., and whose area is, therefore, $12 \times 600 = 7,200$ sq. yd. The cost of paving is, then, $2 \times 7,200 = \$14,400$. Ans.

EXAMPLE 2.—One side of a triangular plot of land is 125 feet long and the perpendicular distance from the opposite vertex to this side is 174.24 feet; it is desired to find a side of a rectangle that has the same area as the triangle and one side 75 feet long.

SOLUTION.—The area of the triangle is $\frac{1}{2} \times 125 \times 174.24 = 10,890$ sq. ft. Then the other side of the rectangle is $10,890 \div 75 = 145.2$ ft. Ans.

EXAMPLE 3.—Divide a triangular plot of land into any number of equal parts by lines from a vertex to the opposite side.

SOLUTION.—Divide the side opposite the vertex through which the lines are to be run into the required number of equal parts and run lines from the vertex of the triangle to the points of division. Then, since the triangles thus formed have equal bases and their vertexes in the same point, they are equivalent. Ans.

EXAMPLE 4.—Divide a given triangle into parts proportional to any given numbers by lines run through a vertex.

SOLUTION.—Let the given triangle be ABC, Fig. 33, and let it be required to divide it into parts proportional to 3, 4, and 5, by lines drawn from the vertex A.

The base BC is divided into parts proportional to the numbers 3, 4, and 5, by dividing it into $3 + 4 + 5 = 12$ equal parts, and then marking the third and the seventh points of division. From the points thus marked, lines are run to the vertex A. Then, by Art. 45,

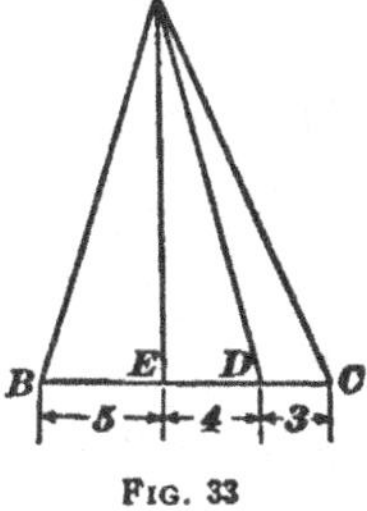

FIG. 33

$$CAD : ADE = 3 : 4$$

and, $$ADE : ABE = 4 : 5 \quad \text{Ans.}$$

EXAMPLES FOR PRACTICE

1. Find the area of a square whose side is 5 feet 9 inches.

Ans. 33.062 sq. ft

2. Find the area of a rhombus whose length is 12.5 feet, and whose height is 9.25 feet.   Ans. 115.62 sq. ft.

3. One side of a room is 16 feet long; if the floor contains 240 square feet, what is the length of the other side?   Ans. 15 ft.

4. In a trapezium two not adjacent sides are 16 and 14 inches, respectively. A diagonal divides the trapezium into two triangles whose altitudes from their vertexes to the given sides as bases are 17 inches and 3 inches, respectively; what is the area of the trapezium?   Ans. 157 sq. in.

5. The base BC of a triangle is 150 chains and the perpendicular from the opposite vertex A to BC is 45 chains; it is desired to divide the triangle into two parts equal in area by a line from A to BC; how far from B is D, the intersection of this line with BC?   Ans. 75 ch.

6. From the mid-point E of the side AB of a parallelogram $ABCD$, lines are drawn to the vertexes D and C and to the mid-point of the side CD; show that these lines divide the parallelogram into four triangles that are equal in area.

7. Find the area of a triangle whose three sides are 13, 14, and 15 feet.   Ans. 84 sq. ft.

8. Find the area of a right triangle whose hypotenuse is 50 feet and one of whose legs is 40 feet.   Ans. 600 sq. ft.

---

50. Area of a Trapezoid.—The area of a trapezoid is equal to one-half the sum of the parallel sides multiplied by the altitude; that is, in Fig. 34, area of trapezoid $ABCD = \frac{1}{2}(AB + DC) \times MN$.

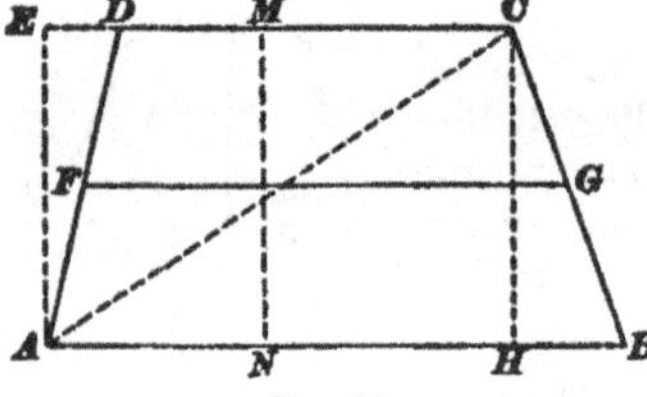

Fig. 34

The area of the trapezoid is equal to the sum of the areas of the two triangles ABC and ADC; hence,

$$ABCD = \frac{1}{2}AB \times CH + \frac{1}{2}DC \times AE$$
$$= \frac{1}{2}AB \times MN + \frac{1}{2}DC \times MN$$
$$= \frac{1}{2}(AB + DC) \times MN$$

Let b_1 = length of lower base;

b_2 = length of upper base;

h = altitude.

Then, the area A of the trapezoid $A\,B\,C\,D$ is

$$A = \tfrac{1}{2}(b_1 + b_2)h$$

51. Since the median line $F\,G$, Fig. 34, joining the mid-points of the non-parallel sides is equal to $\tfrac{1}{2}(A\,B + D\,C)$, the area of a trapezoid is equal to the product of the median line by the altitude.

EXAMPLE.—Divide a plot of ground in the form of a trapezoid into any number of equal parts by lines intersecting the two bases.

SOLUTION.—Divide each of the bases into the same number of equal parts into which the trapezoid is to be divided and run lines through the corresponding points of division. The trapezoids thus formed have equal bases and the same altitude and are, therefore, equal in area. Ans.

EXAMPLES FOR PRACTICE

1. The parallel sides of a trapezoid are 321.51 and 214.24 feet, and the perpendicular distance between them is 171.16 feet; what is the area of the trapezoid?   Ans. 45,849 sq. ft.

2. Find the area of a trapezoid whose parallel sides are 20.5 and 12.25 chains, the perpendicular distance between them being 10.75 chains.   Ans. 17.603 A.

3. The parallel sides of a trapezoidal plot of ground are 400 feet and 360 feet long; the distance between the parallel sides is 100 feet. It is desired to divide this plot into five lots by lines intersecting the parallel sides; what will be the length of the front and the rear of one of the lots?   Ans. 80 ft. and 72 ft.

4. How many square feet are there in a board 12 feet long, 18 inches wide at one end, and 12 inches wide at the other end?
Ans. 15 sq. ft.

---

52. Area of Any Polygon.—The area of any polygon can be found by dividing the polygon into triangles, determining the area of each triangle, and adding the results.

53. Comparison of the Areas of Similar Polygons. The areas of two similar triangles are to each other as the squares of their homologous sides.

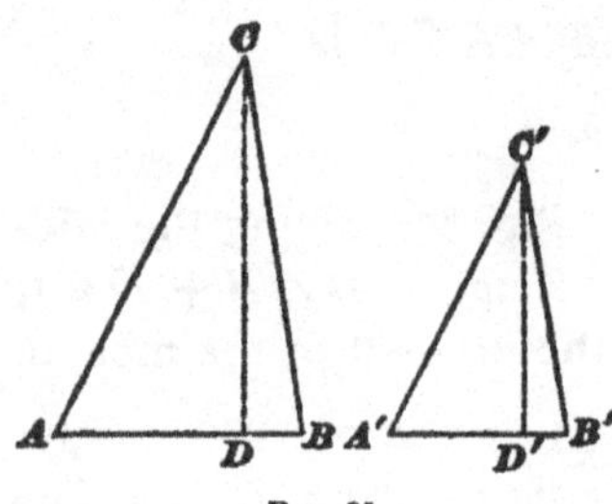

FIG. 35

In Fig. 35,

Area $ABC = \frac{1}{2} AB \times CD$ (1)

Area $A'B'C' = \frac{1}{2} A'B' \times C'D'$ (2)

Dividing equation (1) by equation (2), $\dfrac{ABC}{A'B'C'} = \dfrac{A'B}{A'B'} \times \dfrac{CD}{C'D'}$ (3)

but, by Art. 22, $\dfrac{CD}{C'D'} = \dfrac{AB}{A'B'}$;

hence, substituting in (3) $\dfrac{AB}{A'B'}$ for $\dfrac{CD}{C'D'}$, $\dfrac{ABC}{A'B'C'} = \dfrac{AB}{A'B'} \times \dfrac{AB}{A'B'} = \dfrac{\overline{AB}^2}{\overline{A'B'}^2}$

that is, $\qquad ABC : A'B'C' = \overline{AB}^2 \cdot \overline{A'B'}^2$

54. The areas of two similar triangles are to each other as the squares of any two homologous lines.

55. The areas of two similar polygons are to each other as the squares of their homologous lines.

By Art. **33,** two similar polygons can be divided into the same number of similar triangles. The sums of these triangles will, by Art. **35,** be to each other as any triangle of one polygon is to the corresponding triangle of the other. But these triangles are to each other as the squares of any two homologous lines. Hence, the sum of the triangles, or the polygons, are to each other as the squares of any two homologous lines.

EXAMPLE 1.—Divide a given triangle by a line parallel to the base into parts such that the given triangle shall be to the triangle cut off as $m : n$.

SOLUTION.—Let ABC, Fig. 36, be the given triangle, and ADE be the triangle cut off so that $ABC : ADE = m : n$. By Art. **18,** ADE and ABC are similar; hence, by Art. **53,**

$$ABC : ADE = \overline{AB}^2 : \overline{AD}^2$$

But by the conditions of the problem,

$$ABC : ADE = m : n$$

Therefore, $\overline{AB}^2 : \overline{AD}^2 = m : n$;

whence, $\quad AD = AB\sqrt{\dfrac{n}{m}}$. Ans.

When the triangle ABC is to be divided into two equal parts,

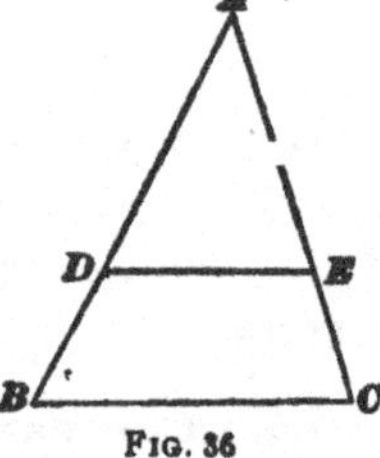

FIG. 36

$$AD = AB\sqrt{\dfrac{1}{2}} = .70711\, AB$$

EXAMPLE 2.—Let the length AB of example 1 be 32 chains and the area of ABC be 25.6 acres; what is the length of AD, if it is desired to make the triangle ADE contain 15 acres?

SOLUTION.—The area ABC is to be to the area of ADE as $25.6 : 15$; hence, $m : n = 25.6 : 15$.

Then, $\qquad AD = 32\sqrt{\dfrac{15}{25.6}} = 24.495$ ch.   Ans.

EXAMPLE 3.—Divide a given triangle ABC, by lines parallel to the base, into n equal parts.

SOLUTION.—Let ABC, Fig. 37, be the triangle, and DE, FG, HI, etc., divide it into n equal parts. Then ADE is one part, AFG is two parts, and so on. Hence,

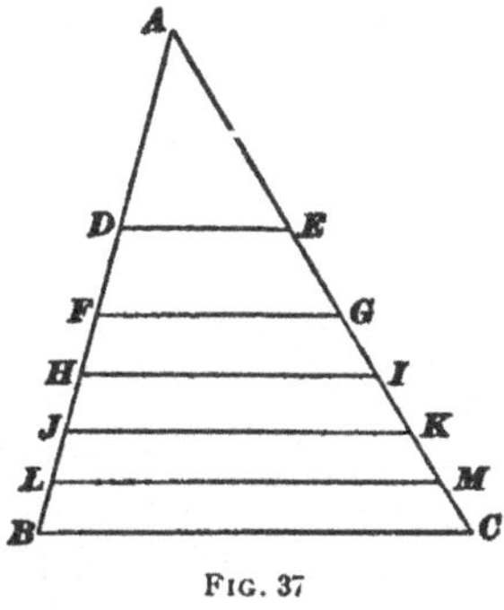

FIG. 37

$$ABC : ADE = n : 1$$
$$ABC : AFG = n : 2; \text{ etc.}$$

Then, by example 1,

$$AD = AB\sqrt{\frac{1}{n}}; \quad AF = AB\sqrt{\frac{2}{n}};$$

$$AH = AB\sqrt{\frac{3}{n}}; \text{ etc. Ans.}$$

EXAMPLE 4.—Two triangles ABC and $A'B'C'$ are similar. The sides of the triangle ABC are: $AB = 10$ inches, $BC = 21$ inches, $AC = 17$ inches, and in the triangle $A'B'C'$ the side $B'C' = 42$ inches; what is the area of the triangle $A'B'C'$?

SOLUTION.—In the triangle ABC, $s = \dfrac{10 + 17 + 21}{2} = 24$.   Then $s - a = 3$, $s - b = 7$, $s - c = 14$, and the area is $\sqrt{24 \times 3 \times 7 \times 14}$ $= 84$ sq. in. By the principle of Art. **53**,

$$\text{area of } A'B'C' : \text{area of } ABC = \overline{B'C'}^2 : \overline{BC}^2,$$

that is, $\qquad$ area of $A'B'C' : 84 = 42^2 : 21^2$

But $\qquad\qquad\qquad 42^2 : 21^2 = 4 : 1$

hence, $\qquad\qquad$ area of $A'B'C' : 84 = 4 : 1$

whence, $\qquad$ area of $A'B'C' = 4 \times 84 = 336$ sq. in. **Ans.**

EXAMPLES FOR PRACTICE

1. Suppose that the sides of the triangle $A'B'C'$ in example 4 of Art. **55** are $A'B' = 20$ inches, $B'C' = 42$ inches, and $C'A' = 34$ inches; show that the answer that is given to the example is correct.

2. The triangles ABC and $A'B'C'$ are similar; being given $BC = 13$ inches, $CA = 14$ inches, $AB = 15$ inches, and $B'C' = 19.5$ inches; find the area of the triangle $A'B'C'$.   Ans. 189 sq. in.

3. Let $A B$, one side of a triangle $A B C$, be 60 chains long, and let it be required to divide, by lines parallel to $B C$, the triangle $A B C$ into five equal parts. (*a*) What are the lengths of the lines $A D$, $A F$, $A H$, and $A T$? (*b*) Let the area of $A B C$ be 120 acres; by means of Art. **53**, prove your results.

Ans. $\begin{cases} A D = 26 \text{ ch. } 83.3 \text{ l.} \\ A F = 37 \text{ ch. } 94.7 \text{ l.} \\ A H = 46 \text{ ch. } 47.6 \text{ l.} \\ A T = 53 \text{ ch. } 66.6 \text{ l.} \end{cases}$

4. Find the lengths of $A D$ and $A F$ when the triangle of example 3 is divided into three parts, whose areas shall be proportional to the numbers 3, 4, and 5.

Ans. $\begin{cases} A D = 30 \text{ ch.} \\ A F = 45 \text{ ch. } 82.6 \text{ l.} \end{cases}$

HINT.—This is the same as if the triangle were divided into $3 + 4 + 5$ equal parts and $A D E$ contained three, and $A F G$, seven of these equal parts.

---

56.   The Theorem of Pythagoras.—In any right triangle, the square described on the hypotenuse is equivalent to the sum of the squares described on the other two sides.

Let $A B C$, Fig. 38, be a right triangle. Draw an equal triangle in the position $C B' C'$, so that $C B'$ will be in the prolongation of $B C$. Construct the squares $A B D E$ and $B' C' F D$ on $A B$ and $B' C'$, respectively. Since $M + N_1$ ($= M + N$) is a right angle, $A C C'$ is also a right angle. Produce $E F$ to A', making $F A' = B A = D E$.

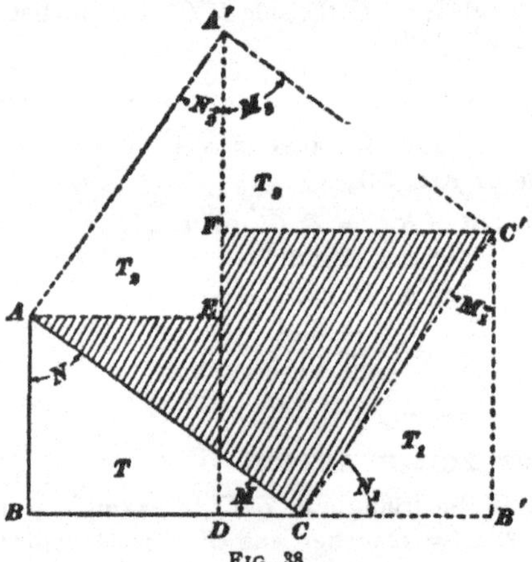

FIG. 38

Then, since $E F$ is the difference between $D F$ and $D E$, or $B C$ and $A B$, $E A' = B C$. Draw $A A'$ and $C' A'$. Each of the right triangles T_2 and T_3 is equal to T, since their legs are respectively equal. The quadrilateral $A C C' A'$, having all its sides equal and a right angle C, is a square—the square on the hypotenuse $A C$. This square is equal to the shaded figure plus the sum of the triangles T_2 and T_3; or to the shaded figure plus twice the triangle T. The sum of the squares $A B D E$ and $B' C' F D$ is equal to the shaded figure plus the sum of the triangles T and T_1, or to the shaded figure plus twice the triangle T. Therefore, square $A C C' A' = $ square $A B D E +$ square $B' C' F D$.

A particular case of the proposition just proved is shown in Fig. 39.

Let c be the hypotenuse, and a and b the other two sides of any right triangle.

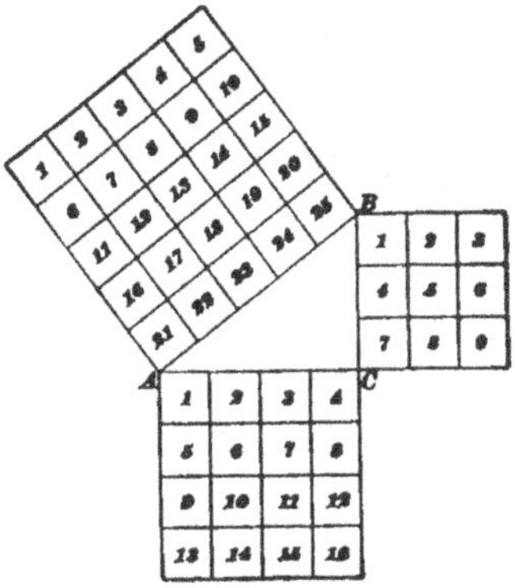

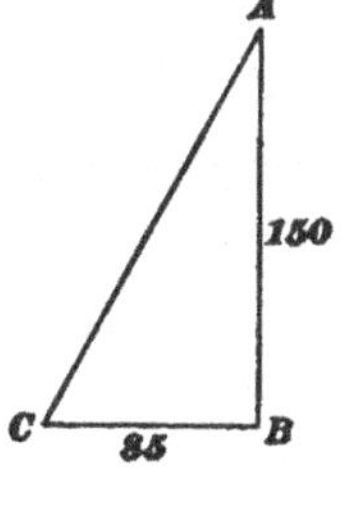

FIG. 39

Then,

$$c^2 = a^2 + b^2 \qquad (1)$$
$$c = \sqrt{a^2 + b^2} \qquad (2)$$
$$a = \sqrt{c^2 - b^2} \qquad (3)$$

Formula **3** may be written

$$a = \sqrt{(c - b)(c + b)} \qquad (4)$$

EXAMPLE 1.—If $AB = 3$ inches and $BC = 4$ inches, what is the length of the hypotenuse AC, Fig. 38?

SOLUTION.—

$$AC = \sqrt{\overline{AB}^2 + \overline{BC}^2}$$
$$= \sqrt{3^2 + 4^2} = \sqrt{25} = 5 \text{ in.} \quad \text{Ans.}$$

EXAMPLE 2.—The side given is 3 inches ($= b$, say), the hypotenuse is 5 inches ($= c$); what is the length of the other side?

SOLUTION.—Applying formula **4**, Art. **56**,

$$a = \sqrt{(5 - 3)(5 + 3)} = \sqrt{16} = 4 \text{ in.} \quad \text{Ans.}$$

Also, $\qquad a = \sqrt{c^2 - b^2} = \sqrt{5^2 - 3^2} = 4 \text{ in.}$ Ans.

EXAMPLE 3.—If, from a church steeple that is 150 feet high, a rope is to be attached at the top and to a stake in the ground 85 feet from its foot (the ground being supposed to be level), what must be the length of the rope?

SOLUTION.—In Fig. 40, AB represents the steeple 150 ft. high; C, a stake 85 ft. from the foot of the steeple; and AC, the rope. Here we have a triangle right-angled at B, of which AC is the hypotenuse. The square of $AC = 85^2 + 150^2$ $= 7{,}225 + 22{,}500 = 29{,}725$. Therefore,

$$AC = \sqrt{29{,}725} = 172.4 \text{ ft., nearly.} \quad \text{Ans.}$$

FIG. 40

EXAMPLE 4.—Referring to Fig. 16, it is required to find the length of the post AB and that of the member BC.

SOLUTION.—Draw BK parallel to ED. Then, $BK = ED = 16$ ft. and $CK = CD - DK = CD - EB = 15 - 12 = 3$ ft. The right triangles AEB and BCK give

$$AB = \sqrt{\overline{AE}^2 + \overline{EB}^2} = \sqrt{16^2 + 12^2} = \sqrt{400} = 20 \text{ ft.}\quad \textbf{Ans.}$$

$$BC = \sqrt{\overline{BK}^2 + \overline{CK}^2} = \sqrt{16^2 + 3^2} = \sqrt{265} = 16.279 \text{ ft.}\quad \textbf{Ans.}$$

EXAMPLES FOR PRACTICE

1. If the two sides about the right angle in a right triangle are 52 and 39 feet long, how long is the hypotenuse?   Ans. 65 ft.

2. A ladder 65 feet long reaches to the top of a house when its foot is 25 feet from the house; how high is the house, supposing the ground to be level?   Ans. 60 ft.

3. The shortest distance from a point to a line is 25 inches; the distances from this point to the extremities of the line are 54 inches and 40 inches, respectively; what is the length of the line?

Ans. 79.08 in.

4. Show that the diagonal of a square is equal to the side multiplied by $\sqrt{2}$.

REGULAR POLYGONS

57. A **regular** polygon is a polygon that has equal sides and equal angles, that is, it is equilateral and equiangular.

58. A circle can be circumscribed about any regular polygon.

Take any three vertexes of the regular polygon $ABCDE$, Fig. 41, as the vertexes A, B, C, and pass a circle through them. Let O be the center of this circle. Join O to A, B, C, D, and E. The polygon being equiangular, the angle $ABC =$ angle BCD. The angles OCB and OBC, being opposite equal sides OC and OB of the triangle OBC, are equal. Hence,

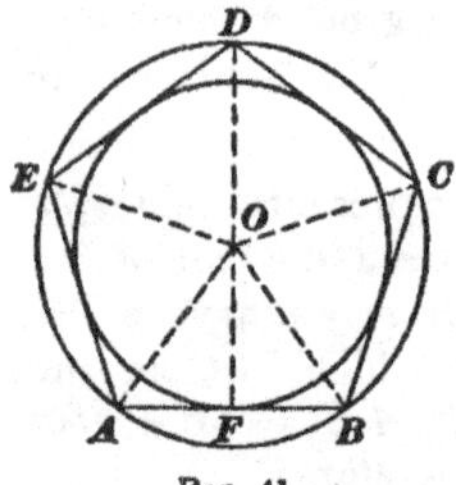

$$ABC - OBC = BCD - OCB$$
or,
$$ABO = OCD$$

FIG. 41

The polygon being equilateral, the sides AB and CD are equal. Hence, the triangles AOB and OCD, having two sides and included angle of one equal to two sides and included angle of the other equal, are equal. Therefore, $OD = OA$, and a circle passing through A,

B, and C must pass through D. In like manner, it can be shown that the circle passes through E.

59. A circle can be inscribed in any regular polygon.

In Fig. 41, OA, OB, OC, OD, and OE, being radii of the circum-scribed circle, are equal and divide the polygon into equal isosceles triangles that have a common vertex O. The altitudes of these equal triangles are equal, hence the perpendicular distances, as OF, from O to each of the sides are the same. Therefore, a circle drawn with O as center and a radius equal to OF will be inscribed in the regular polygon.

60. The **center** of a regular polygon is the common center of the circumscribed and the inscribed circle.

61. The **radius** of a regular polygon is the radius of the circumscribed circle, as OA, Fig. 41.

62. The **apothem** of a regular polygon is the radius of the inscribed circle, as OF, Fig. 41.

63. The **angle at the center** of a regular polygon is the angle included by the radii drawn to the extremities of any side.

64. The angle at the center of any regular polygon is equal to four right angles, or $360°$, divided by the number of the sides.

65. If n is the number of sides of a regular polygon, the sum of its interior angles is $2(n-2)$ right angles (see *Geometry*, Part 1), or, $90° \times 2(n-2) = 180° \times (n-2)$, and, since all the angles are equal, each angle is equal to $\dfrac{180° \times (n-2)}{n} = 180° - \dfrac{360°}{n}$. Since this value depends only on the number of sides, all regular polygons of the same number of sides have the same angles.

66. Regular polygons of the same number of sides are similar; their perimeters are to each other as any two homologous lines, and their areas are to each other as the squares of any two homologous lines.

67. The area of a regular polygon is equal to one-half the product of the perimeter and the apothem.

Let l be the side MN of a regular polygon, Fig. 42, n the number of sides, $p(= nl)$ the perimeter, $a(= OF)$ the apothem, and A the area.   As A is equal to the sum of n triangles, each equal to MON, we have,

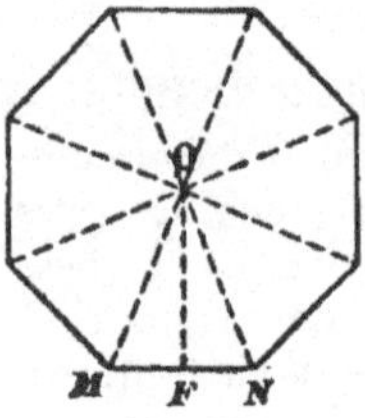

$$A = (\tfrac{1}{2} MN \times OF) \times n = \tfrac{1}{2} l a \times n = \tfrac{1}{2} n l \times a,$$

or, $$A = \tfrac{1}{2} p a$$

EXAMPLE.—Find the area of a regular pentagon whose side is 25 feet and apothem is 17.2 feet.

FIG. 42

SOLUTION.—The figure is a pentagon, hence it has five sides.   The perimeter is 5×25 and the area is $\dfrac{5 \times 25 \times 17.2}{2}$ = 1,075 sq. ft. Ans.

68.   The areas of regular polygons each of whose sides is equal to 1 are given in the following table:

TABLE I

AREAS OF REGULAR POLYGONS

Name	Number of Sides	Area When Side = 1	Name	Number of Sides	Area When Side = 1
Triangle .	3	.4330	Octagon .	8	4.8284
Square . .	4	1.0000	Nonagon .	9	6.1818
Pentagon .	5	1.7205	Decagon .	10	7.6942
Hexagon' .	6	2.5981	Undecagon	11	9.3656
Heptagon .	7	3.6339	Dodecagon	12	11.1960

From the principle of Art. **55**, the following rule is derived:

Rule.—*To find the area of any regular polygon, square the length of a side and multiply by the area of the similar polygon whose side is equal to the unit of length.*

Let A = area; l = length of side of required polygon; a = area of similar polygon whose side is 1; then, by Art. **55**,

$$A : a = l^2 : 1^2$$

whence, $$A = a l^2$$

EXAMPLE.—The side of a regular octagon is 3 inches, find its area.

SOLUTION.—From the table, the area of a regular octagon whose side is 1 in. is 4.8284 sq. in.   Hence, the area of the octagon whose side is 3 in. is 4.8284×3^2 = 43.456 sq. in. Ans.

69. If the vertexes of a regular inscribed polygon are joined to the middle points of the arcs subtended by the sides of the polygon, the joining lines form a regular inscribed polygon of double the number of sides. Thus, the octagon $AFBG$, etc., Fig. 43, is formed by joining the middle points of the arcs subtended by the sides of the square $ABCD$.

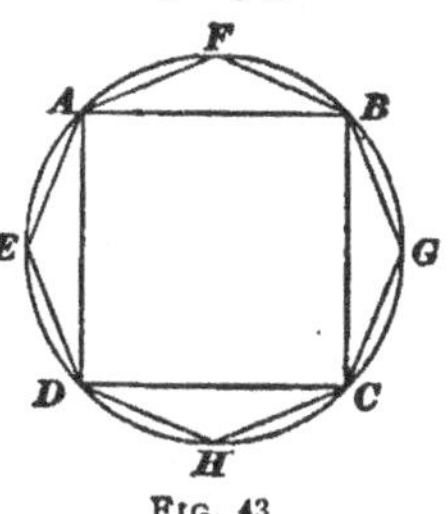

Fig. 43

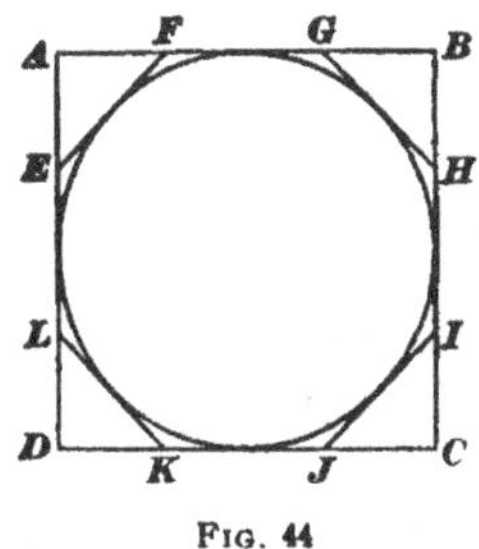

Fig. 44

70. If tangents are drawn at the middle points of the arcs between adjacent points of contact of the sides of a regular circumscribed polygon, a regular circumscribed polygon of double the number of sides is formed. Thus, in Fig. 44, the octagon $EFGH$, etc., is formed by drawing tangents at the middle points of the arcs between adjacent points of contact of the sides of the circumscribed square $ABCD$.

CIRCULAR MEASUREMENTS

THE CIRCLE

LENGTH OF ANY ARC

71. If any two circles are taken, and two regular polygons of the same number of sides are inscribed in them, the perimeters of these polygons are to each other as the radii of the circles (Art. **66**). This relation holds whatever the number of sides of the polygon. Now, it is evident that, as this number increases, the perimeters of the two polygons approach the circumferences of their respective circles. We may, therefore, consider these circumferences as extreme cases of the perimeters of regular polygons, in which the number of sides is increased indefinitely; whence we conclude that the circumferences, also, are to each other as their radii.

If c and c' are the circumferences of any two circles, and r and r' their respective radii, we may write,

$$c : c' = r : r'$$

whence,

$$c : r = c' : r'$$

or,

$$\frac{c}{r} = \frac{c'}{r'}$$

Dividing both numbers by 2, and denoting the diameters by d and d',

$$\frac{c}{2r} = \frac{c'}{2r'}$$

that is,

$$\frac{c}{d} = \frac{c'}{d'}$$

As c and c' are any two circumferences, it is seen that the ratio obtained by dividing any circumference by its diameter is the same for all circumferences. This ratio is usually denoted by the Greek letter π (pronounced *pi*). We have, therefore, for any circle,

$$\frac{c}{d} = \pi$$

whence, $\qquad c = \pi d = 2\pi r$

72. The quantity π can be determined by elementary geometrical methods, which may be found in treatises on geometry; but these methods are very laborious. A much better method is afforded by the theory of series, which is treated in works on trigonometry and the differential calculus. It is found that π cannot be expressed as an exact fraction, either decimal or vulgar. Its value can, however, be calculated to any desired degree of approximation. The following value is approximate to fifteen decimal places:

$$\pi = 3.1415926535897 93 +$$

For nearly all practical purposes, 3.1416 is a sufficiently close value. This value is used very generally, and will be used in this Course, unless otherwise stated. The student should commit it to memory. A value that is often used in rough calculations is $\frac{22}{7}$; it can be used when no more than three significant figures are required in the result.

73. The length of an arc, when the number of degrees in the arc and the radius of the circle are given, may be found as follows:

The length of the arc is evidently the same part of the length of the circumference $(2\pi r)$ as the number of degrees in the arc is of the number of degrees in the whole circumference, or 360°. Thus, if n is the number of degrees in the arc, and l is its length, we shall have,

$$\frac{2\pi r}{l} = \frac{360}{n}$$

whence, $\qquad l = \frac{\pi r n}{180}$

In applying this formula, minutes and seconds should be expressed as fractions of a degree.

EXAMPLE 1.—Find the length of a rope that will go around a wheel or drum 7.5 feet in diameter.

SOLUTION.—The required length is equal to the length c of the circumference of the wheel or drum. Here $d = 7.5$ ft., and, taking
$\pi = 3.1416$, we have, by formula of Art. **71,**
$$c = 3.1416 \times 7.5 = 23.562 \text{ ft.} \quad \text{Ans.}$$
Using $\frac{22}{7}$ for π, the result, to three significant figures, is
$$c = \frac{22}{7} \times 7.5 = 23.6 \text{ ft.} \quad \text{Ans.}$$

EXAMPLE 2.—Find the diameter of a circular race track 1 mile
in length.

SOLUTION.—Here c is given ($= 1$ mi. $= 5,280$ ft.) and the quantity
required is d. From the formula $c = \pi d$, we get
$$d = \frac{c}{\pi} = \frac{5,280}{3.1416} = 1,680.7 \text{ ft.} \quad \text{Ans.}$$

EXAMPLE 3.—What is the length of a railroad circular curve having
a radius of 1,540 feet and subtending an angle at the center equal
to 26° 35′?

SOLUTION.—To apply formula of Art. **73,** we have $r = 1,540$ ft.,
$n = 26\frac{35}{60}° = 26.583°$, nearly. Therefore,
$$l = \frac{3.1416 \times 1,540 \times 26.583}{180} = 714.50 \text{ ft.} \quad \text{Ans.}$$

74. When only the chord $A B$, Fig. 45, of an arc and the
height, or "rise," $C D$ of the segment are known, the following approximate method gives good results. $A C$, the chord
of half the arc, has the value
$$A C = \sqrt{\overline{A D}^2 + \overline{C D}^2} = \sqrt{\left(\frac{A B}{2}\right)^2 + \overline{C D}^2}$$

Then, to find the length of the arc:

RULE.—*From eight times the chord of half the arc, subtract
the chord of the whole arc and divide the remainder by 3.*

That is,
$$\text{arc } A C B = \frac{8 \times A C - A B}{3}$$

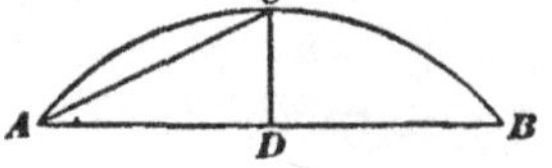

Let c = chord of whole arc;
h = height of segment;
l = length of arc.

FIG. 45

Then, $$A C = \sqrt{\frac{c^2}{4} + h^2} = \frac{1}{2}\sqrt{c^2 + 4 h^2}$$

and $$l = \frac{4 \sqrt{c^2 + 4 h^2} - c}{3}$$

This formula gives the length of an arc less than one-sixth of the circumference correct to four figures, and it gives the length of an arc less than one-third of the circumference correct to three figures.

EXAMPLE.—Find the length of the arc $A\,C\,B$, Fig. 46.

SOLUTION.—In this example, $c = 72$, $h = 8$. Therefore,

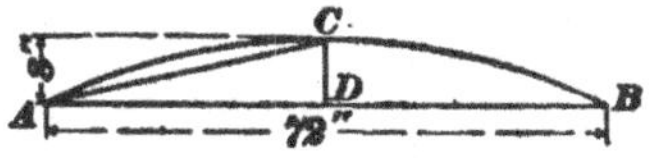

$$l = \frac{4\sqrt{72^2 + 4 \times 8^2} - 72}{3} = 74.34 \text{ in.}$$

Ans.

FIG. 46

75. For very flat arcs, that is, when $\dfrac{h}{c}$ is very small (say not greater than .1), the following approximate formula may be used, the notation being the same as in the preceding article:

$$l = c + \frac{8\,h^2}{3\,c}$$

EXAMPLE 1.—Find the length of the arc $A\,B$, Fig. 46.

SOLUTION.—

$$l = 72 + \frac{8 \times 8^2}{3 \times 72} = 72 + 2.37 = 74.37. \quad \text{Ans.}$$

This is not a very close approximation, because the ratio $\dfrac{h}{c}\left(= \dfrac{8}{72} = \dfrac{1}{9}\right)$ is not very small; however, the approximate value thus found would be close enough for most practical purposes.

EXAMPLE 2.—The chord of a railroad curve is 675 feet long, and the rise (or, ''middle ordinate,'' as the rise is called in railroad work) is 40 feet; what is the length of the curve?

SOLUTION.—Here $c = 675$, $h = 40$, and therefore

$$l = 675 + \frac{8 \times 40^2}{3 \times 675} = 675 + 6.32 = 681.32 \text{ ft.} \quad \text{Ans.}$$

76. Circular Measure of an Angle.—The following equation follows from the formula of Art. **73**:

$$\frac{l}{r} = \frac{\pi n}{180} = \frac{\pi}{180} \times n$$

If we assume the radius to be 1, then

$$l = \frac{\pi}{180} \times n \qquad (1)$$

This equation gives the length of the arc that the angle subtends on a circle whose radius is equal to unity. The length of such arc is called the **circular measure** of the angle, and the angle is often referred to by stating that measure. Thus, an angle of 1.34, circular measure, means an angle that subtends an arc of length 1.34 on a circle whose radius is 1. An angle expressed in circular measure is also said to be expressed **in radians.**

If in equation 1 we make $n = 180°$, we obtain, for the circular measure of 180°, $l = \pi$, that is, 180° is equivalent to π radians. Likewise, 90° is equivalent to $\dfrac{\pi}{2}$ radians, etc.

EXAMPLES FOR PRACTICE

1. Find the distance around the outside of a waterwheel whose outside diameter is 22 feet 8 inches. Ans. 71.21 ft.

2. The wheel of a carriage is observed to turn 375 times in going from a certain place to another; the diameter of the wheel is 3.5 feet; what is the distance between the two places? Ans. 4,123.4 ft.

3. A circular column measures 45.5 inches around the outside; what is its diameter? Ans. 14.483 in.

4. A belt covers an arc of 50° on a pulley whose diameter is 5 feet; what length of the belt is in contact with the pulley? Ans. 2.1817 ft.

5. How long will it take a train to move over a curve subtending an angle of 100°, the radius of the curve being 1,800 feet, and the train going at the rate of 20 miles an hour? Ans. 1.79 min.

6. The length of arc of a circle is equal to the radius; find the number of degrees in the arc. Ans. 57.3° = 57° 18', nearly

7. The chord of a railroad curve is 600 feet long and the middle ordinate is 80 feet; what is the length of the curve? Ans. 628 ft.

AREAS BOUNDED BY CIRCULAR ARCS

77. The area of a circle is equal to one-half the product of its circumference and radius (Art. **67**). This at once follows by considering the circle as an extreme case of a regular polygon.

Let A = area of circle;

c = circumference of circle;

r = radius of circle.

Then, $A = \frac{1}{2}cr$

or, since $c = 2\pi r$,

$$A = \frac{1}{2}\,2\,\pi r \times r$$

or, simplifying,

$$A = \pi r^2 = 3.1416\,r^2 \qquad (1)$$

Writing $\frac{d}{2}$ for r, we obtain for the area in terms of the diameter,

$$A = \frac{\pi d^2}{4} = .7854\,d^2 \qquad (2)$$

These formulas serve likewise to find r or d when A is given. Since $2\pi r = c$, we have

$$r = \frac{c}{2\pi}, \text{ and } \pi r^2 = \pi\left(\frac{c}{2\pi}\right)^2 = \frac{c^2}{4\pi}$$

that is, $A = \dfrac{c^2}{4\pi}$ $\qquad (3)$

This formula gives the area of a circle when its circumference is known.

Example 1.—The steam pressure on a piston is 75 pounds per square inch, and the diameter of the piston is 15 inches; what is the pressure on the whole surface of the piston?

Solution.—The required pressure is evidently seventy-five times the number of square inches in the surface of the piston, or seventy-five times the area A of the piston. Here $d = 15$ in., and formula **2** gives

$$A = .7854 \times 15^2$$

whence the total pressure is

$$75 \times .7854 \times 15^2 = 13{,}254 \text{ lb. } \textbf{Ans.}$$

Example 2.—The distance around a circular park is 2.75 miles; what is the area of the park, in acres?

Solution.—Here c is given equal to 2.75 mi. = (2.75 × 80) ch. Therefore, the area of the park, in square chains, is (formula **3**)

$$\frac{(2.75 \times 80)^2}{4 \times 3.1416}$$

The area, in acres, is one-tenth of this, or

$$\frac{1}{10} \times \frac{(2.75 \times 80)^2}{4 \times 3.1416} = \frac{220^2}{125.664} = 385.15 \text{ A. } \textbf{Ans.}$$

Example 3.—What must be the diameter of a circular sewer pipe that its cross-section may be 12.75 square feet?

Solution.—Solving formula **2** for d,

$$d = \sqrt{\frac{A}{.7854}} = \sqrt{\frac{12.75}{.7854}} = 4.03 \text{ ft. } \textbf{Ans.}$$

EXAMPLES FOR PRACTICE

1   The cable of a suspension bridge measures 40 inches around its circumference; find: (*a*) the diameter *d* of the cable; (*b*) the area *A* of the cross-section.

$$\text{Ans.}\begin{cases}(a)\ d = 12.732 \text{ in.}\\(b)\ A = 127.32 \text{ sq. in}\end{cases}$$

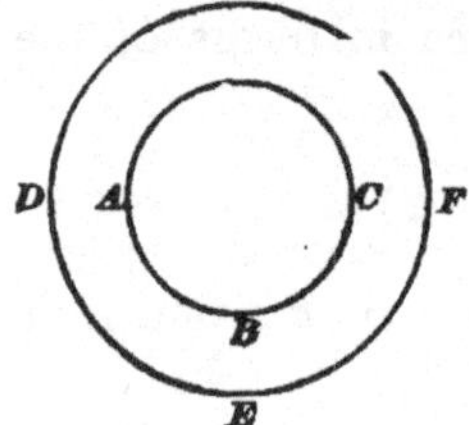

FIG. 47

2.   Find a formula for the area *A* of the space enclosed between two circles *A B C* and *D E F*, Fig. 47, the diameter of the outer circle being *D*, and that of the inner circle *d*.

$$\text{Ans.}\begin{cases}A = \frac{\pi}{4}(D^2 - d^2)\\[2mm]A = \frac{\pi}{4}(D + d)(D - d)\end{cases}$$

3.   What must be the inner diameter of a circular chimney, that its inner cross-section may be 14 square feet?   Ans. 4.2221 ft

4.   The diameter of a circular airway of a mine is 10 feet; find: (*a*) the circumference *c*; (*b*) the area *A* of the cross-section.

$$\text{Ans.}\begin{cases}(a)\ c = 31.416 \text{ ft.}\\(b)\ A = 78.54 \text{ sq. ft.}\end{cases}$$

------

78.   A sector is the same part of a circle as its arc is of the circumference.

Let A = area of circle;

A' = area of sector;

n = number of degrees in arc of sector.

Then, $A' : A = n : 360$

whence, $A' = \dfrac{n A}{360} = \dfrac{\pi r^2 n}{360}$

EXAMPLE.—The angle of a sector of a circle is 75°; the diameter of the circle is 12 inches; what is the area of the sector?

SOLUTION.—The area A of the circle is $12^2 \times .7854$ sq. in. Then the area of the sector is

$$\frac{n A}{360} = \frac{75 \times 12^2 \times .7854}{360} = 23.562 \text{ sq. in.}\quad \text{Ans.}$$

79.   The area of a sector is equal to one-half the product of its base by the radius of the circle.

$$A' = \tfrac{1}{2}\, r\, l$$

If l is the length of the arc, or **base**, of a sector, we have (Art. **73**),

$$l = \frac{\pi\, r\, n}{180}$$

whence,

$$n = \frac{180\, l}{\pi\, r}$$

This value of n substituted in formula of Art. **78** gives

$$A' = \frac{\pi\, r^2}{360} \times \frac{180\, l}{\pi\, r}$$

or, reducing,

$$A' = \tfrac{1}{2}\, r\, l$$

EXAMPLE.—If the radius of an arc is 5 feet and the length of the arc is 4 feet, what is the area of the sector?

SOLUTION.—By formula of Art. **79**,

$$A' = \frac{l\, r}{2} = \frac{4 \times 5}{2} = 10 \text{ sq. ft. Ans.}$$

80. The area of a segment, as ADB, Fig. 48, is evidently equal to the area of the sector $AOBD$ minus the area of the triangle AOB.

EXAMPLE 1.—The diameter of a circle is 10 inches, and the chord of the arc of a segment is 7 inches; what is the area of the segment?

SOLUTION.—In Fig. 48, let $AB = 7$ in. and the diameter $= 10$ in. Then, $OB = 5$ in., and $CB = 3.5$ in. Hence, $OC = \sqrt{5^2 - 3.5^2} = 3.57$ in., and $CD = 5 - 3.57 = 1.43$ in. Then, by formula of Art. **74**,

arc $ADB = \dfrac{4\sqrt{7^2 + 4 \times 1.43^2} - 7}{3} = 7.75$ in.

Hence, area of sector $AOBD = \tfrac{1}{2} \times 5 \times 7.75 = 19.38$ sq. in. The area of the triangle $AOB = \tfrac{1}{2} \times 3.57 \times 7 = 12.50$ sq. in. Therefore, the area of the segment is $19.38 - 12.50 = 6.88$ sq. in. Ans.

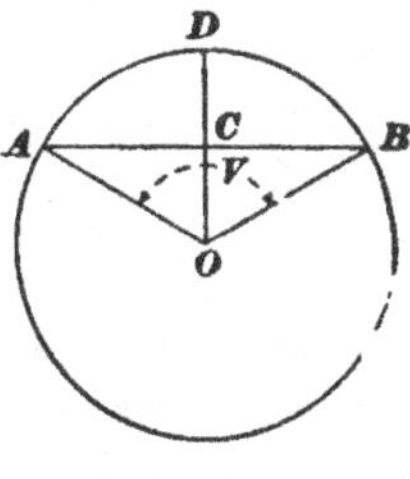

FIG. 48

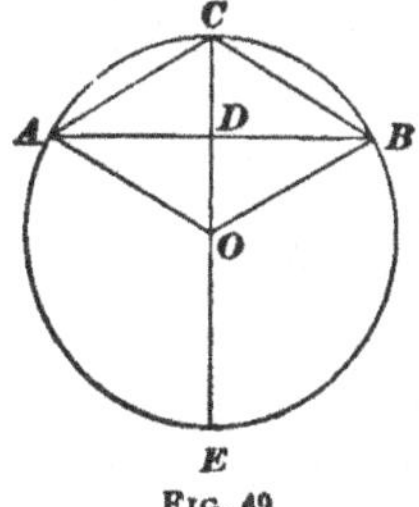

FIG. 49

EXAMPLE 2.—The chord of the arc of a segment is 79 inches and the height of the segment is 20 inches; find the area of the segment.

SOLUTION.—Let $ACBE$, Fig. 49, be the circle; let $AB = 79$ in. and $CD = 20$ in. Then, $AD = \tfrac{1}{2} \times 79$ in. $= 39.5$ in. By Art. **28**,

$$CD : AD = AD : DE$$

or, $\quad 20 : 39.5 = 39.5 : DE$

whence, $\quad DE = 78.01$

Hence, the diameter $= 20 + 78.01 = 98.01$ in., and the radius $= 49$.

Then the arc $ACB = \dfrac{4\cdot\sqrt{79^3} + 4 \times 20^3 - 79}{3} = 91.7$ in.   Hence, the area of sector $AOBC = 91.7 \times \frac{1}{2} \times 49 = 2{,}246.65$ sq. in.   The area of the triangle $AOB = \frac{1}{2} \times 79 \times 29 = 1{,}145.5$ sq. in.   Therefore, the area of the segment $= 2{,}246.65 - 1{,}145.50 = 1{,}101.15$ sq. in.   Ans.

THE ELLIPSE

81.   **An ellipse** is a plane figure bounded by a curved line such that the sum of the distances of any point on that line from two fixed points within is always equal to the length of the line passing through the fixed points and terminating at both ends in the curved line.

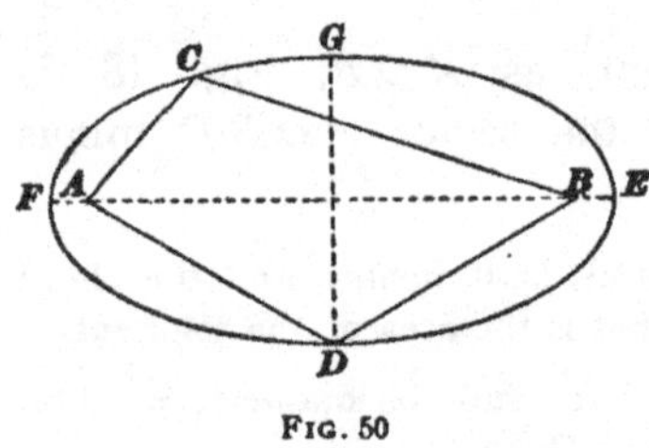

Fig. 50

In Fig. 50, the fixed points are A and B, and if C and D are any two points on the curve, $AC + CB = AD + DB = FE$.   The two fixed points are the **foci.**   The line FE through the foci is the **transverse, or major, axis.**

The line GD, which is the perpendicular bisector of FE, is the **conjugate, or minor, axis.**   The foci may be located from G or D as a center by striking arcs with a radius equal to one-half FE.

82.   There is no simple and exact method of finding the periphery (perimeter) of an ellipse.   The following formula gives values very nearly exact:

Let $C =$ periphery;

$\qquad a =$ half the major axis;

$\qquad b =$ half the minor axis;

$$D = \frac{a - b}{a + b}.$$

Then, $C = \pi(a + b)\dfrac{64 - 3\,D^4}{64 - 16\,D^2}$

EXAMPLE.—What is the periphery of an ellipse whose axes are 10 inches and 4 inches long?

SOLUTION.—In this example, $a = 5$, $b = 2$, $D = \dfrac{5-2}{5+2} = \dfrac{3}{7}$.

Then, $C = 3.1416(5+2)\dfrac{64 - 3(\frac{3}{7})^4}{64 - 16(\frac{3}{7})^2} = 23.013$

Therefore, the periphery is 23.013 in.   Ans.

83. The area of an ellipse is equal to the product of its two semiaxes multiplied by π.

Let a = half the major axis;

  b = half the minor axis;

  A = area.

Then, $A = \pi\,a\,b = 3.1416\,a\,b$

EXAMPLE.—What is the area of an ellipse whose axes are 10 inches and 6 inches?

SOLUTION.—Here, $a = \frac{1}{2} \times 10 = 5$, $b = \frac{1}{2} \times 6 = 3$.

Then, $A = 3.1416 \times 5 \times 3 = 47.124$

Therefore, the area is 47.12 sq. in.   Ans.

EXAMPLES FOR PRACTICE

1. The number of degrees in the angle formed by drawing radii from the center of a circle to the extremities of an arc of the circle is 84; the diameter of the circle is 17 inches; what is the area of the sector?   Ans. 52.96 sq. in.

2. Given the chord of the arc of a segment equal to 24 inches, and the height of the segment equal to 6.5 inches, find: (*a*) the diameter of the circle; (*b*) the area of the segment.   Ans. $\begin{cases} (a) & 28.7 \text{ in.} \\ (b) & 109.5 \text{ sq. in.} \end{cases}$

3. (*a*) What is the perimeter of an ellipse whose axes are 15 inches and 9 inches? (*b*) What is the area?   Ans. $\begin{cases} (a) & 38.29 \text{ in.} \\ (b) & 106.03 \text{ sq. in.} \end{cases}$

4. The base of a sector is 24 inches and the diameter of the circle is 54 inches; what is the area of the sector?   Ans. 324 sq. in.

THE MENSURATION OF SOLIDS

84. A **solid**, or body, has three dimensions: length, breadth, and thickness.

85. The **entire area** of a solid is the area of the whole outside of the solid.

The **convex area** of a solid having one or two flat ends is the same as the entire surface, except that the areas of the ends or bases are not included.

86. The **volume** of a solid is expressed by the number of times that it will contain another volume, called the unit of volume. Instead of the word *volume*, the expression **cubical contents** is frequently used.

THE PRISM AND CYLINDER

87. A **prism** is a solid whose ends are equal polygons in parallel planes, and whose sides are parallelograms.

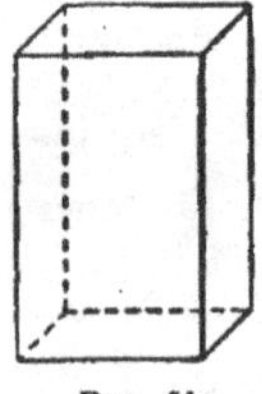

FIG. 51

88. A **parallelopipedon**, Fig. 51, is a prism whose bases (ends) are parallelograms.

89. A **cube**, Fig. 52, is a parallelopipedon whose faces and ends are squares.

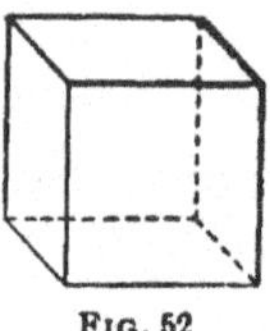

FIG. 52

90. The cube whose edges are equal to the unit of length is taken as the unit of volume when finding the volume of a solid.

Thus, if the unit of length is 1 inch, the unit of volume will be the cube each of whose edges measures 1 inch, or 1 cubic inch; and the number of cubic inches the solid contains will be its volume. If the unit of length is 1 foot, the unit of volume will be 1 cubic foot, etc. Cubic inch, cubic foot, and cubic yard are abbreviated to cu. in., cu. ft., and cu. yd., respectively.

91. Prisms take their names from their bases. Thus, a *triangular prism* is one whose bases are triangles; a *pentagonal prism* is one whose bases are pentagons, etc.

92. A cylinder, Fig. 53, is a round body of uniform diameter with circles for its ends.

93. A **right prism,** or **right cylinder,** is one whose center line (axis) is perpendicular to its bases.

FIG. 53

94. The **altitude** of a prism or cylinder is the perpendicular distance between its two ends.

95. To find the convex area of any right prism, or right cylinder:

Rule.—*Multiply the perimeter of the base by the altitude.*

Let p = perimeter of base;
   h = altitude;
   c = convex area.
Then, $c = p\,h$

EXAMPLE 1.—What is the convex area of a right prism whose base is a square, one side of which is 9 inches, and whose altitude is 16 inches?

SOLUTION.— $9 \times 4 = 36$ in., the perimeter of the base. Applying formula of Art. **95,**

   $c = 36 \times 16 = 576$ sq. in., the convex area. Ans.

To find the entire area, add the areas of the two ends to the convex area.

EXAMPLE 2.—What is the entire area of the parallelopipedon mentioned in the last question?

Solution.—The area of one end is $9^2 = 81$ sq. in. $81 \times 2 = 162$ sq. in., is the area of both ends. $576 + 162 = 738$ sq. in., the entire area of the parallelopipedon. Ans.

Example 3.—What is the entire area of a right cylinder whose base is 16 inches in diameter, and whose altitude is 24 inches?

Solution.— $16 \times 3.1416 = 50.27$ in., or the perimeter (circumference) of the base. $50.27 \times 24 = 1,206.48$ sq. in., the convex area.

$$16^2 \times .7854 \times 2 = 402.12 \text{ sq. in., the area of the ends.}$$

$$1,206.48 + 402.12 = 1,608.6 \text{ sq. in., the entire area. Ans.}$$

96. To find the volume of a prism, or cylinder:

Rule.—*The volume of any prism or cylinder is equal to the area of the base multiplied by the altitude.*

Let A = area of base;

h = altitude;

V = volume.

Then, $$V = A h$$

If the given prism is a cube, the three dimensions are all equal, and the volume equals the cube of one of the edges. Hence, if the volume is given, the length of an edge is found by extracting the cube root.

If the volume and the area of the base are given, the altitude is $h = \dfrac{V}{A}$. If the cylinder or prism is hollow, the volume is equal to the area of the ring or base multiplied by the altitude.

Example 1.—What is the volume of a rectangular prism whose base is 6 inches by 4 inches, and whose altitude is 12 inches?

Solution.—The base of a rectangular prism is a rectangle; hence, $6 \times 4 = 24$ sq. in., the area of the base. Applying formula of Art. **96**,

$$V = 24 \times 12 = 288 \text{ cu. in., or the volume. Ans.}$$

Example 2.—What is the volume of a cube whose edge is 9 inches?

Solution.— $9^3 = 9 \times 9 \times 9 = 729$ cu. in., the volume. Ans.

Example 3.—What is the volume of a cylinder whose base is 7 inches in diameter, and whose altitude is 11 inches?

Solution.— $7^2 \times .7854 = 38.48$ sq. in., the area of the base. Applying formula of Art. **96**,

$$V = 38.48 \times 11 = 423.28 \text{ cu. in., the volume. Ans.}$$

THE PYRAMID AND CONE

97. A pyramid, Fig. 54, is a solid whose base is a polygon, and whose sides are triangles uniting at a common point, called the **vertex.** If the base is a regular polygon, and the sides have the same inclination to the base, the pyramid is a **regular pyramid.**

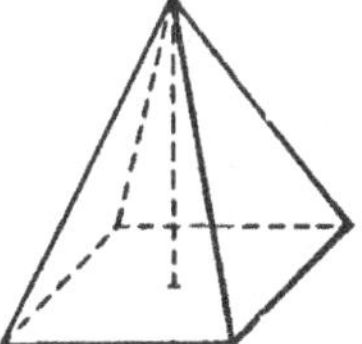

Fig. 54

98. A cone, Fig. 55, is a solid whose base is a circle, and whose convex surface tapers uniformly to a point called the **vertex.**

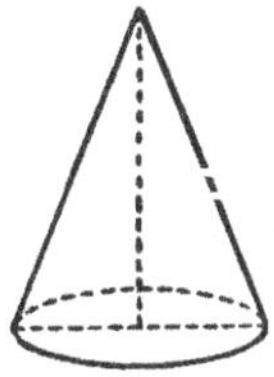

99. The **altitude** of a pyramid or cone is the perpendicular distance from the vertex to the base.

Fig. 55

100. The **slant height** of a regular pyramid is a line drawn from the vertex perpendicular to one of the sides of the base. The slant height of a cone is a straight line drawn from the vertex to the circumference of the base, and lying on the surface of the cone.

101. To find the convex area of a regular pyramid or a cone:

Rule.—*The convex area of a regular pyramid or of a cone is equal to the perimeter of the base multiplied by one-half the slant height.*

Let p = perimeter;
s = slant height;
c = convex area.

Then,
$$c = \frac{p\,s}{2}$$

Example 1.—What is the convex area of a regular pentagonal pyramid, if each side of the base measures 6 inches and the slant height measures 14 inches?

Solution.—The base of the pentagonal pyramid is a pentagon, and consequently it has five sides. $6 \times 5 = 30$ in., or the perimeter of the base. Applying formula of Art. **101,**

$$c = \frac{p\,s}{2} = \frac{30 \times 14}{2} = 210 \text{ sq. in., the convex area.} \quad \textbf{Ans.}$$

EXAMPLE 2.—What is the entire area of a cone whose altitude is 15 inches, and whose base is 16 inches in diameter?

SOLUTION.—The slant height of the cone is the hypotenuse of a right triangle whose legs are the radius of the base and altitude of the cone, respectively. Therefore, the slant height is equal to $\sqrt{15^2 + 8^2} = 17$ in. (Art. **56**). The perimeter of the base is $16 \times 3.1416 = 50.2656$ in. Applying formula of Art. **101**,

$$c = \frac{50.2656 \times 17}{2} = 427.26 \text{ sq. in.}$$

The area of the base is $16^2 \times .7854 = 201.06$ sq. in. The entire area is, therefore, $427.26 + 201.06 = 628.32$ sq. in. Ans.

102. To find the volume of any pyramid or cone:

Rule.—*The volume of any pyramid or cone equals the area of the base multiplied by one-third of the altitude.*

Let A = area of base;

h = altitude;

V = volume.

Then,
$$V = \frac{A\,h}{3}$$

EXAMPLE 1.—What is the volume of a triangular pyramid, each edge of whose base measures 6 inches, and whose altitude is 8 inches?

SOLUTION.—The base is an equilateral triangle; hence, applying the rule of Art. **68**, the area is $6^2 \times .433 = 15.59$ sq. in. Applying formula of Art. **102**,

$$V = \frac{A\,h}{3} = \frac{15.59 \times 8}{3} = 41.57 \text{ cu. in.} \quad \textbf{Ans.}$$

EXAMPLE 2.—What is the volume of a cone whose altitude is 18 inches, and whose base is 14 inches in diameter?

SOLUTION.— $14^2 \times .7854 = 153.94$ sq. in., the area of the base. Applying formula of Art. **102**,

$$V = \frac{A\,h}{3} = \frac{153.94 \times 18}{3} = 923.64 \text{ cu. in., the volume.} \quad \textbf{Ans.}$$

103. It has been stated that the volume of a cone or a pyramid is equal to one-third the product of the area of the base multiplied by the altitude. Similarly, the volume of any solid whose base is a plane figure and which tapers to a point like a cone or a pyramid is equal to one-third of the product of its base and altitude.

EXAMPLE.—Find the volume of an elliptical cone, whose base is an ellipse with diameters 8 inches and 6 inches, and the altitude is 7.5 inches.

SOLUTION.—The area of the ellipse at the base is $3.1416 \times 4 \times 3$ The volume is equal to one-third the product of the area of the base and altitude; that is,

$$V = \frac{1}{3} \times 3.1416 \times 4 \times 3 \times 7.5 = 94.248$$

Hence, the volume is 94.248 cu. in.   Ans.

EXAMPLES FOR PRACTICE

1. Find the volume of a triangular pyramid of which the altitude is 4 inches and the base is an equilateral triangle having each side 3 inches long.   Ans. 5.2 cu. in.

2. Find the weight of a steel bar 16 feet long and 2 inches in diameter, the weight of steel being taken as .28 pound per cubic inch.   Ans. 168.89 lb.

3. What is the entire area of a hexagonal prism 12 inches long, each side of the base being 1 inch long?   Ans. 77.196 sq. in.

4. (a) Find the convex area of a cone whose altitude is 12 inches, and the circumference of whose base is 31.416 inches. (b) Find the volume of the cone.   Ans. $\begin{cases} (a) & 204.2 \text{ sq. in.} \\ (b) & 314.16 \text{ cu. in} \end{cases}$

THE FRUSTUM OF A PYRAMID OR A CONE

104. If a pyramid is cut by a plane parallel to the base, as in Fig. 56, so as to form two parts, the lower part is called a **frustum** of the pyramid.

105. If a cone is cut in a similar manner, as in Fig. 57, the lower part is called a **frustum of the cone.**

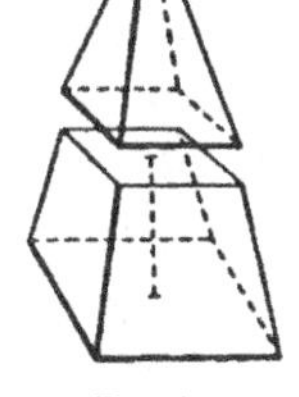

FIG. 56

106. The upper end of a frustum of a pyramid or cone is called the **upper base,** and the lower end the **lower base.** The altitude of a frustum is the perpendicular distance between the bases.

107. To find the convex area of a frustum of a regular pyramid or of a cone:

Rule.—*The convex area of a frustum of a regular pyramid or of a cone equals one-half the sum of the perimeters of its bases multiplied by the slant height of the frustum.*

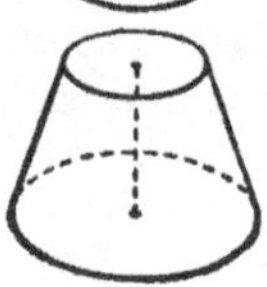

Fig. 57

Let p = perimeter of lower base;
p' = perimeter of upper base;
s = slant height;
c = convex area.

Then, $$c = \left(\frac{p + p'}{2}\right)s$$

EXAMPLE 1.—Given the frustum of a triangular pyramid in which each side of the lower base measures 10 inches, each side of the upper base measures 6 inches, and whose slant height is 9 inches; find the convex area.

SOLUTION.— 10 in. × 3 = 30 in., the perimeter of the lower base. 6 in. × 3 = 18 in., the perimeter of the upper base. Applying formula of Art. **107,**

$$c = \left(\frac{p + p'}{2}\right)s = \frac{30 + 18}{2} \times 9 = 216 \text{ sq. in., the convex area.} \quad \text{Ans.}$$

EXAMPLE 2.—If the diameters of the two bases of a frustum of a cone are 12 inches and 8 inches, respectively, and the slant height is 12 inches, what is the entire area of the frustum?

SOLUTION.— $\dfrac{(12 \times 3.1416) + (8 \times 3.1416)}{2} \times 12 = 376.99$ sq. in., the convex area.

$$8^2 \times .7854 = 50.27 \text{ sq. in.}$$
$$12^2 \times .7854 = 113.1 \text{ sq. in.}$$

113.1 + 50.27 = 163.37 sq. in., the area of the two ends. 376.99 + 163.37 = 540.36 sq. in., the entire area of the frustum. Ans.

108. To find the volume of the frustum of a pyramid or a cone:

Rule.—*Add the areas of the upper base, the lower base, and the square root of the product of the areas of the two bases; multiply this sum by one-third of the altitude.*

Let A = area of lower base;
a = area of upper base;
h = altitude;
V = volume.

Then,
$$V = (A + a + \sqrt{A\,a})\frac{h}{3}$$

Example 1.—Given a frustum of a hexagonal pyramid in which each edge of the lower base measures 8 inches, and each edge of the upper base measures 5 inches, and whose altitude is 14 inches, what is its volume?

Solution.—A hexagonal pyramid is one whose base is a regular hexagon, as shown in Fig. 58. Hence, applying formula of Art. **68**,

$$A = 8^2 \times 2.5981 = 166.28 \text{ sq. in.}$$

In a similar way, the area of the upper base is found to be 64.95 sq. in. Then, applying formula of Art. **108**,

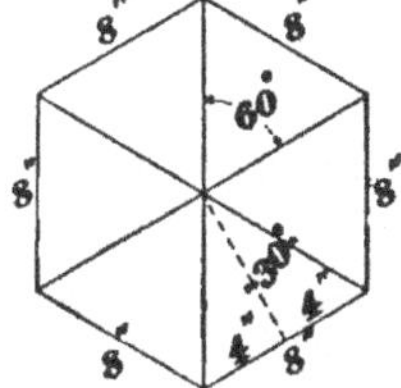

Fig. 58

$$V = (166.28 + 64.95 + \sqrt{166.28 \times 64.95})\frac{14}{3}$$

$$= 335.15 \times \frac{14}{3} = 1{,}564.03 \text{ cu. in., the volume. Ans.}$$

Example 2.—What is the volume of a frustum of a cone whose upper base is 8 inches in diameter, whose lower base is 12 inches in diameter, and whose altitude is 15 inches?

Solution.—The area of the upper base is $8^2 \times .7854 = 50.27$ sq. in. The area of the lower base is $12^2 \times .7854 = 113.1$ sq. in., nearly. The square root of their product is $\sqrt{50.27 \times 113.1} = 75.4$.

Then,
$$V = (50.27 + 113.1 + 75.4)\frac{15}{3}$$

$$= 238.77 \times \frac{15}{3} = 1{,}193.85 \text{ cu. in., the volume. Ans.}$$

THE WEDGE

109. A **wedge**, as here considered, is a solid whose base is a rectangle, two of whose opposite faces are parallel triangles, and two are parallelograms whose intersection is called the edge of the wedge. A wedge may therefore be defined as a triangular prism having one rectangular face, called the base. In Fig. 59, $ABCD$ is the base and EF the edge of the wedge.

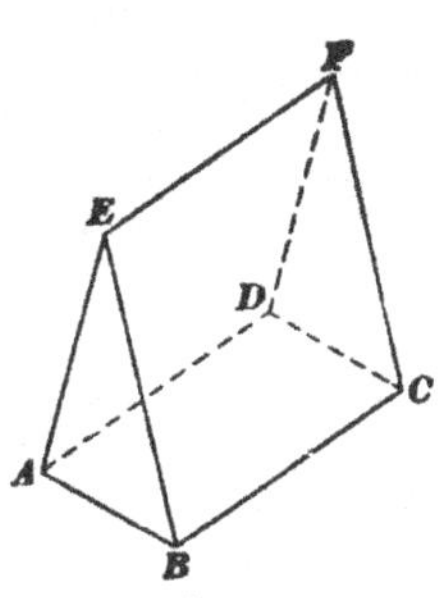

Fig. 59

110. The altitude of a wedge is the perpendicular distance between the base and the opposite edge.

111. To find the volume of a wedge:

Rule.—*The volume of any wedge is equal to the area of the base multiplied by one-half the altitude.*

Let A = area of base;
h = altitude;
V = volume.

Then,
$$V = \frac{A\,h}{2}$$

EXAMPLE.—What is the volume of a wedge whose base is a rectangle 6 feet long and 4 feet wide, and whose altitude is 10 feet?

SOLUTION.—The area of the base is $4 \times 6 = 24$ sq. ft. Applying formula of Art. **111**,
$$V = \frac{24 \times 10}{2} = 120 \text{ cu. ft. } \textbf{Ans.}$$

EXAMPLES FOR PRACTICE

1. Steel weighs .28 pound per cubic inch; find the weight of a steel wedge whose base is a rectangle 3 inches by $1\frac{1}{2}$ inches and whose altitude is 8 inches. *Ans.* 5.04 lb.

2. Find the volume of the frustum of a square pyramid of which the larger base is 15 inches square, the smaller base, 14 inches square, and the altitude, 3 inches. *Ans.* 631 cu. in.

3. A round tank is 8 feet in diameter at the top (inside) and 10 feet at the bottom; if the tank is 12 feet deep, how many gallons will it hold, there being 231 cubic inches in a gallon? *Ans.* 5,734.2 gal.

4. (*a*) What is the convex area of the frustum of a square pyramid whose altitude is 16 inches, one side of whose lower base is 28 inches long, and of the upper base 10 inches? (*b*) What is the volume of the frustum. *Ans.* $\begin{cases} (a) & 1,395.18 \text{ sq. in.} \\ (b) & 6,208 \text{ cu. in.} \end{cases}$

THE SPHERE

FIG. 60

112. A sphere, Fig. 60, is a solid bounded by a uniformly curved surface every point of which is equally distant from a point within, called the **center.**

The word **ball** is commonly used instead of sphere.

113. To find the area of the surface of a sphere

Rule.—*The area of the surface of a sphere equals the square of the diameter multiplied by π.*

Let S = surface;
$\quad d$ = diameter.

Then, $$S = \pi d^2$$

EXAMPLE.—What is the area of the surface of a sphere whose diameter is 14 inches?

SOLUTION.—Applying formula of Art. **113,** $S = 3.1416 \times 14^2$ $= 3.1416 \times 14 \times 14 = 615.75$ sq. in., the area. Ans.

114. To find the volume of a sphere:

Rule.—*The volume of a sphere equals the cube of the diameter multiplied by $\dfrac{\pi}{6}$.*

Let V = volume;
$\quad d$ = diameter.

Then, $$V = \frac{\pi}{6}d^3 = .5236d^3$$

EXAMPLE.—What is the weight of a lead cannon ball 12 inches in diameter, a cubic inch of lead weighing .41 pound?

SOLUTION.—Applying formula of Art. **114,** $V = .5236 \times 12 \times 12 \times 12 = 904.78$ cu. in., the volume of the ball.
$$904.78 \times .41 = 370.96 \text{ lb. Ans.}$$

The volume of a spherical shell, or hollow sphere, is equal to the difference in volume between two spheres having, respectively, the outer and the inner diameter of the shell.

115. To find the diameter of a sphere of known volume:

Rule.—*Divide the volume by .5236 and extract the cube root of the quotient. The result is the diameter.*

$$d = \sqrt[3]{\frac{V}{.5236}} = 1.2407 \sqrt[3]{V}$$

EXAMPLE.—The volume of a sphere is 96.1 cubic inches; what is its diameter?

SOLUTION.—Applying formula of Art. **115,**

$$d = \sqrt[3]{\frac{V}{.5236}} = \sqrt[3]{\frac{96.1}{.5236}} = 1.2407 \sqrt[3]{96.1} = 5.68 \text{ in. Ans.}$$

116. If any solid is cut into two parts by a plane, the surface of either part exposed by the removal of the other part is called a **plane section** of the solid.

Plane sections are divided into three classes: longitudinal sections, cross-sections, and right sections. A **longitudinal section** is any plane section taken lengthwise through the solid. Any other plane section is called a **cross-section.** If the surface exposed by taking a plane section of a solid is perpendicular to the center line of the solid, the section is called a **right section.** The surface exposed by any longitudinal section of a cylinder is a rectangle. The surface exposed by a right section of a cube is a square; of a cylinder or a cone, a circle. An oblique cross-section of a cylinder is an ellipse.

THE CYLINDRICAL RING

117. A **cylindrical ring** is a solid that may be generated by a circle revolving about an external axis in its plane.

118. To find the convex area of a cylindrical ring:

Rule.—*Multiply the circumference of an imaginary cross-section on the line A B, Fig. 61, by the length of the center line D.*

EXAMPLE.—A piece of round iron rod is bent into circular form to make a ring for a chain; if the outside diameter of the ring is 12 inches and the inside diameter is 8 inches, what is its convex area?

SOLUTION.—The diameter of the center circle equals one-half the sum of the inside and outside diameters, $\dfrac{12+8}{2} = 10$, and $10 \times 3.1416 = 31.416$ in., the length of the center line. The radius of the inside circle is 4 in., of the outside circle 6 in.; therefore, the diameter of the cross-section on the line $A\,B$ is 2 in. Then, $2 \times 3.1416 = 6.2832$ in., and $6.2832 \times 31.416 = 197.4$ sq. in., or the convex area. Ans.

119. To find the volume of a cylindrical ring:

Rule.—*The volume will be the same as that of a cylinder whose altitude equals the length of the dotted center line D.*

Fig. 61, and whose base is the same as a cross-section of the ring on the line A B, drawn from the center O. Hence, to find the volume of a cylindrical ring, multiply the area of an imaginary cross-section on a line A B, by the length of the center line D.

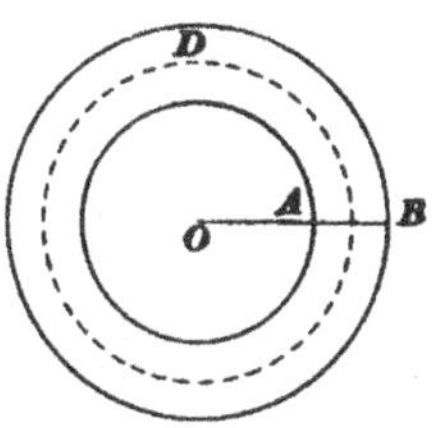

Fig. 61

EXAMPLE.—What is the volume of a cylindrical ring whose outside diameter is 12 inches, and whose inside diameter is 8 inches?

SOLUTION.—The diameter of the center circle equals one-half the sum of the inside and outside diameters, $\dfrac{12+8}{2} = 10$. $10 \times 3.1416 = 31.416$ in., the length of the center line. The radius of the outside circle is 6 in., of the inside circle, 4 in.; therefore, the diameter of the cross-section on the line $A\,B$ is 2 in. Then, $2^2 \times .7854 = 3.1416$ sq. in., the area of the imaginary cross-section; and $3.1416 \times 31.416 = 98.7$ cu. in., the volume. Ans.

EXAMPLES FOR PRACTICE

1. (*a*) What is the area of the surface of a sphere 30 inches in diameter? (*b*) What is the volume of the sphere?

Ans. $\begin{cases} (a) & 2{,}827.44 \text{ sq. in.} \\ (b) & 14{,}137.2 \text{ cu. in.} \end{cases}$

2. (*a*) What is the convex area of a cylindrical ring, the outside diameter of the ring being 10 inches and the inside diameter $7\frac{1}{4}$ inches? (*b*) What is the volume of the ring?

Ans. $\begin{cases} (a) & 107.95 \text{ sq. in.} \\ (b) & 33.734 \text{ cu. in.} \end{cases}$

3. The volume of a sphere is 606.132 cubic inches; what is the convex area of a cone whose slant height is 10 inches, and the diameter of whose base is the same as the diameter of the sphere?

Ans. 164.934 sq. in.

THE PRISMOID

120. A **prismoid** is a solid whose two bases are any polygons in parallel planes, and whose lateral faces may be divided into triangles and trapezoids by lines joining the vertexes of one base with those of the other. Thus, the solid shown in Fig. 62 is a prismoid; its bases are the pentagon $A\,B\,C\,D\,E$ and the quadrilateral $F\,G\,H\,I$, which lie in parallel planes; and its faces are the triangle GBC and the

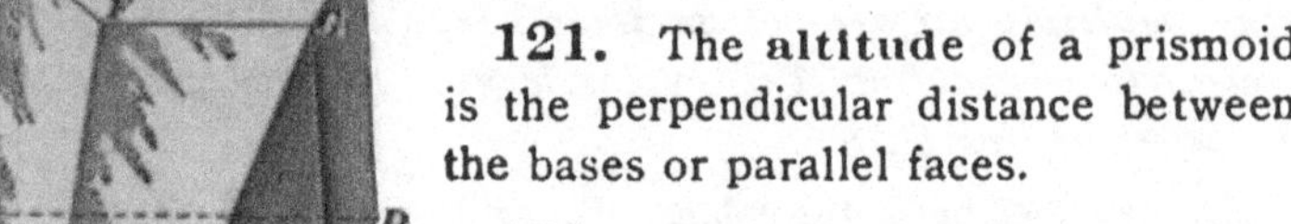

trapezoids $GCDH, HDEI, IEAF,$ and $FABG.$

121. The **altitude** of a prismoid is the perpendicular distance between the bases or parallel faces.

122. The parallel faces or bases of a prismoid are commonly called its **end sections.**

Fig. 62

A prismoid is also defined as a solid having two parallel end faces, and composed of any combination of prisms, wedges, and pyramids, whose common altitude is the perpendicular distance between the parallel faces.

123. The **middle section** of a prismoid is the polygon formed by a plane, parallel to the bases, and cutting the prismoid at equal distances from the two bases or end sections. Thus, polygon $PQRS$ is the middle section of the prismoid shown in Fig. 63.

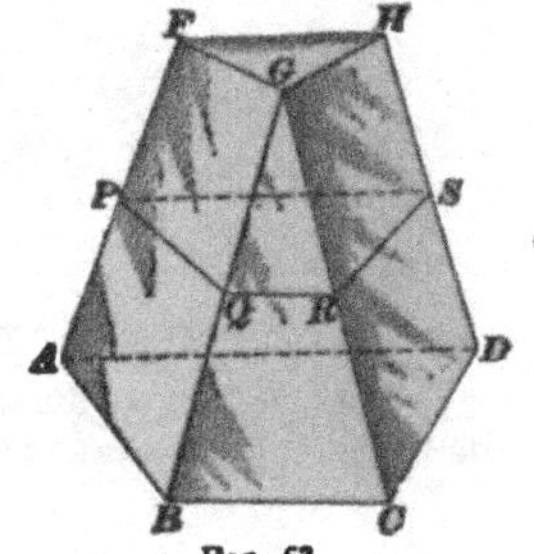

Fig. 63

124. Any dimension of the middle section of a prismoid may be taken equal to one-half the sum of the corresponding dimensions of the two end sections or bases. Thus, in Fig. 63, $PQ = \tfrac{1}{2}(AB + FG)$, $QR = \tfrac{1}{2}BC$, $RS = \tfrac{1}{2}(GH + CD)$, and $SP = \tfrac{1}{2}(HF + DA)$.

125. The area of the middle section of a prismoid may be measured directly, or calculated from its dimensions as determined from the dimensions of the end sections. It is not, in general, equal to one-half the sum of the areas of the bases.

The area of the middle section of a prism is the same as the area of either base; the area of the middle section of a wedge is equal to one-half the area of the base; the area of the middle section of a pyramid is equal to one-fourth the area of the base.

126. To find the volume of a prismoid:

Rule.—*Multiply the sum of the areas of the two end sections plus four times the area of the middle section by one-sixth the altitude.*

Let A = area of one base or end section;
A' = area of opposite base or end section;
M = area of middle section;
h = altitude;
V = volume of prismoid.

Then, $$V = \frac{h}{6}(A + A' + 4M)$$

This formula for finding the volume of a prismoid is known as the **prismoidal formula.** It is theoretically exact for determining the volumes of those solids to which it applies.

The derivation of this formula is as follows:

A prismoid can always be divided into elementary parts that will be prisms, wedges, and pyramids. From formula of Art. **96**, the volume of a prism is $V = Ah$; from formula of Art. **111**, the volume of a wedge is $V = \frac{Ah}{2}$; and from formula of Art. **102**, the volume of a pyramid is $V = \frac{Ah}{3}$. If these expressions are reduced to a common denominator, there will result,

For a prism, $$V = \frac{6Ah}{6} \qquad (1)$$

For a wedge, $$V = \frac{3Ah}{6} \qquad (2)$$

For a pyramid, $$V = \frac{2Ah}{6} \qquad (3)$$

Since any prism is of uniform cross-section throughout its length, every section will have the same area A, and equation (1) may be written

$$V = \frac{6Ah}{6} = \frac{h}{6}(A + A' + 4M)$$

For a wedge, evidently $A' = 0$, and $M = \frac{1}{4}A$. Hence, equation (2) may be written

$$V = \frac{3Ah}{6} = \frac{h}{6}(A + 0 + 2A) = \frac{h}{6}(A + A' + 4M)$$

For a pyramid, $A' = 0$, and $M = \frac{1}{4}A$. Hence, equation (3) may be written

$$V = \frac{2Ah}{6} = \frac{h}{6}(A + 0 + A) = \frac{h}{6}(A + A' + 4M)$$

Each of these formulas is the same as the formula given in this article; which shows that the latter formula applies correctly to the volume of a prism, pyramid, or wedge, and since it applies to each, it applies also to their sum, or the volume of a prismoid.

EXAMPLE.—Find the volume of the prismoid shown in Fig. 64, whose altitude is 14 inches.

SOLUTION.—Let PQR be the middle section. Then,

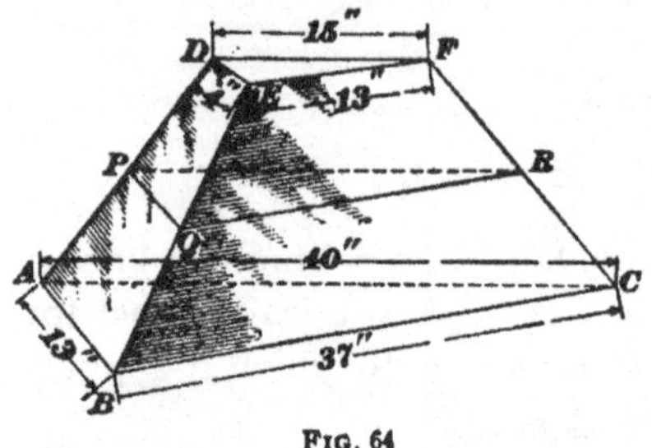

FIG. 64

$PQ = \frac{1}{2}(AB + DE) = \frac{1}{2}(13 + 4) = 8.5$ in.

$QR = \frac{1}{2}(BC + EF) = \frac{1}{2}(37 + 13) = 25$ in.

$RP = \frac{1}{2}(AC + DF) = \frac{1}{2}(40 + 15) = 27.5$ in.

The areas of the triangles are calculated by formula of Art. **47**, which gives the area of $ABC = 240$ sq. in., area of $DEF = 24$ sq. in., and area of $PQR = 105.2$ sq. in., nearly. Hence,

$$V = \frac{14}{6} \times (240 + 24 + 4 \times 105.2) = 1{,}597.9 \text{ cu. in., nearly.   Ans.}$$

127.   A familiar example of a prismoid is a railway cutting where the roadway is a horizontal plane, the side slopes are inclined planes, and the original surface of the ground is more or less inclined and irregular.

For calculating the volume of cuts and fills the prismoidal formula, though theoretically exact, gives results that are only approximate, on account of the inequalities of the surface of the ground. The nearer to each other the cross-sections are taken, the more accurate will be the result.

EXAMPLE 1.—Find, by the prismoidal formula, the volume of the frustum of a square pyramid of which the larger base is 2.5 feet square, the smaller base is 1 foot square, and the altitude is 16 feet.

SOLUTION.—The area of the larger base is $2.5 \times 2.5 = 6.25$ sq. ft.; the area of the smaller base is $1 \times 1 = 1$ sq. ft. The middle section is a square whose side is one-half the sum of the side of the upper and lower base; that is, $\frac{1}{2} \times (2.5 + 1) = 1.75$ ft. The area of the middle section is $1.75^2 = 3.0625$ sq. ft. Applying formula of Art. **126**, the volume of the frustum is

$$\frac{1}{6} \times 16 \times (6.25 + 1 + 4 \times 3.0625) = 52 \text{ cu. ft.   Ans.}$$

EXAMPLE 2.—In a railway cutting 200 feet long, the following are the areas, in square feet, of the cross-sections taken every 50 feet, namely: 2,700, 2,619, 2,556, 2,484, 2,610. What is its volume?

SOLUTION.—The volume between the first and the third cross-section is, by formula of Art. **126,**

$$V = \frac{100}{6}(2,700 + 2,556 + 4 \times 2,619) = 262,200 \text{ cu. ft.}$$

The volume between the third and the fifth section is

$$V = \frac{100}{6}(2,556 + 2,610 + 4 \times 2,484) = 251,700 \text{ cu. ft.}$$

The volume of the cutting is the sum of the volumes of the two prismoids, which is 513,900 cu. ft. = 19,033 cu. yd. Ans.

128. **Average End Areas.**—In practice, the volume of cuts and fills is often calculated by what is known as the **method by average end areas,** or simply as the **end area method.** By this method, the volume of the solid is found by multiplying one-half the sum of the two end areas by the distance between the two sections. Thus, let

A = area of one cross-section;
A' = area of next cross-section;
h = perpendicular distance between **sections;**
V = volume.

Then, $$V = \frac{h}{2}(A + A')$$

Results obtained by this formula are approximate and slightly larger than those given by the prismoidal formula. On account of its simplicity, the average end area formula is much used in practical earth-work calculations. The inequalities of the surface of the ground make it impossible to find the exact volume of a cut or fill, however accurate may be the formula applied.

EXAMPLE.—The areas of two cross-sections of a fill 50 feet apart are 2,700 and 2,619 square feet respectively; find the volume of the section, in cubic yards.

SOLUTION.—In this case, $A = 2,700$; $A' = 2,619$; and $h = 50$; then

$$V = \frac{50}{2}(2,700 + 2,619) = 132,975$$

Hence, the volume is 132,975 cu. ft. = 4,925 cu. yd. Ans.

EXAMPLES FOR PRACTICE

1. Find the volume of a right prismoid whose bases are rectangles that measure 10 inches by 8 inches and 8 inches by 6 inches, and whose height is 40 inches. Ans. 2,533.3 cu. in.

2. A railway cutting is 800 feet in length; the areas, in square yards, of cross-sections taken every 100 feet are: 237, 220, 204, 187, 171, 186, 204, 210, 220. Find the number of cubic yards in the cutting: (*a*) by the prismoidal formula; (*b*) by average end areas.

$$\text{Ans.}\begin{cases}(a)\ 53,633 \text{ cu. yd.}\\(b)\ 53,683 \text{ cu. yd.}\end{cases}$$

3. Find, by the prismoidal formula, the volume of a frustum of a hexagonal pyramid, each side of the lower base being 12 inches; of the upper base, 8 inches; and the altitude being 12 inches. Ans. 3,159.3 cu. in.

4. Find, by the prismoidal formula, the volume of a wedge whose base is a rectangle 15 feet in length and 9 feet in width, and whose altitude is 12 feet. Ans. 810 cu. ft.

PLANE TRIGONOMETRY
(PART 1)

THE TRIGONOMETRIC FUNCTIONS

DEFINITIONS

1. Trigonometric Functions and Trigonometry Defined.—Let A, Fig. 1, be any acute angle; AM and AN, its sides; BC, a perpendicular drawn to the side AN from any point on the side AM; and $B'C'$, a perpendicular drawn to the side AM from any point on the side AN. In the right triangle ABC, one of the vertexes of which is the vertex of the angle A, the hypotenuse AB will be referred to as *the hypotenuse;* the perpendicular BC, opposite the vertex of the angle A, as *the side opposite;* and the leg AC, containing

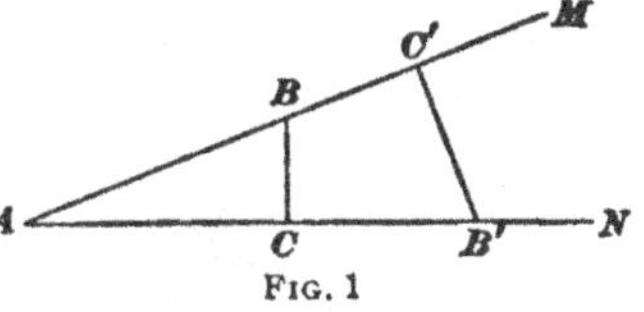

FIG. 1

the vertex of the angle A, as *the side adjacent.* Likewise, in the right triangle $AB'C'$, the hypotenuse is AB'; the side opposite is $B'C'$; and the side adjacent, or the leg containing the vertex of the angle A, is AC'. It should be borne in mind that these terms are used in connection with, or with reference to, the angle A.

The two right triangles ABC and $AB'C'$, having the acute angle A in common, are similar. Therefore,

$$\frac{AB}{AC} = \frac{AB'}{AC'}, \quad \frac{BC}{AB} = \frac{B'C'}{AB'}, \quad \frac{BC}{AC} = \frac{B'C'}{AC'}$$

It will be observed that, from whichever side the perpendicular is drawn, and whatever the point from which it is drawn, the ratio of the hypotenuse to the side adjacent remains unchanged, or is **constant**. The same is true of the ratio of the side opposite to the side adjacent, and, in general, of the ratio of any two of the three lines—hypotenuse, side adjacent, and side opposite. Evidently, these ratios are different for different angles. Thus, if A is 45°, both acute angles B and B' are also 45°; the triangles $A B C$ and $A B' C'$ are isosceles; and therefore

$$\frac{B C}{A C} = \frac{B' C'}{A C'} = 1$$

If A is greater than 45°, $B C$ is greater than $A C$, and the ratio $\dfrac{B C}{A C}$, having its numerator greater than its denominator, is greater than 1.

Confining ourselves to the ratio $\dfrac{B C}{A C}$ of the side opposite to the side adjacent, it is seen that the value of this ratio depends on the magnitude of the angle, and may, therefore, be used for the determination of the angle. Thus, it has just been shown that when the angle is 45° the ratio is equal to 1; hence, if in the solution of a problem it is found that the two legs of a right triangle are equal, or that their ratio is 1, it can be at once concluded that each of the acute angles is 45°.

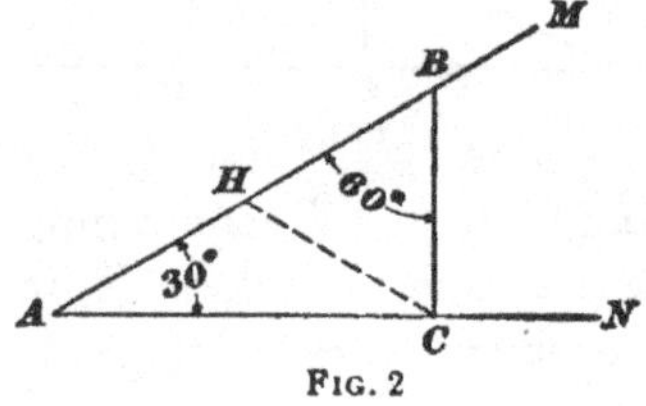

FIG. 2

Consider now an angle A, Fig. 2, of 30°. The right triangle $A B C$ having been constructed, $B C$ is the side opposite and $A C$ the side adjacent. If H is the middle point of the hypotenuse, the line $H C$ is equal to $A H$, or $\dfrac{A B}{2}$; for, if a semicircle is described on $A B$ as a diameter, with $H A$ as a radius, that semicircle must pass through C, since the angle $A C B$ is a right angle. Now, $H C$ being equal to $H B$, the angle $H C B$ is equal to B, or 60°; and, as the sum of the three angles of the triangle $B H C$ is 180°, the angle $B H C$ must be 60°. The triangle $H B C$ being equiangular, it is also equilateral, and therefore $B C = B H = \dfrac{A B}{2}$, and the ratio of the side opposite to the hypotenuse is $\dfrac{BC}{AB}$ $= \dfrac{\frac{1}{2}AB}{AB} = \dfrac{1}{2}$. Suppose, now, that in dealing with a right triangle the hypotenuse is found, by measurement, to be 1,500 feet and one of the sides 750 feet. Since the ratio of 750 to 1,500 is $\dfrac{1}{2}$, we at once conclude that the angle opposite the 750-foot side is 30°, and the other angle of the triangle, 60°.

These illustrations give a general idea of the practical value and use of the ratios under consideration. These ratios are determined for each angle, by methods that will be again referred to further on, and collected together in a table, from which the angle corresponding to any given ratio can be determined. Thus, if in a certain angle the ratio of the opposite side to the hypotenuse is $\dfrac{1}{2}$, this ratio is looked for in the table, where it is found as that belonging to 30°.

In this manner, the value of the angle is determined from the ratio in question, that ratio being obtained from the measured lengths of certain lines.

2. The ratios considered in the preceding article are called **trigonometric functions** of the angle A. In the

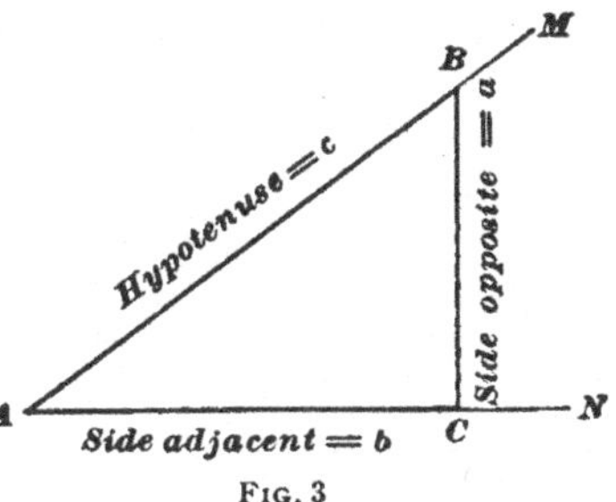

Fig. 3

triangle ABC, Fig. 3, two ratios are obtained by dividing any of three sides by each of the other two. Hence, there are six trigonometric functions of the angle A. This is true of any angle, since A is here used to represent any angle whatever. These functions have very important and useful properties, which make them exceedingly valuable for the solution of geometrical problems by computation.

3. **Trigonometry** is that branch of mathematics that treats of the properties of trigonometric functions and of their application to the solution of triangles.

4. The Sine and the Tangent.—Two of the most important of the trigonometric functions are the ratio of the side opposite to the hypotenuse, and that of the side opposite to the side adjacent; that is, $\frac{a}{c}$ and $\frac{a}{b}$, Fig. 3. They are called, respectively, the **sine** of A and the **tangent** of A. The words *sine* and *tangent* are abbreviated to *sin* and *tan*, respectively, and the expressions *sin A*, *tan A*, are for brevity read *sine A*, *tangent A*, instead of *sine of A*, and *tangent of A*. We have, then,

$$\sin A = \frac{\text{side opposite}}{\text{hypotenuse}} = \frac{a}{c} \qquad (1)$$

$$\tan A = \frac{\text{side opposite}}{\text{side adjacent}} = \frac{a}{b} \qquad (2)$$

If these formulas are fixed in the mind, little difficulty will be experienced in remembering the others that will be given. It should be noticed that the side opposite is the numerator in both ratios. The occurrence of the letter a in both the words *adjacent* and *tangent* will help one to remember which of the two fractions represents the tangent and which the sine.

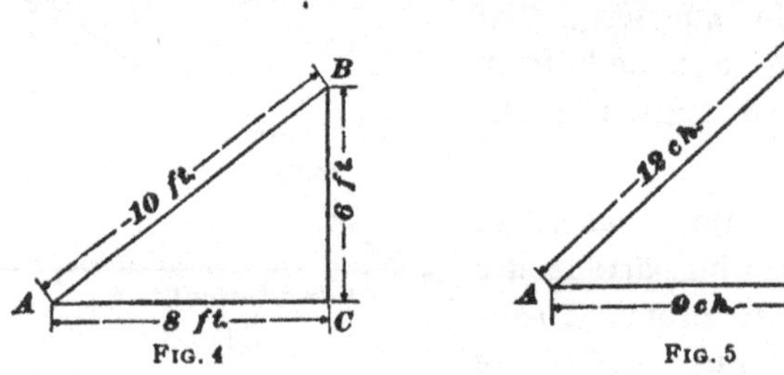

FIG. 4 FIG. 5

EXAMPLE 1.—In the right triangle ABC, Fig. 4, the lengths of the sides are shown; find the sine and the tangent of A.

SOLUTION.—In this case, the hypotenuse $AB = 10$; the side adjacent, $AC = 8$; side opposite, $BC = 6$. These values in formulas 1 and 2 give

$$\sin A = \frac{6}{10} = .6. \text{ Ans.}$$

$$\tan A = \frac{6}{8} = .75. \text{ Ans.}$$

EXAMPLE 2.—In the right triangle ABC, Fig. 5, the hypotenuse is 12 chains, and the side AC is 9 chains; find: (a) the sine and the tangent of A; (b) the sine and the tangent of B.

SOLUTION.—(*a*) For the angle A, we have

$$\text{hypotenuse} \quad A\,B = 12$$
$$\text{side adjacent,} \ A\,C = 9$$

$$\text{side opposite,} \ B\,C = \sqrt{\overline{A\,B}^2 - \overline{A\,C}^2} = \sqrt{12^2 - 9^2} = 7.9372$$

Substituting in formulas **1** and **2**,

$$\sin A = \frac{B\,C}{A\,B} = \frac{7.9372}{12} = .66143. \quad \text{Ans.}$$

$$\tan A = \frac{B\,C}{A\,C} = \frac{7.9372}{9} = .88191. \quad \text{Ans.}$$

(*b*) For angle B, we have

$$\text{hypotenuse} \quad B\,A = 12$$
$$\text{side opposite,} \ A\,C = 9$$
$$\text{side adjacent,} \ B\,C = 7.9372$$

Therefore,
$$\sin B = \frac{A\,C}{A\,B} = \frac{9}{12} = .75. \quad \text{Ans.}$$

$$\tan B = \frac{A\,C}{B\,C} = \frac{9}{7.9372} = 1.1339. \quad \text{Ans.}$$

EXAMPLES FOR PRACTICE

1. In a right triangle $A\,B\,C$ (make a sketch of this triangle), A and B are the two acute angles; the hypotenuse = 40 feet; side opposite B = 15 feet; find: (*a*) sin A and tan A; (*b*) sin B and tan B.

Ans. $\begin{cases} (a) \ \sin A = .92703, \ \tan A = 2.47207 \\ (b) \ \sin B = .37500, \ \tan B = .40452 \end{cases}$

2. From a point on one side of an angle M, a perpendicular is drawn on the other side; it is found that this perpendicular is 12.5 inches long, and that it meets the other side at a distance of 7.75 inches from the vertex; find the sine and the tangent of the angle M. (Make a sketch of this triangle.)

Ans. $\begin{cases} \sin M = .84988 \\ \tan M = 1.61290 \end{cases}$

3. From a point on one side of an angle A distant 10 inches from the vertex, a perpendicular is drawn on the other side; the distance from the vertex to the foot of the perpendicular is 6 inches; find sin A and tan A.

Ans. $\begin{cases} \sin A = .80000 \\ \tan A = 1.33333 \end{cases}$

4. The two acute angles of a right triangle are P and Q; the side opposite P is 150 feet, and that opposite Q is 225 feet; find: (*a*) sin P and tan P; (*b*) sin Q and tan Q.

Ans. $\begin{cases} (a) \ \sin P = .55469, \ \tan P = .66667 \\ (b) \ \sin Q = .83204, \ \tan Q = 1.50000 \end{cases}$

5. The Cosine and Cotangent.—The cosine and cotangent of an angle are, respectively, the sine and the tangent of the complement of the angle. The words *cosine* and *cotangent* are abbreviated to *cos* and *cot*, respectively, and the expressions *cos A*, *cot A* are read *cosine A*, *cotangent A*. Denoting any angle by A, its complement is $90° - A$; therefore, according to the definitions just given,

$$\cos A = \sin (90° - A) \qquad (1)$$
$$\cot A = \tan (90° - A) \qquad (2)$$

Since the complement of $90° - A$ is A, it also follows that

$$\cos (90° - A) = \sin A \qquad (3)$$
$$\cot (90° - A) = \tan A \qquad (4)$$

With reference to the angle B, Fig. 3, BC is the side adjacent and AC the side opposite. Therefore, by formulas 1 and **2**, Art. **4**,

$$\sin B = \frac{b}{c}, \ \tan B = \frac{b}{a}$$

and therefore, since A is the complement of B,

$$\cos A = \sin B = \frac{b}{c}$$

$$\cot A = \tan B = \frac{b}{a}$$

or, again referring to the angle A, which is the angle under consideration,

$$\cos A = \frac{\text{side adjacent}}{\text{hypotenuse}} \qquad (5)$$

$$\cot A = \frac{\text{side adjacent}}{\text{side opposite}} \qquad (6)$$

The student will, after some practice, become familiar with these formulas. Whenever he forgets them, he should refer to the definitions of the cosine and cotangent, which will at once enable him to write down the formulas, provided that he remembers those for the sine and the tangent.

6. The Secant and Cosecant.—The secant of an angle is the reciprocal of the cosine of the angle; that is, 1 divided by the cosine.

The word *secant* is abbreviated to *sec.* According to the definition, we have

$$\sec A = \frac{1}{\cos A} \qquad (1)$$

It follows that

$$\cos A = \frac{1}{\sec A} \qquad (2)$$

7. The **cosecant** of an angle is the secant of the complement of the angle. The abbreviations *cosec* and *csc* are used for *cosecant.* According to the definition, we have

$$\csc A = \sec (90° - A) \qquad (1)$$

Since A is the complement of $90° - A$, we have also

$$\csc (90° - A) = \sec A \qquad (2)$$

By means of formula **1**, Art. **6**, this relation may be written

$$\csc A = \sec (90° - A) = \frac{1}{\cos (90° - A)}$$

or, since $\cos (90° - A) = \sin A$ (formula **3**, Art. **5**),

$$\csc A = \frac{1}{\sin A} \qquad (3)$$

Therefore, the cosecant of an angle may also be defined as the reciprocal of the sine. Notice very particularly that

$$secant = \text{reciprocal of } cosine$$
$$cosecant = \text{reciprocal of } sine$$

From formula **3** above follows

$$\sin A = \frac{1}{\csc A} \qquad (4)$$

8. Cofunctions and Complementary Functions. The functions cosine, cotangent, and cosecant are sometimes called **cofunctions** of the angle considered; while the sine, tangent, and secant are called **fundamental functions.** As has been explained, the cofunctions of an angle are the corresponding fundamental functions of the complement of the angle. Thus, the cosine of A is the sine of $90° - A$; the cotangent of A is the tangent of $90° - A$; etc.

A fundamental function and its corresponding cofunction are called **complementary functions** of each other. The sine, for example, is the complementary function of the cosine; and the cosine is the complementary function of the sine.

EXAMPLE 1.—Find: (a) the cosine of the angle A, Fig. 5; (b) the cotangent; (c) the secant; (d) the cosecant.

SOLUTION.—(a) The cosine of A is equal to the sine of B, or

$$\frac{A\,C}{A\,B} = \frac{9}{12} = .75. \quad \text{Ans.}$$

(b) The cotangent of A is equal to the tangent of B, or (see example 2, Art. 4)

$$\frac{A\,C}{B\,C} = \frac{9}{7.9372} = 1.1339. \quad \text{Ans.}$$

(c) The secant of A is 1 divided by cos A, or

$$1 \div \frac{9}{12} = \frac{12}{9} = 1.33333. \quad \text{Ans.}$$

(d) The cosecant of A is 1 divided by sin A, or

$$1 \div \frac{B\,C}{A\,B} = \frac{A\,B}{B\,C} = \frac{12}{7.9372} = 1.51187. \quad \text{Ans.}$$

EXAMPLE 2.—Find the functions of 30°.

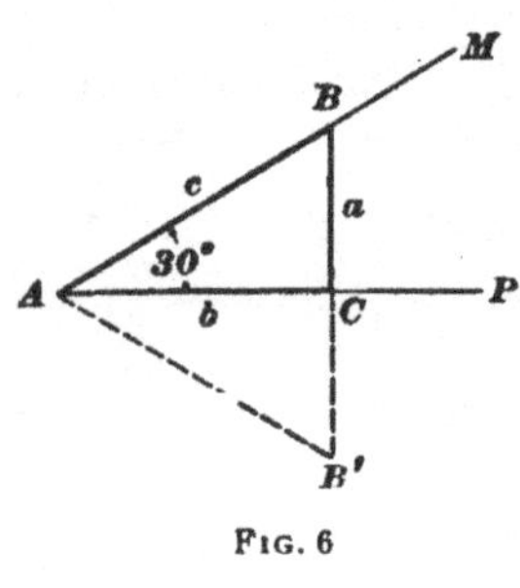

FIG. 6

SOLUTION.—Let the angle $M\,A\,P$, Fig. 6, be 30°. Draw $B\,C$ perpendicular to $A\,P$, produce it to B', making $C\,B' = C\,B$, and draw $A\,B'$. The triangle $B\,A\,B'$ thus formed is isosceles, and angle $C\,A\,B' = C\,A\,B = 30°$. Therefore, $B\,A\,B' = 30° + 30° = 60°$. Also, angle $B = 90° - 30° = 60°$; and angle $B' = $ angle $B = 60°$. As the three angles of $A\,B\,B'$ are equal, the sides are also equal, and $c = B\,B' = 2\,a$. Now, the figure gives,

$$b = \sqrt{c^2 - a^2} = \sqrt{(2\,a)^2 - a^2} = \sqrt{3\,a^2} = a\sqrt{3}$$

Bearing these values in mind, we have

$$\sin 30° = \frac{a}{c} = \frac{a}{2\,a} = \frac{1}{2}. \quad \text{Ans.}$$

$$\tan 30° = \frac{a}{b} = \frac{a}{a\sqrt{3}} = \frac{1}{\sqrt{3}} = \frac{\sqrt{3}}{3}. \quad \text{Ans.}$$

$$\cos 30° = \frac{b}{c} = \frac{a\sqrt{3}}{2\,a} = \frac{\sqrt{3}}{2}. \quad \text{Ans.}$$

$$\cot 30^\circ = \frac{b}{a} = \frac{a\sqrt{3}}{a} = \sqrt{3}. \quad \textbf{Ans.}$$

$$\sec 30^\circ = \frac{1}{\cos 30^\circ} = 1 \div \frac{\sqrt{3}}{2} = \frac{2}{\sqrt{3}} = \frac{2}{3}\sqrt{3}. \quad \textbf{Ans.}$$

$$\csc 30^\circ = \frac{1}{\sin 30^\circ} = 1 \div \frac{1}{2} = 2. \quad \textbf{Ans.}$$

Note.—It is only in a few cases that the values of the trigonometric functions of an angle can be derived by elementary principles, as above. The general method for determining the functions of any angle is comparatively complicated, and is beyond the scope of this work. The trigonometric functions of any angle can be obtained from a table, as will be presently explained.

EXAMPLES FOR PRACTICE

1. The acute angles of a right triangle are B and C; the side opposite B is 1,200 feet; and that opposite C is 1,500 feet; find the fundamental functions of B, and from them the cofunctions of C.

Ans. $\begin{cases} \sin B = .62471,\ \tan B = .8,\ \sec B = 1.2806 \\ \cos C = .62471,\ \cot C = .8,\ \csc C = 1.2806 \end{cases}$

2. From example 2, Art. 8, derive the functions of 60° ($= 90^\circ - 30^\circ$).

Ans. $\begin{cases} \sin 60^\circ = \dfrac{\sqrt{3}}{2},\ \tan 60^\circ = \sqrt{3},\ \cos 60^\circ = \dfrac{1}{2} \\[2mm] \cot 60^\circ = \dfrac{\sqrt{3}}{3},\ \sec 60^\circ = 2,\ \csc 60^\circ = \dfrac{2}{3}\sqrt{3} \end{cases}$

3. Given $\sin A = \dfrac{2}{3}$ and $\cos B = \dfrac{4}{5}$, find $\csc A$ and $\sec B$.

Ans. $\begin{cases} \csc A = 1.5 \\ \sec B = 1.25 \end{cases}$

4. Find the trigonometric functions of 45°. (Notice that here the side opposite is equal to the side adjacent. Denote the hypotenuse by c, and express the other two sides in terms of c.)

Ans. $\begin{cases} \sin 45^\circ = \cos 45^\circ = \frac{1}{2}\sqrt{2} \\ \tan 45^\circ = \cot 45^\circ = 1 \\ \sec 45^\circ = \csc 45^\circ = \sqrt{2} \end{cases}$

9. The Versed Sine and Coversed Sine.—The versed sine (*vers*) of an angle is 1 minus the cosine; and the coversed sine (*covers*) is 1 minus the sine.

$$\text{vers } A = 1 - \cos A \qquad \textbf{(1)}$$
$$\text{covers } A = 1 - \sin A \qquad \textbf{(2)}$$

These two functions are not much used, except in railroad work.

10. Summing Up.—The foregoing definitions are summed up in the table given below, which contains the expressions for the functions of the angle A, Fig. 7, in terms of the hypotenuse c, the side opposite, a, and the side adjacent, b.

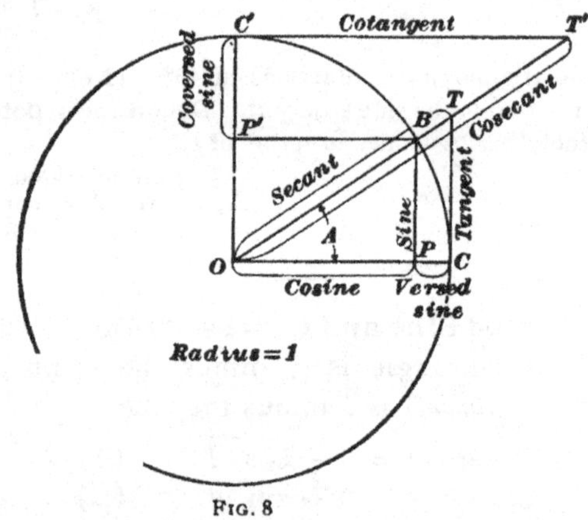

Fig. 7

TABLE I

Function	sin	tan	cos	cot	sec	csc	vers	covers
Value . . .	$\dfrac{a}{c}$	$\dfrac{a}{b}$	$\dfrac{b}{c}$	$\dfrac{b}{a}$	$\dfrac{c}{b}$	$\dfrac{c}{a}$	$1-\dfrac{b}{c}$	$1-\dfrac{a}{c}$

The ratios $\dfrac{c}{b}$ and $\dfrac{c}{a}$ for the secant and cosecant are obtained from the formulas $\sec A = 1 \div \cos A = 1 \div \dfrac{b}{c} = \dfrac{c}{b}$, $\csc A = 1 \div \sin A = 1 \div \dfrac{a}{c} = \dfrac{c}{a}$.

11. Representation of the Trigonometric Functions by Lines.—Let A, Fig. 8, be any angle. From its

Fig. 8

vertex O, describe a circle of radius 1; or, otherwise, describe any circle and take its radius as unity. This circle intersects the sides of the angle at B and C. Draw the tangent CT, meeting OB produced at T; the radius OC' perpendicular to OC; the lines BP and BP' perpendicular to OC and OC', respectively; and the tangent $C'T'$, meeting OB produced at T'.

Since the angle A is measured by the arc CB, the trigonometric functions of the angle are said to be likewise the trigonometric functions of the arc. It is, for instance, immaterial whether we say that 1 is the tangent of an angle of $45°$ or of an arc of $45°$.

In the figure constructed as just explained, the trigonometric functions of the angle A, or of the arc CB, may be represented by lines, as marked. For, in the right triangle OPB, in which BP, OP, and OB are, respectively, the side opposite, the side adjacent, and the hypotenuse, we have

$$\sin A = \frac{BP}{OB}, \cos A = \frac{OP}{OB}$$

or, since $OB = 1$,

$$\sin A = \frac{BP}{1} = BP, \cos A = \frac{OP}{1} = OP$$

In the triangle OCT, in which CT and OC are, respectively, the side opposite and the side adjacent, and OT is the hypotenuse,

$$\tan A = \frac{CT}{OC} = \frac{CT}{1} = CT$$

$$\sec A = \frac{OT}{OC} = \frac{OT}{1} = OT$$

By the same reasoning, it can be shown that $C'T'$ and OT' are, respectively, the tangent and the secant of the angle $C'OT'$, or the cotangent and the cosecant of A, since $C'OT'$ is the complement of A.

Let the student verify that, according to the definitions of the versed sine and coversed sine, these functions are represented by PC and $P'C'$, respectively.

RELATIONS AMONG THE FUNCTIONS OF
AN ANGLE

12. Method of Marking a Triangle.—The triangle ABC, Fig. 7, has the angles marked by the capital letters A, B, and C and the sides opposite these angles marked by the small letters a, b, and c, respectively. This method of marking a triangle is very useful and convenient, as it points out at once the relative position of the sides and the angles. In a right triangle, the right angle is usually designated by C. In the figures that follow, when only the angles are marked, the sides opposite are taken as marked by the small letters corresponding to the capital letters that mark the angles.

13. Relation Between Tangent and Cotangent. In Fig. 7,

$$\tan A = \frac{a}{b}, \cot A = \frac{b}{a}$$

Multiplying these equations together gives

$$\tan A \times \cot A = \frac{a}{b} \times \frac{b}{a} = 1$$

whence,

$$\cot A = \frac{1}{\tan A}$$

$$\tan A = \frac{1}{\cot A}$$

That is, the tangent and cotangent are each the reciprocal of the other. This is a very important relation, and should be committed to memory, together with those given in the two articles following.

14. Tangent and Cotangent in Terms of Sine and Cosine.—In Fig. 7,

$$\sin A = \frac{a}{c}, \cos A = \frac{b}{c}$$

Dividing these equations member by member gives

$$\frac{\sin A}{\cos A} = \frac{a}{c} \div \frac{b}{c} = \frac{a}{b}$$

that is, since $\frac{a}{b} = \tan A$,

$$\tan A = \frac{\sin A}{\cos A} \qquad (1)$$

Also, because the cotangent is the reciprocal of the tangent,

$$\cot A = \frac{\cos A}{\sin A} \qquad (2)$$

15. Relations Between the Squares of Certain Functions.—A power of a trigonometric function is indicated by writing the exponent immediately after the abbreviation used for the function. Thus, the square of the sine of A, or of $\sin A$, is written $\sin^2 A$, and read *sine square A*. Similarly, the cube of $\tan A$ is written $\tan^3 A$, and read *tangent cube A*, etc.

In the right triangle ABC, Fig. 7, we have

$$a^2 + b^2 = c^2$$

Dividing both members of this equality by c^2 gives

$$\frac{a^2}{c^2} + \frac{b^2}{c_2} = 1$$

that is, $\qquad \sin^2 A + \cos^2 A = 1 \qquad (1)$

Again, dividing both members of the equation $c^2 = a^2 + b^2$ by b^2,

$$\frac{c^2}{b^2} = \frac{a^2}{b^2} + 1 = 1 + \frac{a^2}{b^2}$$

that is, $\qquad \sec^2 A = 1 + \tan^2 A \qquad (2)$

Similarly, if both members of the equation $c^2 = a^2 + b^2$ are divided by a^2,

$$\frac{c^2}{a^2} = 1 + \frac{b^2}{a^2}$$

that is, $\qquad \csc^2 A = 1 + \cot^2 A \qquad (3)$

16. To Express Any Function in Terms of Any Other Function.—In the triangle ABC, Fig. 7, we have

$$a^2 + b^2 = c^2 \qquad (1)$$

Dividing both members of this equation by c^2 gives

$$\frac{a^2}{c^2} + \frac{b^2}{c^2} = 1 \qquad (2)$$

From these two equations, any of the six ratios $\dfrac{a}{b}, \dfrac{a}{c}, \dfrac{b}{a}, \dfrac{b}{c},$

$\dfrac{c}{a}, \dfrac{c}{b}$ can be found when one of them is given. If, for instance, $\dfrac{a}{c}$ is given, $\dfrac{c}{a}$ is obtained by dividing 1 by $\dfrac{a}{c}$; $\dfrac{b}{c}$, by solving equation (2) for $\dfrac{b^2}{c^2}$ and taking the square root; $\dfrac{c}{b}$, by taking the reciprocal of the value just found for $\dfrac{b}{c}$. To find $\dfrac{a}{b}$, divide both members of equation (1) by b^2, which gives

$$\frac{a^2}{b^2} + 1 = \frac{c^2}{b^2}$$

whence, multiplying through by $\dfrac{b^2}{c^2}$,

$$\frac{b^2}{c^2}\left(\frac{a^2}{b^2} + 1\right) = \frac{c^2}{b^2} \times \frac{b^2}{c^2} = 1$$

and hence, dividing through by $\dfrac{a^2}{b^2} + 1$,

$$\frac{b^2}{c^2} = \frac{1}{\dfrac{a^2}{b^2} + 1}$$

Substituting this value of $\dfrac{b^2}{c^2}$ in equation (2) and solving for $\dfrac{a}{b}$, the latter ratio is obtained in terms of $\dfrac{a}{c}$.

Table II gives the relation between any two functions of any angle A.

TABLE II

RELATIONS BETWEEN THE FUNCTIONS OF AN ANGLE

In Terms of	$\sin A$	$\cos A$	$\tan A$	$\cot A$	$\sec A$	$\csc A$
$\sin A =$	$\sin A$	$\sqrt{1 - \cos^2 A}$	$\dfrac{\tan A}{\sqrt{1 + \tan^2 A}}$	$\dfrac{1}{\sqrt{1 + \cot^2 A}}$	$\dfrac{\sqrt{\sec^2 A - 1}}{\sec A}$	$\dfrac{1}{\csc A}$
$\cos A =$	$\sqrt{1 - \sin^2 A}$	$\cos A$	$\dfrac{1}{\sqrt{1 + \tan^2 A}}$	$\dfrac{\cot A}{\sqrt{1 + \cot^2 A}}$	$\dfrac{1}{\sec A}$	$\dfrac{\sqrt{\csc^2 A - 1}}{\csc A}$
$\tan A =$	$\dfrac{\sin A}{\sqrt{1 - \sin^2 A}}$	$\dfrac{\sqrt{1 - \cos^2 A}}{\cos A}$	$\tan A$	$\dfrac{1}{\cot A}$	$\sqrt{\sec^2 A - 1}$	$\dfrac{1}{\sqrt{\csc^2 A - 1}}$
$\cot A =$	$\dfrac{\sqrt{1 - \sin^2 A}}{\sin A}$	$\dfrac{\cos A}{\sqrt{1 - \cos^2 A}}$	$\dfrac{1}{\tan A}$	$\cot A$	$\dfrac{1}{\sqrt{\sec^2 A - 1}}$	$\sqrt{\csc^2 A - 1}$
$\sec A =$	$\dfrac{1}{\sqrt{1 - \sin^2 A}}$	$\dfrac{1}{\cos A}$	$\sqrt{1 + \tan^2 A}$	$\dfrac{\sqrt{1 + \cot^2 A}}{\cot A}$	$\sec A$	$\dfrac{\csc A}{\sqrt{\csc^2 A - 1}}$
$\csc A =$	$\dfrac{1}{\sin A}$	$\dfrac{1}{\sqrt{1 - \cos^2 A}}$	$\dfrac{\sqrt{1 + \tan^2 A}}{\tan A}$	$\sqrt{1 + \cot^2 A}$	$\dfrac{\sec A}{\sqrt{\sec^2 A - 1}}$	$\csc A$

TRIGONOMETRIC TABLES

TABLES OF NATURAL FUNCTIONS

17. To facilitate calculations, tables of the trigonometric functions are used. The tables give values for the sines, cosines, tangents, and cotangents of angles from $0°$ to $90°$. The values of the secant and cosecant are not generally given in tables; they are obtained by dividing 1 by the cosine and the sine, respectively, according to formula **1**, Art. **6**, and formula **3**, Art. **7**.

There are two kinds of trigonometric tables; namely, the table of *natural functions* and the table of *logarithmic functions*. The table of natural functions gives the actual values of the functions, while the table of logarithmic functions gives the logarithms of the functions. It may be remarked that, except in making a table, the values of the functions are never calculated directly because the process is so long and laborious that it would require considerable time to calculate even the value of one function of an angle; nor is there a simple method of calculating the angle corresponding to a given function.

NOTE.—In all that follows, the number of seconds by which an angle exceeds a whole number of degrees and minutes will be referred to as *the odd seconds*, or *the number of odd seconds*, or simply *the number of seconds* in the angle; while the expression *total number of seconds* will be applied to the number obtained by reducing the degrees and minutes to seconds, and adding the odd seconds. Thus, the odd seconds, or, for shortness, the seconds in $34° 36' 16''$ are 16; while the total number of seconds is the number of seconds in $34°$, plus the number of seconds in $36'$, plus 16; that is, $34 \times 60 \times 60 + (36 \times 60) + 16 = 124{,}576$. A similar notation will be used with regard to minutes. The explanations that follow refer to the *Trigonometric Tables* used with this Course.

18. To Find the Natural Functions of an Angle Less Than 45° and Containing No Odd Seconds.—The required function is found in the double column marked at the top with the given number of degrees, in the subdivision of that column headed by the name of the given function, and horizontally opposite the number in the left-hand column (marked ′) that expresses the number of odd minutes in the angle. When the function considered is a sine or a cosine, it is taken from the table headed Natural Sines and Cosines; when a tangent or cotangent, from the table headed Natural Tangents and Cotangents.

EXAMPLE.—Find the natural functions of an angle of 37° 23′.

SOLUTION.—On page 30 of the table headed Natural Sines and Cosines, the double column headed 37° is found. Looking in the left-hand minute column for 23 (number of odd minutes in the given angle), and glancing along the horizontal row to the right of 23, the number .60714 is found in the single column marked Sine under 37°; and the number .79459 is found in the column marked Cosine. Therefore,

$$\sin 37° 23′ = .60714. \quad \text{Ans.}$$
$$\cos 37° 23′ = .79459. \quad \text{Ans.}$$

The tangent and cotangent are taken in a similar manner from the table headed Natural Tangents and Cotangents, page 39. The results are:

$$\tan 37° 23′ = .76410. \quad \text{Ans.}$$
$$\cot 37° 23′ = 1.30873. \quad \text{Ans.}$$

EXAMPLES FOR PRACTICE

Verify the following values:

(a) $\sin 39° 55′ = .64167$; $\cos 39° 55′ = .76698$; $\tan 39° 55′ = .83662$; $\cot 39° 55′ = 1.19528$.

(b) $\tan 16° 32′ = .29685$; $\cos 16° 32′ = .95865$; $\sec 16° 32′ = 1.04313$; $\csc 16° 32′ = 3.51407$.

(c) $\cot 43° 2′ = 1.07112$; $\csc 43° 2′ = 1.46537$; $\tan 43° 2′ = .93360$; $\cos 43° 2′ = .73096$.

---

19. To Find the Natural Functions of an Angle Greater Than 45° and Containing No Odd Seconds. The required function is found in the double column marked at the bottom with the given number of degrees, in the subdivision of that column having at the bottom the name of the given function, and horizontally opposite the number in the right-hand column (marked ′) that expresses the odd minutes in the angle. It will be observed that the number of degrees at the bottom of the pages decrease as the pages increase, and that the number of minutes in the right-hand column increase from bottom to top.

EXAMPLE.—Find the functions of 53° 43′.

SOLUTION.—The double column marked 53° at the bottom is found on page 30 of Natural Sines and Cosines. Looking along the horizontal row determined by the number 43 in the right-hand minute column, the number .80610 is found in the single column marked Sine at the bottom, and the number .59178 in the single column marked Cosine at the bottom, these two columns forming the double column marked 53° at the bottom. Therefore,

$$\sin 53° 43′ = .80610. \quad \text{Ans.}$$
$$\cos 53° 43′ = .59178. \quad \text{Ans.}$$

The tangent and cotangent are similarly taken from page 39 of Natural Tangents and Cotangents. The results are:

$$\tan 53° 43′ = 1.36217. \quad \text{Ans.}$$
$$\cot 53° 43′ = .73413. \quad \text{Ans.}$$

EXAMPLES FOR PRACTICE

Verify the following values:

(a) $\sin 67° 45′ = .92554$; $\cos 67° 45′ = .37865$; $\tan 67° 45′ = 2.44433$; $\cot 67° 45′ = .40911$.

(b) $\cot 74° 3′ = .28580$; $\csc 74° 3′ = 1.04004$; $\sin 74° 3′ = .96150$.

(c) $\cos 48° 9′ = .66718$; $\cot 48° 9′ = .89567$; $\csc 48° 9′ = 1.34248$.

---

20. To Find the Natural Functions of an Angle Containing Odd Seconds.—The method of solving this problem by means of the table is founded on the following principle, which applies within the limits of approximation with which the table is constructed:

If several angles are taken within an interval not greater than 1′; that is, so that the difference between the greatest and the smallest shall not exceed 1′, the ratio of the difference between any two of these angles to the difference between any other two is the same as the ratio obtained by dividing the difference between the values of any trigonometric function for the first pair of angles, by the difference between the values of the same function for the second pair of angles. For instance, if the angles 43° 46′ 32″, 43° 46′ 34″, 43° 46′ 40″, and 43° 47′ are taken between 43° 46′ and 43° 47′, then

$$\frac{43^\circ\ 47' - 43^\circ\ 46'\ 40''}{43^\circ\ 46'\ 34'' - 43^\circ\ 46'\ 32''} = \frac{\sin 43^\circ\ 47' - \sin 43^\circ\ 46'\ 40''}{\sin 43^\circ\ 46'\ 34'' - \sin 43^\circ\ 46'\ 32''}$$

In general, if A, B, C, D are any angles within an interval of $1'$, then

$$\frac{A - B}{C - D} = \frac{\sin A - \sin B}{\sin C - \sin D} = \frac{\cos A - \cos B}{\cos C - \cos D}$$

$$= \frac{\tan A - \tan B}{\tan C - \tan D} = \frac{\cot A - \cot B}{\cot C - \cot D}$$

Similarly,

$$\frac{A - B}{B - C} = \frac{\sin A - \sin B}{\sin B - \sin C} = \frac{\cos A - \cos B}{\cos B - \cos C}, \text{ etc.}$$

Let A be the number of degrees and minutes in any angle, and s the number of odd seconds. Then the angle, which will be represented by $A + s''$, lies between A and $A + 1'$ or between A and $A + 60''$. For instance, if the angle is $25^\circ\ 15'\ 37''$, it lies between $25^\circ\ 15'$, which is represented by A, and $25^\circ\ 16'$, which is $25^\circ\ 15' + 1'$, or $A + 1'$, or $A + 60''$. In this case s represents $37''$. From the principle stated above we have,

$$\frac{(A + 60'') - A}{(A + s'') - A} = \frac{\sin (A + 60'') - \sin A}{\sin (A + s'') - \sin A}$$

or,

$$\frac{60}{s} = \frac{\sin (A + 1') - \sin A}{\sin (A + s'') - \sin A}$$

whence, solving this equation for $\sin (A + s'')$,

$$\sin (A + s'') = \sin A + [\sin (A + 1') - \sin A]\frac{s}{60} \qquad (1)$$

Similarly,

$$\tan (A + s'') = \tan A + [\tan (A + 1') - \tan A]\frac{s}{60} \qquad (2)$$

For the cosine, we have

$$\cos (A + s'') = \cos A + [\cos (A + 1') - \cos A]\frac{s}{60}$$

but, since the cosine of an angle decreases as the angle increases, $\cos A$ is greater than $\cos (A + 1')$, and therefore it is better to write the formula thus,

$$\cos (A + s'') = \cos A - [\cos A - \cos (A + 1')]\frac{s}{60} \qquad (3)$$

Similarly,

$$\cot (A + s'') = \cot A - [\cot A - \cot (A + 1')]\frac{s}{60} \qquad (4)$$

The functions of A and $A + 1'$ can be readily taken from the table, as explained in the preceding articles, and from them the functions of $A + s''$ are determined by the formulas just given, or by the following rule, which states in words what the formulas express in symbols:

Rule.—*Find, in the table, the sine, cosine, tangent, or cotangent corresponding to the degrees and minutes in the angle.*

For the seconds, find the difference between this value and the value of the sine, cosine, tangent, or cotangent of an angle 1 minute greater; multiply this difference by a fraction whose numerator is the number of seconds in the given angle and whose denominator is 60.

If the sine or tangent is sought, add this correction to the value first found; if the cosine or cotangent is sought, subtract the correction.

Example.—Find: (a) the sine of $56° 43' 17''$; (b) the cosine; (c) the tangent; and (d) the cotangent.

Solution.—(a) Here $A = 56° 43'$, $s = 17$, $A + 1' = 56° 44'$.

$$\sin (A + 1') = \sin 56° 44' = .83613$$
$$\sin A = \sin 56° 43' = .83597$$
$$\text{Difference} = .00016$$
$$\times \frac{17}{60}$$
$$.00005, \text{ nearly}$$

Adding this product to $\sin A$, we have
$$\sin 56° 43' 17'' = .83597 + .00005 = .83602. \quad \text{Ans.}$$

(b)
$$\cos A = \cos 56° 43' = .54878$$
$$\cos (A + 1') = \cos 56° 44' = .54854$$
$$\text{Difference} = .00024$$
$$\times \frac{17}{60}$$
$$.00007, \text{ nearly}$$

Subtracting this product from $\cos A$, we have
$$\cos 56° 43' 17'' = .54878 - .00007 = .54871. \quad \text{Ans.}$$

(c)
$$\tan (A + 1') = \tan 56° 44' = 1.52429$$
$$\tan A = \tan 56° 43' = 1.52332$$
$$\text{Difference} = .00097$$
$$\times \frac{17}{60}$$
$$.00027, \text{ nearly}$$

Adding this product to tan A, we have

$$\tan 56° 43' 17'' = 1.52332 + .00027 = 1.52359. \textbf{ Ans.}$$

(*d*)
$$\cot A = \cot 56° 43' = .65646$$
$$\cot (A + 1') = \cot 56° 44' = .65604$$
$$\text{Difference} = .00042$$
$$\times \frac{17}{60}$$
$$.00012, \text{ nearly}$$

Subtracting this product from cot A, we have

$$\cot 56° 43' 17'' = .65646 - .00012 = .65634. \textbf{ Ans.}$$

EXAMPLES FOR PRACTICE

Verify the following values:

(*a*) sin 18° 54′ 45″ = .32412; tan 18° 54′ 45″ = .34262.
(*b*) cos 34° 17′ 18″ = .82621; cot 34° 17′ 18″ = 1.46659.
(*c*) sin 72° 26′ 20″ = .95340; cot 72° 26′ 20″ = .31647.
(*d*) cos 65° 6′ 9″ = .42100; tan 65° 6′ 9″ = 2.15457.
(*e*) sin 80° 0′ 3″ = .98481; cot 80° 0′ 3″ = .17631.
(*f*) tan 14° 14′ 14″ = .25373; cos 14° 14′ 14″ = .96928.

---

21. **To Find the Angle Corresponding to a Given Function, When the Function Is in the Table.**—This case does not present any difficulty. Having found the given function in the table, the degrees in the angle are taken from the top or the bottom, and the minutes from the left- or the right-hand column, according as the name of the function is at the top or at the bottom of the page.

EXAMPLE 1.—The sine of an angle is .47486; what is the angle?

SOLUTION.—Glancing down the columns marked Sine in the table of Natural Sines and Cosines, .47486 is found (on page 28) in the column headed 28°. The number of minutes, 21, is found in the left-hand minute column, horizontally opposite .47486. Therefore, .47486 = sin 28° 21′. Ans.

EXAMPLE 2.—Find the angle whose cosine is .27032.

SOLUTION.—Looking in the columns marked Cosine at the top of the page, the given cosine is not found; hence, the angle is greater than 45°. Consequently, looking in the columns marked Cosine at the bottom of the page, .27032 is found (on page 26) in the double column marked 74° at the bottom, and in the horizontal row beginning with 19 in the right-hand minute column. Therefore, the angle whose cosine is .27032 is 74° 19′; or, .27032 = cos 74° 19′. Ans.

EXAMPLE 3.—Find the angle whose tangent is 2.15925.

SOLUTION.—On searching the table of Natural Tangents, the given tangent is found to belong to an angle greater than 45°, so that it must be looked for in the column marked Tangent at the bottom. It is found in the column having 65° at the bottom and opposite 9′ in the right-hand minute column. Therefore, 2.15925 = tan 65° 9′. Ans.

EXAMPLE 4.—Find the angle whose cotangent is .43412.

SOLUTION.—From the table of Natural Cotangents, it is found that this value is less than the cotangent of 45°, so it must be found in the column marked Cotangent at the bottom. Looking there, it is found in the column having 66° at the bottom, and opposite 32′, in the right-hand column of minutes. Therefore, the angle whose cotangent is .43412 is 66° 32′, or .43412 = cot 66° 32′. Ans.

EXAMPLES FOR PRACTICE

1. Find the angle whose sine is .47486.   Ans. 28° 21′
2. Find the angle whose cosine is .74353.   Ans. 41° 58′
3. Find the angle whose tangent is 2.06247.   Ans. 64° 8′
4. Find the angle whose cotangent is 1.20665.   Ans. 39° 39′
5. Find the angle whose sine is .76903.   Ans. 50° 16′
6. Find the angle whose tangent is 9.93101.   Ans. 84° 15′

22. To Find the Angle Corresponding to a Given Function, When the Function Is Not in the Table. Since the table includes the functions of all angles containing no odd seconds, a function not found in the table must correspond to an angle having odd seconds. Let the odd seconds that are to be determined be denoted by s, and the degrees and minutes by A, as in Art. **20.** Now, two consecutive functions including the given function can always be found in the table; that is, two consecutive functions of which one is greater and the other less than the given function. The required angle must, therefore, lie between the two angles corresponding to these two consecutive functions, and its number of degrees and minutes, A, is the number of degrees and minutes in the smaller of the two angles. The larger angle is $A + 1′$, or $A + 60″$, while the required angle is $A + s″$. Having determined A, it only remains to determine the number of odd seconds, or s. This is done by means of the following formulas, obtained by solving for s the formulas found in Art. **20.**

If the given function is a sine or tangent,

$$s = \frac{\sin (A + s'') - \sin A}{\sin (A + 1') - \sin A} \times 60 \qquad (1)$$

$$s = \frac{\tan (A + s'') - \tan A}{\tan (A + 1') - \tan A} \times 60 \qquad (2)$$

If the given function is a cosine or cotangent,

$$s = \frac{\cos A - \cos (A + s'')}{\cos A - \cos (A + 1')} \times 60 \qquad (3)$$

$$s = \frac{\cot A - \cot (A + s'')}{\cot A - \cot (A + 1')} \times 60 \qquad (4)$$

Observe that, although $A + s''$ is not known, its sine, cosine, etc., as the case may be, is known, or given. Thus, if the problem is to find the angle whose cotangent is .97888, we have cot $(A + s'') = .97888$.

The foregoing formulas lead to the following general rule for finding the angle corresponding to a given function:

Rule.—*Find the difference of the two numbers in the table between which the given function lies, and use that difference as the denominator of a fraction.*

Find the difference between the function belonging to the smaller angle and the given function, and use that difference as the numerator of the fraction mentioned above. Multiply this fraction by 60. The result will be the number of seconds to be added to the smaller angle in order to obtain the required angle.

Example 1.—Find the angle whose sine is .57698.

Solution.—Looking in the table of Natural Sines, in the columns marked Sine, it is found that the given sine lies between .57691 (= sin 35° 14′) and .57715(= sin 35° 15′). The difference between them is .57715 − .57691 = .00024. The difference between the sine of the smaller angle, or .57691, and the given sine, or .57698, is .57698 − .57691 = .00007. Then, $\dfrac{.00007}{.00024} \times 60 = \dfrac{7}{24} \times 60 = 18''$, nearly, and the required angle is 35° 14′ 18″; or, .57698 = sin 35° 14′ 18″. Ans.

Note.—In practice, only the significant figures of the differences forming the terms of the function are used, the decimal point being dispensed with. Thus, .57715 − .57691 = 24, it being understood that this means 24 units of the fifth decimal order, or .00024.

EXAMPLE 2.—Find the angle whose cosine is .27052.

SOLUTION.—Looking in the table of Cosines, the given cosine is found to belong to a greater angle than 45° and therefore it must be looked for in the columns marked Cosine at the bottom of the page. It is found between the numbers .27060($= \cos 74° 18'$) and .27032($= \cos 74° 19'$). The difference between the two numbers is .27060 − .27032 = 28 units of the fifth order. The cosine of the smaller angle, or 74° 18', is .27060, and the difference between this and the given cosine is .27060 − .27052 = 8 units of the fifth order. Hence, $\frac{8}{28} \times 60 = 17''$; and, therefore, .27052 = cos 74° 18' 17''. Ans.

EXAMPLE 3.—Find the angle whose tangent is 2.15841.

SOLUTION.— 2.15841 falls between 2.15760($= \tan 65° 08'$) and 2.15925 ($= \tan 65° 9'$). The difference between these numbers is 2.15925 − 2.15760 = 165 units of the fifth order; 2.15841 − 2.15760 = 81 units of the fifth order. Hence, $\frac{81}{165} \times 60 = 30''$, nearly, and therefore 2.15841 = tan 65° 8' 30''. Ans.

EXAMPLE 4.—Find the angle whose cotangent is 1.26342.

SOLUTION.— 1.26342 falls between 1.26395($= \cot 38° 21'$) and 1.26319 ($= \cot 38° 22'$). The difference between these numbers is 1.26395 − 1.26319 = .00076. Also, 1.26395 − 1.26342 = .00053. $\frac{53}{76} \times 60 = 42''$, and therefore 1.26342 = cot 38° 21' 42''. Ans.

EXAMPLES FOR PRACTICE

1. Find: (*a*) the sine of 48° 17'; (*b*) the cosine; (*c*) the tangent.

Ans. $\begin{cases} (a) & .74644 \\ (b) & .66545 \\ (c) & 1.12172 \end{cases}$

2. Find: (*a*) the sine of 13° 11' 6''; (*b*) the cosine; (*c*) the tangent.

Ans. $\begin{cases} (a) & .22810 \\ (b) & .97364 \\ (c) & .23427 \end{cases}$

3. Find: (*a*) the sine of 72° 0' 2''; (*b*) the cosine; (*c*) the tangent.

Ans. $\begin{cases} (a) & .95106 \\ (b) & .30901 \\ (c) & 3.07778 \end{cases}$

4. (*a*) Of what angle is .26489 the sine? (*b*) Of what angle is it the cosine?

Ans. $\begin{cases} (a) & 15° 21' 37'' \\ (b) & 74° 38' 23'' \end{cases}$

5. (*a*) Of what angle is .688 the sine? (*b*) Of what angle is it the cosine? (*c*) Of what angle is it the tangent?

Ans. $\begin{cases} (a) & 43° 28' 20'' \\ (b) & 46° 31' 40'' \\ (c) & 34° 31' 40'' \end{cases}$

TABLE OF LOGARITHMIC FUNCTIONS

23. The student is already familiar with the use of the table of logarithms of numbers. As stated in Art. **17,** a **table of logarithmic functions** is a table containing the logarithms of the natural functions, these logarithms being, for convenience, called **logarithmic functions.** Thus, the logarithm of the sine of an angle is referred to as the **logarithmic sine** of the angle.

The connection between the tables can be seen from the following:

From table of natural functions, cot 44° . . . . . = 1.03553
From table of logarithms, log 1.03553 . . . . . = .01516
From table of logarithmic functions, log cot 44° = .01516

Few tables give the logarithmic secants and cosecants. These logarithmic functions may be obtained from the relations,

$$\sec A = \frac{1}{\cos A}, \; \csc A = \frac{1}{\sin A}$$

which give,

log sec A = — log cos A, log csc A = — log sin A

That is, instead of adding the logarithmic secant or cosecant, the logarithmic cosine or sine, respectively, may be subtracted. Likewise, instead of subtracting the logarithmic secant, the logarithmic cosine may be added, and instead of subtracting the logarithmic cosecant, the logarithmic sine may be added.

24. **Description of the Table.**—The table of logarithmic functions contains for every minute the logarithms, to five decimal places, of the trigonometric sines, cosines, tangents, and cotangents of angles from 0° to 90°. From 0° to 45°, the degrees are placed at the top of the page and the minutes in the column headed ′ on the left. From 45° to 90°, the degrees are at the bottom of the page, the minutes in the last whole column at the right, and the name of the trigonometric function is placed at **the** bottom of the column.

This arrangement is similar to that in the table of natural functions. It will be observed that the numbers of degrees at the top of the pages increase in the order of the pages from 0° to 44°, while those at the bottom decrease from 89° to 44°.

The general description of the table will be better understood by referring to one of its pages. Take, for instance, the page marked 11° at the top and 78° at the bottom. The first column on the left (marked ′) contains the natural numbers from 1 to 60. These numbers represent minutes. Horizontally opposite to these numbers, and in the columns marked at the top log sin, log tan, etc., are printed the logarithmic functions, each function being in the same horizontal line as the number of minutes by which the corresponding angle exceeds 11°. Thus, the logarithmic tangent of 11° 39′, which is $\overline{1}.31425$, is found in the column marked log tan at the top, and in the same horizontal line as the number 39 in the left-hand column. Similarly, the number $\overline{1}.99072$, being in the column marked at the top log cos, and in the same horizontal line as 48 in the left-hand column, is the logarithmic cosine of 11° 48′. In some tables, several mantissas are printed under and to the right of the same characteristic, and are understood to belong with that characteristic. Thus, in the logarithm just considered, only the mantissa .99072 is printed, the characteristic being the same as the first one found above that mantissa.

The last column but one (marked ′ at the bottom) contains the natural numbers from 1 to 60, increasing from bottom to top. It will be observed that any angle determined by the number of degrees at the bottom (78 in this case) and any number of minutes in the right-hand minute column, is the complement of the angle determined by the number of degrees at the top (11 in this case) and the number of minutes in the left-hand minute column, horizontally opposite the number of minutes in the right-hand minute column. Thus, the number 18 in the right-hand minute column is horizontally opposite the number 42 in the left-hand column, and we have, 78° 18′ + 11° 42′ = 90°. Therefore, since the fundamental functions of an angle are equal to the cofunctions of its complement,

$$\sin 11° \; 42' = \cos 78° \; 18'$$
$$\cot 11° \; 42' = \tan 78° \; 18', \text{ etc.}$$

and $\qquad \log \sin 11° \; 42' = \log \cos 78° \; 18', \text{ etc.}$

For this reason, the notation log tan is written at the bottom of the column headed log cot, to indicate that the logarithms in this column are the logarithmic tangents of angles whose number of degrees is the number (78 in this case) at the bottom of the page, and whose number of minutes is opposite those logarithms in the right-hand minute column. Similarly, the columns marked log sin, log tan, and log cos at the top are marked, respectively, log cos, log cot, and log sin at the bottom.

25. After the column marked log sin there is a column marked d. This column contains the differences, expressed in units of the fifth decimal order, between the consecutive logarithmic sines given in the sine column. Thus, referring to the page headed 11°, the first number in the d-column following the sine column is 65; it will be observed that this number is opposite the space between the logarithmic sines $\overline{1}.28125$ and $\overline{1}.28060$, and is the difference, in units of the fifth decimal order, or expressed in hundred thousandths, between these two logarithmic sines. These differences are called **tabular differences.** Similar differences are printed in the column marked d after the cosine column, and in the column marked c. d. between the tangent and the cotangent column. The notation c. d. means *common difference*, as the differences between the successive logarithmic tangents are the same as those between the corresponding cotangents, although obtained by reversing the order in which the functions are subtracted; that is to say, $\log \tan A - \log \tan B = \log \cot B - \log \cot A$.

The tabular differences for the cosines are not given in the first ten pages, both for want of space and because they are so small that they can be readily determined by mental subtraction.

The use of the tabular differences, the use and contents of the column marked p. p. in all pages but the first three, and the peculiarities and applications of these first three pages of the table will be explained further on.

26. To Find the Logarithmic Functions of an Angle Having No Odd Seconds.

Rule.—For an angle less than 45°, look for the degrees at the top of the page and for the minutes in the column (marked ') at the left of the page on which the number of degrees is found. Then look across the page along the horizontal row containing the given number of minutes, into the column headed by the name of the function whose logarithm is required. The desired logarithm is found in this row and column.

For an angle between 45° and 90°, find the degrees at the bottom of the page and the minutes in the column (marked ') at the right of the page. Then look across the page, along the horizontal row containing the given number of minutes, into the column marked at the bottom with the name of the function whose logarithm is to be found. The row and column thus determined contain the desired logarithm.

EXAMPLE 1.—Find the logarithmic sine and the logarithmic tangent of 15° 24′.

SOLUTION.—On the page marked 15° at the top, in the column headed log sin, and in the same horizontal row with 24, the number $\overline{1}.42416$ is found; and in the column headed log tan, the number $\overline{1}.44004$ is found. Hence,

$$\log \sin 15° 24′ = \overline{1}.42416. \quad \text{Ans.}$$
$$\log \tan 15° 24′ = \overline{1}.44004. \quad \text{Ans.}$$

EXAMPLE 2.—Find the logarithmic tangent and cosine of 73° 10′.

SOLUTION.—As 73 is greater than 45, it is found at the bottom of the page. Looking for the number of minutes (10′) in the right-hand minute column, and following the horizontal row determined by this number into the column marked log tan at the bottom, the number .51920 is found. Likewise, the number $\overline{1}.46178$ is found in the column marked log cos at the bottom, and horizontally opposite the number 10 in the right-hand minute column. Therefore,

$$\log \tan 73° 10′ = \ .51920. \quad \text{Ans.}$$
$$\log \cos 73° 10′ = \overline{1}.46178. \quad \text{Ans.}$$

EXAMPLES FOR PRACTICE

1. Find: (*a*) the logarithmic cosine of 36° 58′; (*b*) the logarithmic tangent.

Ans. $\begin{cases} (a) & \overline{1}.90254 \\ (b) & \overline{1}.87659 \end{cases}$

2. Find: (*a*) the logarithmic tangent of 23° 39′; (*b*) the logarithmic cotangent.

Ans. $\begin{cases} (a) & \overline{1}.64140 \\ (b) & .35860 \end{cases}$

3. Find: (*a*) the logarithmic sine of 79° 45′; (*b*) the logarithmic cosine.

Ans. $\begin{cases} (a) & \overline{1}.99301 \\ (b) & \overline{1}.25028 \end{cases}$

4. Find: (*a*) the logarithmic tangent of 46° 59′; (*b*) the logarithmic cotangent.

Ans. $\begin{cases} (a) & .03009 \\ (b) & \overline{1}.96991 \end{cases}$

---

27. To Find the Logarithmic Functions of an Angle Containing an Odd Number of Seconds.—Let the number of degrees and minutes in an angle any of whose logarithmic functions is required be denoted by A, and the number of odd seconds by s. Thus, if the angle is 37° 43′ 19″, A will equal 37° 43′, and s will equal 19″; also, $A + 1′$, or $A + 60″$, will equal 37° 43′ + 1′, or 37° 44′. (See Art. **20.**) Since the table gives the logarithmic functions of any angle containing no odd seconds, the logarithmic functions of A and $A + 1′$ may be readily found, as explained in the last article. Let these logarithmic functions be denoted by l and $l′$, respectively, and the required logarithmic function by L. In the general theory of logarithms, treated in advanced works on mathematics, it is shown that if two consecutive angles (as 37° 43′ and 37° 44′) are taken from the table, the difference between any logarithmic function of the greater and the same logarithmic function of the smaller angle is to the difference between the same logarithmic function of any intermediate angle (as 37° 43′ 19″) and the same function of the smaller angle, as the difference between the greater and the smaller angle is to the difference between the intermediate and the smaller angle. If the notation $F(A)$, read *function of A*, is employed to denote any logarithmic function of an angle A, we have, writing $A + 60″$ instead of $A + 1′$,

$$\frac{F(A + 60'') - F(A)}{F(A + s) - F(A)} = \frac{(A + 60'') - A}{(A + s) - A} = \frac{60}{s}$$

that is,
$$\frac{l' - l}{L - l} = \frac{60}{s}$$

whence,
$$L - l = (l' - l)\frac{s}{60}$$

and
$$L = l + (l' - l)\frac{s}{60}$$

The difference between l' and l, being the difference between two consecutive logarithmic functions, may be taken from the column of tabular differences in the table. (See Art. 25.) Denoting the tabular difference $l' - l$ by D, the preceding equation becomes

$$L = l + D \times \frac{s}{60}$$

It should be observed that, since the sine and the tangent increase with the angle, while the cosine and cotangent decrease as the angle increases, $l' - l$ is positive or negative according as the functions considered are fundamental functions (sine, tangent) or cofunctions (cosine, cotangent). In the latter case, D in the formula should be treated as negative; that is, the product $D \times \frac{s}{60}$ should be subtracted from l

It should also be borne in mind that the tabular difference D is expressed in units of the fifth order of decimals, or hundred thousandths. Thus, if the number of seconds s is 15, and the tabular difference is 36, the quantity to be added to l is $.00036 \times \frac{15}{60} = .00009.$

If $l = \bar{1}.59812$, the work is arranged as follows:
$$l = \bar{1}.59812$$
$$D \times \frac{s}{60} = \quad\quad 9$$
$$\overline{\qquad\qquad}$$
$$L = \bar{1}.59821$$

When, as in this case, the product $D \times \frac{s}{60}$ is small, it can readily be added or subtracted mentally. Only the significant figures of D (those given in the d-column) are used, it being understood that the result expresses units of the fifth order of decimals. Thus, instead of writing D = .00036, and $D \times \frac{s}{60}$ = .00036 $\times \frac{s}{60}$, the following abbreviated notation is used: $D = 36$; $D \times \frac{s}{60} = 36 \times \frac{s}{60}$, the latter product expressing decimal units of the fifth order, or hundred thousandths.

The foregoing formula indicates the process by which the logarithmic functions of an angle containing odd seconds are obtained. It may be stated in words as follows:

Rule.—*Drop the seconds, and find the logarithmic function of the remaining angle. Find the tabular difference between this logarithmic function and the same function of the angle next higher in the table. Multiply this tabular difference by the number of seconds in the angle and divide the product by 60. Add this result to or subtract it from the logarithm found, according as the logarithm to be determined is that of a fundamental function or that of a cofunction. The result thus obtained is the required logarithmic function.*

Example 1.—Find: (a) the logarithmic sine of 15° 40′ 32″; (b) the logarithmic cosine.

Solution.—(a) Dropping the seconds, 15° 40′ is obtained, whose logarithmic sine, found as in Art. **25**, is $\bar{1}.43143$; that is, $l = \bar{1}.43143$. Opposite the space between this logarithm and the following, and in the column marked d, is found the tabular difference 45($= D$). Applying the formula given in Art. **27**,

$$L = \bar{1}.43143 + .00045 \times \tfrac{32}{60}$$
$$l = \bar{1}.43143$$
$$D \times \frac{s}{60} = 45 \times \frac{32}{60} = \qquad 24$$
$$L = \overline{\bar{1}.43167}$$

that is,   log sin 15° 40′ 32″ = $\bar{1}.43167$. Ans.

In practice, it is not necessary to write all the figures of l before adding the correction $D \times \frac{s}{60}$. Having found the value of l in the table, one places and keeps the finger on that value and calculates the correction $D \times \frac{s}{60}$. In the majority of cases, this correction can be added mentally to l. Thus, in the example just explained, the correction is 24, which, being mentally added to the number 43 formed by the last two figures of l, gives 67 as the last two figures of L. The other figures of L are the same as those of l.

(*b*) The logarithmic cosine of 15° 40′ is $\bar{1}.98356 (= l)$. Horizontally opposite the space between this logarithm and the following, the tabular difference $4 (= D)$ is found in the column marked d on the right of the cosine column. As the function under consideration is a cofunction, the correction $D \times \dfrac{s}{60}$ must be subtracted for l. We have, then,

$$l = \bar{1}.98356$$

$$D \times \frac{s}{60} = 4 \times \frac{32}{60} = \qquad 2, \text{ to the nearest unit}$$

$$L = \overline{\bar{1}.98354}$$

Therefore, $\qquad \log \cos 15° 40′ 32″ = \bar{1}.98354.$ Ans.

In practice, the correction 2 would be subtracted mentally, without previously writing the value of l.

EXAMPLE 2.—Find the logarithmic tangent of 63° 39′ 27″.

SOLUTION.—Dropping the seconds, and referring to the page marked 63° at the bottom, the logarithmic tangent of 63° 39′ is found to be $.30512 (= l)$. Since in this case the angles increase from bottom to top, the tabular difference to be used is that horizontally opposite the space between the logarithm just taken and the one immediately above it in the column (that is, .30543). This difference is 31, printed in the column marked c. d. on the left of the cotangent column. We have, therefore,

$$l = .30512$$

$$\frac{s}{60} \times D = \frac{27}{60} \times 31 = \qquad 14, \text{ to the nearest unit}$$

$$L = \overline{.30526}$$

Therefore, $\qquad \log \tan 63° 39′ 27″ = .30526.$ Ans.

EXAMPLE 3.—Find the logarithmic cotangent of 54° 8′ 9″.

SOLUTION.—Dropping the seconds, the value of l is found to be $\bar{1}.85913$. The tabular difference in the c. d. column and horizontally opposite the space between this logarithm and the one immediately above it is 26. As the cotangent is a cofunction, the correction $\dfrac{s}{60} \times D$ is to be subtracted from l. Then,

$$l = \bar{1}.85913$$

$$\frac{s}{60} \times D = \frac{9}{60} \times 26 = \qquad 4$$

$$L = \overline{\bar{1}.85909}$$

Therefore, $\qquad \log \cot 54° 8′ 9″ = \bar{1}.85909.$ Ans.

EXAMPLES FOR PRACTICE

1. Find the logarithmic sine, tangent, and cosine of 33° 21′ 46′ .

$$\text{Ans.}\begin{cases}\log \sin = \bar{1}.74032 \\ \log \tan = \bar{1}.81852 \\ \log \cos = \bar{1}.92179\end{cases}$$

2. Find the logarithmic sine and cotangent of 23° 3′ 17″.

$$\text{Ans.}\begin{cases}\log \sin = \bar{1}.59286 \\ \log \cot = .37100\end{cases}$$

3. Find the logarithmic tangent and cosine of 49° 12′ 12″.

$$\text{Ans.}\begin{cases}\log \tan = .06395 \\ \log \cos = \bar{1}.81516\end{cases}$$

4. Find the logarithmic sine, tangent, and cosine of 72° 52′ 49″.

$$\text{Ans.}\begin{cases}\log \sin = \bar{1}.98031 \\ \log \tan = .51143 \\ \log \cos = \bar{1}.46890\end{cases}$$

5. Find the logarithmic sine and cotangent of 81° 38′ 28″.

$$\text{Ans.}\begin{cases}\log \sin = \bar{1}.99536 \\ \log \cot = \bar{1}.16712\end{cases}$$

6. Find the logarithmic tangent and cosine of 65° 0′ 47″.

$$\text{Ans.}\begin{cases}\log \tan = .33159 \\ \log \cos = \bar{1}.62574\end{cases}$$

7. Find the logarithmic secant and cosecant of 59° 0′ 9″.

$$\text{Ans.}\begin{cases}\log \sec = .28819 \\ \log \csc = .06692\end{cases}$$

---

28. Use of the Column of Proportional Parts.—The method described in the preceding article can be applied to any table of logarithmic functions. Some tables, however, among them the table furnished with this Course, contain a column giving the products of the tabular differences by the fractions $\frac{6}{60}$, $\frac{7}{60}$, $\frac{8}{60}$, $\frac{9}{60}$, $\frac{10}{60}$, $\frac{20}{60}$, $\frac{30}{60}$, $\frac{40}{60}$, and $\frac{50}{60}$. These products are called **proportional parts**, and are given in the right-hand column (marked p. p. at the top) of each page, beginning with 3°. The tabular differences are here printed in heavy figures. Under each tabular difference are given the products of it by $\frac{6}{60}$, $\frac{7}{60}$, etc., the number of sixtieths being printed horizontally opposite the product, on the left of a vertical line. Thus, referring to the right-hand column of the page marked 13° at the top, the numbers 54, 53, 52, printed in heavy type, are tabular differences. The number 27, directly under 54, and horizontally opposite **the**

number 30 on the left of the vertical line, is the product of 54 by $\frac{30}{60}$. Likewise, 17.3, found under 52, and horizontally opposite 20, is the product of 52 by $\frac{20}{60}$. The proportional parts for 1, 2, 3, 4, 5 are obtained from those for 10, 20, 30, etc., by moving the decimal point one place to the left. Thus, the proportional part for 20, under the tabular difference 52, is 17.3, as just explained. The proportional part for 2, that is, the product of 52 by $\frac{2}{60}$, is 1.73.

In the first three pages of the logarithmic table, no proportional parts are given, the use of these pages being different from that of the others. In pages 45, 46, and 47, not all the tabular differences are given in the p. p. column, owing to want of space; but the proportional part for any tabular difference is easily obtained by means of the proportional parts for digits given at the bottom of the p. p. column. Referring, for example, to page 45, the tabular difference 215, which is found in the c. d. column, does not appear in the p. p. column. If we wish to find the product of 215 by $\frac{30}{60}$, we look in the p. p. column for the tabular difference next lower than 215, which is 212. Horizontally opposite 30, and under 212, we find 106; that is, $212 \times \frac{30}{60} = 106$. As 215 $= 212 + 3$, we must add to the product just found (106), the product of $3 \times \frac{30}{60}$. This is taken from the column headed 3 near the bottom of the p. p. column: there we find 1.5 horizontally opposite 30; that is, $3 \times \frac{30}{60} = 1.5$. Therefore, $215 \times \frac{30}{60} = 106 + 1.5 = 107.5$. The addition of these two products can usually be effected mentally.

The correction $D \times \frac{s}{60}$ to be applied to l in order to find L (formula of Art. **27**) is found from the table of proportional parts as follows:

Rule.—*Having found the tabular difference D, look for this difference in the column of proportional parts. If this difference is found in that column and the number of seconds is a digit greater than 5 or a digit followed by a cipher, look for it on the left of the vertical line under D; the correction is then found horizontally opposite this number, and directly under D. If the* *number of seconds is a digit less than 6, add a cipher, find the proportional part corresponding to the resulting number, and move the decimal point one place to the left. If the number of seconds consists of two significant digits (as 39), find the correction for the first digit followed by a cipher, and that for the second digit, and add the two corrections. (Thus, if the number of seconds is 43, the correction is found by adding the corrections for 40 and 3.)*

If the tabular difference D is not found in the p. p. column (which may happen only on pages 45 to 47), take, as just explained, the proportional part corresponding to the next lower tabular difference found in the p. p. column; then, from the digit columns found at the bottom of the p. p. column, find the proportional part corresponding to the difference between D and the tabular difference just used. Add the two proportional parts thus found.

EXAMPLE 1.—Find: (a) the logarithmic tangent of $22° 17' 8''$; (b) the logarithmic cosine.

SOLUTION.—(a) Dropping the seconds, we find log tan 22° 17′ $= \bar{1}.61256(= l)$; $D = 36$. Turning to the column of proportional parts, 36 is found in heavy type near the top of the page. Following the horizontal row that begins with 8 (number of seconds) at the left of the vertical line under 36, we find in that row, and directly under 36, the correction 4.8, which may be called 5, as there are no other numbers to be combined with it. Therefore,

$$l = \bar{1}.61256$$
$$\frac{s}{60} \times D = \text{p. p.} = \qquad 5$$
$$L = \bar{1}.61261$$

That is,   log tan 22° 17′ 8″ $= \bar{1}.61261$. Ans.

(b) $l = \log \cos 22° 17' = \bar{1}.96629$; $D = 5$. Looking for the column headed 5 among the proportional parts, the correction .7 (or say 1) is found directly under 5 and horizontally opposite 8. Therefore,

$$l = \bar{1}.96629$$
$$\frac{s}{60} \times D = \text{p. p.} = \qquad 1$$
$$L = \bar{1}.96628$$

That is,   log cos 22° 17′ 8″ $= \bar{1}.96628$. Ans.

EXAMPLE 2.—Find the logarithmic sine of $3° 18' 9''$.

SOLUTION.— $l = \sin 3° 18' = \bar{2}.76015$; $D = 219$. The difference 219 is not found in the p. p. column; the tabular difference in the p. p. column next lower is 216. Under 216, and horizontally opposite 9, is found 32.4. The difference between 219 and 216 is 3. Looking for 3 in the digit columns at the bottom of the p. p. column, .5 is found under 3, and horizontally opposite 9. Therefore, $219 \times \frac{9}{60} = 32.4 + .5 = 33$, nearly.

$$l = \bar{2}.76015$$
$$219 \times \frac{9}{60} = \quad 33$$
$$L = \bar{2}.76048$$

That is, $\log 3° 18' 9'' = \bar{2}.76048$. Ans.

EXAMPLE 3.—Find: (a) the logarithmic tangent of 53° 47′ 04″; (b) the logarithmic cosine.

SOLUTION.—(a) $l = \log \tan 53° 47' = .13529$; $D = 26$; the proportional part for 40, under D, that is, under 26, is 17.3; the proportional part for 4 is $\frac{17.3}{10}$, or 2, nearly.

$$l = .13529$$
$$26 \times \frac{4}{60} = \quad 2$$
$$L = .13531$$

That is, $\log \tan 53° 47' 4'' = .13531$. Ans.

(b) $l = \log \cos 53° 47' = \bar{1}.77147$; $D = 17$. The number horizontally opposite 40, in the column headed 17 among the proportional parts, is 11.3; the proportional part for 4 is, therefore, $\frac{11.3}{10} = 1$, nearly.

$$l = \bar{1}.77147$$
$$17 \times \frac{4}{60} = \quad 1$$
$$L = \bar{1}.77146$$

That is, $\log \cos 53° 47' 4'' = \bar{1}.77146$. Ans.

EXAMPLE 4.—To find the logarithmic cotangent of 72° 35′ 47″.

SOLUTION.— $l = \log \cot 72° 35' = \bar{1}.49652$; $D = 45$. Looking among the proportional parts for the column headed 45, the correction for 40 is found to be 30, and that for 7 is found to be 5.3. Therefore,

$$l = \bar{1}.49652$$

p. p. for 40 = 30.0
p. p. for 7 = 5.3
p. p. for 47 =   35

$$L = \bar{1}.49617$$

That is, $\log \cot 72° 35' 47'' = \bar{1}.49617$. Ans.

In practice, it would not be necessary to write down the corrections 30 and 5.3, which would be added mentally. The same remark applies to all similar cases.

EXAMPLES FOR PRACTICE

1. Find the logarithmic sine and cotangent of $9° 39' 17''$.

$$\text{Ans.}\begin{cases}\log \sin = \bar{1}.22456 \\ \log \cot = .76924\end{cases}$$

2. Find the logarithmic sine, tangent, and cosine of $39° 8' 52''$.

$$\text{Ans.}\begin{cases}\log \sin = \bar{1}.80025 \\ \log \tan = \bar{1}.91065 \\ \log \cos = \bar{1}.88959\end{cases}$$

3. Find the logarithmic cotangent and cosecant of $80° 3' 46''$.

$$\text{Ans.}\begin{cases}\log \cot = \bar{1}.24352 \\ \log \csc = .00657\end{cases}$$

4. Find the logarithmic sine, secant, and tangent of $49° 0' 54''$.

$$\text{Ans.}\begin{cases}\log \sin = \bar{1}.87788 \\ \log \sec = .18319 \\ \log \tan = .06107\end{cases}$$

5. Find the logarithmic tangent and cosine of $4° 2' 4''$.

$$\text{Ans.}\begin{cases}\log \tan = \bar{2}.84838 \\ \log \cos = \bar{1}.99892\end{cases}$$

---

29. To Find the Angle Corresponding to Any Logarithmic Function When the Given Function Is Found in the Table.—In this case, the angle, which contains no odd seconds, is found as follows:

Rule.—*Find the given logarithm in the column marked by the name of the function whose logarithm is given. Then, if the name of the given function is at the top of the column, the number of degrees in the angle is that at the top of the page, and the number of minutes is horizontally opposite the logarithm, in the left-hand minute column. If the name of the function is at the foot of the column, the number of degrees in the angle is that at the foot of the page, and the number of minutes is in the right-hand minute column, horizontally opposite the given logarithm.*

In searching the table for a given logarithm, it should be borne in mind that the logarithmic sines and tangents increase, and the cosines and cotangents decrease, from $0°$ to $90°$. Therefore, in the columns marked log sin and log tan at the top, the logarithms increase, and in the columns headed log cos and log cot the logarithms decrease, from the first to the last page. The sines and tangents continue to increase, and the cosines and cotangents to decrease, from the last page to the first, in the columns marked with the names of these functions, respectively, at the bottom. Thus, the last page contains, in the column headed log sin, the logarithmic sines of the angles between 44° and 45°. The sines are continued in the column marked log sin at the bottom, which contains the logarithmic sines of the angles between 45° and 46°; the preceding page contains the sines of angles between 46° and 47°, etc. Here the logarithmic sines increase from bottom to top, and in the inverse order of the pages.

When looking for a given logarithmic sine, open the table at random. Glance at both of the sine columns, that is, the column marked log sin at the top and the column marked log sin at the bottom, and compare the logarithms in them with the given logarithm. If the given logarithm is less than those found in the column marked log sin at the top, said given logarithm must be in that column, but in a preceding page. If the given logarithm is greater than those in the column marked log sin at the bottom, said given logarithm must be in that column, but in a preceding page. If neither of these is the case, the given logarithm must be in a subsequent page. Turn a few pages forwards or backwards, as the case may be, and repeat the operation. The comparison of the two columns, however, is not usually necessary after the first three figures of the given logarithm have been found in one of them, as that logarithm is then found in that column, and can be readily seen among the logarithms beginning with those three figures.

Proceed exactly in the same manner when the given function is a cosine; that is, treat the cosine as though it were a sine; but, having found the given logarithm, treat it as that of a cosine and take the angle accordingly.

As the tangents of angles less than 45° are less than 1, their logarithmic tangents have negative characteristics, and as the tangents of angles greater than 45° are greater than 1, their logarithmic tangents have positive characteristics. Therefore, a logarithmic tangent should be looked for in the column marked log tan at the top or at the bottom, according as its characteristic is negative or positive.   For a logarithmic cotangent, the rule should be reversed.

Example 1.—Find the angle whose logarithmic sine is $\bar{1}.57669$.

Solution.—Opening the table at random, say at the page marked 36° at the top, it is at once seen that the logarithms in the column marked log sin at the top are greater than the given logarithm.   This logarithm must, therefore, be in that column, but in a preceding page. Turning the pages backwards, a few at a time, the given logarithm is found on page 64, among those logarithms whose first three figures are $\bar{1}.57$.   As the name of the function is at the head of the column, the number of degrees (22) is taken from the top of the page, and that of minutes (10) from the left-hand minute column.   Therefore, the angle whose logarithmic sine is $\bar{1}.57669$ is 22° 10′, or $\bar{1}.57669 = \log \sin 22° 10′$.

Suppose that the table had first been opened at page 56.   Since the given logarithm is greater than those in the column marked log sin at the top and less than those in the column marked log sin at the bottom (or log cos at the top), the given logarithm is to be found in a subsequent page.   Suppose also that, turning the pages forwards, a few at a time, we come to page 63, and find the first three figures ($\bar{1}.57$) of the given logarithm in the column marked log sin at the top. Then, without consulting the other column, we follow the former column to the bottom, and into the next page, where we find the given logarithm, and take the corresponding angle as before.

Example 2.—To find the angle whose logarithmic sine is $\bar{1}.89810$.

Solution.—Open the table at random, say at page 73.   Since the given logarithm is greater than those in the column marked log sin at the top, and less than those in the column marked log sin at the bottom, it must be found in a subsequent page.   Suppose that we turn next to page 85.   We see at once that the given logarithm is greater than those in the column headed log sin, and also than those in the column marked log sin at the bottom.   Therefore, it must be in the latter column in some preceding page.   Turning the pages backwards, we find the first three figures ($\bar{1}.89$) of the given logarithm on page 79, and among the logarithms to which these three figures are common, we find $\bar{1}.89810$.   As this is a logarithmic sine, and the name sine is at the bottom of the column, the degrees in the corresponding angle are taken from the bottom of the page, and the minutes from the right-hand minute column.   Therefore, 52° 16′ is the angle whose logarithmic sine is $\bar{1}.89810$; that is, $\bar{1}.89810 = \log \sin 52° 16′$.   Ans.

Example 3.—Find the angle whose logarithmic cosine is $\bar{1}.86924$.

Solution.—Treating this as though it were a logarithmic sine, it is found, as explained above, on page 84, in the column marked log sin at the bottom.   Since the name cosine is at the top of the column, the required angle is 42° 16′.   That is, $\bar{1}.86924 = \log \cos 42° 16′$.   **Ans.**

EXAMPLE 4.—Find the angle whose logarithmic cotangent is .15639.

SOLUTION.—As the characteristic is positive, the logarithm should be looked for in the column marked log cot at the top.   After looking in a few pages, the first three figures (0.15) of the logarithm are found on page 76, and among them is found the given logarithm.   The name of the function being at the head of the column, the degrees in the angle are taken from the top of the page, and the minutes from the left-hand minute column.   Therefore, $.15639 = \log \cot 34° 54′$.   **Ans.**

EXAMPLES FOR PRACTICE

1.   Find the angle whose logarithmic sine is $\bar{1}.57885$.   Ans. 22° 17′

2.   Find the angle whose logarithmic sine is $\bar{1}.66731$.   Ans. 27° 42′

3.   Find the angle whose logarithmic sine is $\bar{2}.93740$.   Ans. 4° 58′

4.   Find the angle whose logarithmic sine is $\bar{1}.98345$.   Ans. 74° 17′

5.   Find the angle whose logarithmic cosine is $\bar{1}.92086$.

Ans. 33° 33′

6.   Find the angle whose logarithmic cosine is $\bar{1}.57232$. Ans. 68° 4′

7.   Find the angle whose logarithmic cosine is $\bar{1}.84949$. Ans. 45° 0′

8.   Find the angle whose logarithmic tangent is $\bar{1}.97649$.

Ans. 43° 27′

9.   Find the angle whose logarithmic cotangent is $\bar{2}.89274$.

Ans. 85° 32′

10.   Find the angle whose logarithmic tangent is .67377.

Ans. 78° 2′

11.   Find the angle whose logarithmic cotangent is .35517.

Ans. 23° 49′

12.   Find the angle whose logarithmic tangent is 1.28060.

Ans. 87° 0′

---

30.   To Find the Angle Corresponding to a Given Logarithmic Function When the Function Is Not in the Table.—*Without the Use of Proportional Parts.*—From the formula given in Art. **27**, the following may be obtained:

$$s = \frac{(L - l) \times 60}{D} = \frac{(L - l) \times 60}{l' - l}$$

Therefore, if the function L is given and it is found to lie between the consecutive logarithms l and l', the corresponding angle $A + s$ is that corresponding to l increased by the number of seconds determined by the formula just given. It will be remembered (see Art. **27**) that l and l' are, respectively, the logarithmic functions of two angles (A and $A + 1'$) differing by one minute. If the function is a fundamental function (sine or tangent) l' is greater than l; and since L lies between l and l', L is also greater than l; therefore, both $L - l$ and $l' - l$ are positive. If the function is a cofunction, l is greater than l', and also greater than L; therefore, both $L - l$ and $l' - l$ are negative, and $\dfrac{L - l}{l' - l}$ is positive. In such case, however, it is better to write this fraction in the form $\dfrac{l - L}{l - l'}$.

From the formula and the explanations just given, the following rule is derived for finding the angle corresponding to any given logarithmic function:

Rule.—*Find in the table the two consecutive logarithmic functions between which the given function lies. The degrees and minutes in the smaller of the angles corresponding to these two functions are the degrees and minutes in the required angle.*

Find the difference between the given function and that of the smaller angle; multiply that difference by 60, and divide the product by the tabular difference between the two functions in the table. The result will be the number of odd seconds in the required angle.

As the tabular difference is expressed in units of the fifth decimal order, the difference $L - l$ should be likewise expressed. Thus, if $L = \bar{1}.25198$, and $l = \bar{1}.25168$, the difference $L - l$ will be called 30.

EXAMPLE 1.—Find the angle whose logarithmic sine is $\bar{1}.47867$ ($= L$).

SOLUTION.—The first three figures of the given logarithm are always found in the table, and this makes it easy to determine the functions between which the given logarithm lies. Searching the sine columns of the table, it is found that $\bar{1}.47867$ lies between $\bar{1}.47854 (= l)$ and

$\bar{1}.47894(=l')$ on page 59. The smaller of the two angles corresponding to these two logarithms is $17° 31'(= A)$. Now, $L - l = 13, l' - l$ (tabular difference taken from table) $= 40$. Therefore,

$$s = \frac{13 \times 60}{40} = 19.5'', \text{ or, say, } 20''$$

and $\qquad A + s = 17° 31' + 20'' = 17° 31' 20''$

that is, $\qquad \bar{1}.47867 = \log \sin 17° 31' 20''.$ Ans.

EXAMPLE 2.—Find the angle whose logarithmic tangent is .27743 $(= L)$.

SOLUTION.—As the characteristic is positive, the logarithms between which L lies should be looked for in the column marked log tan at the bottom. These two logarithms are $.27738(= l)$ and $.27769(= l')$. The smaller angle corresponds to .27738, and is $62° 10'(= A)$. Also,

$$L - l = 5, l' - l(= D) = 31$$

$$A + s = A + \frac{5 \times 60}{31} = 62° 10' + 10'', \text{ nearly, } = 62° 10' 10''$$

that is, $\qquad .27743 = \log \tan 62° 10' 10''.$ Ans.

EXAMPLE 3.—Find the angle whose logarithmic cotangent is $\bar{1}.85899(= L)$.

SOLUTION.— L is found to lie between $\bar{1}.85887(= l')$ and $\bar{1}.85913$ $(= l)$. It will be noticed that here l is the greater, and l' the smaller of the two logarithms. Angle corresponding to $l = 54° 8'(= A)$.

$$l = \bar{1}.85913$$
$$L = \bar{1}.85899$$
$$l - L = 14; l - l' = 26$$
$$A + s = 54° 8' + \frac{14 \times 60}{26} = 54° 8' 32'', \text{ nearly}$$

that is, $\qquad \bar{1}.85899 = \log \cot 54° 8' 32''.$ Ans.

EXAMPLES FOR PRACTICE

1. Find the angle whose logarithmic sine is $\bar{1}.45566$.

    Ans. $16° 35' 27''$

2. Find the angle whose logarithmic tangent is $\bar{1}.33471$.

    Ans. $12° 11' 44''$

3. Find the angle whose logarithmic sine is $\bar{1}.89798$.

    Ans. $52° 14' 42''$

4. Find the angle whose logarithmic cosine is $\bar{1}.67412$.

    Ans. $61° 49' 23''$

5. Find the angle whose logarithmic cosine is $\bar{1}.92386$.

    Ans. $32° 56' 45''$

6. Find the angle whose logarithmic cotangent is .54139.

Ans. 16° 2′ 20″

7. Find the angle whose logarithmic tangent is $\overline{1}$.86712.

Ans. 36° 22′ 7″

8. Find the angle whose logarithmic cosine is $\overline{1}$.99785.

Ans. 5° 42′ 0′′

9. Find the angle whose logarithmic cotangent is $\overline{1}$.12345.

Ans. 82° 25′ 52″

---

31. *With the Use of Proportional Parts.*—Having found the degrees and minutes in the angle as in the preceding case, the number s of odd seconds may be conveniently found from the column of proportional parts. In order to facilitate the explanations that follow, the proportional parts corresponding to the tabular difference 105 are here copied from page 48 of the table. It will, therefore, be assumed that the value of D is 105, and, for what is said below, the student should refer to these proportional parts. Such being the case, the formula given at the beginning of the preceding article may be written,

	105
6	10.5
7	12.3
8	14.0
9	15.8
10	17.5
20	35.0
30	52.5
40	70.0
50	87.5

$$s = \frac{(L - i) \times 60}{105}$$

The value $L - l$, which is the difference between the given logarithm and the logarithm of the degrees and minutes (A) in the required angle, is readily determined, as already explained. It is only necessary to repeat that, if the function is a cofunction, $l - L$ should be used instead of $L - l$. Since the numbers on the right of the vertical line are the products of $\frac{105}{60}$ by the numbers on the left, it follows that the numbers on the left are the products of those on the right by $\frac{60}{105}$. Thus, $52.5 = \frac{105}{60} \times 30$, and $30 = 52.5 \times \frac{60}{105}$ $= \frac{52.5 \times 60}{105}$. Therefore, if $L - l$ is found among the numbers directly under 105, the value of s is the number on the left of the vertical column horizontally opposite $L - l$. For example, if $L - l = 35$, then $s = 20''$. If $L - l = 16$, then

$s = 9''$, the number 9 being opposite 15.8, which, to the nearest unit, may be called 16.

It will be remembered that the proportional parts opposite 10, 20, 30, 40, 50, when divided by 10 (that is, when the period is moved one place to the left), give the products of $\frac{10.5}{60}$ by 1, 2, 3, 4, and 5. From those parts we may, therefore, find by inspection the products of $\frac{10.5}{60}$ by all the digits from 1 to 9; and, in what follows, we shall proceed as if the products 1.75, 3.50, 5.25, 7.00, 8.75 of $\frac{10.5}{60}$ by 1, 2, 3, 4, and 5 were actually printed in the table opposite those digits; that is, it will be assumed that the proportional parts run in this order: 1.75, 3.50, 5.25, 7.00, 8.75, 10.5, 12.3, 14.0, etc., up to 87.5, the corresponding numbers on the left being, 1, 2, 3, 4, 5, 6, 7, 8, 9, 10, 20, 30, 40, 50. The proportional parts 1.75, 3.50, 5.25, 7.00, 8.75 will be referred to as proportional parts found in the table, corresponding to 1, 2, 3, 4, and 5 seconds, respectively.

This being understood, the number s of odd seconds in the angle is determined as follows:

Rule.—*Find $l, l', L - l$, and $l' - l$ (= tabular difference, or D), as before. Look for the tabular difference D in the column of proportional parts. Look for $L - l$ in the column of proportional parts directly under D. If $L - l$ is found there, the number horizontally opposite it on the left of the vertical line is the required number of seconds s. If $L - l$ is not found under D, take the proportional part next lower, which call p. Find the difference between $L - l$ and p, and look among the proportional parts under D for this difference, or the part nearest to it, whether higher or lower. Call this part p'. Add the numbers horizontally opposite p and p' on the left of the vertical line. The result will be the required number of seconds s.*

EXAMPLE 1.—Find the angle whose logarithmic tangent is $\bar{1}.42822 (= L)$.

SOLUTION.— $l = 1.42805$, $A = 15° 0'$, $L - l = 17$, $D = 51$. Looking in the column marked p. p. for 51, the number $17 (= L - l)$ is found under it, horizontally opposite the number 20 on the left of the vertical column. Therefore, $s = 20''$, and
$$\bar{1}.42822 = \log \tan 15° 0' 20''. \quad \textbf{Ans.}$$

EXAMPLE 2.—Find the angle whose logarithmic cosine is $\bar{1}.52783(=L)$.

SOLUTION.— $l = \bar{1}.52811$, $A = 70° 17'$, $l - L = 28$, $D = 36$. The proportional part under 36 next lower than 28 is 24; $28 - 24 = 4$; the proportional part nearest 4 is 4.2; the number horizontally opposite 24 is 40; and the number horizontally opposite 4.2 is 7; hence, $s = 40 + 7 = 47''$, and therefore
$$\bar{1}.52783 = \log \cos 70° 17' 47''. \quad \text{Ans.}$$

EXAMPLE 3.—Find the angle whose logarithmic sine is $\bar{1}.66191(L)$.

SOLUTION.— $l = \bar{1}.66173$; $A = 27° 19'$; $L - l = 18$; $D = 24$. Looking in the p. p. column for 24, the proportional part next lower than 18 is $16(= p)$, horizontally opposite which is 40. $18 - p = 18 - 16 = 2$. This difference is found among the proportional parts in the table (since it is the same as 20 with the decimal point moved one place to the left), and corresponds to $5'' \left(= \dfrac{50}{10}\right)$. Therefore, $s = 40 + 5 = 45''$, and
$$\bar{1}.66191 = \log \sin 27° 19' 45''. \quad \text{Ans.}$$

EXAMPLE 4.—Find the angle whose logarithmic cotangent is $\bar{1}.00375(=L)$.

SOLUTION.— $l = \bar{1}.00427$; $A = 84° 14'$; $l - L = 52$; $D = 126$. The proportional part under 126 next lower than 52 is $42(= p)$, which corresponds to $20''$; $52 - 42 = 10$. The proportional part nearest to 10 is $10.50 \left(= \dfrac{105.0}{10}\right)$, which corresponds to $5'' \left(= \dfrac{50}{10}\right)$. Therefore, $s = 20'' + 5'' = 25''$, and
$$\bar{1}.00375 = \log \cot 84° 14' 25''. \quad \text{Ans.}$$

EXAMPLES FOR PRACTICE

1. Find the angle whose logarithmic sine is $\bar{1}.78988$. Ans. 38° 3' 20''

2. Find the angle whose logarithmic tangent is $\bar{1}.78540$.
Ans. 31° 23' 15''

3. Find the angle whose logarithmic sine is $\bar{1}.77777$.
Ans. 36° 49' 56''

4. Find the angle whose logarithmic cosine is $\bar{1}.87341$.
Ans. 41° 39' 21''

5. Find the angle whose logarithmic cotangent is .31789.
Ans. 25° 41' 9''

6. Find the angle whose logarithmic cosine is $\bar{1}.34567$.
Ans. 77° 11' 38''

7.   Find the angle whose logarithmic cotangent is $\overline{1}.00381$.

Ans. 84° 14′ 22″

8.   Find the angle whose logarithmic tangent is 1.00300.

Ans. 84° 19′ 42″

9.   Find the angle whose logarithmic sine is $\overline{2}.99001$.

Ans. 5° 36′ 30″

---

32.   Tabular Values Increased by 10.—To avoid calculating with negative characteristics, they may be made positive by increasing them by 10.   Thus, log sin 27° may be given as 9.65705 instead of $\overline{1}.65705$.   The true logarithm is, therefore, 9.65705 − 10; the − 10 is usually not written, but is implied.   In many books this method is used for the logarithms of trigonometric functions.   In applying such logarithms to the solution of a problem, the characteristic in the final result must be corrected to agree with the conditions of the problem.

GENERAL PRINCIPLE OF INTERPOLATION

33.   It has been explained in some of the preceding articles how to determine the natural or the logarithmic functions of any angle containing an odd number of seconds, and therefore, not found in the table; also, how to find the angle corresponding to a given function, when that function is not in the table but lies between two values given in the table. The operation by which such intermediate values are determined from a table is called **interpolation.**   The values that are actually given in the table are called **tabular values.**   For example, in the table of logarithmic functions already described are found all angles that lie between 0° and 90° and contain no odd seconds, and also the logarithmic sines, cosines, etc. of such angles; those are all tabular values.   Angles containing odd seconds are not in the table, nor are their logarithmic functions.   Both these angles and their functions are intermediate values, and it is in connection with them that interpolation is used.

34. The general principle of interpolation, to be explained presently, is of the utmost importance, and of great value to the engineer, whose work requires the frequent use of tables of various kinds. That principle, although only approximately true, applies to nearly all tables with which the engineer has to deal, and the student should endeavor to make himself thoroughly familiar with it.

Let a table be constructed on the general type shown on the margin, the left-hand column containing values of a quantity X, and the right-hand column corresponding values of some quantity whose values depend on the values of X. Thus, the values of X may be the natural numbers 1, 2, 3, 4, etc., and the corresponding values of F may be the logarithms or the square roots of those numbers; or the values of X may be angles, and those of F may be sines, cosines, etc., either natural or logarithmic. So far

X	F
—	—
—	—
—	—
x_1	f_1
x	f
x_2	f_2
—	—

as the principle of interpolation is concerned, it is immaterial what kind of quantity is represented by X, and what kind of quantity is tabulated under F. It should be stated, however, that the principle applies only to tables in which the differences between consecutive values of X and the differences between the corresponding values of F do not vary very rapidly.

Let x_1 and x_2, as shown in the above general form, be two consecutive values of X given in the table, and f_1 and f_2 the corresponding values of F. Let x be a value of X lying between x_1 and x_2, and f the corresponding value of F. Neither x nor f is in the table, but one of them is given, and the problem is to find the other by interpolation. For instance, if the table is one of natural tangents in which the angles increase by whole minutes, x_1 and x_2 may be, respectively, 31° 42′ and 31° 43′, and f_1 and f_2 their corresponding tangents; while x may be any angle between 31° 42′ and 31° 43′, and f its tangent. Either x may be given to find f; or f may be given to find x.

The quantity by which the tabular value x_1 must be algebraically increased in order to obtain x will be called the **increment** of x_1, and denoted by $i(x_1)$, read *increment of x_1* (mathematicians use the notation $\varDelta x_1$, read *delta x_1*). We have, then,

$$x = x_1 + i(x_1) \qquad (1)$$

Using a similar notation for f_1,

$$f = f_1 + i(f_1) \qquad (2)$$

If x is given, $i(x_1)$ may be assumed as given, since $i(x_1) = x - x_1$. Then $i(f_1)$ is determined by interpolation, as explained below, and f is found from formula **2**. Similarly, if f is given, $i(f_1)$ is likewise given, and x is found by interpolation.

The difference, as $x_2 - x_1$, of two consecutive values of X, will be called the **interval** of X; and that between two consecutive values of F, the interval of F. The notation $I(x_1)$. read *interval of x_1*, will be used to denote the interval $x_2 - x_1$. Similarly, $I(f_1)$ will denote the interval $f_2 - f_1$.

The principle of interpolation is this: *The increments $i(x_1)$ and $i(f_1)$ are to each other as the corresponding intervals $I(x_1)$ and $I(f_1)$; or, algebraically,*

$$\frac{i(x_1)}{i(f_1)} = \frac{I(x_1)}{I(f_1)} \qquad (3)$$

This formula is very easily remembered on account of its symmetry. The following, derived from it, serve, respectively, to find $i(f_1)$ when x is given, and $i(x_1)$ when f is given:

$$i(f_1) = I(f_1) \times \frac{i(x_1)}{I(x_1)} \qquad (4)$$

$$i(x_1) = I(x_1) \times \frac{i(f_1)}{I(f_1)} \qquad (5)$$

The last two formulas may be stated in the form of a general principle, as follows: *Either increment is equal to the corresponding interval multiplied by the ratio of the other increment to the other interval.* It is easy to remember what the numerator of this ratio is, by noticing that the ratio is always less than 1, and that, since the increment is always less than the interval, the former must be the numerator and the latter the denominator. It should be noted that $i(x_1)$, $i(f_1)$, $I(x_1)$, and $I(f_1)$ may be expressed in any convenient units, it being understood that $i(f_1)$, as determined from formula **4**, is in the same units as $I(f_1)$; and that $i(x_1)$, as determined from formula **5**, is in the same units as $I(x_1)$. Thus, if the values of f_1 and f_2 in the table are, respectively, 4.3476 and 4.3463, then, $I(f_1) = f_2 - f_1 = .0013$, or, if one ten-thousandth is taken as the unit, we may write $I(f_1) = 13$. The value of $i(f_1)$, determined from formula **4**, must be understood to express ten-thousandths. For instance, if $\dfrac{i(x_1)}{I(x_1)} = .3$, then, $i(f_1) = 13 \times .3 = 3.9$ (ten-thousandths) $= 4$ (ten-thousandths), nearly.

The value of f is then found thus,

$$f_1 = 4.3463$$
$$i(f_1) = \underline{\quad\quad 4}$$
$$f = 4.3467$$

Usually, the correction $i(f_1)$ can be added to f_1 mentally, in order to find f.

EXAMPLE 1.—Find the logarithm of 57,846 by means of a five-place table giving the logarithms of numbers consisting of four figures.

SOLUTION.—Only the mantissas will be considered, since the characteristics are determined by inspection. The given number lies between $57,840(=x_1)$ and $57,850(= x_2)$, whose logarithms are, respectively, $.76223(= f_1)$ and $.76230(= f_2)$. We have, therefore, expressing $f_2 - f_1$, or $I(f_1)$, in units of the fifth order.

$$x = 57846 \qquad\qquad f_2 = .76230$$
$$x_1 = 57840 \qquad\qquad f_1 = .76223$$
$$i(x_1) = \overline{\quad 6} \qquad\qquad I(f_1) = \overline{\quad 7}$$
$$I(x_1) = x_2 - x_1 = 10$$

Then (formula **4**),

$$i(f_1) = 7 \times \frac{6}{10} = 4.2 = 4, \text{ nearly}$$

and
$$f = \begin{cases} f_1 \\ + i(f_1) \end{cases} = \begin{cases} .76223 \\ +4 \end{cases} = .76227. \text{ Ans.}$$

EXAMPLE 2.—Find, by means of a five-place table, the number the mantissa of whose logarithm is .47693.

SOLUTION.—Here $f(=.47693)$ lies between the tabular values $.47683(=f_1)$ and $.47698(=f_2)$, which are, respectively, the logarithms of $29,980(=x_1)$ and $29,990(=x_2)$. We have, then,

$$
\begin{array}{ll}
f_2 = .47698 & x_2 = 29,990 \\
f = .47693 & x_1 = 29,980 \\
f_1 = .47683 & I(x_1) = \overline{10} \\
\end{array}
$$

$$
\begin{array}{l}
I(f_1) = f_2 - f_1 = \overline{15} \\
i(f_1) = f - f_1 = 10 \\
\end{array}
$$

Then (formula 5),

$$i(x_1) = 10 \times \frac{10}{15} = 7, \text{ nearly}$$

and $\qquad x = x_1 + i(x_1) = 29,980 + 7 = 29,987. \text{ Ans.}$

This gives the significant figures of the number. The decimal point should be placed according to the characteristic of the given logarithm.

EXAMPLE 3.—Find the angle whose natural tangent is $.56781(=f)$ by means of a table giving the natural tangents of angles varying by minutes.

SOLUTION.—Here f is found to lie between $.56769(= \tan 29° 35'$ $= f_1)$ and $.56808(= \tan 29° 36' = f_2)$. Expressing $x_2 - x_1$, or $I(x_1)$, in seconds, we have

$$
\begin{array}{ll}
x_2 = 29° 36' & f_2 = .56808 \\
x_1 = 29° 35' & f = .56781 \\
I(x_1) = \overline{60''} & f_1 = .56769 \\
& I(f_1) = \overline{39} \\
& i(f_1) = 12 \\
\end{array}
$$

Then (formula 5),

$$i(x_1) = 60'' \times \frac{12}{39} = 18'', \text{ nearly}$$

and $\qquad x = x_1 + i(x_1) = 29° 35' 18''. \text{ Ans.}$

EXAMPLE 4.—In Searles' field book is given a table of lengths of arcs for different degrees of curvature. Part of it is as follows (lengths in feet):

Degree of Curve $(=X)$	Length of Arc for One Station $(=F)$
10° 10′	100.131
10° 20′	100.136
10° 30′	100.140

Find the length of the arc between two stations for a 10° 26′ curve

SOLUTION.—Here we have, $x = 10° 26'$, which lies between $10° 20'$ ($=x_1$) and $10° 30 (= x_2)$. Expressing $I(x_1)$ and $i(x_1)$ in minutes, and $I(f_1)$ and $i(f_1)$ in thousandths, we have $I(x_1) = 10$, $i(x_1) = 6$, $I(f_1) = 140 - 136 = 4$.

Therefore (formula 4),

$$i(f_1) = 4 \times \frac{6}{10} = 2, \text{ nearly}$$

and $$f = f_1 + i(f_1) = \left\{ \begin{matrix} 100.136 \\ +2 \end{matrix} \right\} = 100.138. \quad \text{Ans.}$$

In all simple cases like this the operations can be performed mentally and very rapidly.

EXAMPLES FOR PRACTICE

1. From the following table, find, by interpolation, the cube root of 347.3 and that of 349.7.

Number	Cube Root
347	7.0271
348	7.0338
349	7.0406
350	7.0473

Ans. $\left\{ \begin{matrix} 7.0291 \\ 7.0453 \end{matrix} \right.$

2. Find, from the following table, the diameter of a circle whose circumference is 63.57318.

Diameter	Circumference
20.1	63.14601
20.2	63.46017
20.3	63.77433

Ans. 20.236

SOLUTION OF RIGHT TRIANGLES

35.   Fundamental Equations.—Let $A\,B\,C$, Fig. 9, be a right triangle, in which A, B, and C are the angles and a, b, and c are the lengths of the sides, c being the hypotenuse. Since A and B are complementary angles, we have

$$\sin A = \cos B \qquad \tan A = \cot B$$
$$\cos A = \sin B \qquad \cot A = \tan B$$

Also, from the definitions of the trigonometric functions,

$$\sin A = \frac{a}{c}, \ \tan A = \frac{a}{b}, \ \cos B = \sin A = \frac{a}{c}, \ \cot A = \frac{b}{a};$$

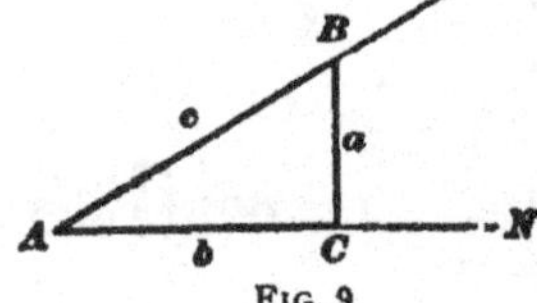

whence, expressing the value of a from each of these equations,

$$a = c \sin A \qquad (1)$$
$$a = b \tan A \qquad (2)$$
$$a = c \cos B \qquad (3)$$
$$a = b \cot B \qquad (4)$$

Fig. 9

From formulas **1** and **3**, the following values are found for c:

$$c = \frac{a}{\sin A} = a \csc A \qquad (5)$$

$$c = \frac{a}{\cos B} = a \sec B \qquad (6)$$

Finally, from geometry,

$$c^2 = a^2 + b^2 \qquad (7)$$

Of the trigonometric formulas just given, it is only necessary to commit to memory formulas **1** and **2,** as the others are immediate consequences of these.   These two formulas may be stated in words thus:

Either leg of a right triangle is equal to the hypotenuse multiplied by the sine, or to the other leg multiplied by the tangent, of the opposite angle.

It should be observed that, since a is either leg whose opposite angle is A, and adjacent angle B, the letters a and b may be interchanged in the preceding formulas, provided that A and B are likewise interchanged. Thus, by interchanging a and b, A and B in formulas **1** and **5**, we obtain,

$$b = c \sin B, \quad c = \frac{b}{\sin B} = b \csc B$$

36. **Solution of a Right Triangle.**—In general, when some of the parts of a triangle are given, the process of determining the others is called **solving the triangle,** or **the solution of the triangle.** The latter expression is applied also to the triangle determined in accordance with the given data.

In order to solve a right triangle, two parts, one at least of which should be a side, must be known in addition to the right angle. The two parts may be either (1) one side and one of the acute angles, or (2) two sides.

37. **Case I.**—*Given a Side and an Acute Angle.* The other acute angle is found from the relation $A + B = 90°$, and the other two sides by means of formulas **1** to **7**, Art. **35,** as illustrated by the following examples:

EXAMPLE 1.—In Fig. 10, the length of the hypotenuse $A B$ of the right triangle $A C B$, right-angled at C, is 24 feet, and the angle A is 29° 31′; find the sides $A C$ and $B C$, and the angle B.

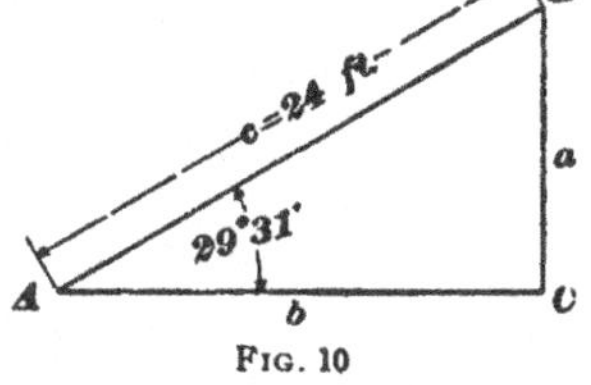

FIG. 10

NOTE.—When working examples of this kind, make a sketch and mark the known parts, as shown in the figure.

SOLUTION WITHOUT LOGARITHMS.— $B = 90° - A = 90° - 29° 31'$ $= 60° 29'$. By formula **3,** Art. **35,** interchanging a and b, and A and B,

$b = c \cos A = 24 \cos 29° 31' = 24 \times .87021 = 20.89$ ft., nearly.

By formula **1,** Art 35,

$a = 24 \sin 29° 31' = 24 \times .49268 = 11.82$ ft., nearly.

$$\text{Ans.} \begin{cases} B = 60° 29' \\ A\,C = 20.89 \text{ ft.} \\ B\,C = 11.82 \text{ ft.} \end{cases}$$

SOLUTION BY LOGARITHMS.—By formulas 3 and 1, Art. 35,
$$b = 24 \cos 29° 31' \qquad (1)$$
$$a = 24 \sin 29° 31' \qquad (2)$$

LOGARITHMS FOR (1)

$$\log 24 = 1.38021$$
$$\log \cos 29° 31' = \bar{1}.93963$$
$$\log b = \overline{1.31984}$$
$$b = 20.89$$

LOGARITHMS FOR (2)

$$\log 24 = 1.38021$$
$$\log \sin 29° 31' = \bar{1}.69256$$
$$\log a = \overline{1.07277}$$
$$a = 11.82$$

In working examples of this kind, the two logarithmic functions should be taken from the table at the same time. It saves time and space to arrange the operations as follows:

$$\log a = 1.07277;\ a = 11.82$$
$$\log \sin 29° 31' = \bar{1}.69256$$
$$\log 24 = 1.38021$$
$$\log \cos 29° 31' = \bar{1}.93963$$
$$\log b = \overline{1.31984};\ b = 20.89.\quad \text{Ans.}$$

The logarithm of 24 is written first, and then the logarithms of the sine and cosine, one over, the other under, log 24, the addition being performed upwards in one case and downwards in the other.

EXAMPLE 2.—One leg of a right triangle $A\,C\,B$, Fig. 11, is 37 feet 7 inches long; the angle opposite is 25° 33′ 7″; what are the lengths of the hypotenuse and the side adjacent, and what is the other angle?

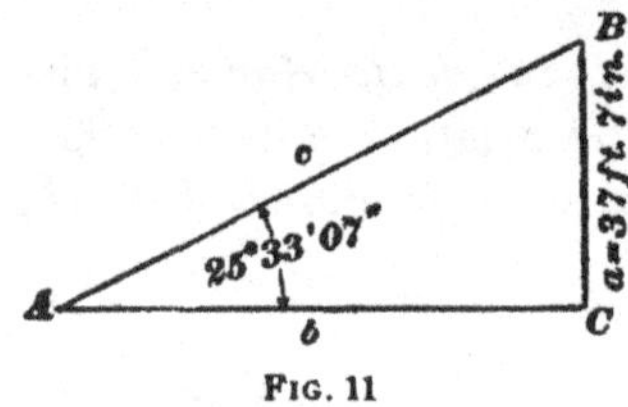

FIG. 11

SOLUTION WITHOUT LOGARITHMS.
$B = 90° - 25° 33' 07'' = 64° 26' 53''$.
Reducing 37 ft. 7 in. to ft., we have,
$a = 37.583$ ft., nearly.

By formula 5, Art. 35,
$$c = \frac{37.583}{\sin 25° 33' 07''} = \frac{37.583}{.43133} = 87.133 \text{ ft., nearly.}$$

By formula 4, Art. 35, interchanging a and b, and A and B,
$$b = a \cot A = 37.583 \times 2.09166 = 78.611 \text{ ft., nearly.}$$

$$\text{Ans.}\begin{cases} B = 64° 26' 53'' \\ A\,C = 78.611 \\ A\,B = 87.133 \text{ ft} \end{cases}$$

SOLUTION BY LOGARITHMS.—As before,
$$c = \frac{37.583}{\sin 25° 33' 7''}$$

Also,
$$b = 37.583 \cot 25° 33' 7''$$
$$\log b = 1.89548;\ b = A\,C = 78.611 \text{ ft.}$$
$$\log \cot 25° 33' 7'' = .32049$$
$$\log 37.583 = 1.57499$$
$$\log \sin 25° 33' 7'' = \bar{1}.63481$$
$$\log c = \overline{1.94018};\ c = A\,B = 87.132 \text{ ft.}\quad \textbf{Ans.}$$

It is to be noted that the value of $A\,B$ given by logarithms is different in the fifth figure from the result given by natural functions. This is due to the fact that in using five-place tables the results can be depended on to be correct to only four figures, and to have a very close approximation to the fifth figure.

NOTE.—In the majority of cases, the solution by logarithms is far more expeditious than the solution by natural functions. The student is strongly advised to form the habit of solving all trigonometric problems by means of logarithms and the logarithmic functions, whenever these functions can be used.

38. **Case II.**—*Given Two Sides.* If the given sides are the two legs a and b, A is found from formula **2**, Art. **35**, and B, from the relation $A + B = 90°$. To find c, formula **7**, Art. **35**, may be used; but, unless a and b are convenient numbers to square, it is preferable to determine c by formula **5**, Art. **35**, after having determined A.

If the given sides are the hypotenuse c and one leg, say a, the

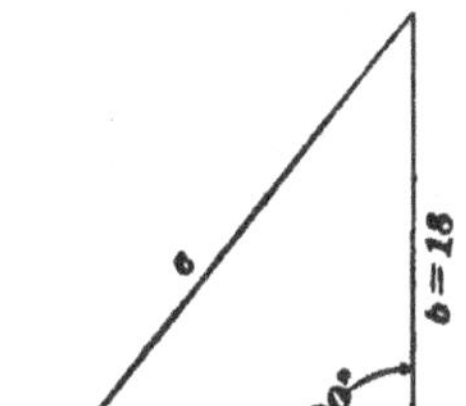

FIG. 12

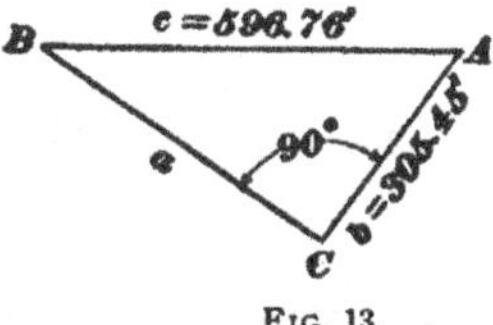

FIG. 13

angle A is found by formula **1**, Art. **35**, B from the relation $A + B = 90°$, and b from either formula **4**, or formula **7**, Art. **35**. The latter gives

$$b = \sqrt{c^2 - a^2}$$

Unless c and a are convenient numbers to square, the quantity under the radical should be replaced by the product $(c + a)\,(c - a)$, and then

$$\log b = \tfrac{1}{2}\,[\log\,(c + a) + \log\,(c - a]$$

from which b can be readily determined.

EXAMPLE 1.—Given a and b as shown in Fig. 12, to find $A, B,$ and c

SOLUTION.—Formula 2, Art. 35,

$$\tan A = \frac{a}{b} = \frac{15}{18} = \frac{5}{6} = .83333$$
$$A = 39°\ 48'\ 20''$$
$$B = 90° - 39°\ 48'\ 20'' = 50°\ 11'\ 40''$$

Formula 5, Art. 35,

$$c = \frac{15}{\sin A} = \frac{15}{\sin 39°\ 48'\ 20''}$$

$$\log 15 = 1.17609$$
$$\log \sin 39°\ 48'\ 20'' = \overline{1}.80630$$
$$\log c = 1.36979;\ c = 23.431$$

Otherwise,

$$c = \sqrt{15^2 + 18^2} = \sqrt{(3 \times 5)^2 + (3 \times 6)^2} = 3\sqrt{5^2 + 6^2} = 3\sqrt{61} = \textbf{23.431.}$$

Ans.

EXAMPLE 2.—The hypotenuse c and the leg b having the values shown in Fig. 13, find the acute angles and the leg a.

SOLUTION.—By formula 3, Art. 35, interchanging a and b, A and B,

$$\cos A = \frac{b}{c} = \frac{305.45}{596.76}$$

$\log 305.45 = 2.48494$	$90° = 89°\ 59'\ 60''$
$\log 596.76 = 2.77580$	$A = 59°\ 12'\ 46''$
$\log \cos A = \overline{1}.70914;\ A = 59°\ 12'\ 46''$	$B = 30°\ 47'\ 14''$

Formula 2, Art. 35,

$$a = 305.45 \tan 59°\ 12'\ 46''$$
$$\log 305.45 = 2.48494$$
$$\log \tan 59°\ 12'\ 46'' = .22489$$
$$\log a = 2.70983;\ a = 512.66$$

Otherwise, $a = \sqrt{c^2 - b^2} = \sqrt{(c + b)(c - b)}$

$c + b = 902.21$	$\log (c + b) = 2.95531$
$c = 596.76$	$\log (c - b) = 2.46435$
$b = 305.45$	$2)5.41966$
$c - b = 291.31$	$\log a = 2.70983;\ a = 512.66$

$$\text{Ans.}\begin{cases} A = 59°\ 12'\ 46'' \\ B = 30°\ 47'\ 14'' \\ a = 512.66 \text{ ft.} \end{cases}$$

EXAMPLES FOR PRACTICE

1. In a right triangle ACB, right-angled at C (let the student make a sketch), the hypotenuse $AB = 40$ inches and angle $A = 28°$ $14'\ 14''$; solve the triangle.

$$\text{Ans.}\begin{cases} \text{Angle } B = 61°\ 45'\ 46'' \\ AC = 35.239 \text{ in.} \\ BC = 18.925 \text{ in.} \end{cases}$$

2. In a right triangle ACB, right-angled at C, the side BC = 10 feet 4 inches; if angle $A = 26°\ 59'\ 6''$, what are the other parts?

$$\text{Ans.}\begin{cases} \text{Angle } B = 63°\ 0'\ 54'' \\ AB = 22 \text{ ft. } 9\tfrac{1}{4} \text{ in., nearly} \\ AC = 20 \text{ ft. } 3\tfrac{1}{4} \text{ in., nearly} \end{cases}$$

3. In a right triangle $A\,C\,B$, the hypotenuse $A\,B = 60$ feet and the side $A\,C = 22$ feet; solve the triangle.

Ans. $\begin{cases} \text{Angle } A = 68° \ 29' \ 22'' \\ \text{Angle } B = 21° \ 30' \ 38'' \\ B\,C = 55.821 \text{ ft.} \end{cases}$

4. In a right triangle $A\,C\,B$, right-angled at C, side $A\,C = .364$ foot and side $B\,C = .216$ foot; solve the triangle.

Ans. $\begin{cases} \text{Angle } A = 30° \ 41' \ 6'' \\ \text{Angle } B = 59° \ 18' \ 54'' \\ A\,B = .423 \text{ ft.} \end{cases}$

___

PRACTICAL EXAMPLES

39. When an object is viewed by an observer, the object may be either above or below a horizontal plane passing through the observer's eye. The angle made with this plane by the line of sight, that is, by the line from the observer's eye to the object, is called an **angle of elevation** if the object is above that plane; an **angle of depression** if the object is below that plane. The object is said to be seen at an angle of elevation or at an angle of depression according as it is above or below the plane in question. For example, a lighthouse is seen from a ship at sea at an angle of elevation, while the ship is seen from the lighthouse at an angle of depression.

EXAMPLE 1.—The angle of elevation of the top of a vertical cliff, CB, Fig. 14, at a point 100 feet from its base, is 36° 50'; find the height of the cliff.

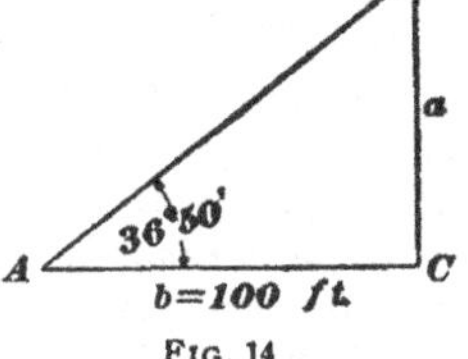

SOLUTION.—By formula **2**, Art. **35**, required height $= a = 100 \times \tan 36° 50' = 100 \times .74900 = 74.9$ ft. Ans.

EXAMPLE 2.—A statue is placed on the top of a column. At a point on the ground 130 feet from the base of the column, the angle of elevation of the top of the statue and that of the column are 43° 38' and 40° 58', respectively; find the height of the statue and column. (Let the student make a sketch.)

SOLUTION.—Let h = height of column;

h' = height of column and statue.

Then, $\tan 40° 58' = \dfrac{h}{130}$

Whence, $h = 130 \times \tan 40° 58' = 112.875$

Also, $\tan 43° 38' = \dfrac{h'}{130}$

Whence, $h' = 130 \times \tan 43° 38' = 123.942$

Therefore, the height of the column is 112.875 ft.   Ans.

The height of the statue is $123.942 - 112.875 = 11.07$ ft.   Ans.

EXAMPLE 3.—The top and bottom of a lighthouse $L\,T$, Fig. 15,

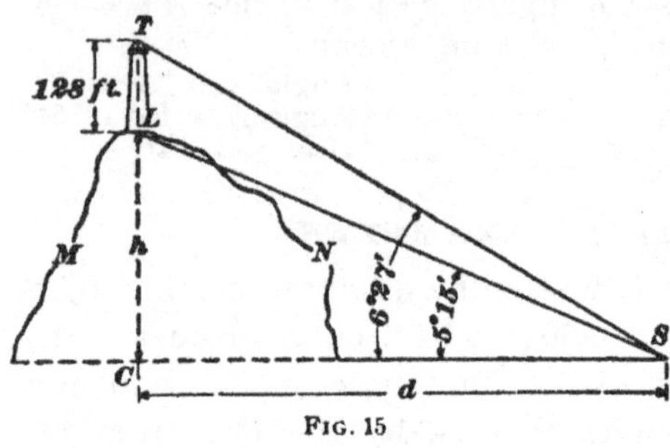

located on a hill $M\,N$, are observed from a ship S with a sextant. It is found that the angles of elevation of T and L are, respectively, $6° 27'$ and $5° 15'$. If the height of the lighthouse is 128 feet, and the surface of the sea is assumed to be plane, what are: (a) the height $h(= C\,L)$ of the

FIG. 15

hill above sea level? (b) the horizontal distance $d(= S\,C)$ of the ship from the lighthouse?

SOLUTION.—(a) In the right triangles $L\,C\,S$ and $T\,C\,S$, we have

$$d = h \cot 5° 15', \text{ and } d = (h + 128) \cot 6° 27'$$

Equating the two values of d,

$$h \cot 5° 15' = (h + 128) \cot 6° 27'$$

whence,

$$h = \frac{128 \cot 6° 27'}{\cot 5° 15' - \cot 6° 27'} = \frac{128 \times 8.84551}{10.8829 - 8.84551} = 555.72. \quad \text{Ans.}$$

(b) From (a),

$$d = h \cot 5° 15' = 555.72 \cot 5° 15' = 6,047.8 \text{ ft.} \quad \text{Ans.}$$

EXAMPLE 4.—In Fig. 16, $P_1\,T_1$ is the track of a railroad that curves into a circular arc $T_1\,M\,T_2$ at T_1. The chord $T_1\,T_2$ of the whole arc is found, by measurement, to be 764.7 feet, and the chord $T_1\,M$ of half the arc, 393.2 feet. Find: (a) the external angle I between $P_2\,T_2$ and $P_1\,T_1$ produced; (b) the radius $r(= C\,T_1)$ of the curve $T_1\,M\,T_2$.

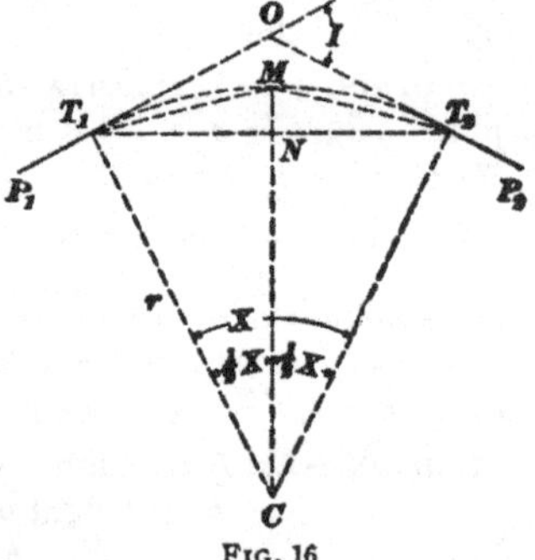

FIG. 16

SOLUTION.—(a) Draw $C\,T_1$, $C\,M$, and $C\,T_2$, as shown. Since $P_1\,O$ and $P_2\,O$ are tangent to the circle, the angles $O\,T_1\,C$ and $O\,T_2\,C$ are right angles; and as the sum of the angles in the quadrilateral $O\,T_1\,C\,T_2$ is four right angles, we must have $X + T_1\,O\,T_2 = 2$ right angles $= 180°$; we have also, $I + T_1\,O\,T_2 = 180°$; therefore, $I = X$. The line $C\,M$ bisects both the angle X and the chord $T_1\,T_2$. As the angle $M\,T_1\,T_2$ is measured by one-half the arc $M T_2$, it is equal to one-half of $M C T_2$, or to $\frac{1}{4} X$. The right tri-angle $M T_1 N$ gives

$$\cos \frac{1}{4} X \, (= \cos M T_1 N) = \frac{T_1 N}{T_1 M} = \frac{\frac{1}{2} T_1 T_2}{T_1 M} = \frac{382.35}{393.2}$$

whence, by either logarithms or natural functions (logarithms are far preferable in this case),

$$\frac{1}{4}X = 13° \, 29' \, 20''; \quad I(= X) = 4 \times 13° \, 29' \, 20'' = 53° \, 57' \, 20''. \quad \text{Ans.}$$

(b) In the right triangles $C T_1 N$,

$$r(= C T_1) = \frac{T_1 N}{\sin \frac{1}{2} X} = \frac{382.35}{\sin 26° \, 58' \, 40''} = 842.83 \text{ ft. Ans.}$$

EXAMPLE 5.—Fig. 17 is a cross-section of a dam, the dimensions being as shown. The batter of the face $A B$ is 30 in 100, or .3.   Find: (a) the width $w_1(= A B)$ of the face; (b) the batter of the back $C D$; (c) the width $w_2(= C D)$ of the back.

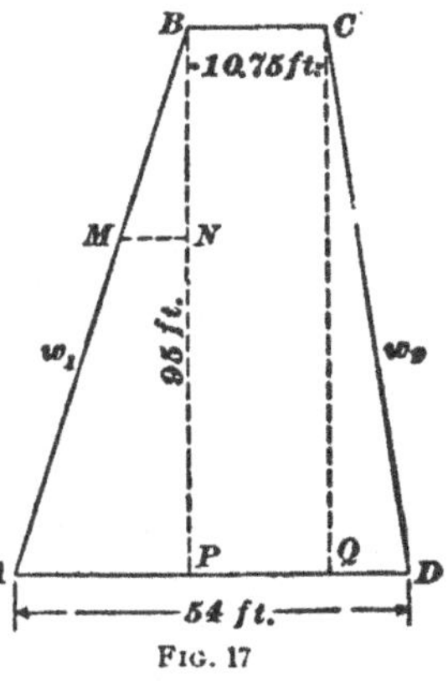

FIG. 17

NOTE.—By the batter of one of the sides of an inclined wall is meant the rate at which that side deviates from the vertical. Thus, in Fig. 17, the side $B A$ deviates from the vertical by the amount $M N$ in the vertical distance $B N$, or by the amount $A P$ in the distance $B P$. Either of the ratios $\dfrac{M N}{B N}$ or $\dfrac{A P}{B P}$ expresses the batter of the wall. A batter of 30 in 100 is the same as $\dfrac{30}{100}$, or .3. It will be noticed that the batter is equal to the tangent of the inclination of the side of the wall to the vertical.

SOLUTION.—(a) As just explained, $\tan A B P = \text{batter} = .3$; whence $A B P = 16° \, 41' \, 58''$.   The triangle $A B P$ gives,

$$w_1 = \frac{B P}{\cos A B P} = \frac{95}{\cos 16° \, 41' \, 58''} = 99.182 \text{ ft. Ans.}$$

(b)   The triangle $A P B$ gives,

$$A P = P B \tan A B P = 95 \times .3 = 28.5 \text{ ft.}$$

From the figure,

$$Q D = A D - A P - P Q = 54 - 28.5 - 10.75 = 14.75 \text{ ft.}$$

The triangle $C Q D$ gives,

$$\text{batter of } C D = \tan Q C D = \frac{Q D}{Q C} = \frac{14.75}{95} = .15526$$

or, say, 15.5 in 100; also, $Q C D = 8° \, 49' \, 31''$. Ans.

$$(c) \quad w_2 = \frac{C Q}{\cos Q C D} = \frac{95}{\cos 8° \, 49' \, 31''} = 96.139 \text{ ft. Ans.}$$

EXAMPLE 6.—Fig. 18 represents a derrick; the dimensions being as shown, determine: (a) the inclination A of the boom $Q R$ to the verti-cal; (b) the inclination M of the rod $P R$ to the vertical; (c) the point U at which the guy rope $P U$ must be tied, that it may make an angle of 60° with the horizontal; (d) the length $P U$ of the guy rope.

SOLUTION.—(*a*) The triangle RQS gives,

$$\sin A = \frac{QS}{QR} = \frac{28}{42} = \frac{2}{3};\text{ whence, } A = 41° \ 48' \ 38''. \quad \textbf{Ans.}$$

(*b*) The same triangle gives,

$$RS = \sqrt{(42 + 28)(42 - 28)} = \sqrt{70 \times 14} = 31.305$$

The triangle PTR gives,

$$RT = RS - ST = RS - QP = 31.305 - 11.5 = 19.805 \text{ ft.}$$

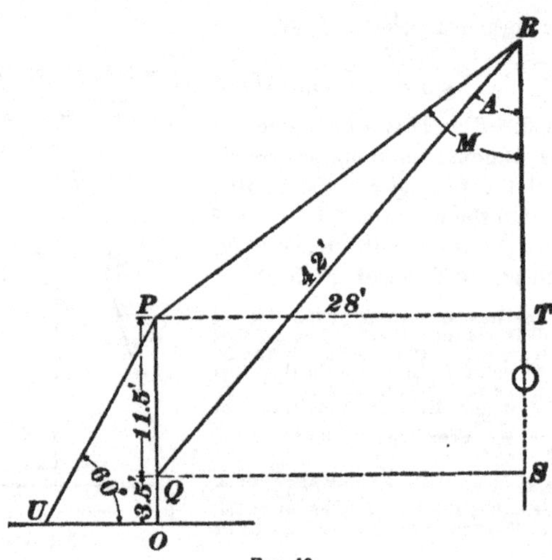

FIG. 18

$$\tan M = \frac{PT}{RT} = \frac{28}{19.805};\text{ whence, } M = 54° \ 43' \ 38''. \quad \textbf{Ans.}$$

(*c*) In the triangle POU,

$$OU = OP \cot 60° = 15 \cot 60° = 8.660 \text{ ft.} \quad \textbf{Ans.}$$

(*d*) In the same triangle,

$$PU = \frac{OP}{\sin 60°} = \frac{15}{\sin 60°} = 17.320 \text{ ft.} \quad \textbf{Ans.}$$

EXAMPLES FOR PRACTICE

1. In order to determine the distance CB, Fig. 19, across an intervening stream, a line CA, at right angles to CB, was measured; the angle CAB was also measured, and found to be 50° 16'. If CA = 100 feet, what is the distance CB? Ans. CB = 120.31 ft.

2. A ship was observed from the top of a lighthouse under an angle of depression of 50°; if the top of the lighthouse is 250 feet above sea level, what was the horizontal distance of the ship from the lighthouse? **Ans. 209.78 ft.**

3. From two points P_1, P_2, Fig. 20, assumed to be on the same horizontal line, the angles of elevation of the top O of a column were found to be as shown. If $P_1 P_2 = 300$ feet, and the points P_1 and P_2 are 9 feet higher than the base of the column, find: (a) the height h ($= OH$) of the column; (b) the horizontal distance d from P_1 to the axis of the column.
Ans. $\begin{cases} (a)\ h = 366.77 \text{ ft.} \\ (b)\ d = 292.31 \text{ ft.} \end{cases}$

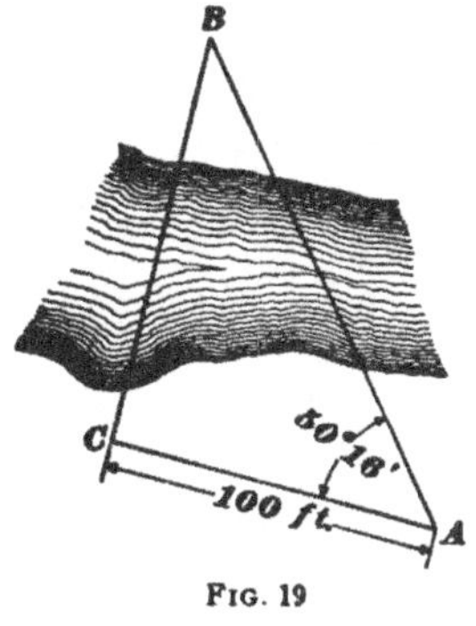

FIG. 19

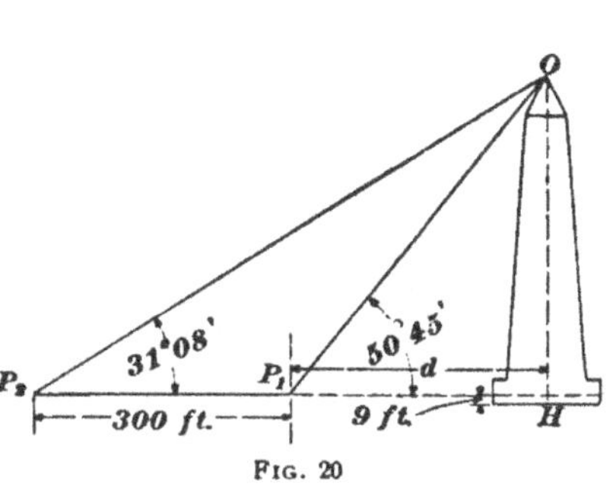

FIG. 20

4. A water pipe has a grade of 5.5 per cent. (that is, the pipe drops or rises 5.5 feet in every hundred feet measured horizontally); find: (a) the inclination of the pipe to the horizontal; (b) the length of pipe required for a horizontal distance of 2,764 feet.
Ans. $\begin{cases} (a)\ 3° 8' 54'' \\ (b)\ 2,768.2 \text{ ft.} \end{cases}$

5. The face $A\,B$, Fig. 17, and back $C\,D$ of a dam 80 feet high are to have a batter of 26 and 12 in 100, respectively; if the base $A\,D$ is to be 45 feet wide, find: (a) the angles A and D at the base; (b) the width $B\,C$ of the top.
Ans. $\begin{cases} (a)\ A = 75° 25' 33'',\ D = 83° 9' 26'' \\ (b)\ B\,C = 14.6 \text{ ft.} \end{cases}$

6. Show that the base of an isosceles triangle is equal to twice one of the equal sides multiplied by the sine of one-half the vertical angle (angle opposite base).

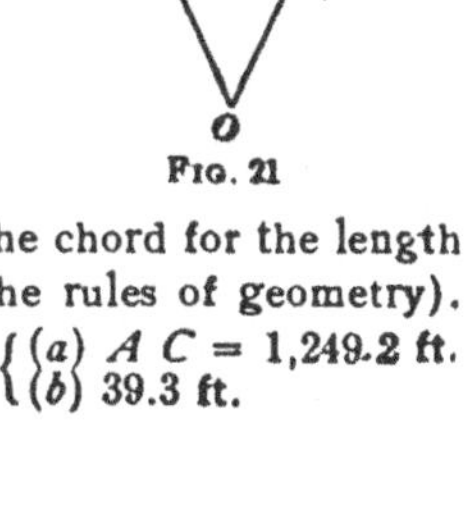

7. A railroad curve $A\,B\,C$, Fig. 21, radius 1,500 feet, subtends a central angle of 49° 13'. (a) Find the length of the chord $A\,C$. (b) What will be the error in taking the length of the chord for the length of the arc? (Determine the latter length by the rules of geometry).
Ans. $\begin{cases} (a)\ A\,C = 1,249.2 \text{ ft.} \\ (b)\ 39.3 \text{ ft.} \end{cases}$

FIG. 21

PLANE TRIGONOMETRY

(PART 2)

———

LOGARITHMIC FUNCTIONS OF SMALL ANGLES

1. Angles less than 3° are of comparatively rare occurrence in practice. When, however, they do occur, and they contain odd seconds, their logarithmic sines, tangents, and cotangents cannot be accurately determined by the general formulas and rules given in *Plane Trigonometry*, Part 1. These functions are found from a special table, which covers the first three pages of the general table of logarithmic functions furnished with this Course. These pages differ from the others in several respects, namely:

(*a*) The column of seconds on the left, marked ″ at the top, gives the total number of seconds in all angles between 0° and 3°, at intervals of 1 minute. Thus, on page 43, the number 6,360 in the column of seconds is horizontally opposite 46 in the minute column, and is, therefore, the total number of seconds in 1° 46′.

(*b*) The column headed S T, between the sine and the tangent column, contains the values of $\log \tan A - \log A''$, and $\log \sin A - \log A''$ for all values of A between 0° and 3°, varying from minute to minute; A'' is the total number of seconds in the angle A. The first four figures of these differences are common to the tangent and the sine and are printed near the head of the column; the other two figures are printed under S for the sine and under T for the tangent. The two figures corresponding to any angle are horizontally opposite the total number of seconds in the

§ 10

angle, this total number of seconds being given in the left-hand column. Thus, for 1° 45′ (= 6,300″), the value of S, or of log sin 1° 45′ − log 6,300, is $\bar{6}$.68551; and the value of T, or of log tan 1° 45′ − log 6,300, is $\bar{6}$.68571.

(*c*) Next to the cotangent column, there is a column marked C, containing the values of $-T$. The first four figures of these values are common to all angles between 0° and 3°, and are printed but once; the other two are printed horizontally opposite the number of seconds in the corresponding angles. Thus, for 1° 51′ (= 6,660″), the value of C is 5.31427. The values of S, T, and C will here be referred to as **corrections**.

2.   To Find the Logarithmic Sine or Tangent of an Angle Between 0° and 3°.—If there are no odd seconds in the angle, the logarithm may be at once taken from the table, as in *Plane Trigonometry*, Part 1. Here it will be assumed that the angle contains a number of odd seconds. Let the angle be denoted by A, and the total number of seconds in it by A''; that is, let A'' be the angle reduced to seconds.   (See Art. **1**.)

Rule.—*Open the table at the page headed by the number of degrees in the given angle. Look in the minute column for the number of minutes nearest (whether greater or less) to the number of odd minutes and seconds in the given angle. (Thus, if the given angle is 2° 36′ 40″, look for 2° 37′; if the given angle is 2° 36′ 21″, look for 2° 36′.) Take from the column headed S T the correction horizontally opposite the number of minutes found as just described, using the correction under S for the sine, and that under T for the tangent. Look in the column of seconds at the left of the page for the number horizontally opposite the number of minutes in the given angle, and to it add the number of odd seconds in that angle. The result will be the total number of seconds (A'') in the given angle. Find the logarithm of this number of seconds from the table of logarithms of numbers. Add to this logarithm the correction found as above. The result will be the required logarithmic sine or tangent, according to the correction used.*

EXAMPLE 1.—To find the logarithmic sine of 1° 3′ 45″ (= A).

SOLUTION.—Opening the table at page 43 (headed 1°), we look for 4′ in the minute column, since 3′ 45″ is nearer to 4′ than to 3′. Horizontally opposite 4, and in the column headed S T, the sine correction $\bar{6}.68555$ (= S) is found. We now look in the minute column for the number of minutes (3) in the given angle; horizontally opposite it in the left-hand column is the number 3,780, number of seconds in 1° 3′; adding 45″, we obtain 3,825 (= A'') for the total number of seconds in the given angle.

$$\log A'' = \log 3{,}825 = 3.58263$$
$$S = \bar{6}.68555$$
$$\log \sin A = \bar{2}.26818$$

that is, $\log \sin 1° \, 3′ \, 45″ = \bar{2}.26818$.   Ans.

EXAMPLE 2.—To find the logarithmic tangent of 2° 36′ 17″.

SOLUTION.—On page 44, the correction for the tangent, opposite 36′, is $\bar{6}.68587$ (= T). Number of seconds opposite 36′ in the left-hand column, 9,360; $A'' = 9{,}360 + 17 = 9{,}377$.

$$\log 9{,}377 = 3.97206$$
$$T = \bar{6}.68587$$
$$\log \tan 2° \, 36′ \, 17″ = \bar{2}.65793. \quad \text{Ans.}$$

3. To Find the Logarithmic Cotangent of an Angle Between 0° and 3°.

Rule.—*Find C, A'', and log A'' exactly as in the last article, C being taken from the correction column next to the cotangent column. Subtract log A'' from C. The result will be the required logarithmic cotangent.*

EXAMPLE.—To find the logarithmic cotangent of 1° 52′ 37″.

SOLUTION.—On page 43, the correction under C, and horizontally opposite 53′, is 5.31427; $A'' = 6{,}720 + 37 = 6{,}757$.

$$C = 5.31427$$
$$\log A'' = \log 6{,}757 = 3.82975$$
$$C - \log A'' = 1.48452$$

that is, $\log \cot 1° \, 52′ \, 37″ = 1.48452$.   Ans.

4. To Find the Logarithmic Tangent, Cosine, or Cotangent of an Angle Between 87° and 90°.—These functions also are to be taken from the first three pages of the table of logarithmic functions. The simplest way to proceed is to subtract the angle from 90° and look for the corresponding complementary function as explained in Arts. **2** and **3.** Thus, log cos 88° 55′ 38″ is obtained by looking for log sin (90° − 88° 55′ 38″) = log sin 1° 4′ 22″.

EXAMPLES FOR PRACTICE

1. Find the logarithmic sine of 1° 6′ 19″.   Ans. $\overline{2}$.28532
2. Find the logarithmic sine of 0° 2′ 41″.   Ans. $\overline{4}$.89240
3. Find the logarithmic tangent of 2° 56′ 57″.   Ans. $\overline{2}$.71196
4. Find the logarithmic cotangent of 1° 30′ 18″.   Ans. 1.58049
5. Find the logarithmic cosine of 88° 50′ 49″.   Ans. $\overline{2}$.30370
6. Find the logarithmic tangent of 89° 3′ 9″.   Ans. 1.78151
7. Find the logarithmic cotangent of 88° 0′ 25″.   Ans. $\overline{2}$.54157

------

5.   To Find the Angle Corresponding to a Given Logarithmic Function, When the Function Lies Between Two of the Functions in the First Three Pages of the Table.—I. *Sine and Tangent.*—As explained in Art. **1,** log sin $A = S +$ log A''; therefore,

$$\log A'' = \log \sin A - S \qquad (1)$$

Likewise, when log tan A is given,

$$\log A'' = \log \tan A - T \qquad (2)$$

From these formulas is derived the following

Rule.—*Find in the table the logarithm nearest to the given one. Take the correction horizontally opposite this logarithm, and subtract it from the given logarithm. The result will be the logarithm of the total number of seconds* (A'') *in the given angle. Find the number corresponding to this logarithm, and reduce it to degrees, minutes, and seconds.*

It is here assumed that the given function lies between two functions in the column marked log sin or log tan, as the case may be, at the top. If the names of the functions are at the bottom, the sine should be treated as in *Plane*

Trigonometry, Part 1; the tangent should be treated as if it were a cotangent, according to the directions to be given presently, and when the angle corresponding to that cotangent is found, it should be subtracted from 90°.

II. *Cotangent.*—Since log cot $A = C -$ log A'' (Art. **3**), we have

$$\log A'' = C - \log \cot A \qquad (3)$$

From this formula is derived the following

Rule.—*Find in the table the logarithmic function nearest the given cotangent. Take from the C column the correction horizontally opposite the logarithm just found, and from it subtract the given logarithmic cotangent. The result will be the logarithm of the total number of seconds in the angle.*

Here, as before, it is assumed that the given cotangent lies between two of those marked log cot at the top.   If it lies between two logarithms in the column marked log cot at the bottom, it should be treated as if it were a tangent, and having found the angle corresponding to this tangent, it should be subtracted from 90° to obtain the required angle.

III.   *Cosine.*

Rule.—*If the given cosine lies between two of those in the column headed log cos, apply the general rule given in Plane Trigonometry, Part 1.   If it lies between two of the logarithms in the column marked log cos at the bottom, treat it as if it were a sine, find the angle corresponding to that sine as above, and subtract the result from 90°.*

Example 1.—To find the angle whose logarithmic tangent is $\overline{2}.32803$.

Solution.—The logarithmic tangent nearest to $\overline{2}.32803$ is $\overline{2}.32711$, found in the column headed log tan on page 43.   The T correction horizontally opposite $\overline{2}.32711$ is $\overline{6}.68564$.

$$\log \tan A = \overline{2}.32803$$
$$T = \overline{6}.68564$$
$$\overline{}$$
$$\log A'' = 3.64239$$

From the table of logarithms of numbers,

$$A'' = 4{,}389'' = 1° \ 13' \ 9''. \text{ Ans.}$$

Example 2.—To find the angle whose logarithmic cotangent is 2.49567.

SOLUTION.—The nearest logarithmic cotangent found in the table is 2.49488. The number opposite this logarithm in the C column is 5.31442.

$$C = 5.31442$$
$$\log \cot A = 2.49567$$
$$\log A'' = 2.81875;$$
$$A'' = 659'' = 0°\ 10'\ 59''.\quad \text{Ans.}$$

NOTE.—Angles are here given to the nearest whole second.

EXAMPLE 3.—To find the angle whose logarithmic cosine is $\bar{2}.63723$.

SOLUTION.—The nearest logarithm, $\bar{2}.63678$, is found on page 44, in the column headed log sin. The given function is, therefore, to be treated as if it were a logarithmic sine, and the angle A, corresponding to this sine is to be subtracted from $90°$ to obtain the required angle A. The correction horizontally opposite $\bar{2}.63678$, in the S column, is $\bar{6}.68544$.

$$\log \sin A_1 = \bar{2}.63723$$
$$S = \bar{6}.68544$$
$$\log A_1'' = 3.95179;$$
$$A_1 = 8{,}949'' = 2°\ 29'\ 9''$$
$$A = 90° - 2°\ 29'\ 9'' = 87°\ 30'\ 51''.\quad \text{Ans.}$$

EXAMPLES FOR PRACTICE

Verify the following values:

(a) $\bar{2}.17645 = \log \sin 0°\ 51'\ 37''$ (e) $\bar{2}.48790 = \log \cot 88°\ 14'\ 19''$

(b) $\bar{3}.94316 = \log \sin 0°\ 30'\ 10''$ (f) $2.47608 = \log \cot 0°\ 11'\ 29''$

(c) $\bar{2}.65783 = \log \cos 87°\ 23'\ 36''$ (g) $1.31009 = \log \tan 87°\ 11'\ 48''$

(d) $\bar{2}.58349 = \log \tan 2°\ 11'\ 41''$ (h) $\bar{3}.95377 = \log \cos 89°\ 29'\ 6''$

6. Use of the Column of Seconds for Obtaining the Angle Corresponding to a Given Function.—In order to avoid confusing the student by too many rules, the reduction of A'' to degrees, minutes, and seconds was, in the preceding articles, effected by the ordinary rules of arithmetic, without any reference to the table. The following is a more expeditious method:

Let the given function lie between the functions of two consecutive angles, A_1 and $A_1 + 1'$. Then, the degrees and minutes in the required angle are those in A_1, and may be at once written down. The number in the column of seconds on the left, horizontally opposite the number of minutes in A_1, gives the total number of seconds in A_1. Denoting that number by A_1'' and the number of odd seconds in the required angle by s, we have

$$s = A'' - A_1''$$

EXAMPLE.—To find the angle whose logarithmic tangent is $\bar{2}.30217$.

SOLUTION.—The given function lies between $\bar{2}.29629$ and $\bar{2}.30263$. The angle corresponding to the first of these two functions is $1°\ 8'$ $(= A_1)$; $A_1'' = 4,080''$.

$$\log \tan A = \bar{2}.30217$$
$$T = \bar{6}.68563$$
$$\log A'' = \overline{3.61654}; \quad A'' = 4,136$$
$$s = A'' - A_1'' = 4,136 - 4,080 = 56''$$
$$A = A_1 + s = 1°\ 8'\ 56''. \quad \text{Ans.}$$

The subtraction $A'' - A_1''$ can usually be effected mentally.

EXAMPLES FOR PRACTICE

Apply the method just described to the Examples for Practice given after Art. 4.

GENERAL TRIGONOMETRIC FORMULAS

ANGLES AND THEIR TRIGONOMETRIC FUNCTIONS

7. Angle of Any Magnitude.—In trigonometry, an angle is considered as being generated by a straight line turning about one of its ends, which is the vertex of the angle. In this motion, any point in the turning line describes a circular arc, whose number of degrees is the measure of the angle. The turning line is called the **generating line**. The position that this line occupies before it begins to turn, and from which arcs are measured, is called the **initial**

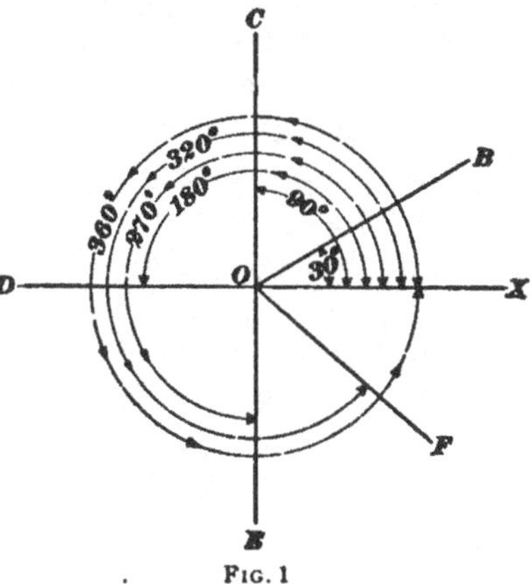

FIG. 1

line, or the **initial position** of the generating line; and the position it occupies after turning through a certain angle is called the **final position**. In Fig. 1, for example, the initial position of the generating line is OX. The turning is supposed to take place about the point O and in a direction opposite to that in which the hands of a clock move. When the line turns from the position OX to the final positions OB, OC, OD, OE, OF, it generates angles o' $30°$, $90°$, $180°$, $270°$, $320°$, respectively, as indicated or the figure. If the line makes a complete turn, so that its final position coincides with its initial position OX, the angle generated is $360°$.

8. Positive and Negative Angles.—When an angle is described by a line turning in a direction contrary to that

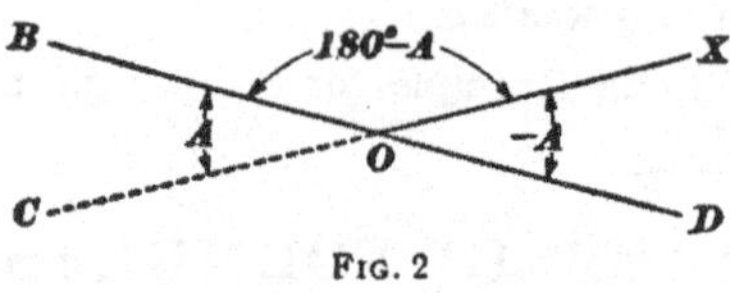

FIG. 2

in which the hands of a watch move, the angle is considered **positive**; if described in the opposite direction, it is considered **negative**. Referring to Fig. 2, the angle XOB, whose supplement is A, may be regarded as having been described in any of the following manners:

(*a*) By turning the generating line about O from the position OX in the positive direction through $(180 - A)$ degrees to the position OB.

(*b*) By turning the generating line about O in a positive direction through an angle of $180°$, when it will be in the position OC, and then turning it back from OC in the negative direction through the angle $-A$ (negative, because turned in the negative direction) into the position OB.

(*c*) By turning the generating line about O in the negative direction through the angle $-A$, into the position OD, and then turning it back in the positive direction through $180°$ into the position OB.

It is to be noticed that, however the angle $(180° - A)$ may be regarded as described, the resulting angle XOB is the same.

9. Quadrants.—Let OX, Fig. 3, be the initial position of the generating line, and OM_1, OM_2, OM_3, OM_4 final positions, determining, respectively, the angles A_1, A_2, A_3, A_4, all measured from OX upwards and toward the left. Producing XO and drawing through O a perpendicular YY' to OX, the plane of the figure is divided into four right angles, called **quadrants**. Taking them in order, following the direction in which positive angles are reckoned, they are distinguished as follows: XOY is the **first quadrant**; YOX', the **second quadrant**; $X'OY'$, the **third quadrant**; and $Y'OX$, the **fourth quadrant**.

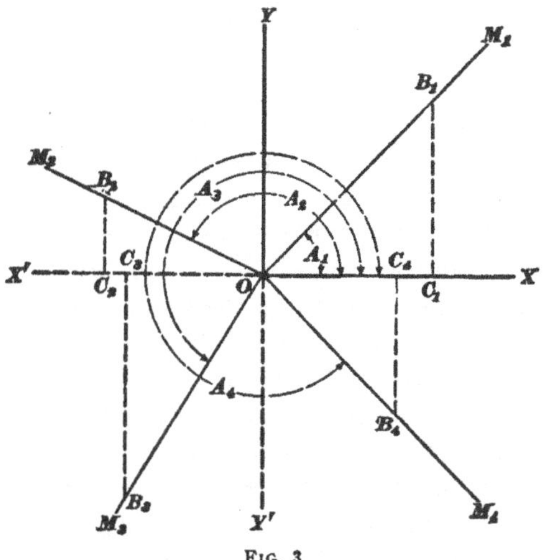

Fig. 3

10. Trigonometric Functions of Any Angle.—In the definitions given in *Plane Trigonometry*, Part 1, only acute angles were considered. Referring to Fig. 3, in which $B_1 C_1$ is perpendicular to OX, the trigonometric functions of the acute angle A_1 were defined by the following equations:

$$\sin A_1 = \frac{\text{side opposite}}{\text{hypotenuse}} = \frac{B_1 C_1}{O B_1} \qquad \tan A_1 = \frac{\text{side opposite}}{\text{side adjacent}} = \frac{B_1 C_1}{O C_1}$$

$$\cos A_1 = \frac{\text{side adjacent}}{\text{hypotenuse}} = \frac{O C_1}{O B_1} \qquad \cot A_1 = \frac{\text{side adjacent}}{\text{side opposite}} = \frac{O C_1}{B_1 C_1}$$

$$\sec A_1 = \frac{\text{hypotenuse}}{\text{side adjacent}} = \frac{O B_1}{O C_1} \qquad \csc A_1 = \frac{\text{hypotenuse}}{\text{side opposite}} = \frac{O B_1}{B_1 C_1}$$

These formulas serve as the definitions of the trigonometric functions of any angle; that is, the sine of any angle is the ratio of the side opposite to the hypotenuse; the **tangent** is the ratio of the side opposite to the side adjacent, etc.   But, in order that these definitions may be correct, it is necessary to apply to them some algebraic rules relating to signs.

In Fig. 3, the hypotenuse used for the determination of the functions of A_1 is any portion OB_1 of the side OM_1, which is the final position of the generating line. From B_1, a perpendicular B_1C_1 is drawn on the initial line OX, thus determining the right triangle OB_1C_1. The length of the perpendicular B_1C_1, which is the side opposite the vertex of the angle, is the distance of B_1 above the initial line OX, and the length of the adjacent side OC_1 is the distance of the point B_1 to the right of the vertex, measured along the initial line; or, what is the same thing, OC_1 is the distance of B_1C_1 from the vertex, measured toward the right.

Consider now the angle XOM_2, or A_2, in which the final position OM_2 of the generating line lies in the second quadrant.   As before, the hypotenuse to be used in the definitions of the trigonometric functions of A_2 is any portion OB_2 of the side OM_2, which is the final position of the generating line. As before, also, a perpendicular from B_2 is drawn on the initial line OX; but, in this case, the perpendicular falls on OX produced.   In the right triangle OB_2C_2, the perpendicular B_2C_2 is the side opposite the vertex of the angle A_2, and OC_2 is the side adjacent.   It should be noted very particularly that the terms *side opposite* and *side adjacent* are used to describe the positions of the legs of the right triangle with reference to the *vertex* of the angle considered, not to the angle itself.   Thus, B_2C_2 is not opposite the angle A_2, but opposite the vertex O of that angle.   The length of the side opposite, B_2C_2, measures the distance of B_2 above the initial line; and the length of OC_2, or the side adjacent, measures the distance of the opposite side B_2C_2 to the left of the vertex; or, in the language of algebra, it may be said that $-OC_2$ is the distance of B_2C_2 to the *right* of O.

Having defined the cosine of any angle as the ratio of the side adjacent to the hypotenuse, and the side adjacent as the distance of the side opposite from the vertex, measured toward the right of the vertex, it is necessary, when the side opposite is to the left of the vertex, to consider its distance from the vertex, or the side adjacent, as negative. This is in accordance with the general principle of algebra, that, if distances counted in one direction are treated as positive, distances in the opposite direction must be treated as negative. In the triangle OB_2C_2, therefore, OC_2 should be treated as negative, and therefore, the cosine of A_2 is $\dfrac{-OC_2}{OB_2}$.

Considering now the angle A_3, the hypotenuse is, as above, any portion OB_3 of the side OM_3, which is the final position of the generating line. From B_3, the perpendicular B_3C_3 on the initial line (produced) is drawn, and thus a right triangle is determined, in which B_3C_3 is the side opposite, and OC_3 the side adjacent. As previously explained, OC_3 should be treated as negative. The opposite side B_3C_3, which is the distance of B_3 below the initial line, should also be treated as negative; for if distances above the initial line are treated as positive, those below the initial line must be treated as negative.

Finally, in the angle A_4, which terminates in the fourth quadrant, OC_4, the side adjacent, is positive, while B_4C_4, the side opposite, is negative.

The foregoing explanations may be summed up as follows: The side opposite is positive or negative according as the hypotenuse is above or below the initial line. The side adjacent is positive or negative according as it extends toward the right or toward the left of the vertex. The hypotenuse is always positive.

11. Algebraic Signs of the Functions.—Referring again to Fig. 3, it will be observed that, for any angle, as A_1, terminating in the first quadrant, both the side adjacent and the side opposite, or OC_1 and B_1C_1, are positive, and therefore all the functions are positive; for any angle, as A_2, terminating in the second quadrant, the side adjacent, or OC_2, is negative, and the side opposite, or B_2C_2, is positive. Therefore,

$$\sin A_2 = \frac{+ B_2 C_2}{+ O B_2}, \text{ positive} \qquad \tan A_2 = \frac{+ B_2 C_2}{- O C_2}, \text{ negative}$$

$$\cos A_2 = \frac{- O C_2}{+ O B_2}, \text{ negative} \qquad \sec A_2 = \frac{+ O B_2}{- O C_2}, \text{ negative}$$

$$\csc A_2 = \frac{+ O B_2}{+ B_2 C_2}, \text{ positive}$$

The signs of the functions of angles terminating in the third and in the fourth quadrant are similarly determined. The results are tabulated below.

TABLE I

Function	Quadrant			
	First	Second	Third	Fourth
	Sign of Function			
Sine	+	+	−	−
Cosine	+	−	−	+
Tangent	+	−	+	−
Cotangent	+	−	+	−
Secant	+	−	−	+
Cosecant	+	+	−	−

12. Trigonometric Functions of 0° and 90°.—In the right triangles ACB, Fig. 4, the hypotenuse AB may

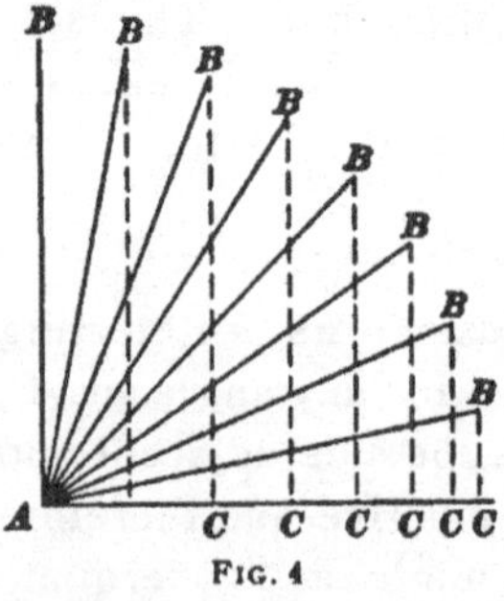

Fig. 4

be taken to have any value whatever. It is evident that BC, the side opposite, decreases as the angle $\angle AB$ decreases, and becomes zero when the angle becomes zero; and that BC coincides with the hypotenuse AB when the angle CAB is 90°. Again, AC, the adjacent side, increases as the angle decreases, and is equal to the hypotenuse AB when the angle CAB is 0°. Also, AC becomes zero when CAB is 90°. Now, from the definitions of the trigonometric functions,

$$\sin CAB = \frac{\text{side opposite}}{\text{hypotenuse}}, \text{ whence} \begin{cases} \sin\ 0° = \dfrac{0}{AB} = 0 \\[2ex] \sin\ 90° = \dfrac{AB}{AB} = 1 \end{cases}$$

$$\cos CAB = \frac{\text{side adjacent}}{\text{hypotenuse}}, \text{ whence} \begin{cases} \cos\ 0° = \dfrac{AB}{AB} = 1 \\[2ex] \cos 90° = \dfrac{0}{AB} = 0 \end{cases}$$

In like manner,

$$\tan 0° = \frac{0}{AC} = 0 \qquad\qquad \tan 90° = \frac{BC}{0} = \infty$$

$$\cot 0° = \frac{AC}{0} = \infty \qquad\qquad \cot 90° = \frac{0}{CB} = 0$$

NOTE.—The cotangent of CAB is equal to $\dfrac{CA}{CB}$. Now, as the angle decreases, the side CB becomes less and less, and it is evident that, as the denominator of a fraction becomes less and less, the numerator remaining the same, the value of the fraction increases. As the denominator decreases indefinitely, the value of the fraction increases indefinitely, and when the value of the fraction exceeds any known quantity, however great, it is said to be infinite. The sign ∞ is used to express an infinite number.

13. Functions of $(180° - A)$.—Let XOM, Fig. 5, be any angle, and $A\,(= MOX')$ its supplement. Draw OM' making with OX an angle equal to A, as shown. Take any part OB of OM for the hypotenuse, and draw BC perpendicular to OX produced; draw BB' parallel to

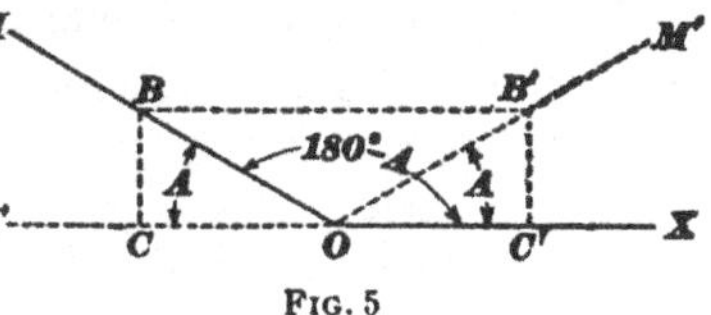

FIG. 5

OX, and $B'C'$ perpendicular to OX. Then, $BC = B'C'$; $OB = OB'$; $OC = -OC'$ (Art. **10**); and, by the definitions of the functions,

$$\sin XOM = \frac{BC}{OB} = \frac{B'C'}{OB'} = \sin A$$

$$\cos XOM = \frac{OC}{OB} = \frac{-OC'}{OB'} = -\cos A$$

that is, $\sin(180° - A) = \sin A$ **(1)**

$\cos(180° - A) = -\cos A$ **(2)**

Similarly, $\tan (180^\circ - A) = - \tan A$ (3)

$$\cot (180^\circ - A) = - \cot A \qquad (4)$$

These relations are especially useful for finding the logarithmic functions of angles greater than 90°, since these functions are arithmetically equal to those of the supplements of the angles; that is, when signs are disregarded, any function cf an angle and that of its supplement are equal. For example, $\sin 105^\circ = \sin (180^\circ - 105^\circ) = \sin 75^\circ$; $\cos 105^\circ = - \cos (180^\circ - 105^\circ) = - \cos 75^\circ$.

14. Functions of $(90^\circ + A)$.—By formula 1 of Art. **13,**
$\sin (90^\circ + A) = \sin [180^\circ - (90 + A)] = \sin (90^\circ - A)$
or, since $\sin (90^\circ - A) = \cos A$,

$$\sin (90^\circ + A) = \cos A \qquad (1)$$

The following formulas may be derived in a similar manner:

$$\tan (90^\circ + A) = - \cot A \qquad (2)$$

$$\cos (90^\circ + A) = - \sin A \qquad (3)$$

$$\cot (90^\circ + A) = - \tan A \qquad (4)$$

15. Functions of Negative Angles.—The complement of an angle is the algebraic difference between the angle and 90°. If the angle is greater than 90°, its complement is negative. Thus, the complement of 95° is $90^\circ - 95^\circ = - 5^\circ$. The cofunctions of an angle are the corresponding fundamental functions of its complement, whether that complement be positive or negative. Thus, $\cos 85^\circ = \sin (90^\circ - 85^\circ) = \sin 5^\circ$; $\cos 95^\circ = \sin (90^\circ - 95^\circ) = \sin (- 5^\circ)$. Similarly, $\sin 95^\circ = \cos (90^\circ - 95^\circ) = \cos (- 5^\circ)$. It is, therefore, necessary to know how to determine the functions of negative angles.

If $90^\circ + A$ is any angle, its complement is $90^\circ - (90^\circ + A) = - A$; and, therefore,

$$\cos (90^\circ + A) = \sin (- A), \quad \cot (90^\circ + A) = \tan (- A)$$
$$\sin (90^\circ + A) = \cos (- A), \quad \tan (90^\circ + A) = \cot (- A)$$

whence, replacing the values of cos $(90° + A)$, cot $(90° + A)$, etc. from the preceding article,

$$\sin (- A) = - \sin A \qquad (1)$$
$$\tan (- A) = - \tan A \qquad (2)$$
$$\cos (- A) = \cos A \qquad (3)$$
$$\cot (- A) = - \cot A \qquad (4)$$

ADDITION OF ANGLES

16. To Express the Sine or Cosine of the Sum or Difference of Two Angles in Terms of the Sine and Cosine of the Angles.—The following formulas are fundamental; being of frequent occurrence, they are very important, and should be committed to memory:

$$\sin (A + B) = \sin A \cos B + \cos A \sin B \qquad (1)$$
$$\cos (A + B) = \cos A \cos B - \sin A \sin B \qquad (2)$$
$$\sin (A - B) = \sin A \cos B - \cos A \sin B \qquad (3)$$
$$\cos (A - B) = \cos A \cos B + \sin A \sin B \qquad (4)$$

NOTE.—The derivation of these formulas is given in the Appendix at the end of this Section, under the Roman numeral I. That Appendix contains this and a few other demonstrations that are comparatively laborious and may be found irksome by some. They are not essential to the understanding of the formulas, and the student is not required to learn them. He is, however, advised to peruse them carefully, as they are good exercises in the handling and transforming of both algebraic and trigonometric expressions.

These formulas are not used, as they seem to imply, to determine the sine or the cosine of the sum or difference of two angles, when the sine and cosine of those angles are given. They can be used for this purpose, but there would be no advantage in so doing. Their main value consists in their application to transforming complicated trigonometric expressions into simpler ones. The student will often have occasion to employ them in this manner. In order that he may have an idea of this application of the formulas, **two** examples are given here.

EXAMPLE 1.—To determine the angle A from the relation

$$\frac{\sin(A + 28°)}{\sin A} = .95$$

SOLUTION.—Applying formula 1, we have

$$\frac{\sin(A + 28°)}{\sin A} = \frac{\sin A \cos 28° + \cos A \sin 28°}{\sin A}$$

$$= \frac{\sin A \; \text{os } 28°}{\sin A} + \frac{\cos A \sin 28°}{\sin A} = \cos 28° + \cot A \sin 28°$$

replacing $\dfrac{\cos A}{\sin A}$ by its equal cot A (see *Plane Trigonometry*, Part 1). Substituting this value of the quotient $\dfrac{\sin(A + 28°)}{\sin A}$ in the given equation, we have,

$$\cos 28° + \cot A \sin 28° = .95$$

whence $\cot A = \dfrac{.95 - \cos 28°}{\sin 28°} = \dfrac{.95 - .88295}{.46947} = .14282$

and, therefore, $A = 81° 52' 19''$. Ans.

EXAMPLE 2.—To transform the expression tan A + tan B into the expression $\dfrac{\sin(A + B)}{\cos A \cos B}$.

NOTE.—Transformations of this kind are very often useful, when logarithms are employed. Thus, if tan A + tan B were to be multiplied by 39.578, it would be necessary first to find the natural tangent of A, then that of B, add the two together, take the logarithm of the sum thus obtained, and add this logarithm to that of 39.578. If, however, the expression $\dfrac{\sin(A + B)}{\cos A \cos B}$ is used, the logarithms of sin $(A + B)$, cos A cos B can be taken from the table, and the operation performed without having recourse to natural functions, which are often inconvenient.

SOLUTION.—We have (*Plane Trigonometry*, Part 1),

$$\tan A + \tan B = \frac{\sin A}{\cos A} + \frac{\sin B}{\cos B} = \frac{\sin A \cos B + \cos A \sin B}{\cos A \cos B}$$

According to formula 1, the numerator of this last fraction is equal to sin $(A + B)$. Therefore,

$$\tan A + \tan B = \frac{\sin(A + B)}{\cos A \cos B}. \text{ Ans.}$$

17. Sine and Cosine of 2 A and of $\frac{1}{2} A$.—From the formulas for the sine and cosine of the sum of two angles, the following are deduced:

$$\sin 2A = 2 \sin A \cos A \qquad (1)$$

$$\cos 2A = \cos^2 A - \sin^2 A \qquad (2)$$

$$\cos 2A = 1 - 2 \sin^2 A \qquad (3)$$

$$\cos 2A = 2 \cos^2 A - 1 \qquad (4)$$

$$\sin A = 2 \sin \tfrac{1}{2} A \cos \tfrac{1}{2} A \qquad (5)$$

$$\cos A = \cos^2 \tfrac{1}{2} A - \sin^2 \tfrac{1}{2} A \qquad (6)$$

$$\cos A = 1 - 2 \sin^2 \tfrac{1}{2} A \qquad (7)$$

$$\cos A = 2 \cos^2 \tfrac{1}{2} A - 1 \qquad (8)$$

As in the case of formulas **1** to **4, Art. 16,** these formulas are used mainly for the purposes of transformation. They are very simply derived as follows:

When B is made equal to A, formula **1, Art. 16,** becomes

$$\sin (A + A) = \sin A \cos A + \cos A \sin A$$

that is, $\sin 2 A = 2 \sin A \cos A$

Similarly, formula **2, Art. 16,** becomes

$$\cos (A + A) = \cos A \cos A - \sin A \sin A$$

that is, $\cos 2 A = \cos^2 A - \sin^2 A$

Formula **3** follows from this, by writing $1 - \sin^2 A$ instead of $\cos^2 A$ (since $\sin^2 A + \cos^2 A = 1$); and formula **4,** by writing $1 - \cos^2 A$ instead of $\sin^2 A$.

Formulas **1** to **4** give the sine and cosine of twice any angle in terms of the sine and cosine of the angle. If the angle is denoted by $\tfrac{1}{2} A$, twice the angle will be A, and formulas **1** to **4** take the forms of formulas **5** to **8.**

OBLIQUE TRIANGLES

FUNDAMENTAL PRINCIPLES

NOTE.—For the general method of marking and naming the sides and angles of a triangle, see *Plane Trigonometry*, Part 1.

18. Principle of Sines.—*In any triangle, the sides are proportional to the sines of the opposite angles.* That is,

$$\frac{a}{b} = \frac{\sin A}{\sin B}, \quad \frac{a}{c} = \frac{\sin A}{\sin C}, \quad \frac{b}{c} = \frac{\sin B}{\sin C}$$

Let $A B C$, Fig. 6, be any triangle and p the perpendicular from C on the opposite side. Then, in (a), the right triangles $A C D$ and $B C D$ give, respectively,

$$p = b \sin A, \; p = a \sin B$$

whence, putting the two values of p equal to each other,

$$a \sin B = b \sin A$$

and, therefore, dividing by $b \sin B$,

$$\frac{a}{b} = \frac{\sin A}{\sin B}$$

In (b), the right triangles ACD and BCD give, respectively, $p = b \sin A,\ p = a \sin CBD$

whence, $a \sin CBD = b \sin A$

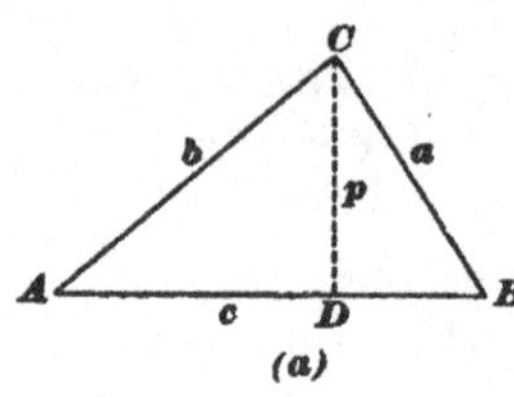

FIG. 6

But, as $CBD = 180° - B$, we may write $\sin B$ instead of $\sin CBD$ (Art. **13**), and, therefore,

$$a \sin B = b \sin A$$

whence, as before,

$$\frac{a}{b} = \frac{\sin A}{\sin B} \qquad (1)$$

By drawing a perpendicular from B on AC, and reasoning in the same manner, it may be shown that

$$\frac{a}{c} = \frac{\sin A}{\sin C} \qquad (2)$$

Similarly, $\dfrac{b}{c} = \dfrac{\sin B}{\sin C}$

By transforming equation (1), we obtain

$$\frac{a}{\sin A} = \frac{b}{\sin B}$$

and by a similar transformation of equation (2),

$$\frac{a}{\sin A} = \frac{c}{\sin C}$$

We have, therefore,

$$\frac{a}{\sin A} = \frac{b}{\sin B} = \frac{c}{\sin C}$$

The principle of sines may, then, be stated in this **form:** *In every triangle, the quotient obtained by dividing the length of any side by the sine of the opposite angle is the same, whatever the side taken.*

This quotient is called the **modulus** of the triangle, and will here be denoted by M. The modulus can be found when any of the sides and the opposite angle are known.

The principle of sines is one of the most important in trigonometry, and both forms in which it is stated in this article should be committed to memory.

19. The Cosine Principle.—*In any triangle, the square of one side is equal to the sum of the squares of the other two sides minus twice the product of these two sides and the cosine of their included angle.* That is (Fig. 6),

$$a^2 = b^2 + c^2 - 2\,bc\cos A$$
$$b^2 = a^2 + c^2 - 2\,ac\cos B$$
$$c^2 = a^2 + b^2 - 2\,ab\cos C$$

These formulas are derived in Appendix II.

20. Principle of Tangents.—*The sum of any two sides of a triangle is to their difference as the tangent of half the sum of the opposite angles is to the tangent of half their difference.* That is (Fig. 6),

$$\frac{a+b}{a-b} = \frac{\tan\frac{1}{2}(A+B)}{\tan\frac{1}{2}(A-B)}$$

The derivation of this formula is given in Appendix III. The student should have no difficulty in committing the formula to memory, as its symmetry makes it very easy to remember.

SOLUTION OF OBLIQUE TRIANGLES

21. The solution of oblique triangles is treated under four cases:

Case I: Given Two Sides and the Included Angle. Let a, b, and C, Fig. 6, be given and A, B, and c be required. Of the two methods given below, the first is preferable in most cases.

First Method.—From the formula in Art. **20,** the following is readily derived:

$$\tan\tfrac{1}{2}(A-B) = \frac{a-b}{a+b}\tan\tfrac{1}{2}(A+B) \qquad (1)$$

Now, since $A + B + C = 180°$, we have also,

$$A + B = 180° - C; \text{ and } \tfrac{1}{2}(A + B) = \tfrac{1}{2}(180° - C)$$
$$= 90° - \tfrac{1}{2}C$$

Therefore, $\tfrac{1}{2}C$ is the complement of $\tfrac{1}{2}(A + B)$, and hence, $\tan \tfrac{1}{2}(A + B) = \cot \tfrac{1}{2}C$. Substituting this value in equation (1), the following formula is derived:

$$\tan \tfrac{1}{2}(A - B) = \frac{a - b}{a + b} \cot \tfrac{1}{2}C \qquad (1)$$

If the student remembers the formula in Art. **20,** or the principle of tangents, he will have no difficulty in remembering this formula, which is derived from the formula in Art. **20,** by simply writing $\cot \tfrac{1}{2}C$ instead of $\tan \tfrac{1}{2}(A + B)$.

From this formula $\tfrac{1}{2}(A - B)$ can be found. Let this value of $\tfrac{1}{2}(A - B)$ be denoted by D. We have also, as explained above, $\tfrac{1}{2}(A + B) = \tfrac{1}{2}(180° - C) = 90° - \tfrac{1}{2}C$.

$$\tfrac{1}{2}(A + B) = 90° - \tfrac{1}{2}C \qquad (2)$$
$$\tfrac{1}{2}(A - B) = D \qquad (3)$$

Adding equations (2) and (3) gives

$$A = (90° - \tfrac{1}{2}C) + D$$

Subtracting equation (3) from (2) gives

$$B = (90° - \tfrac{1}{2}C) - D$$

Knowing A and B, the side c may be found from the relation (Art. **18**),

$$\frac{c}{\sin C} = \frac{a}{\sin A}, \text{ which gives } c = \frac{a \sin C}{\sin A}$$

It is, however, more convenient to find c from the following formula, the derivation of which is given in Appendix IV:

$$c = \frac{(a - b) \cos \tfrac{1}{2}C}{\sin \tfrac{1}{2}(A - B)} \qquad (2)$$

It will be noticed that, for calculating $\tan \tfrac{1}{2}(A - B)$, the logarithms of $(a - b)$ and $\cot \tfrac{1}{2}C$ have to be found. The logarithm of $\cos \tfrac{1}{2}C$ may be taken out of the table at the same time as that of $\cot \tfrac{1}{2}C$. Also, when the angle $\tfrac{1}{2}(A - B)$ is taken from the table, its logarithmic sine should be taken at the same time. This greatly simplifies the application of formula **2.**

Second Method.—The third side c can be found directly from the formula in Art. **19**, which gives

$$c = \sqrt{a^2 + b^2 - 2\,a\,b\,\cos C}$$

Then, by the principle of sines,

$$\sin A = \frac{a \sin C}{c}, \quad \sin B = \frac{b \sin C}{c}$$

This method is of value when the only required part is the side c, especially if a and b are convenient numbers to square.

EXAMPLE 1.—In a triangle, $a = 17$ feet, $b = 12$ feet, and the included angle $C = 59°\ 23'$. To find the other parts of the triangle.

SOLUTION.—Here $\frac{1}{2}\,C = 29°\ 41'\ 30''$; $a + b = 17 + 12 = 29$, and $a - b = 17 - 12 = 5$. Then, by the first method,

$$\tan \tfrac{1}{2}\,(A - B) = \frac{5}{29} \times \cot 29°\ 41'\ 30''$$

log 5 = .69897	log 5 = .69897
log 29 = 1.46240	log cos 29° 41′ 30″ = $\overline{1}$.93887
$\overline{1}$.23657	.63784
log cot 29° 41′ 30″ = .24397	log sin D = $\overline{1}$.46154
log tan ½(A − B) = $\overline{1}$.48054	log c = 1.17630
D = ½(A − B) = 16° 49′ 25″;	c = 15.007. Ans.

$$A = (90° - 29°\ 41'\ 30'') + 16°\ 49'\ 25'' = 77°\ 7'\ 55''. \quad \text{Ans.}$$
$$B = (90° - 29°\ 41'\ 30'') - 16°\ 49'\ 25'' = 43°\ 29'\ 5''. \quad \text{Ans.}$$

EXAMPLE 2.—Given $a = 10$, $b = 15$, and $C = 60°$; to find c.

SOLUTION.—By the second method,

$$c = \sqrt{10^2 + 15^2 - 2 \times 10 \times 15 \cos 60°}$$
$$= \sqrt{325 - 300 \times .5} = \sqrt{175} = 13.229 \text{ ft.} \quad \text{Ans.}$$

EXAMPLES FOR PRACTICE

1. Given $a = 37.46$ feet, $b = 59.17$ feet, and $C = 69°\ 13'$; find A, B, and c.

$$\text{Ans.} \begin{cases} A = 37°\ 21'\ 30'' \\ B = 73°\ 25'\ 30'' \\ c = 57.72 \text{ ft.} \end{cases}$$

2. Two sides of a triangle are, respectively, 687.64 and 319.58 feet long, and their included angle is $47°\ 15'\ 8''$; find the other two angles and the third side.

$$\text{Ans.} \begin{cases} \text{Angles, } 106°\ 14'\ 56'' \text{ and } 26°\ 29'\ 56'' \\ \text{Third side} = 525.97 \end{cases}$$

3. Given $c = 4$ chains, $a = 6$ chains, and $B = 45°\ 18'$; find b.

$$\text{Ans. } b = 4.271 \text{ ch.}$$

4. Given $b = 43.16$ chains, $c = 51.29$ chains, and $A = 35°\ 8'\ 10''$; find B, C, and a.

$$\text{Ans.}\begin{cases} B = 57°\ 13'\ 20'' \\ C = 87°\ 38'\ 30'' \\ a = 29.544 \text{ ch.} \end{cases}$$

---

22. Case II: Given a Side and Two Angles.—Let c, A, and B be known, to find a, b, and C. The angle $C = 180° - A - B$. By the principle of sines,

$$\frac{a}{\sin A} = \frac{c}{\sin C}, \text{ whence } a = \frac{c}{\sin C}\sin A$$

Similarly,
$$b = \frac{c}{\sin C}\sin B$$

Since $\dfrac{c}{\sin C}$ is the modulus of the triangle (Art. **18**), these formulas may be thus stated: *Any side of a triangle is equal to the modulus of the triangle multiplied by the sine of the angle opposite that side.*

EXAMPLE.—Given $a = 98.48$, $B = 60°\ 45'$, and $C = 39°\ 15'$; to find b, c, and A.

SOLUTION.— $A = 180° - (60°\ 45' + 39°\ 15') = 80°$. Ans.

$$M = \frac{98.48}{\sin 80°}; \quad b = \frac{98.48}{\sin 80°}\sin 60°\ 45'; \quad c = \frac{98.48}{\sin 80°}\sin 39°\ 15'$$

$$\log 98.48 = 1.99335 \qquad\qquad \log b = 1.94076;\ b = 87.248.\ \text{Ans.}$$
$$\log \sin 80° = \overline{1}.99335$$
$$\log \sin 60°\ 45' = \overline{1}.94076$$
$$\log M = 2.00000 \qquad\qquad \log M = 2.00000$$
$$\log \sin 39°\ 15' = \overline{1}.80120$$
$$\log c = \overline{1}.80120;\ c = 63.27.\ \text{Ans.}$$

NOTE.—Attention is called to the convenient way in which the work is here arranged. Having determined $\log M$, this logarithm is copied, and then one of the logarithms to be added to it is written above it, the other under it, the addition being performed upwards in one case, and downwards in the other.

EXAMPLES FOR PRACTICE

1. Given $a = 45.39$ feet, $B = 38°\ 12'$, and $C = 11°\ 11'\ 34''$; find A, b, and c.

$$\text{Ans.}\begin{cases} A = 130°\ 36'\ 26'' \\ b = 36.973 \text{ ft.} \\ c = 11.605 \text{ ft.} \end{cases}$$

2. Given $c = 101.11$ chains, $C = 55°\ 55'\ 55''$, and $A = 10°\ 10'\ 10''$; find B, a, and b.

$$\text{Ans.}\begin{cases} B = 113°\ 53'\ 55'' \\ a = 21.551 \text{ ch.} \\ b = 111.59 \text{ ch.} \end{cases}$$

23. Case III: Given Three Sides.—Let a, b, and c be given, to find A, B, and C.

First Method.—The angles can be found directly from the cosine formulas (Art. **19**), which, being solved for cos A, cos B, and cos C, respectively, give

$$\left.\begin{array}{c} \cos A = \dfrac{b^2 + c^2 - a^2}{2\,bc} \\[2mm] \cos B = \dfrac{a^2 + c^2 - b^2}{2\,ac} \\[2mm] \cos C = \dfrac{a^2 + b^2 - c^2}{2\,ab} \end{array}\right\} \qquad (1)$$

These formulas are to be used when the numbers a, b, c are convenient to square; otherwise, they are too cumbersome, and those given below for the functions of half the angles should be employed. It is necessary to apply the formulas in determining only two of the angles, as the third follows from the relation $A + B + C = 180°$. As a check, however, the formulas should be applied to the third angle also.

It should be borne in mind that, if the cosine of an angle is found to be negative, this implies that the angle is obtuse (Art. **13**). In such case, the cosine is treated as positive, and the corresponding angle taken from the table is subtracted from 180° to obtain the required angle. Thus, if cos $A = -.97030$, we look for the angle whose cosine is $+.97030$, which is 14°. Then, $A = 180° - 14° = 166°$.

Example.—Given $a = 4$ inches, $b = 5$ inches, and $c = 7$ inches; to find A, B, and C.

Solution.— $\cos A = \dfrac{b^2 + c^2 - a^2}{2\,bc} = \dfrac{5^2 + 7^2 - 4^2}{2 \times 5 \times 7} = \dfrac{58}{70} = .82857,$

and, therefore, $A = 34° 2' 53''$. Ans.

$\cos B = \dfrac{a^2 + c^2 - b^2}{2\,ac} = \dfrac{4^2 + 7^2 - 5^2}{2 \times 4 \times 7} = \dfrac{40}{56} = .71429$

and, therefore, $B = 44° 24' 54''$. Ans.

$C = 180° - A - B = 101° 32' 13''$. Ans.

As a check, we have

$\cos C = \dfrac{a^2 + b^2 - c^2}{2\,ab} = \dfrac{4^2 + 5^2 - 7^2}{2 \times 4 \times 5} = -\dfrac{8}{40} = -.20000$

The angle whose cosine is .20000 is 78° 27′ 47″. Therefore, $C = 180°$ $- 78° 27′ 47″ = 101° 32′ 13″$.

Second Method.—As said before, this method is to be applied when the operations required by formula **1** involve too much labor, which happens when the lengths of the given sides consist of three or more significant figures—the usual case. If the sum of the sides is denoted by $2s$, or half their sum by s, the angles A, B, C may be found by the following formulas, which are derived in Appendix V:

$$\left. \begin{aligned} \tan \tfrac{1}{2} A &= \sqrt{\frac{(s-b)(s-c)}{s(s-a)}} \\ \tan \tfrac{1}{2} B &= \sqrt{\frac{(s-a)(s-c)}{s(s-b)}} \\ \tan \tfrac{1}{2} C &= \sqrt{\frac{(s-a)(s-b)}{s(s-c)}} \end{aligned} \right\} \qquad (2)$$

$$\left. \begin{aligned} \cos \tfrac{1}{2} A &= \sqrt{\frac{s(s-a)}{bc}} \\ \cos \tfrac{1}{2} B &= \sqrt{\frac{s(s-b)}{ac}} \\ \cos \tfrac{1}{2} C &= \sqrt{\frac{s(s-c)}{ab}} \end{aligned} \right\} \qquad (3)$$

For angles differing but little from 90° (say between 85° and 90°), use the cosine formulas **3**; in all other cases, the tangent formulas **2**.

We have also,

$$\sin \tfrac{1}{2} A = \sqrt{\frac{(s-b)(s-c)}{bc}} \qquad (4)$$

with similar formulas for $\sin \tfrac{1}{2} B$ and $\sin \tfrac{1}{2} C$. These formulas are of value for deriving the tangent formulas **2**, as well as for deriving an expression for the area of a triangle when the sides are given. They may also be used instead of the tangent formulas **2** for the determination of the angles, but the latter are preferable.

EXAMPLE.—In the triangle ABC, $a = 567$ feet, $b = 736$ feet, and $c = 264$ feet; to find the angles A, B, and C.

SOLUTION.—The tangent formulas will be used.

To find A

$$
\begin{aligned}
a &= 567 \\
b &= 736 \\
c &= 264 \\
\hline
2\,s &= 1{,}567 \\
s &= 783.5 \\
s - a &= 216.5 \\
s - b &= 47.5 \\
s - c &= 519.5
\end{aligned}
$$

$\log (s - c) = 2.71559$
$\log (s - b) = 1.67669$
$ 4.39228$

$\log s = 2.89404$
$\log (s - a) = 2.33546$
$ 5.22950$
$2)\overline{1}.16278$

$\log \tan \tfrac{1}{2} A = \overline{1}.58139$

$\tfrac{1}{2} A = 20° 52' 38''$, $A = 41° 45' 16''$.   Ans.

To find B

$\log (s - a) = 2.33546$
$\log (s - c) = 2.71559$
$ 5.05105$

$\log s = 2.89404$
$\log (s - b) = 1.67669$
$ 4.57073$

$2)0.48032$

$\log \tan \tfrac{1}{2} B = 0.24016$
$\tfrac{1}{2} B = 60° 5' 29''$; $B = 120° 10' 58''$
Ans.

To find C

$\log (s - a) = 2.33546$
$\log (s - b) = 1.67669$
$ 4.01215$

$\log s = 2.89404$
$\log (s - c) = 2.71559$
$ 5.60963$

$2)\overline{2}.40252$

$\log \tan \tfrac{1}{2} C = \overline{1}.20126$
$\tfrac{1}{2} C = 9° 1' 54''$; $C = 18° 3' 48''$
Ans.

To check, add the angles:

$$
\begin{aligned}
&41° \ 45' \ 16'' \\
&120 \ \ \ 10 \ \ \ 58 \\
&18 \ \ \ 3 \ \ \ 48 \\
\hline
&180° \ 00' \ 2''
\end{aligned}
$$

The triangle closes within 2 sec. This error is due to the use of five-place tables, and to the fact that the angle in each case was taken out to the nearest second.

EXAMPLES FOR PRACTICE

1. Given $a = 1$ mile, $b = 2$ miles, and $c = 1.5$ miles; find A, B, and C. (Use first method.)

$$\text{Ans.} \begin{cases} A = 28° 57' 17'' \\ B = 104° 28' 39'' \\ C = 46° 34' 4'' \end{cases}$$

2. Given $a = 50$ chains, $b = 30$ chains, and $c = 45$ chains; find A, B, and C. (Use first method.)

$$\text{Ans.} \begin{cases} A = 80° 56' 36'' \\ B = 36° 20' 7'' \\ C = 62° 43' 17'' \end{cases}$$

3.   Given a = 63.47 feet, b = 89.36 feet, and c = 109.83 feet; find
A, B, and C (Use second method.)

$$\text{Ans.} \begin{cases} A = 35°\ 18'\ 10'' \\ B = 54°\ 27'\ 2'' \\ C = 90°\ 14'\ 50'' \end{cases}$$

4.   Given a = 2,354 feet, b = 3,115 feet, and c = 836.6 feet; find
A, B, and C.   (Use second method.)

$$\text{Ans.} \begin{cases} A = 21°\ 7'\ 24'' \\ B = 151°\ 31'\ 8'' \\ C = 7°\ 21'\ 30'' \end{cases}$$

**24.   Case IV: Given Two Sides and the Angle
Opposite One of Them.**—In the triangle ABC, let a, b,
and A be given, to find B, C, and c. The angle B or C is
found by means of the principle of sines; thus,

$$\frac{a}{\sin A} = \frac{b}{\sin B}, \text{ whence } \sin B = \frac{b \sin A}{a}$$

Then, $C = 180° - A - B$, and $c = \dfrac{a}{\sin A} \sin C$

When the data are given as above, without any further
restrictions, there may be two triangles that will answer the
given conditions; and the problem is said to have two solutions. For here the angle B is determined from its sine,
and as every sine corresponds to two supplementary angles,
either of these angles may be taken. Thus, if sin B is found
to be .64746, the corresponding angle may be either 40° 21'
or 180° − 40° 21' = 139° 39', since these angles both have
the same sine (Art. **13**).

The same result is obtained from geometrical consider-
ations. On any line AX, Fig. 7, construct an angle equal

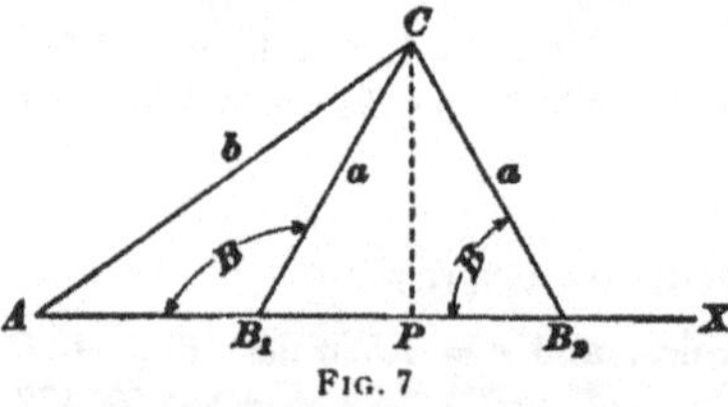

to the given angle A, and on
its side AC take AC equal
to one of the given sides b.
From C as a center, with a
radius equal to the side a,
describe an arc. This arc
will generally cut AX at two

FIG. 7

points, B_1 and B_2, and either of the triangles ACB_1 or ACB_2
will answer the conditions of the problem, for they both contain the given sides b and a, and the angle A opposite a.

The problem will have but one solution in the following
cases:

1.   If $a = b \sin A$.   For in this case a will be equal to the perpendicular CP, Fig. 7, and the arc described from C will touch AX at P only.

2.   If $a = b$.   For in this case the angles A and B must be equal, and therefore both acute, since a triangle cannot have two obtuse angles. In this case B_1 coincides with A in Fig. 7, since $CB_1 = CA$.

3.   When a is greater than b.   For in this case A must be greater than B,

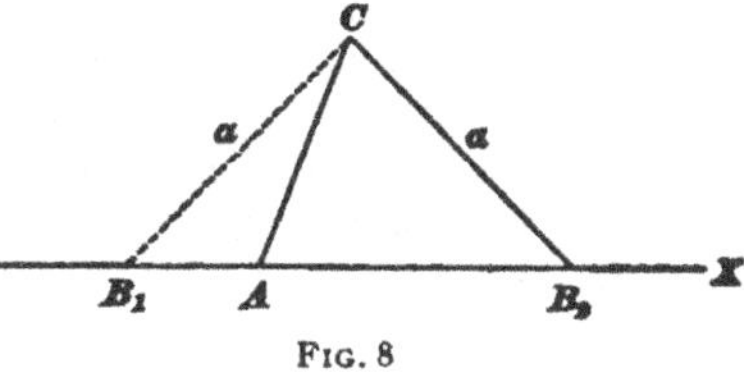

FIG. 8

and the latter angle must therefore be acute.   This is shown by Fig. 8; the arc described from C cuts AX produced at B_1, and CB_1, although equal to a, is not opposite A ($= CAB_2$).

When a is less than $b \sin A$, the problem is impossible.   For then a is less than CP, Fig. 7, and the arc does not cut AX at all.   This is also shown by the formula $\sin B = \dfrac{b \sin A}{a}$, which would give $\sin B$ a value greater than 1, which is an impossible value, for no sine can be greater than 1.

EXAMPLE.—Given $a = 273$ feet, $b = 392$ feet, and $A = 37°\ 14'$; to find B, C, and c.

SOLUTION.—Here a is less than b, and, unless $\sin B$ is found to be greater than 1 (in which case the problem is impossible), there are two solutions.

$$\sin B = \frac{b \sin A}{a} = \frac{392 \times \sin 37°\ 14'}{273}; \quad B = \begin{cases} 60°\ 19'\ 17'', \text{ or} \\ 180° - 60°\ 19'\ 17'' \\ = 119°\ 40'\ 43''. \text{ Ans.} \end{cases}$$

$$C = \begin{cases} 180° - 37°\ 14' - 60°\ 19'\ 17'' = 82°\ 26'\ 43'', \text{ or} \\ 180° - 37°\ 14' - 119°\ 40'\ 43'' = 23°\ 5'\ 17''. \text{ Ans.} \end{cases}$$

$$c = \frac{a}{\sin A}\sin C = \frac{273}{\sin 37°\ 14'}\sin\begin{cases} 82°\ 26'\ 43'', \text{ or} \\ 23°\ 5'\ 17'' \end{cases}\begin{cases} = 447.27 \text{ ft., or} \\ 176.93 \text{ ft. Ans.} \end{cases}$$

PRACTICAL EXAMPLES

EXAMPLE 1.—The distance between two points A and B, Fig. 9, is 360.38 feet, the angles from A and B to a station C are found, with a transit, to be, respectively, $62°\ 17'$ and $39°\ 51'$.   What are the distances of C from A and B?

SOLUTION.— $C = 180° - 62° 17' - 39° 51' = 77° 52'$. Modulus (M) of triangle $= \dfrac{360.38}{\sin 77° 52'}$. Then (Art. **18**),

$$a = \frac{360.38}{\sin 77° 52'} \sin 62° 17' = 326.32 \text{ ft.} \quad \textbf{Ans.}$$

$$b = \frac{360.38}{\sin 77° 52'} \sin 39° 51' = 236.2 \text{ ft.} \quad \textbf{Ans.}$$

EXAMPLE 2.—The distances of a fort C from two other forts A and B are as marked in Fig. 10; the lines of sight from C to A and B make an angle of 53° 8′ 16″. What is the distance between the two forts A and B?

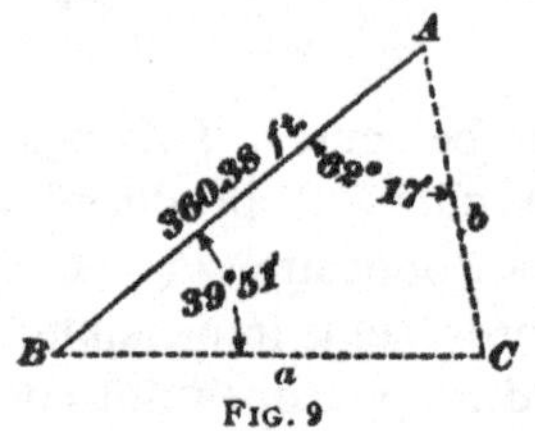

FIG. 9

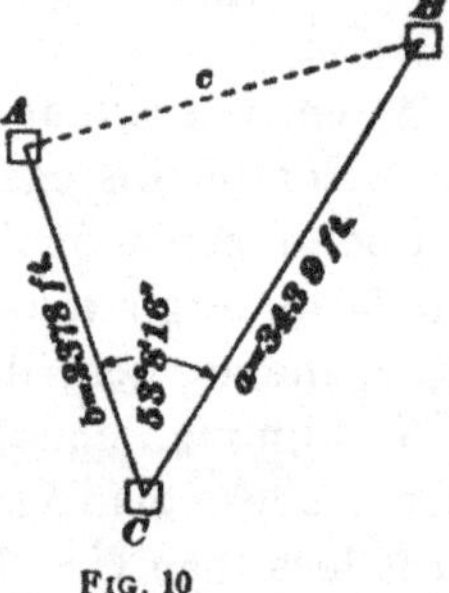

FIG. 10

SOLUTION.—The two sides and the included angle are given, and formulas 1 and 2, Art. **21**, will be applied. It is not necessary to determine the angles A and B, for they are not required. Formula 1, Art. **21**,

$$\tan \tfrac{1}{2}(A - B) = \frac{a - b}{a + b} \cot \tfrac{1}{2} C$$

$$= \frac{3{,}439 - 2{,}378}{3{,}439 + 2{,}378} \cot \frac{53° 8' 16''}{2}$$

$$= \frac{1{,}061}{5{,}817} \cot 26° 34' 8''$$

$$\tfrac{1}{2}(A - B) = 20° 2' 20''$$

Formula 2, Art. **21**,

$$c = AB = \frac{(a - b) \cos \tfrac{1}{2} C}{\sin \tfrac{1}{2}(A - B)}$$

$$= \frac{1{,}061 \cos 26° 34' 8''}{\sin 20° 2' 20''}$$

$$= 2{,}769.4 \text{ ft.} \quad \textbf{Ans.}$$

EXAMPLE 3.—A weight W, Fig. 11, is to be hung from a pulley sliding freely on the rope $O Q P$. The length of the rope is l, and its ends are fastened at two points

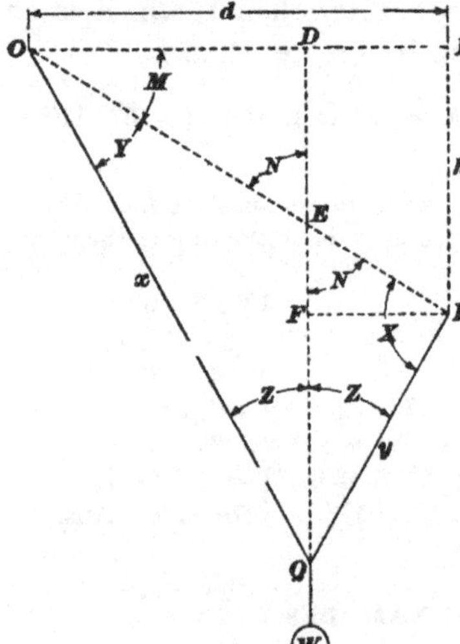

FIG. 11

O and P, whose horizontal distance is d and whose vertical distance is h, as shown. It being proved in mechanics that the pulley will rest in equilibrium when the vertical line WQ bisects the angle OQP, what are the lengths $x\,(=OQ)$ and $y\,(=PQ)$ of the two segments of the rope for which that condition obtains?

Note.—This problem is given here as an illustration of the many problems that occur in practice requiring the exercise of some ingenuity and the performance of some transformations, both algebraical and trigonometrical, before the required results are obtained.

Solution.—Let QD be a vertical line through Q. According to the data, this line makes equal angles with OQ and QP. These angles are denoted by Z. The angles made by OQ and PQ with OP are denoted by Y and X, respectively. The line OR is horizontal, and PR vertical

As OR and RP are known, the right triangle OPR gives

$$\tan M = \frac{h}{d}$$

Also, in the triangle OED, $N = 90° - M$.

The angles M and N may, therefore, be assumed to be known.

Drawing PF parallel to RO, we have ,

$$d\,(=OR) = OD + DR = OD + PF$$

or, substituting the values of OD and PF from the triangles ODQ and PFQ,

$$d = x \sin Z + y \sin Z = (x+y) \sin Z = l \sin Z$$

whence,
$$\sin Z = \frac{d}{l}$$

Having found Z, we have
$$X = 180° - (N+Z) \text{ (triangle } PEQ)$$
$$Y = N - Z \text{ (triangle } OEQ)$$

The modulus of the triangle OPQ is
$$\frac{OP}{\sin OQP} = \frac{OP}{\sin 2Z} = \frac{d \div \cos M}{\sin 2Z} = \frac{d}{\cos M \sin 2Z} = \frac{d}{\sin N \sin 2Z}$$

Therefore (Art. 18), $x = \dfrac{d}{\sin N \sin 2Z} \sin X$

or, substituting the value of X, and noticing that $\sin[180° - (N+Z)]$ $= \sin(N+Z)$,

$$x = \frac{d}{\sin N \sin 2Z} \sin(N+Z). \qquad \text{Ans.}$$

Likewise,

$$y = \frac{d}{\sin N \sin 2Z} \sin(N-Z). \qquad \text{Ans.}$$

EXAMPLES FOR PRACTICE

1. Find the distance MN across the lake from the data shown in Fig. 12.

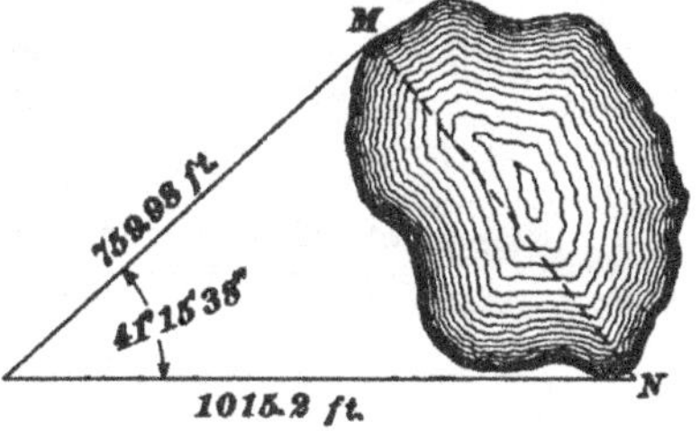

Fig. 12

Ans. $MN = 669.51$ ft.

2. The angles from two stations M and N, Fig. 13, to two inaccessible

points P and Q being as shown, and the distance MN being 550 feet find the distance PQ.

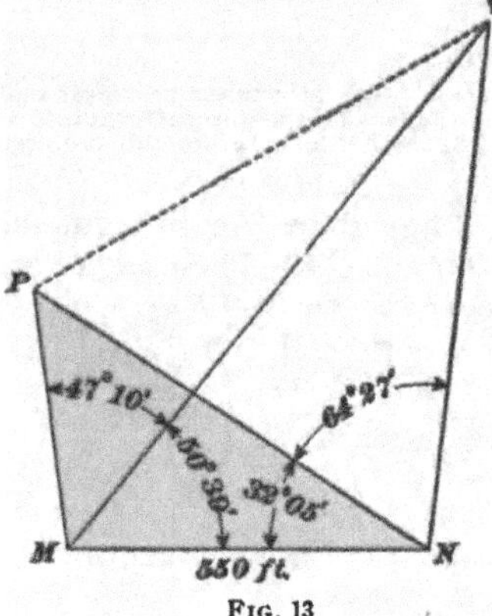

Fig. 13

HINT.—First calculate MP, then MQ, and finally PQ.

Ans. $PQ = 799.7$ ft.

3. In Fig. 14, the sides AB and DE were measured and the angles were turned as marked. Find the lengths of the sides BC, CA, CF, AF, CD, FD, EF.

$$\text{Ans.}\begin{cases} BC = 677.92 \text{ ft.} \\ CA = 1{,}065.8 \text{ ft.} \\ CF = 905.46 \text{ ft.} \\ AF = 703.1 \text{ ft.} \\ CD = 696.83 \text{ ft.} \\ FD = 1{,}019.7 \text{ ft.} \\ EF = 687.97 \text{ ft.} \end{cases}$$

4. Two observers on the same side of a steeple, and in the same vertical plane with it, are 100 feet apart, and find that the angles of elevation are 26° 28′ and 49° 14′. What is the height of the steeple?

Ans. 87.225 ft.

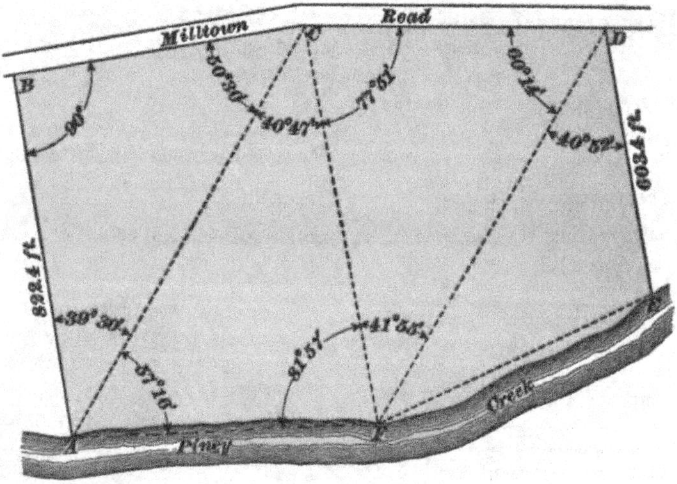

Fig. 14

5. Find the altitude h and the lengths of the sides AB and CD of the trapezoid $ABCD$, Fig. 15.

$$\text{Ans.}\begin{cases} h = 62.22 \text{ ft.} \\ AB = 87.56 \text{ ft.} \\ CD = 64.579 \text{ ft.} \end{cases}$$

6. The connecting-rod AB, Fig. 16, of an engine is 9 feet 3 inches, and the crank-arm CB is $10\frac{1}{2}$ inches; the figure shows the crank after

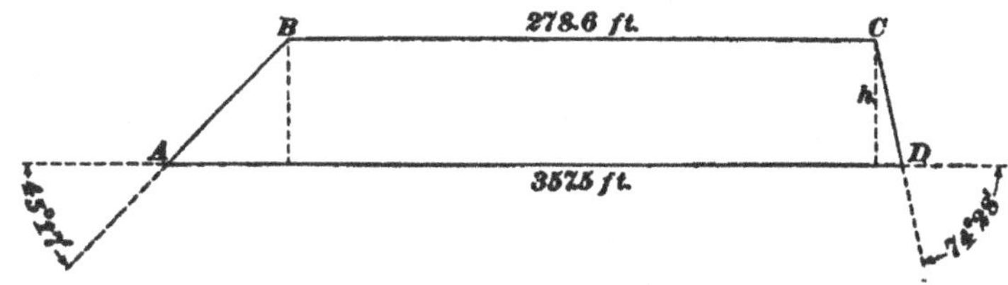

FIG. 15

it has performed one-eighth of a revolution, starting from the position CB'. Find: (a) the inclination M of the connecting-rod to the axis

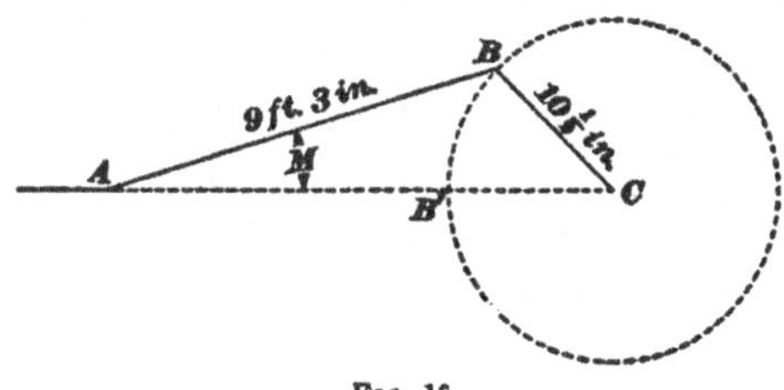

FIG. 16

ot the piston rod, which is in line with CA; (b) the distance AC of the joint A from the center of the crank-circle.

Ans. $\begin{cases} (a) \ \ M = 3°\ 50'\ 7'' \\ (b) \ \ AC = 9 \text{ ft. } 10\frac{1}{2} \text{ in., nearly} \end{cases}$

AREAS

LAND MEASURE

25. In surveying the public lands of the United States and Canada, all linear measurements are made with the surveyors' chain, also known as Gunter's chain, from the name of the inventor. This chain is 66 feet in length and contains 100 links, each 7.92 inches long. In private surveys, the foot is commonly taken as the unit of linear measure, and small land areas are expressed in square feet.

Land areas of considerable extent in the countries mentioned are generally expressed in acres. Fractional parts of an acre, which formerly were expressed in roods, square rods or perches, and square links, are now expressed decimally by nearly all surveyors. Thus, 40.35 acres is written instead of 40 acres, 1 rood, and 16 square rods.

Tables of linear and square measure are given in *Arithmetic*, and to those tables the student is referred for detailed information regarding the subject. The following table gives the relative values of the units of area used in land surveying in the countries referred to above. As already stated, the square foot and acre are now the units most commonly employed.

TABLE OF LAND MEASURE

```
1 square yard (sq. yd.)  .  =    9  square feet (sq. ft.)
1 square rod* (sq. rd)   .  =  30¼ square yards = 272¼ square feet
1 square chain (sq. ch.) .  =   16  square rods = 4,356 square feet
1 acre (A.) . . . . . . .  =   10  square chains = 43,560 square feet
1 rood (R.) . . . . . . .  =   40  square rods = 10,890 square feet
1 acre  . . . . . . . . .  =    4  roods = 160 square rods
1 square mile (sq. mi.)     = 640  acres = 6,400 square chains
1 township (Tp.) . . . .  =   36  square miles = 23,040 acres (app )
```

*Sometimes called a perch or pole. and designated by the abbreviation P.

As will be observed, there are 10 square chains in an acre. In order, therefore, to reduce to acres any number of square chains, it is sufficient to move the decimal point one place toward the left, which is equivalent to dividing by 10. It must also be borne in mind that, since there are 100 links in 1 chain, links are usually expressed decimally as hundredths of a chain. Thus, 6.72 chains is written instead of 6 chains 72 links.

EXAMPLE 1.—A rectangular piece of land is 1,060 feet in length by 820 feet in breadth; what is its area: (a) in acres and decimals? (b) in acres, roods, and perches?

SOLUTION.— (a) $1,060 \times 820 = 869,200$ sq. ft.; $869,200 \div 43,560 = 19.954$ A. Ans.

(b) .954 A. $= .954 \times 4 = 3.816$ R.; .816 R. is equal to $.816 \times 40 = 32.64$ P. Hence, the area is 19 A. 3 R. 32.64 P. Ans.

EXAMPLE 2.—A rectangular piece of land is 12 chains and 6 links (12.06 chains) in length by 8 chains and 55 links (8.55 chains) in breadth; what is its area: (a) in acres and decimals? (b) in acres, roods, and perches?

SOLUTION.— (a) $12.06 \times 8.55 = 103.11$ sq. ch.; $103.11 \div 10 = 10.311$ A. Ans.

(b) .311 A. $= .311 \times 4 = 1.244$ R.; .244 R. is equal to $.244 \times 40 = 9.76$ P. Hence, the area is 10 A. 1 R. 9.76 P. Ans.

EXAMPLES FOR PRACTICE

1. A rectangular piece of land is 1,190 feet in length by 700 feet in breadth; what is its area: (a) in acres and decimals? (b) in acres, roods, and perches?

Ans. $\begin{cases} (a)\ 19.123\ \text{A.} \\ (b)\ 19\ \text{A. 0 R. 19.7 P.} \end{cases}$

2. A rectangular piece of land is 525 feet long by 250 feet wide, what is its area: (a) in acres and decimals? (b) in acres, roods, and perches?

Ans. $\begin{cases} (a)\ 3.013\ \text{A.} \\ (b)\ 3\ \text{A. 0 R. 2.08 P.} \end{cases}$

3. A rectangular piece of land is 15 chains and 65 links in length by 8 chains and 16 links in breadth; what is its area: (a) in acres and decimals? (b) in acres, roods, and perches?

Ans. $\begin{cases} (a)\ 12.77\ \text{A.} \\ (b)\ 12\ \text{A. 3 R. 3.2 P.} \end{cases}$

AREÁS OF POLYGONS

THE TRIANGLE

NOTE.—In all that follows, the area of any figure under consideration will be designated by S, unless otherwise stated.

26. Given the Base and Altitude.—Any of the sides of a triangle may be taken as the base, the altitude being the length of the perpendicular drawn on the base from the vertex of the opposite angle. In Fig. 17, b is taken as the base,

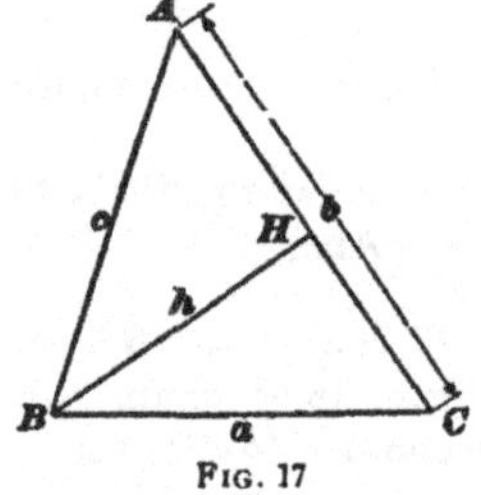

FIG. 17

and the perpendicular BH, denoted by h, is the altitude.

It was shown in *Geometry*, Part 2, that the area of a triangle, when the base b and altitude h are known, is given by the formula

$$S = \tfrac{1}{2} b h$$

27. Given Two Sides and the Included Angle.—Let b, c, and A, Fig. 17, be given. In the right triangle ABH, we have $h = c \sin A$. The substitution of this value of h in the formula in Art. **26** gives

$$S = \tfrac{1}{2} b c \sin A$$

In words, *the area of a triangle is equal to one-half the product of any two sides and the sine of their included angle.*

EXAMPLE.—Two of the sides of a triangular field are 39.47 and 59.23 chains, respectively, and their included angle is 65° 10′ 40″. To find the contents of the field, in acres.

SOLUTION.—By the formula, S (square chains) $= \tfrac{1}{2} \times 39.47 \times 59.23$ sin 65° 10′ 40″ $= 1,060.9$ sq. ch.; whence, dividing by 10 (Art. **25**),

$$S \text{ (acres)} = 106.09 \text{ A.} \quad \text{Ans.}$$

28. Given One Side and Two Angles.—The other angle may be at once found by subtracting the sum of the two given angles from 180°. It may, therefore, be assumed that the three angles are known. Let b, Fig. 17, be the given side. From Art. **22**, the value of c is equal to the modulus of the triangle multiplied by sin C, or,

$$c = \frac{b}{\sin B} \sin C$$

Substituting this value in the formula in Art. **27,** we obtain

$$S = \frac{b^2 \sin A \sin C}{2 \sin B}$$

29. The formula in Art. **28** is convenient when logarithmic functions are employed. For the use of natural functions, the following is preferable:

In the right triangles ABH and CBH, Fig. 17, we have,

$$AH = h \cot A, \quad CH = h \cot C$$

whence, adding these two equations,

$$AH + CH = h \cot A + h \cot C$$

that is, $b = h(\cot A + \cot C)$

and, therefore, $h = \dfrac{b}{\cot A + \cot C}$ (1)

This formula is useful and should be committed to memory. It may be stated in words thus: *The altitude of a triangle is equal to the base divided by the sum of the cotangents of the adjacent angles.*

By substituting, in the formula in Art. **26,** the value of h given in formula **1,** we obtain

$$S = \frac{b^2}{2(\cot A + \cot C)}   \quad (2)$$

In words, *the area of a triangle is equal to the square of any side divided by twice the sum of the cotangents of the angles adjacent to that side.*

EXAMPLE.—One side of a triangular field is 127.64 chains, and the adjacent angles are 46° 15′ and 60° 41′. To find the area.

SOLUTION BY LOGARITHMIC FUNCTIONS.—Here, $b = 127.64$, $A = 46° 15′$, $C = 60° 41′$, and $B = 180° - 46° 15′ - 60° 41′ = 73° 4′$. Formula of Art. **28,**

$$S = \frac{127.64^2 \sin 46° 15′ \sin 60° 41′}{2 \sin 73° 4′}$$

$$= 5{,}363.4 \text{ sq. ch.} = 536.34 \text{ A. Ans.}$$

SOLUTION BY NATURAL FUNCTIONS.—By formula **2,**

$$S = \frac{127.64^2}{2(\cot 46° 15′ + \cot 60° 41′)} = \frac{127.64^2}{2(.95729 + .56156)}$$

$$= \frac{127.64^2}{3.0377} = 5{,}363.4 \text{ sq. ch.} = 536.34 \text{ A. Ans.}$$

NOTE.—Even if natural functions are used, the division is advantageously performed by means of logarithms.

EXAMPLES FOR PRACTICE

1. Two sides of a triangular field are 3,760 and 2,757 feet, respectively, and their included angle is 54° 13′ 13″. What is the area of the field, in acres?　　　　　　　　　　　　Ans. S = 96.534 A.

2. One side of a triangle is 96.34 chains; the opposite angle is 49° 10′, and one of the adjacent angles, 69° 45′ 30″. What is the area of the triangle, in acres?　　　　　　　　Ans. S = 503.69 A.

3. One side of a triangle is 8.93 inches, and the adjacent angles are 34° 16′ and 17° 37′ 18″. What is the area of the triangle?

Ans. S = 8.638 sq. in.

4. Two sides of a triangle are 17 and 25 feet, respectively, and the included angle is 76° 13′. What is the area of the triangle?

Ans. S = 206.38 sq. ft.

---

30. Given the Three Sides.—Let a, b, and c, Fig. 17, be given, and denote $\frac{1}{2}(a + b + c)$ by s. The area S of the triangle is given by the following formula, which is derived in Appendix VI:

$$S = \sqrt{s(s - a)(s - b)(s - c)}$$

EXAMPLE.—The sides of a triangular tract are 1,634.6 (= a, say), 978.28 (= b, say), and 2,176.4 (= c, say) feet, respectively; to find the area, in acres.

SOLUTION.—The work may be conveniently arranged as shown below. The numbers in marks of parenthesis indicate the order in which the several quantities are set down. In (6), s is placed above a, b, c in order to facilitate the subtractions. The differences $s - a$, $s - b$, $s - c$ are written, as the subtractions are performed, horizontally opposite a, b, and c, respectively.

```
(6)    s = 2,394.64
(1)    a = 1,634.60        (7) s - a =    760.04
(2)    b =   978.28        (8) s - b = 1,416.36
(3)    c = 2,176.40        (9) s - c =    218.24
(4) 2 s = 4,789.28
(5)    s = 2,394.64
(10) log  s       = 3.37924
(11) log (s - a) = 2.88083
(12) log (s - b) = 3.15117
(13) log (s - c) = 2.33893
                2)11.75017
        log S = 5.87509;  S = 750,050 sq. ft. = 17.22 A.  Ans.
```

EXAMPLES FOR PRACTICE

1. Find the area of a triangular tract whose sides are 54.36, 73.19, and 101.76 chains, respectively.   Ans. $S = 192.26$ A.

2. Find the area of a triangular plate whose sides are 17.12, 12.75, and 8.95 inches, respectively.   Ans. $S = 55.646$ sq. in.

THE TRAPEZOID

31. Notation.—In Fig. 18, the bases, or parallel sides, of the trapezoid $ABCD$ are denoted by b_1 and b_2; the altitude, by h; and the sides AD and BC, by a and c, respectively. The angles will be designated by the letters A, B, C, D at the vertexes. The line DB' is

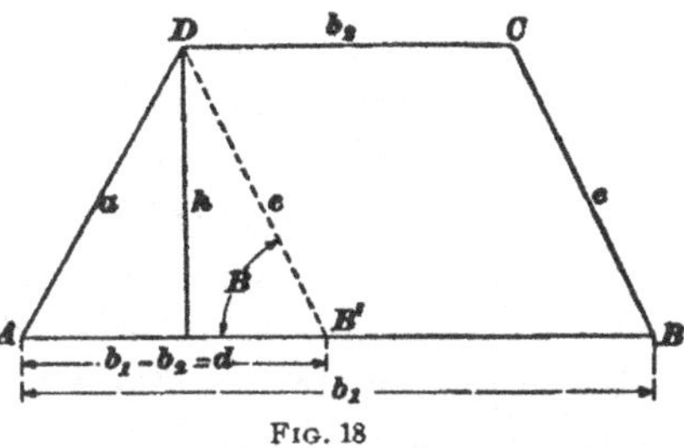

FIG. 18

drawn through D parallel to CB, thus forming a parallelogram in which $B'B = DC = b_2$, and $DB' = CB = c$. Also, angle $DB'A = B$, and $AB' = AB - B'B = b_1 - b_2$. For some purposes, it is convenient to represent this difference by a single letter d, as shown in the figure.

32. Given the Bases and the Altitude.—As shown in *Geometry*, Part 2, the area of a trapezoid is equal to one-half the product of the altitude by the sum of the bases; that is,
$$S = \tfrac{1}{2}(b_1 + b_2)h$$

33. Given the Bases and the Angles Adjacent to One of Them.—Let b_1, b_2, A, and B, Fig. 18, be given. In the triangle ADB' we have (formula **1**, Art. **29**),
$$h = \frac{b_1 - b_2}{\cot A + \cot B}$$

If this value of h is substituted in the formula of Art. **32**, the result is,
$$S = \frac{(b_1 - b_2)(b_1 + b_2)}{2(\cot A + \cot B)} \qquad (1)$$

As the product of the sum of two quantities by their difference is equal to the difference between the squares of the quantities, $(b_1 - b_2)(b_1 + b_2)$ is equal to $b_1{}^2 - b_2{}^2$; and, there-fore, formula **1** may also be written:

$$S = \frac{b_1{}^2 - b_2{}^2}{2(\cot A + \cot B)} \qquad (2)$$

For the use of logarithmic functions, formula **1** may be transformed into the following (see Appendix VII):

$$S = \frac{(b_1 - b_2)(b_1 + b_2) \sin A \sin B}{2 \sin (A + B)} \qquad (3)$$

In the application of these formulas, the student should bear in mind that the cotangent of an angle greater than 90° is negative, and numerically equal to the cotangent of the supplement of the angle; also, that the sine of an angle greater than 90° is equal to the sine of its supplement. Thus, $\cot 105° = - \cot (180° - 105°) = - \cot 75° = - .26795$; and $\sin 105° = \sin 75° = .96593$.

EXAMPLE 1.—The two bases of a trapezoid are 350 and 137 chains, respectively; the angles adjacent to the longer base are 75° 10′ and 63° 54′. What is the area of the trapezoid?

SOLUTION BY NATURAL FUNCTIONS.—Let $350 = b_1$, $137 = b_2$, $A = 75° 10′$, $B = 63° 54′$. As b_1 and b_2 are not convenient numbers to square, formula **1**, which is better adapted to logarithmic work, will be used.

$$S = \frac{(350 - 137)(350 + 137)}{2(\cot 75° 10′ + \cot 63° 54′)} = \frac{213 \times 487}{2(.26483 + .48989)} = 68{,}721 \text{ sq. ch.}$$
$$= 6{,}872.1 \text{ A.} \quad \text{Ans.}$$

SOLUTION BY LOGARITHMIC FUNCTIONS.—By formula **3**,

$$S = \frac{(350 - 137)(350 + 137) \sin 75° 10′ \sin 63° 54′}{2 \sin 139° 4′}$$

or, replacing $\sin 139° 4′$ by $\sin (180° - 139° 4′) = \sin 40° 56′$,

$$S = \frac{213 \times 487 \sin 75° 10′ \sin 63° 54′}{2 \sin 40° 56′} = 68{,}721 \text{ sq. ch.} = 6{,}872.1 \text{ A.} \quad \text{Ans.}$$

EXAMPLE 2.—The bases of a trapezoid are 100 and 70 feet, the angles adjacent to the shorter base being 52° 47′ and 143° 14′. What is the area of the trapezoid?

SOLUTION.—Since the bases are parallel, the two angles adjacent to each of the non-parallel sides are supplementary. Thus, in Fig. 18, $A + D = 180°$, $B + C = 180°$; and, therefore, $A = 180° - D$; $B = 180° - C$. Let $52° 47′ = D$, $143° 14′ = C$. Then,

$$A = 180° - 52° \ 47' = 127° \ 13'$$
$$B = 180° - 143° \ 14' = 36° \ 46'$$
$$\cot A = -\cot (180° - 127° \ 13') = -\cot 52° 47' = -.75950$$
$$\cot B = \cot 36° \ 46' = 1.33835$$

Formula **2,**

$$S = \frac{100^2 - 70^2}{2(+1.33835-.7595)} = \frac{5,100}{1.1577} = 4,405.3 \text{ sq. ft.} \quad \textbf{Ans.}$$

EXAMPLES FOR PRACTICE

1. The bases of a trapezoidal tract are 78.63 and 54.71 chains, respectively; the angles adjacent to the longer base are 55° 18′ and 62° 53′. Find the area, in acres.   Ans. $S = 132.4$ A.

2. Find the number of square feet in a trapezoidal cross-section of a canal 40 feet wide at the bottom, 65 feet wide at the top, and whose non-parallel sides are inclined to the horizontal at an angle of 50°. (The dimensions across the top and bottom are measured horizontally.)
          Ans. $S = 782.09$ sq. ft.

3. The two bases of a trapezoid are 10.25 and 18.76 inches, respectively; one of the angles adjacent to the shorter base is 76° 45′ 10″, and the angle diagonally opposite is 66° 8′ 9″; find the area of the trapezoid.   (Use logarithmic functions.)   Ans. $S = 596.4$ sq. in.

---

34. **Given the Four Sides.**—If the difference between the two bases added to the sum of the non-parallel sides is denoted by $2\,s$; that is, if the expression $\frac{1}{2}(a + c + d)$, Fig. 18, is denoted by s, the area of the trapezoid is given by the following formula (see Appendix VIII):

$$S = \frac{b_1 + b_2}{d}\sqrt{s(s - a)(s - c)(s - d)}$$

EXAMPLE FOR PRACTICE

The bases of a trapezoidal field are 136.43 and 210.18 chains, respectively; one of the non-parallel sides is 96.73 chains, and the other 164.37 chains. Find the area of the tract, in acres.
          Ans. $S = 864.97$ A.

THE REGULAR POLYGON

35. Given the Number of Sides and the Radius.
Let MN, Fig. 19, be one of the sides of a regular polygon
of n sides; O, the center, and r the
radius, of the circumscribed circle
(called also the center and radius,
respectively, of the polygon); and
A, the angle at the center subtended
by a side of the polygon. The length
of the side MN will be denoted by l.

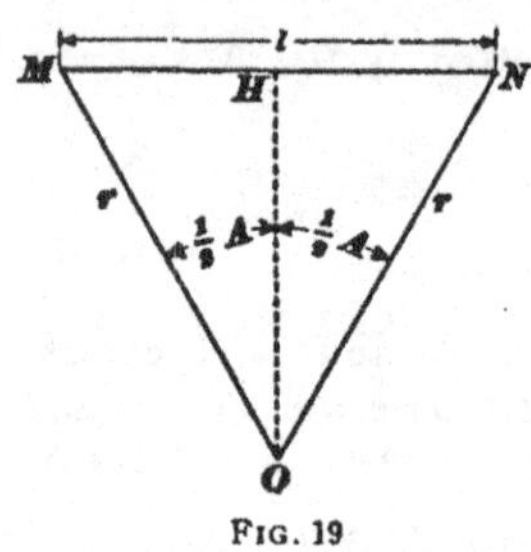

FIG. 19

Let n and r be given, to find the
area S of the polygon and the length l
of each of its sides. From *Geometry*, Part 2, the angle MON,
or A, is found by dividing $360°$ by the number of sides in the
polygon; that is,

$$A = \frac{360°}{n}$$

The area of the triangle MON is (Art. **27**) $\frac{1}{2}OM \times ON$
$\sin MON$, or $\frac{1}{2}r \times r \sin A = \frac{1}{2}r^2 \sin A = \frac{1}{2}r^2 \sin \frac{360°}{n}$.
Since the polygon consists of n triangles equal to MON, its
area S is equal to n times the area of MON; that is,

$$S = n \times \tfrac{1}{2}r^2 \sin \frac{360°}{n}$$

or $\qquad\qquad S = \tfrac{1}{2}nr^2 \sin \dfrac{360°}{n} \qquad$ **(1)**

In the right triangle MOH, we have,

$$MH = r \sin \frac{A}{2}$$

or, since MH is one-half of MN, or of l,

$$\frac{l}{2} = r \sin \frac{A}{2}$$

whence, multiplying by 2,

$$l = 2r \sin \frac{A}{2}$$

Finally, $\dfrac{A}{2} = \tfrac{1}{2}\dfrac{360°}{n} = \dfrac{180°}{n}.$ By the substitution of this value in the expression for l just found, we get, finally,

$$l = 2\,r \sin \frac{180°}{n} \qquad (2)$$

36. When the Number of Sides and Their Common Length Are Given.—Let n and l, Fig. 19, be given, to find the radius r and the area S. The radius is found by solving formula **2**, Art. **35**, for r, which gives,

$$r = \frac{l}{2 \sin \dfrac{180°}{n}} \qquad (1)$$

In the triangle MOH, we have,

$$OH = MH \cot \tfrac{1}{2} A = \frac{MN}{2} \cot \tfrac{1}{2} A$$

The area of MON is $\tfrac{1}{2} MN \times OH$. Writing instead of OH the value just found,

$$\tfrac{1}{2} MN \times \frac{MN}{2} \cot \tfrac{1}{2} A = \frac{MN^2}{4} \cot \tfrac{1}{2} A$$

$$= \frac{l^2}{4} \cot \tfrac{1}{2} A = \frac{l^2}{4} \cot \frac{180°}{n}$$

Multiplying this by n, we obtain, for the area of the polygon,

$$S = \frac{n\,l^2}{4} \cot \frac{180°}{n} \qquad (2)$$

Example 1.—Find the area, and also the length of the side, of a regular decagon inscribed in a 15-inch circle.

Solution.—In practice, it is usual to refer to a circle by its diameter, and so a 15-in. circle is a circle whose diameter is 15 in. We have, therefore, $r = \dfrac{15}{2} = 7.5$, $n = 10$, $\dfrac{360°}{n} = \dfrac{360°}{10} = 36°$, $\dfrac{180°}{n} = 18°$, and formulas **1** and **2**, Art. **35**, give

$$S = \tfrac{1}{4} \times 10 \times 7.5^2 \sin 36° = 165.32 \text{ sq. in. Ans.}$$
$$l = 2 \times 7.5 \sin 18° = 4.635 \text{ in. Ans.}$$

Example 2.—Each of the sides of an octagonal park is 150 feet; what is the area of the park, in acres?

Solution.—Here $l = 150$ ft., $n = 8$, $\dfrac{180°}{n} = \dfrac{180°}{8} = 22\tfrac{1}{2}° = 22°\ 30'$, and formula **2**, Art. **36**, gives,

$$S = \tfrac{1}{4} \times 8 \times 150^2 \cot 22° \ 30' = 2 \times 22{,}500 \cot 22° \ 30' = (45{,}000 \cot$$

$$22° \ 30') \text{ sq. ft.} = \frac{45{,}000 \cot 22° \ 30'}{43{,}560} \text{ A.} = 2.494 \text{ A.} \quad \textbf{Ans.}$$

EXAMPLES FOR PRACTICE

1. Find the side and area of an equilateral triangle inscribed in a 20-inch circle.

$$\text{Ans.} \begin{cases} l = 17.321 \text{ in.} \\ S = 129.9 \text{ sq. in.} \end{cases}$$

2. What must be the length of the side and the radius of a regular pentagon, that its area may be 46.97 square feet?

$$\text{Ans.} \begin{cases} l = 5.225 \text{ ft.} \\ r = 4.445 \text{ ft.} \end{cases}$$

3. An eight-sided drive is to be built around a circular park 1,500 feet in diameter, the drive to be 15 feet wide, with its outer corners on the circumference of the park. Find: (*a*) the length of each of the sides of the outer boundary of the drive; (*b*) the length of each of the sides of the inner boundary; (*c*) the cost of paving the drive with asphalt, at \$2.25 per square yard; (*d*) the difference between the exact area of the drive and the approximate area found by assuming the polygonal boundaries to coincide with the circumferences of their respective circumscribed circles.

$$\text{Ans.} \begin{cases} (a)\ 574.02 \text{ ft.} \\ (b)\ 561.60 \text{ ft.} \\ (c)\ \$17{,}025 \\ (d)\ 844 \text{ sq. yd.} \end{cases}$$

OTHER POLYGONS

37. The area of any polygon can be determined by dividing the polygon into triangles, and measuring in each triangle whatever parts are necessary for the determination of its area. The parts to be measured depend on special conditions and on the instruments used. The polygon may be divided into triangles either by diagonals or by lines drawn from a convenient interior point to the different vertexes. Illustrations of these methods of division will be given in connection with surveying. When the area is to be determined from a plat, the base and altitude of each triangle are usually the most convenient parts to measure.

AREAS BOUNDED BY IRREGULAR OUTLINES

AREA INCLUDED BETWEEN A STRAIGHT LINE AND AN IRREGULAR CURVE

38. By Selected Ordinates.—Let it be required to determine the area between the curve DC and the straight line AB, Fig. 20. A very convenient method is to draw perpendiculars on AB from the points of the curve at which its direction changes appreciably, and to consider the portion of the curve between two consecutive perpendiculars to be a straight line. The

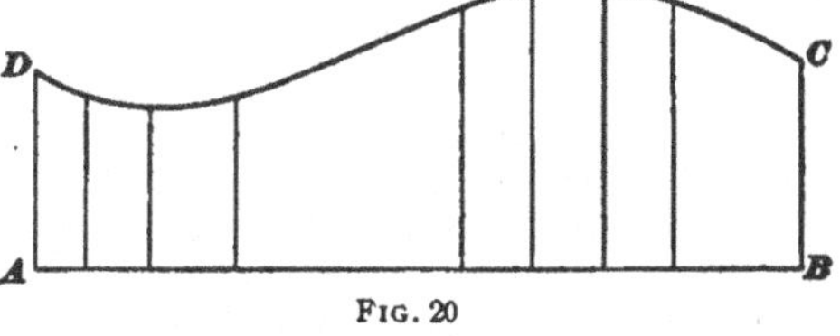

FIG. 20

figure is then treated as if divided into a number of trapezoids, whose areas can be computed by the rules of geometry. The perpendiculars are called **ordinates**. Both the lengths of the ordinates and the distances between every two conseeutive ordinates should be measured. The area of any of the (approximate) trapezoids into which the figure is thus divided is equal to one-half the sum of the two ordinates enclosing it multiplied by the distance between them. It should be understood that both this rule and those given further on relating to the same subject are only approximate. Since the bounding curve is irregular, that is, does not follow any mathematical law, no exact formula can be found for the area.

EXAMPLE.—Referring to Fig. 20, suppose that, beginning at the left of the figure, the successive ordinates measure 15, 13, 12, 13.5, 20, 21.5, 22, 20, and 16 feet, respectively, and that the successive distances between the offsets, from left to right, measure 7.5, 10, 15, 41, 10.5, 11.5, 11.5, and 21 feet, respectively; what is the area of the surface?

SOLUTION.—The area of the figure is approximately equal to the sum of the areas of the trapezoids into which it is divided, and the area of each trapezoid is equal to one-half the sum of its parallel sides multiplied by the perpendicular distance between them. Therefore, the area of the figure is equal to

$$\frac{15+13}{2} \times 7.5 + \frac{13+12}{2} \times 10 + \frac{12+13.5}{2} \times 15 + \frac{13.5+20}{2} \times 41$$

$$+ \frac{20+21.5}{2} \times 10.5 + \frac{21.5+22}{2} \times 11.5 + \frac{22+20}{2} \times 11.5 + \frac{20+16}{2} \times 21$$

$$= 2{,}195.5 \text{ sq. ft.   Ans.}$$

39.   Trapezoidal Rule: Sigma Notation.—In order to facilitate the calculations, the ordinates are often measured at regular intervals along the straight line, as shown in Fig. 21.   The area $ABCD$ included between the straight line and the irregular boundary can then be more easily calculated by what is commonly known as the **trapezoidal rule**.   This

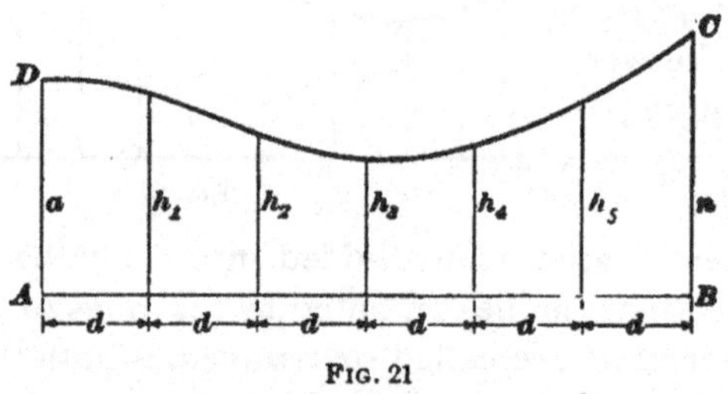

FIG. 21

is merely a rule for calculating the combined area of a series of trapezoids that have the same altitude, the areas being combined for convenience of calculation.   The result given by this rule is closer the smaller the distance between the ordinates. The rule is as follows:

Rule.—*Add together one-half the two end ordinates and all the intermediate ordinates, and multiply the sum by the common distance between the ordinates.*

Let a = first ordinate;

n = last ordinate;

h_1, h_2, h_3 = intermediate ordinates;

d = common distance between ordinates;

S = area of surface.

Then, $S = [\frac{1}{2}(a + n) + h_1 + h_2 + h_3 + \ldots]d$

This expression may be put in a simpler form by using the **sigma notation**, which is as follows: As will be noticed, all the intermediate ordinates are denoted by h, different subscripts being used to indicate different values of h. We may, therefore, write the value of S thus,

$$S = [\tfrac{1}{2}(a + n) + \text{sum of all values of } h]d$$

Instead of the phrase *sum of all values of h*, the expression Σh, read *sigma h*, is used. The symbol Σ is the Greek letter sigma, corresponding to English S, and is very commonly used, as here, to indicate the addition of several quantities of the same character, denoted by a single symbol; hence, the name **sign of summation,** which also is often given to that letter.

By using the sigma notation, the value of S may be written

$$S = \left(\frac{a + n}{2} + \Sigma h\right)d$$

EXAMPLE.—If the ordinates from the straight line $A\,B$ to the curved boundary $D\,C$, Fig. 21, are 19, 18, 14, 12, 13, 17, and 23 links, respectively, and are at equal distances of 50 links, what is the area included between the curved boundary and the straight line?

SOLUTION.—Area $A\,B\,C\,D = \left(\dfrac{19 + 23}{2} + 18 + 14 + 12 + 13 + 17\right)$ $\times 50 = 4{,}750$ sq. li. Ans.

40. Simpson's Rule.—The foregoing rule assumes that all the small figures into which the area is divided are perfect trapezoids, which assumption always involves more or less error, since the irregular boundary is in nearly all cases an irregular curve. When the offsets are taken at

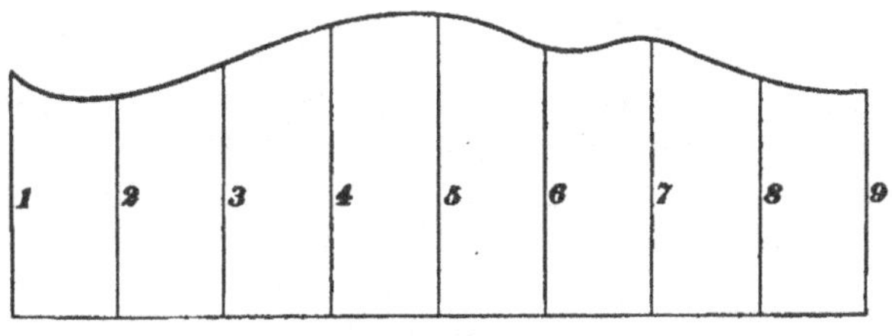

FIG. 22

regular intervals, the following rule, known as **Simpson's one-third rule,** gives a closer approximation. In applying this rule, the base line must be divided into an even number of equal parts; the ordinates measured at the points of division are numbered consecutively, as shown in Fig. 22.

Rule.—*Divide the base line into an even number of equal parts, and at the points of division erect ordinates terminating in the curve. Number the ordinates 1, 2, 3, etc., from left to right, including those at the ends of the base. Add together the end ordinates, four times the sum of all intermediate even-numbered ordinates, and twice the sum of all intermediate odd-numbered ordinates; multiply the total sum by one-third the common distance between adjacent ordinates.*

This rule has been used extensively; it can be expressed by a formula as follows:

Let h_2 = any intermediate even-numbered ordinate;

h_3 = any intermediate odd-numbered ordinate;

and let all other quantities be represented by the same letters as in the preceding article. Then,

$$S = (a + n + 4 \Sigma h_2 + 2 \Sigma h_3) \frac{d}{3}$$

The notation will be readily understood by reference to Fig. 23. The expression $4 \Sigma h_2$ means four times the sum of all the ordinates h_2, or, in other words, four times the sum of all the even-numbered ordinates.

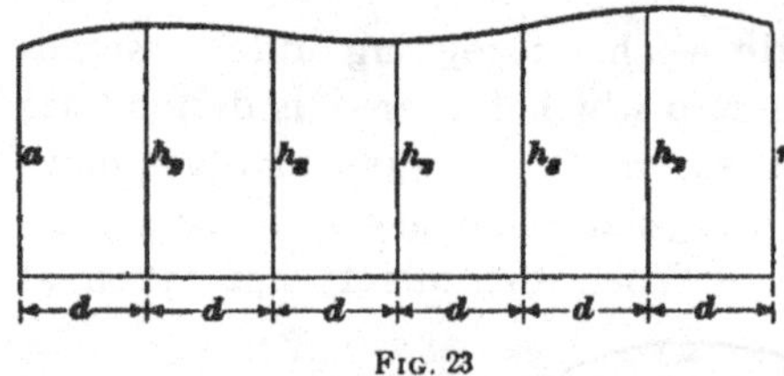

FIG. 23

EXAMPLE.—What is the area $A\,B\,C\,D$, Fig. 21, by Simpson's rule, using the same values as in the example in Art. **39**?

SOLUTION.— $S = [19 + 23 + 4(18 + 12 + 17) + 2(14 + 13)] \times \frac{40}{3}$
$= 4{,}733$ sq. li. Ans.

EXAMPLES FOR PRACTICE

1. A figure included between a straight base line, a curve, and two perpendiculars to the base at the ends has nine ordinates, including the two end perpendiculars, whose lengths are 43, 48, 39, 50, 41, 32, 37, 31, and 22 feet, respectively; the common distance between the ordinates is 60 feet. Find the area: (*a*) by the trapezoidal rule; (*b*) by Simpson's rule.

Ans. $\begin{cases} (a) & 18{,}630 \text{ sq. ft.} \\ (b) & 18{,}860 \text{ sq. ft.} \end{cases}$

2. In order to determine the area included between an irregular boundary, a straight base line, and two perpendiculars to the base at the ends, eight ordinates, including the two end perpendiculars, are measured from the straight line to the boundary. The ordinates are found to measure 16, 18, 12, 13, 15, 17, 19, and 20.5 feet, and the successive distances between them are found to measure 7.8, 10, 15, 20, 12, 40, and 5 feet, respectively. What is the area of the surface?

Ans. 1,760.9 sq. ft.

3. A surface lying between a straight base line and a curve is limited by two perpendiculars to the base line at the ends; the base line is divided into eight parts 50 feet each, and at the points of division ordinates are measured. The lengths of the successive ordinates, including the two end perpendiculars, are 10, 25, 38, 49, 58, 65, 70, 73, and 74 feet, respectively. Find the area of the surface: (a) by the trapezoidal rule; (b) by Simpson's rule.   Ans. $\begin{cases} (a) \ 21,000 \text{ sq. ft.} \\ (b) \ 21,067 \text{ sq. ft.} \end{cases}$

AREA BOUNDED BY AN IRREGULAR CURVE

41. By Ordinates.—Suppose that it is required to find the area enclosed by the heavy irregular curve shown in Fig. 24. A broken line *AEFMGHIA* is drawn around

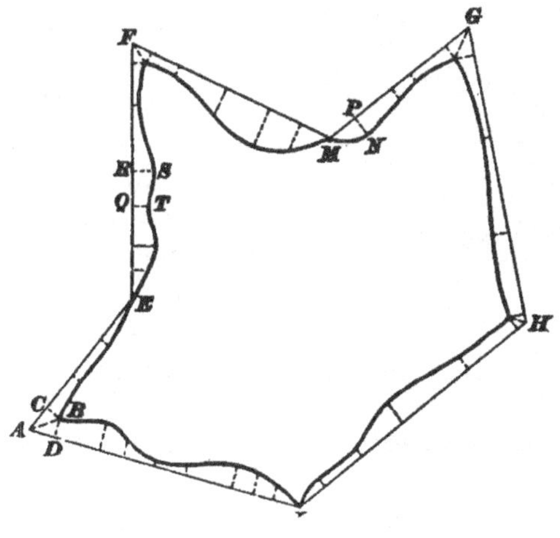

FIG. 24

the curved boundary line, and as close to it as convenient. Ordinates to the straight lines thus drawn are measured from the points where the direction of the curved boundary changes materially, as shown. The area of the polygon

AEFMGHIA is çalculated by one of the methods explained in preceding articles, and from it is subtracted the sum of the areas included between the curved boundary and the broken line, calculated as in Art. **39.**

At such corners as *A*, the triangles *ABC* and *ABD* are computed from the measured bases *AC* and *AD* and the altitudes *BC* and *BD*. All the quadrilaterals, as *QRST*, are treated as trapezoids; and such three-sided figures as *MPN*, as triangles. The process is so simple that it does not require any further explanation.

42. By the Planimeter.—The most convenient way to find the area of a plane surface having an irregular boundary is by the **planimeter**. There are several forms of planimeters; the one most commonly used is the **polar planimeter** (see Fig. 25). As will be seen from the illustration, this instrument has two arms *ij* and *gh* connected by a hinge joint. The point *e* at the end of the bar *ij* is called the **anchor**

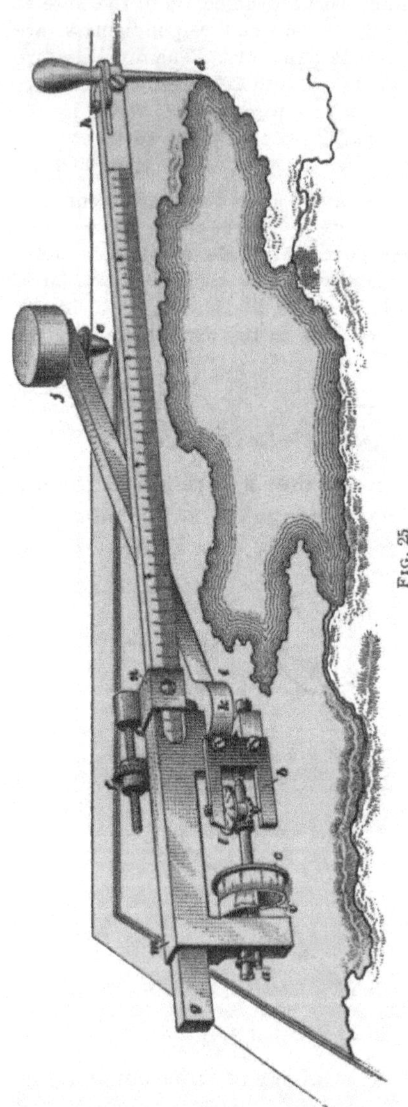

point; it remains stationary while the point d, called the pointer or tracer, at the end of the bar $g\,h$ is moved over the outline of the figure whose area is to be determined. The movement of the pointer d causes the wheel c on the opposite end of the bar to roll on the paper; this wheel is called the **measuring wheel** or **counter wheel**. The graduated bar $g\,h$ can be adjusted by sliding it in or out through the socket m in the top of the frame. This bar is clamped by means of a clamp screw, a part of which is shown back of the small movable socket n, and is set at the exact length required by means of the thumbscrew f. The bar $i\,j$ is of fixed length; it is pivoted at k, the junction of the two bars. The measuring wheel c is mounted on the main axis $a\,b$, which is parallel with the bar $g\,h$. The complete revolutions of the wheel c are read on **the** disk l, and the fractional parts of revolutions are read on the wheel c and the vernier v, the tenths and hundredths being read on the wheel itself, and the thousandths on the vernier.

To use the planimeter, the anchor point e is fixed on the paper or drawing board, preferably outside the figure to be measured, the pointer d is placed on some point in the periphery of the figure, and a reading of the wheel c is taken. The point d is then moved carefully around the periphery of the figure, in a clockwise direction, or from left to right, to the point of beginning. A second reading of the wheel c is then taken, and the difference between the two readings is the number of revolutions of the wheel. If the wheel is set to read zero, the number of revolutions is given directly by the second reading.

If the anchor point is outside the area to be measured, the distance traversed by the wheel, or the product of the number of revolutions by the circumference of the wheel, in inches, multiplied by the length of the bar $n\,h$, in inches, is the area, in square inches, bounded by the path of the pointer d.

If the anchor point is inside the area, the product just referred to must be added to the area of the **zero circle,** whose radius is equal to $\sqrt{p^2 + q^2 + 2\,pr}$, p being the length of the arm $n\,h$; r, the distance from the center of the wheel c to the center of the joint k; and q, the length of the bar $k\,j$. The bar $g\,h$ is generally set at such a length that ten times the number of revolutions of the wheel c is the area measured. This area is the actual area of the figure measured, and the area *represented* by the figure is determined from the scale of the plat. The area given by the planimeter, in square inches, must be multiplied by the square of the scale of the plat, in order to get the area sought. Thus, if the plat has been drawn to a scale of 50 feet to an inch, each square inch of the plat is equivalent to $50 \times 50 = 2,500$ square feet of area.

Suppose that the area bounded by the irregular line in Fig. 25, as measured by the planimeter, is 2.535 square inches, and that the scale of the plat is 100 feet to an inch; then the area represented by a square inch of the plat is $100 \times 100 = 10,000$ square feet, and the area represented by the closed figure is $10,000 \times 2.535 = 25,350$ square feet.

Full directions for using the planimeter are usually furnished by the maker.

APPENDIX: DERIVATION OF FORMULAS

I—FORMULAS 1 TO 4 OF ART. 16

Let $R\,O\,Q$, Fig. 26, be any angle A, and $Q\,O\,S$ any angle B. Then, $A + B = R\,O\,S$. From any point P on $O\,S$, draw $P\,N$ and $P\,M$, perpendicular, respectively, to $O\,R$ and $O\,Q$. Draw $M\,K$ parallel to $O\,R$ and therefore perpendicular to $P\,N$; also, $M\,L$ perpendicular to $O\,R$. The angles $M\,P\,K$ and $R\,O\,Q$, having their sides perpendicular each to each, are equal. Now,

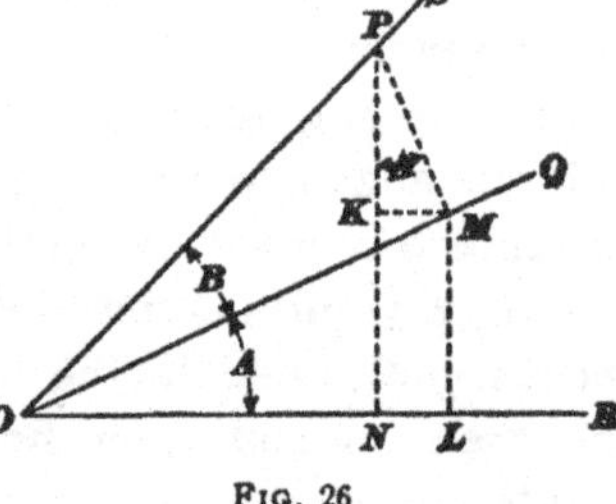

Fig. 26

$$\sin(A+B) = \frac{NP}{OP} = \frac{NK + KP}{OP} = \frac{ML}{OP} + \frac{KP}{OP} = \frac{OM \sin A}{OP} + \frac{PM \cos A}{OP}$$

(triangles $M L O$ and $P M K$) $= \sin A \dfrac{O M}{O P} + \cos A \dfrac{P M}{O P} = \sin A \cos B$
$+ \cos A \sin B$ (triangle $O P M$)

This is formula **1.**

Also,
$$\cos (A - B) = \sin [90° - (A - B)] = \sin [(90° - A) + B]$$
or, by formula **1,**
$$\cos (A - B) = \sin (90° - A) \cos B + \cos (90° - A) \sin B$$
$$= \cos A \cos B + \sin A \sin B$$
which is formula **4.**

Formula **3** follows from this; for
$$\sin (A - B) = \cos [90° - (A - B)] = \cos [(90° + B) - A]$$
$$= \cos (90° + B) \cos A + \sin (90° + B) \sin A$$
or, because $\cos (90° + B) = - \sin B$, and $\sin (90° + B) = \cos B$ (Art. 14),
$$\sin (A - B) = - \sin B \cos A + \cos B \sin A = \sin A \cos B - \cos A \sin B.$$
Finally, applying this formula,
$$\cos (A + B) = \sin [90° - (A + B)] = \sin [(90° - A) - B]$$
$$= \sin (90° - A) \cos B - \cos (90° - A) \sin B$$
$$= \cos A \cos B - \sin A \sin B$$
which is formula **2.**

II—FORMULAS OF ART. 19

Referring to Fig. 6 (a) and (b), Art. **18,**
$$a^2 = p^2 + B D^2 \qquad (1)$$
In (a), $B D = c - A D$, whence $\overline{B D}^2 = c^2 - 2 c \times A D + \overline{A D}^2$.

In (b), $B D = A D - c$, whence $\overline{B D}^2 = \overline{A D}^2 - 2 c \times A D + c^2$.
Substituting this value of $B D$ in equation (1),
$$a^2 = p^2 + \overline{A D}^2 + c^2 - 2 c \times A D \qquad (2)$$
But $p^2 + \overline{A D}^2 = b^2$, and $A D = b \cos A$; therefore,
$$a^2 = b^2 + c^2 - 2 b c \cos A$$
When the angle opposite the side is obtuse, as B in Fig. 6 (b), the
same reasoning leads to the relation,
$$b^2 = a^2 + c^2 + 2 a c \times \cos C B D$$
the second member of which becomes $a^2 + c^2 - 2 a c \cos B$, when
$\cos C B D$ is replaced by its equal $- \cos B$ (Art. **13**).

III—FORMULAS OF ART. 20

Let $A\,B\,C$, Fig. 27, be any triangle. As usual, the angles of the tri-

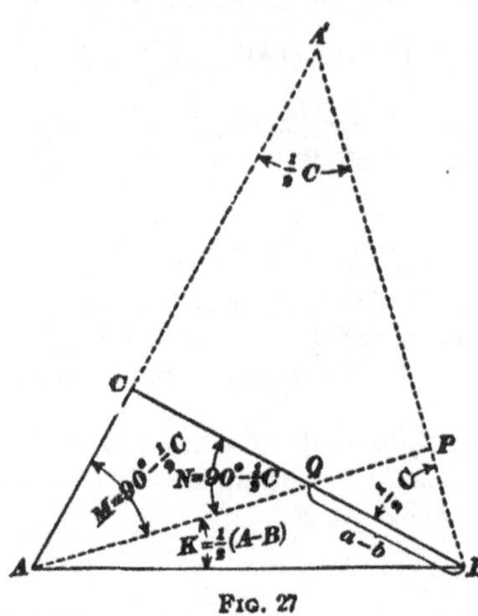

FIG. 27

angle will be denoted by A, B, C, and the opposite sides by a, b, c, respectively; that is, angle $C\,A\,B = A$, $B\,C = a$, etc. Produce $A\,C$ to A', making $C\,A' = B\,C = a$. Draw $B\,A'$, and $A\,P$ perpendicular to it, meeting $B\,C$ at Q.

Since $B\,C = C\,A'$, the triangle $B\,C\,A'$ is isosceles, and, therefore, the angles $C\,A'\,B$ and $C\,B\,A'$ are equal. The sum of these two angles, or twice either of them, is equal to the external angle $B\,C\,A$, or C, and therefore each of these two angles is equal to $\frac{1}{2} C$. In the right triangle $A\,P\,A'$, the angle M, being the complement of A', is equal to $90° - \frac{1}{2} C$.

We have also,

$$K = A - M = A - (90° - \tfrac{1}{2}C),$$

or, since $C = 180° - (A + B) = 180° - A - B$,

$$K = A - [90° - \tfrac{1}{2}(180° - A - B)] = \tfrac{1}{2}(A - B)$$

The angle N being external to the triangle $A\,Q\,B$, we have

$$N = K + B = \tfrac{1}{2}(A - B) + B = \tfrac{1}{2}(A + B)$$
$$= \tfrac{1}{2}(180° - C) = 90° - \tfrac{1}{2}C = M$$

Therefore, the triangle $A\,Q\,C$ is isosceles, and $Q\,C = A\,C = b$; and, consequently, $B\,Q = a - b$.

The right triangle $A\,B\,P$ gives,

$$\tan \tfrac{1}{2}(A - B) = \frac{B\,P}{A\,P}$$

or, writing the values of $B\,P$ and $A\,P$ from the triangles $B\,Q\,P$ and $A\,P\,A'$,

$$\tan \tfrac{1}{2}(A - B) = \frac{B\,Q \,\cos \tfrac{1}{2} C}{A\,A' \,\sin \tfrac{1}{2} C} = \frac{B\,Q}{A\,A'} \cot \tfrac{1}{2} C$$

that is,

$$\tan \tfrac{1}{2}(A - B) = \frac{a - b}{a + b} \cot \tfrac{1}{2} C \qquad (1)$$

Now, $\tfrac{1}{2} C = \tfrac{1}{2}[180° - (A + B)] = 90° - \tfrac{1}{2}(A + B)$, and therefore, $\cot \tfrac{1}{2} C = \tan \tfrac{1}{2}(A + B)$. By substituting this value in equation (1), and transforming, the formula in Art. 20 is obtained.

IV—FORMULA 2 OF ART. 21

This formula is derived from Fig. 27 as follows: In the triangle BPQ,

$$BP = BQ \cos \tfrac{1}{2} C = (a - b) \cos \tfrac{1}{2} C \qquad (1)$$

and, in the triangle ABP,

$$c\,(= AB) = \frac{BP}{\sin \tfrac{1}{2}(A - B)}$$

which becomes formula 2 when BP is replaced by its value (1).

V—FORMULAS 2 TO 4 OF ART. 23

We have (formula 8, Art. 17),

$$2 \cos^2 \tfrac{1}{2} A = 1 + \cos A$$

or, substituting the value of $\cos A$ from formula 1, Art. 23,

$$2 \cos^2 \tfrac{1}{2} A = 1 + \frac{b^2 + c^2 - a^2}{2\,bc} = \frac{2\,bc + b^2 + c^2 - a^2}{2\,bc} = \frac{(b + c)^2 - a^2}{2\,bc}$$

or, remembering that the difference between the squares of two numbers is equal to their sum multiplied by their difference,

$$2 \cos^2 \tfrac{1}{2} A = \frac{(b + c + a)(b + c - a)}{2\,bc} \qquad (1)$$

Now, since $a + b + c = 2s$, we have, subtracting $2a$ from both members, $b + c - a = 2s - 2a = 2(s - a)$. Likewise, $a + b - c = 2(s - c)$, and $a + c - b = 2(s - b)$. Substituting these values in equation (1),

$$2 \cos^2 \tfrac{1}{2} A = \frac{2s \times 2(s - a)}{2\,bc} = \frac{2s(s - a)}{bc}$$

whence, $$\cos \tfrac{1}{2} A = \sqrt{\frac{s(s - a)}{bc}} \qquad (2)$$

which is formula 3, Art. 23.

Likewise (formula 7, Art. 17),

$$2 \sin^2 \tfrac{1}{2} A = 1 - \cos A = 1 - \frac{b^2 + c^2 - a^2}{2\,bc}$$

$$= \frac{2\,bc - b^2 - c^2 + a^2}{2\,bc} = \frac{a^2 - (b^2 - 2\,bc + c^2)}{2\,bc} = \frac{a^2 - (b - c)^2}{2\,bc}$$

$$= \frac{(a + b - c)(a - b + c)}{2\,bc} = \frac{2(s - c) \times 2(s - b)}{2\,bc} = \frac{2(s - b)(s - c)}{bc}$$

whence, $$\sin \tfrac{1}{2} A = \sqrt{\frac{(s - b)(s - c)}{bc}} \qquad (3)$$

which is formula 4, Art. 23.

Formula 2 is obtained by dividing equation (3) by equation (2).

VI—FÓRMULA OF ART. 30

Formulas 3 and 4 of Art. 23 are:

$$\sin \tfrac{1}{2} A = \sqrt{\frac{(s-b)(s-c)}{bc}} \qquad (1)$$

$$\cos \tfrac{1}{2} A = \sqrt{\frac{s(s-a)}{bc}} \qquad (2)$$

Also (formula 5, Art. 17),

$$\sin A = 2 \sin \tfrac{1}{2} A \cos \tfrac{1}{2} A \qquad (3)$$

Substituting in equation (3) the values of $\sin \tfrac{1}{2} A$ and $\cos \tfrac{1}{2} A$ from equations (1) and (2),

$$\sin A = 2 \sqrt{\frac{(s-b)(s-c)}{bc}} \sqrt{\frac{s(s-a)}{bc}} = 2 \sqrt{\frac{s(s-a)(s-b)(s-c)}{b^2 c^2}}$$

$$= 2 \frac{\sqrt{s(s-a)(s-b)(s-c)}}{bc}$$

Substituting this value in formula of Art. 27,

$$S = \sqrt{s(s-a)(s-b)(s-c)}$$

VII—FORMULA 3 OF ART. 33

We have, since $\cot = \dfrac{\cos}{\sin}$,

$$\frac{1}{\cot A + \cot B} = \frac{1}{\dfrac{\cos A}{\sin A} + \dfrac{\cos B}{\sin B}} = \frac{\sin A \sin B}{\sin B \cos A + \cos B \sin A}$$

$$= \frac{\sin A \sin B}{\sin (A + B)}$$

By substituting this value in formula 1, we obtain

$$S = \frac{(b_1 - b_2)(b_1 + b_2) \sin A \sin B}{2 \sin (A + B)}$$

VIII—FORMULA OF ART. 34

Let the area of the triangle $A D B'$, Fig. 18, be denoted by T, and that of the parallelogram $B C D B'$ by P. Then,

$$S = P + T \qquad (1)$$

Now, $\qquad\qquad P = b_2 h, \ T = \tfrac{1}{2} d h$

Dividing the first of these equations by the second,

$$\frac{P}{T} = \frac{b_2}{\tfrac{1}{2} d} = \frac{2 b_2}{d} = \frac{2 b_2}{b_1 - b_2}$$

whence, $\qquad\qquad P = \dfrac{2 b_2}{b_1 - b_2} T$

Substituting this value of P in equation (1),

$$S = \frac{2b_2}{b_1 - b_2}\,T + T = \left(\frac{2b_2}{b_1 - b_2} + 1\right)T = \frac{b_1 + b_2}{b_1 - b_2}\,T = \frac{b_1 + b_2}{d}\,T \tag{2}$$

Let $\tfrac{1}{2}(a + c + d) = s$. Then (formula of Art. 30),

$$T = \sqrt{s(s - a)(s - c)(s - d)}$$

and, substituting this value in equation (2),

$$S = \frac{b_1 + b_2}{d}\,\sqrt{s(s - a)(s - c)(s - d)}$$

TABLE OF TRIGONOMETRIC FORMULAS

The principal formulas occurring in the text, and others that can be readily derived from these, are tabulated in the following pages for convenient reference. As these formulas, which include those for the solution of triangles, are here systematically classified and arranged, the student will find this table useful in the solution of all kinds of problems requiring the application of trigonometry. He is advised to refer to it often, so as to become familiar with its contents and use.

FORMULAS DEFINING THE TRIGONOMETRIC FUNCTIONS

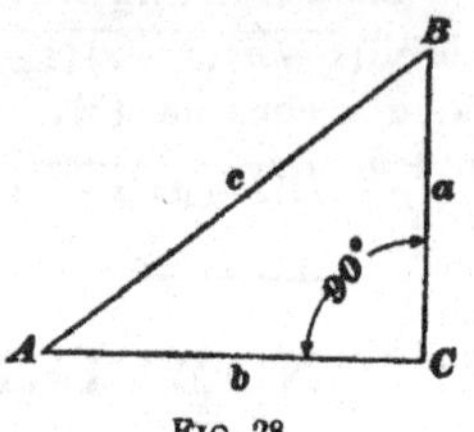

FIG. 28

1. $\sin A = \dfrac{a}{c}$

2. $\tan A = \dfrac{a}{b}$

3. $\cos A = \sin (90° - A) = \dfrac{b}{c}$

4. $\cot A = \tan (90° - A) = \dfrac{b}{a}$

5. $\sec A = \dfrac{c}{b}$

6. $\csc A = \sec (90° - A) = \dfrac{c}{a}$

7. $\text{vers } A = 1 - \cos A = 1 - \dfrac{b}{c}$

8. $\text{covers } A = \text{vers } (90° - A) = 1 - \sin A = 1 - \dfrac{a}{c}$

FUNCTIONS OF 0° AND 90°

9. $\sin 0° = 0$ 15. $\sin 90° = 1$
10. $\tan 0° = 0$ 16. $\tan 90° = \infty$
11. $\cos 0° = 1$ 17. $\cos 90° = 0$
12. $\cot 0° = \infty$ 18. $\cot 90° = 0$
13. $\sec 0° = 1$ 19. $\sec 90° = \infty$
14. $\csc 0° = \infty$ 20. $\csc 90° = 1$

FUNCTIONS OF NEGATIVE ANGLES

21. $\sin (- A) = - \sin A$ 24. $\cot (- A) = - \cot A$
22. $\tan (- A) = - \tan A$ 25. $\sec (- A) = \sec A$
23. $\cos (- A) = \cos A$ 26. $\csc (- A) = - \csc A$

FUNCTIONS OF 90° + A

27. $\sin (90° + A) = \cos A$ 30. $\cot (90° + A) = -\tan A$
28. $\tan (90° + A) = -\cot A$ 31. $\sec (90° + A) = -\csc A$
29. $\cos (90° + A) = -\sin A$ 32. $\csc (90° + A) = \sec A$

FUNCTIONS OF 180° − A AND OF 180° + A

33. $\sin (180° - A) = \sin A$
34. $\tan (180° - A) = -\tan A$
35. $\cos (180° - A) = -\cos A$
36. $\cot (180° - A) = -\cot A$
37. $\sec (180° - A) = -\sec A$
38. $\csc (180° - A) = \csc A$
39. $\sin (180° + A) = -\sin A$
40. $\tan (180° + A) = \tan A$
41. $\cos (180° + A) = -\cos A$
42. $\cot (180° + A) = \cot A$
43. $\sec (180° + A) = -\sec A$
44. $\csc (180° + A) = -\csc A$

FUNCTIONS OF 360° − A AND OF 360° + A

45. $\sin (360° - A) = -\sin A$ 51. $\sin (360° + A) = \sin A$
46. $\tan (360° - A) = -\tan A$ 52. $\tan (360° + A) = \tan A$
47. $\cos (360° - A) = \cos A$ 53. $\cos (360° + A) = \cos A$
48. $\cot (360° - A) = -\cot A$ 54. $\cot (360° + A) = \cot A$
49. $\sec (360° - A) = \sec A$ 55. $\sec (360° + A) = \sec A$
50. $\csc (360° - A) = -\csc A$ 56. $\csc (360° + A) = \csc A$

FUNCTIONS OF $(A + B)$ AND OF $(A - B)$

57. $\sin (A + B) = \sin A \cos B + \cos A \sin B$
58. $\sin (A - B) = \sin A \cos B - \cos A \sin B$
59. $\cos (A + B) = \cos A \cos B - \sin A \sin B$
60. $\cos (A - B) = \cos A \cos B + \sin A \sin B$
61. $\tan (A + B) = \dfrac{\tan A + \tan B}{1 - \tan A \tan B}$
62. $\tan (A - B) = \dfrac{\tan A - \tan B}{1 + \tan A \tan B}$

FUNCTIONS OF $2A$ AND OF $\frac{1}{2}A$

63. $\sin 2A = 2 \sin A \cos A$

64. $\cos 2A = \cos^2 A - \sin^2 A$

65. $\cos 2A = 2 \cos^2 A - 1$

66. $\cos 2A = 1 - 2 \sin^2 A$

67. $\tan 2A = \dfrac{2 \tan A}{1 - \tan^2 A}$

68. $\sin \frac{1}{2} A = \sqrt{\dfrac{1 - \cos A}{2}}$

69. $\cos \frac{1}{2} A = \sqrt{\dfrac{1 + \cos A}{2}}$

70. $\tan \frac{1}{2} A = \sqrt{\dfrac{1 - \cos A}{1 + \cos A}}$

71. $\tan \frac{1}{2} A = \dfrac{1 - \cos A}{\sin A}$

SUMS AND DIFFERENCES OF FUNCTIONS

72. $\sin A + \sin B = 2 \sin \frac{1}{2}(A + B) \cos \frac{1}{2}(A - B)$

73. $\sin A - \sin B = 2 \sin \frac{1}{2}(A - B) \cos \frac{1}{2}(A + B)$

74. $\cos A + \cos B = 2 \cos \frac{1}{2}(A + B) \cos \frac{1}{2}(A - B)$

75. $\cos A - \cos B = 2 \sin \frac{1}{2}(A + B) \sin \frac{1}{2}(B - A)$

76. $\tan A + \tan B = \dfrac{\sin (A + B)}{\cos A \cos B}$

77. $\tan A - \tan B = \dfrac{\sin (A - B)}{\cos A \cos B}$

78. $\sin^2 A - \sin^2 B = \sin (A + B) \sin (A - B)$

79. $\cos^2 A - \cos^2 B = \sin (A + B) \sin (B - A)$

80. $\cos^2 A - \sin^2 B = \cos (A + B) \cos (A - B)$

RELATIONS AMONG THE FUNCTIONS OF AN ANGLE

$\sin A =$	$\tan A =$	$\cos A =$	$\cot A =$	$\sec A =$	$\csc A =$
81. $\dfrac{\tan A}{\sqrt{1+\tan^2 A}}$	86. $\dfrac{\sin A}{\sqrt{1-\sin^2 A}}$	91. $\sqrt{1-\sin^2 A}$	96. $\dfrac{\sqrt{1-\sin^2 A}}{\sin A}$	101. $\dfrac{1}{\sqrt{1-\sin^2 A}}$	106. $\dfrac{1}{\sin A}$
82. $\sqrt{1-\cos^2 A}$	87. $\dfrac{\sqrt{1-\cos^2 A}}{\cos A}$	92. $\dfrac{1}{\sqrt{1+\tan^2 A}}$	97. $\dfrac{1}{\tan A}$	102. $\sqrt{1+\tan^2 A}$	107. $\dfrac{\sqrt{1+\tan^2 A}}{\tan A}$
83. $\dfrac{1}{\sqrt{1+\cot^2 A}}$	88. $\dfrac{1}{\cot A}$	93. $\dfrac{\cot A}{\sqrt{1+\cot^2 A}}$	98. $\dfrac{\cos A}{\sqrt{1-\cos^2 A}}$	103. $\dfrac{1}{\cos A}$	108. $\dfrac{1}{\sqrt{1-\cos^2 A}}$
84. $\dfrac{\sqrt{\sec^2 A-1}}{\sec A}$	89. $\sqrt{\sec^2 A-1}$	94. $\dfrac{1}{\sec A}$	99. $\dfrac{1}{\sqrt{\sec^2 A-1}}$	104. $\dfrac{\sqrt{1+\cot^2 A}}{\cot A}$	109. $\sqrt{1+\cot^2 A}$
85. $\dfrac{1}{\csc A}$	90. $\dfrac{1}{\sqrt{\csc^2 A-1}}$	9. $\dfrac{\sqrt{\csc^2 A-1}}{\csc A}$	100. $\sqrt{\csc^2 A-1}$	105. $\dfrac{\csc A}{\sqrt{\csc^2 A-1}}$	110. $\dfrac{\sec A}{\sqrt{\sec^2 A-1}}$

FORMULAS FOR THE SOLUTION OF RIGHT TRIANGLES

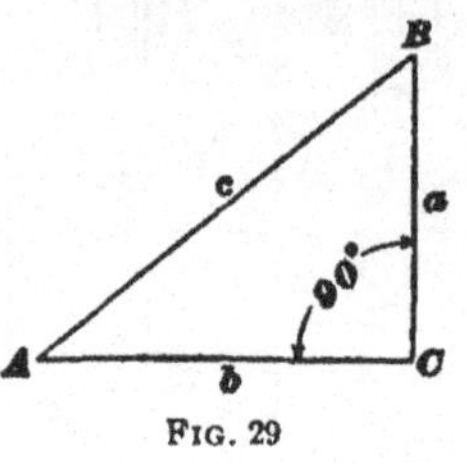

Fig. 29

Given	Required	Formula
a, A	B, b, c	111. $B = 90° - A$ 112. $b = a \cot A$ 113. $c = \dfrac{a}{\sin A} = a \csc A$
a, B	A, b, c	114. $A = 90° - B$ 115. $b = a \tan B$ 116. $c = \dfrac{a}{\cos B} = a \sec B$
c, A	B, a, b	117. $B = 90° - A$ 118. $a = c \sin A$ 119. $b = c \cos A$
a, b	A, B, c	120. $\tan A = \dfrac{a}{b}$ 121. $\tan B = \dfrac{b}{a}$, or $B = 90° - A$ 122. $c = \sqrt{a^2 + b^2}$ 123. $c = \dfrac{a}{\sin A} = a \csc A$
a, c	A, B, b	124. $\sin A = \dfrac{a}{c}$ 125. $\cos B = \dfrac{a}{c}$, or $B = 90° - A$ 126. $b = \sqrt{c^2 - a^2} = \sqrt{(c + a)(c - a)}$ 127. $b = a \cot A$

FORMULAS FOR THE SOLUTION OF OBLIQUE TRIANGLES

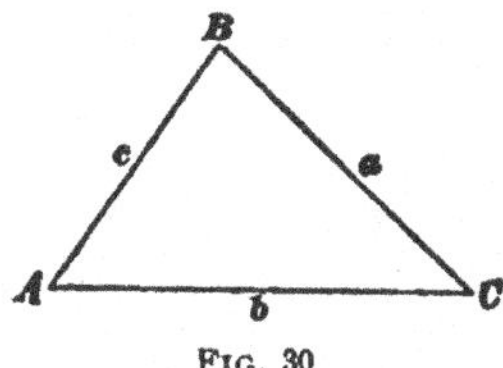

F$_{\text{IG}}$. 30

Given	Required	Formulas
a, b, C	A, B, c	128. $\begin{cases} \tan \tfrac{1}{2}(A - B) = \dfrac{a - b}{a + b}\cot \tfrac{1}{2}C \\ A = (90° - \tfrac{1}{2}C) + \tfrac{1}{2}(A - B) \\ B = (90° - \tfrac{1}{2}C) - \tfrac{1}{2}(A - B) \end{cases}$ 129. $c = \dfrac{(a - b)\cos \tfrac{1}{2}C}{\sin \tfrac{1}{2}(A - B)}$ 130. $c = \sqrt{a^2 + b^2 - 2\,a\,b\,\cos C}$
c, A, B	C, a, b	131. $C = 180° - (A + B)$ 132. $a = \dfrac{c}{\sin C}\sin A$ 133. $b = \dfrac{c}{\sin C}\sin B$
a, b, A	B, C, c	134. $\sin B = \dfrac{b}{a}\sin A$ 135. $C = 180° - A - B$ 136. $c = \dfrac{a}{\sin A}\sin C$
a, b, c $\tfrac{1}{2}(a + b + c) = s$	A	137. $\tan \tfrac{1}{2}A = \sqrt{\dfrac{(s - b)(s - c)}{s(s - a)}}$ 138. $\cos \tfrac{1}{2}A = \sqrt{\dfrac{s(s - a)}{b\,c}}$ 139. $\cos A = \dfrac{b^2 + c^2 - a^2}{2\,b\,c}$

NATURAL TRIGONOMETRIC FUNCTIONS

′	0°		1°		2°		3°		4°		′
	Sine	Cosine	Sine	Cosine	Sine	Cosine	Sine	Cosine	Sine	Cosine	
0	.00000	1.	.01745	.99985	.03490	.99939	.05234	.99863	.06976	.99756	60
1	.00029	1.	.01774	.99984	.03519	.99938	.05263	.99861	.07005	.99754	59
2	.00058	1.	.01803	.99984	.03548	.99937	.05292	.99860	.07034	.99752	58
3	.00087	1.	.01832	.99983	.03577	.99936	.05321	.99858	.07063	.99750	57
4	.00116	1.	.01862	.99983	.03606	.99935	.05350	.99857	.07092	.99748	56
5	.00145	1.	.01891	.99982	.03635	.99934	.05379	.99855	.07121	.99746	55
6	.00175	1.	.01920	.99982	.03664	.99933	.05408	.99854	.07150	.99744	54
7	.00204	1.	.01949	.99981	.03693	.99932	.05437	.99852	.07179	.99742	53
8	.00233	1.	.01978	.99980	.03723	.99931	.05466	.99851	.07208	.99740	52
9	.00262	1.	.02007	.99980	.03752	.99930	.05495	.99849	.07237	.99738	51
10	.00291	1.	.02036	.99979	.03781	.99929	.05524	.99847	.07266	.99736	50
11	.00320	.99999	.02065	.99979	.03810	.99927	.05553	.99846	.07295	.99734	49
12	.00349	.99999	.02094	.99978	.03839	.99926	.05582	.99844	.07324	.99731	48
13	.00378	.99999	.02123	.99977	.03868	.99925	.05611	.99842	.07353	.99729	47
14	.00407	.99999	.02152	.99977	.03897	.99924	.05640	.99841	.07382	.99727	46
15	.00436	.99999	.02181	.99976	.03926	.99923	.05669	.99839	.07411	.99725	45
16	.00465	.99999	.02211	.99976	.03955	.99922	.05698	.99838	.07440	.99723	44
17	.00495	.99999	.02240	.99975	.03984	.99921	.05727	.99836	.07469	.99721	43
18	.00524	.99999	.02269	.99974	.04013	.99919	.05756	.99834	.07498	.99719	42
19	.00553	.99998	.02298	.99974	.04042	.99918	.05785	.99833	.07527	.99716	41
20	.00582	.99998	.02327	.99973	.04071	.99917	.05814	.99831	.07556	.99714	40
21	.00611	.99998	.02356	.99972	.04100	.99916	.05844	.99829	.07585	.99712	39
22	.00640	.99998	.02385	.99972	.04129	.99915	.05873	.99827	.07614	.99710	38
23	.00669	.99998	.02414	.99971	.04159	.99913	.05902	.99826	.07643	.99708	37
24	.00698	.99998	.02443	.99970	.04188	.99912	.05931	.99824	.07672	.99705	36
25	.00727	.99997	.02472	.99969	.04217	.99911	.05960	.99822	.07701	.99703	35
26	.00756	.99997	.02501	.99969	.04246	.99910	.05989	.99821	.07730	.99701	34
27	.00785	.99997	.02530	.99968	.04275	.99909	.06018	.99819	.07759	.99699	33
28	.00814	.99997	.02560	.99967	.04304	.99907	.06047	.99817	.07788	.99696	32
29	.00844	.99996	.02589	.99966	.04333	.99906	.06076	.99815	.07817	.99694	31
30	.00873	.99996	.02618	.99966	.04362	.99905	.06105	.99813	.07846	.99692	30
31	.00902	.99996	.02647	.99965	.04391	.99904	.06134	.99812	.07875	.99689	29
32	.00931	.99996	.02676	.99964	.04420	.99902	.06163	.99810	.07904	.99687	28
33	.00960	.99995	.02705	.99963	.04449	.99901	.06192	.99808	.07933	.99685	27
34	.00989	.99995	.02734	.99963	.04478	.99900	.06221	.99806	.07962	.99683	26
35	.01018	.99995	.02763	.99962	.04507	.99898	.06250	.99804	.07991	.99680	25
36	.01047	.99995	.02792	.99961	.04536	.99897	.06279	.99803	.08020	.99678	24
37	.01076	.99994	.02821	.99960	.04565	.99896	.06308	.99801	.08049	.99676	23
38	.01105	.99994	.02850	.99959	.04594	.99894	.06337	.99799	.08078	.99673	22
39	.01134	.99994	.02879	.99959	.04623	.99893	.06366	.99797	.08107	.99671	21
40	.01164	.99993	.02908	.99958	.04653	.99892	.06395	.99795	.08136	.99668	20
41	.01193	.99993	.02938	.99957	.04682	.99890	.06424	.99793	.08165	.99666	19
42	.01222	.99993	.02967	.99956	.04711	.99889	.06453	.99792	.08194	.99664	18
43	.01251	.99992	.02996	.99955	.04740	.99888	.06482	.99790	.08223	.99661	17
44	.01280	.99992	.03025	.99954	.04769	.99886	.06511	.99788	.08252	.99659	16
45	.01309	.99991	.03054	.99953	.04798	.99885	.06540	.99786	.08281	.99657	15
46	.01338	.99991	.03083	.99952	.04827	.99883	.06569	.99784	.08310	.99654	14
47	.01367	.99991	.03112	.99952	.04856	.99882	.06598	.99782	.08339	.99652	13
48	.01396	.99990	.03141	.99951	.04885	.99881	.06627	.99780	.08368	.99649	12
49	.01425	.99990	.03170	.99950	.04914	.99879	.06656	.99778	.08397	.99647	11
50	.01454	.99989	.03199	.99949	.04943	.99878	.06685	.99776	.08426	.99644	10
51	.01483	.99989	.03228	.99948	.04972	.99876	.06714	.99774	.08455	.99642	9
52	.01513	.99989	.03257	.99947	.05001	.99875	.06743	.99772	.08484	.99639	8
53	.01542	.99988	.03286	.99946	.05030	.99873	.06773	.99770	.08513	.99637	7
54	.01571	.99988	.03316	.99945	.05059	.99872	.06802	.99768	.08542	.99635	6
55	.01600	.99987	.03345	.99944	.05088	.99870	.06831	.99766	.08571	.99632	5
56	.01629	.99987	.03374	.99943	.05117	.99869	.06860	.99764	.08600	.99630	4
57	.01658	.99986	.03403	.99942	.05146	.99867	.06889	.99762	.08629	.99627	3
58	.01687	.99986	.03432	.99941	.05175	.99866	.06918	.99760	.08658	.99625	2
59	.01716	.99985	.03461	.99940	.05205	.99864	.06947	.99758	.08687	.99622	1
60	.01745	.99985	.03490	.99939	.05234	.99863	.06976	.99756	.08716	.99619	0
′	Cosine	Sine	Cosine	Sine	Cosine	Sine	Cosine	Sine	Cosine	Sine	′
	89°		88°		87°		86°		85°		

′	5° Sine	5° Cosine	6° Sine	6° Cosine	7° Sine	7° Cosine	8° Sine	8° Cosine	9° Sine	9° Cosine	′
0	.08716	.99619	.10453	.99452	.12187	.99255	.13917	.99027	.15643	.98769	60
1	.08745	.99617	.10482	.99449	.12216	.99251	.13946	.99023	.15672	.98764	59
2	.08774	.99614	.10511	.99446	.12245	.99248	.13975	.99019	.15701	.98760	58
3	.08803	.99612	.10540	.99443	.12274	.99244	.14004	.99015	.15730	.98755	57
4	.08831	.99609	.10569	.99440	.12302	.99240	.14033	.99011	.15758	.98751	56
5	.08860	.99607	.10597	.99437	.12331	.99237	.14061	.99006	.15787	.98746	55
6	.08889	.99604	.10626	.99434	.12360	.99233	.14090	.99002	.15816	.98741	54
7	.08918	.99602	.10655	.99431	.12389	.99230	.14119	.98998	.15845	.98737	53
8	.08947	.99599	.10684	.99428	.12418	.99226	.14148	.98994	.15873	.98732	52
9	.08976	.99596	.10713	.99424	.12447	.99222	.14177	.98990	.15902	.98728	51
10	.09005	.99594	.10742	.99421	.12476	.99219	.14205	.98986	.15931	.98723	50
11	.09034	.99591	.10771	.99418	.12504	.99215	.14234	.98982	.15959	.98718	49
12	.09063	.99588	.10800	.99415	.12533	.99211	.14263	.93978	.15988	.98714	48
13	.09092	.99586	.10829	.99412	.12562	.99208	.14292	.98973	.16017	.98709	47
14	.09121	.99583	.10858	.99409	.12591	.99204	.14320	.98969	.16046	.98704	46
15	.09150	.99580	.10887	.99406	.12620	.99200	.14349	.98965	.16074	.98700	45
16	.09179	.99578	.10916	.99402	.12649	.99197	.14378	.98961	.16103	.98695	44
17	.09208	.99575	.10945	.99399	.12678	.99193	.14407	.98957	.16132	.98690	43
18	.09237	.99572	.10973	.99396	.12706	.99189	.14436	.98953	.16160	.98686	42
19	.09266	.99570	.11002	.99393	.12735	.99186	.14464	.98948	.16189	.98681	41
20	.09295	.99567	.11031	.99390	.12764	.99182	.14493	.98944	.16218	.98676	40
21	.09324	.99564	.11060	.99386	.12793	.99178	.14522	.98940	.16246	.98671	39
22	.09353	.99562	.11089	.99383	.12822	.99175	.14551	.98936	.16275	.98667	38
23	.09382	.99559	.11118	.99380	.12851	.99171	.14580	.98931	.16304	.98662	37
24	.09411	.99556	.11147	.99377	.12880	.99167	.14608	.98927	.16333	.98657	36
25	.09440	.99553	.11176	.99374	.12908	.99163	.14637	.98923	.16361	.98652	35
26	.09469	.99551	.11205	.99370	.12937	.99160	.14666	.98919	.16390	.98648	34
27	.09498	.99548	.11234	.99367	.12966	.99156	.14695	.98914	.16419	.98643	33
28	.09527	.99545	.11263	.99364	.12995	.99152	.14723	.98910	.16447	.98638	32
29	.09556	.99542	.11291	.99360	.13024	.99148	.14752	.98906	.16476	.98633	31
30	.09585	.99540	.11320	.99357	.13053	.99144	.14781	.98902	.16505	.98629	30
31	.09614	.99537	.11349	.99354	.13081	.99141	14810	.98897	.16533	.98624	29
32	.09642	.99534	.11378	.99351	.13110	99137	.14838	.98893	.16562	.98619	28
33	.09671	.99531	.11407	.99347	.13139	.99133	.14867	.98889	.16591	.98614	27
34	.09700	.99528	.11436	.99344	.13168	.99129	.14896	.98884	.16620	.98609	26
35	.09729	.99526	.11465	.99341	.13197	.99125	.14925	.98880	.16648	.98604	25
36	.09758	.99523	.11494	.99337	.13226	.99122	.14954	.98876	.16677	.98600	24
37	.09787	.99520	.11523	.99334	.13254	.99118	.14982	.98871	.16706	.98595	23
38	.09816	.99517	.11552	.99331	.13283	.99114	.15011	.98867	.16734	.98590	22
39	.09845	.99514	.11580	.99327	.13312	.99110	.15040	.98863	.16763	.98585	21
40	.09874	.99511	.11609	.99324	.13341	.99106	.15069	.98858	.16792	.98580	20
41	.09903	.99508	.11638	.99320	.13370	.99102	.15097	.98854	.16820	.98575	19
42	.09932	.99506	.11667	.99317	.13399	.99098	.15126	.98849	.16849	.98570	18
43	.09961	.99503	.11696	.99314	.13427	.99094	.15155	.98845	.16878	.98565	17
44	.09990	.99500	.11725	.99310	.13456	.99091	.15184	.98841	.16906	.98561	16
45	.10019	.99497	.11754	.99307	.13485	.99087	.15212	.98836	.16935	.98556	15
46	.10048	.99494	.11783	.99303	.13514	.99083	.15241	.98832	.16964	.98551	14
47	.10077	.99491	.11812	.99300	.13543	.99079	.15270	.98827	.16992	.98546	13
48	.10106	.99488	.11840	.99297	.13572	.99075	.15299	.98823	.17021	.98541	12
49	.10135	.99485	.11869	.99293	.13600	.99071	.15327	.98818	.17050	.98536	11
50	.10164	.99482	.11898	.99290	.13629	.99067	.15356	.93814	.17078	.98531	10
51	.10192	.99479	.11927	.99286	.13658	.99063	.15385	.98809	.17107	.98526	9
52	.10221	.99476	.11956	.99283	.13687	.99059	.15414	.98805	.17136	.98521	8
53	.10250	.99473	.11985	.99279	.13716	.99055	.15442	.98800	.17164	.98516	7
54	.10279	.99470	.12014	.99276	.13744	.99051	.15471	.98796	.17193	.98511	6
55	.10308	.99467	.12043	.99272	.13773	.99047	.15500	.98791	.17222	.98506	5
56	.10337	.99464	.12071	.99269	.13802	.99043	.15529	.98787	.17250	.98501	4
57	.10366	.99461	.12100	.99265	.13831	.99039	.15557	.98782	.17279	.98496	3
58	.10395	.99458	.12129	.99262	.13860	.99035	.15586	.98778	.17308	.98491	2
59	.10424	.99455	.12158	.99258	.13889	.99031	.15615	.98773	.17336	.98486	1
60	.10453	.99452	.12187	.99255	.13917	.99027	.15643	.98769	.17365	.98481	0
′	Cosine	Sine	Cosine	Sine	Cosine	Sine	Cosine	Sine	Cosine	Sine	′
	84°		83°		82°		81°		80°		

′	10°		11°		12°		13°		14°		′
	Sine	Cosine	Sine	Cosine	Sine	Cosine	Sine	Cosine	Sine	Cosine	
0	.17365	.98481	.19081	.98163	.20791	.97815	.22495	.97437	.24192	.97030	60
1	.17393	.98476	.19109	.98157	.20820	.97809	.22523	.97430	.24220	.97023	59
2	.17422	.98471	.19138	.98152	.20848	.97803	.22552	.97424	.24249	.97015	58
3	.17451	.98466	.19167	.98146	.20877	.97797	.22580	.97417	.24277	.97008	57
4	.17479	.98461	.19195	.98140	.20905	.97791	.22608	.97411	.24305	.97001	56
5	.17508	.98455	.19224	.98135	.20933	.97784	.22637	.97404	.24333	.96994	55
6	.17537	.98450	.19252	.98129	.20962	.97778	.22665	.97398	.24362	.96987	54
7	.17565	.98445	.19281	.98124	.20990	.97772	.22693	.97391	.24390	.96980	53
8	.17594	.98440	.19309	.98118	.21019	.97766	.22722	.97384	.24418	.96973	52
9	.17623	.98435	.19338	.98112	.21047	.97760	.22750	.97378	.24446	.96966	51
10	.17651	.98430	.19366	.98107	.21076	.97754	.22778	.97371	.24474	.96959	50
11	.17680	.98425	.19395	.98101	.21104	.97748	.22807	.97365	.24503	.96952	49
12	.17708	.98420	.19423	.98096	.21132	.97742	.22835	.97358	.24531	.96945	48
13	.17737	.98414	.19452	.98090	.21161	.97735	.22863	.97351	.24559	.96937	47
14	.17766	.98409	.19481	.98084	.21189	.97729	.22892	.97345	.24587	.96930	46
15	.17794	.98404	.19509	.98079	.21218	.97723	.22920	.97338	.24615	.96923	45
16	.17823	.98399	.19538	.98073	.21246	.97717	.22948	.97331	.24644	.96916	44
17	.17852	.98394	.19566	.98067	.21275	.97711	.22977	.97325	.24672	.96909	43
18	.17880	.98389	.19595	.98061	.21303	.97705	.23005	.97318	.24700	.96902	42
19	.17909	.98383	.19623	.98056	.21331	.97698	.23033	.97311	.24728	.96894	41
20	.17937	.98378	.19652	.98050	.21360	.97692	.23062	.97304	.24756	.96887	40
21	.17966	.98373	.19680	.98044	.21388	.97686	.23090	.97298	.24784	.96880	39
22	.17995	.98368	.19709	.98039	.21417	.97680	.23118	.97291	.24813	.96873	38
23	.18023	.98362	.19737	.98033	.21445	.97673	.23146	.97284	.24841	.96866	37
24	.18052	.98357	.19766	.98027	.21474	.97667	.23175	.97278	.24869	.96858	36
25	.18081	.98352	.19794	.98021	.21502	.97661	.23203	.97271	.24897	.96851	35
26	.18109	.98347	.19823	.98016	.21530	.97655	.23231	.97264	.24925	.96844	34
27	.18138	.98341	.19851	.98010	.21559	.97648	.23260	.97257	.24954	.96837	33
28	.18166	.98336	.19880	.98004	.21587	.97642	.23288	.97251	.24982	.96829	32
29	.18195	.98331	.19908	.97998	.21616	.97636	.23316	.97244	.25010	.96822	31
30	.18224	.98325	.19937	.97992	.21644	.97630	.23345	.97237	.25038	.96815	30
31	.18252	.98320	.19965	.97987	.21672	.97623	.23373	.97230	.25066	.96807	29
32	.18281	.98315	.19994	.97981	.21701	.97617	.23401	.97223	.25094	.96800	28
33	.18309	.98310	.20022	.97975	.21729	.97611	.23429	.97217	.25122	.96793	27
34	.18338	.98304	.20051	.97969	.21758	.97604	.23458	.97210	.25151	.96786	26
35	.18367	.98299	.20079	.97963	.21786	.97598	.23486	.97203	.25179	.96778	25
36	.18395	.98294	.20108	.97958	.21814	.97592	.23514	.97196	.25207	.96771	24
37	.18424	.98288	.20136	.97952	.21843	.97585	.23542	.97189	.25235	.96764	23
38	.18452	.98283	.20165	.97946	.21871	.97579	.23571	.97182	.25263	.96756	22
39	.18481	.98277	.20193	.97940	.21899	.97573	.23599	.97176	.25291	.96749	21
40	.18509	.98272	.20222	.97934	.21928	.97566	.23627	.97169	.25320	.96742	20
41	.18538	.98267	.20250	.97928	.21956	.97560	.23656	.97162	.25348	.96734	19
42	.18567	.98261	.20279	.97922	.21985	.97553	.23684	.97155	.25376	.96727	18
43	.18595	.98256	.20307	.97916	.22013	.97547	.23712	.97148	.25404	.96719	17
44	.18624	.98250	.20336	.97910	.22041	.97541	.23740	.97141	.25432	.96712	16
45	.18652	.98245	.20364	.97905	.22070	.97534	.23769	.97134	.25460	.96705	15
46	.18681	.98240	.20393	.97899	.22098	.97528	.23797	.97127	.25488	.96697	14
47	.18710	.98234	.20421	.97893	.22126	.97521	.23825	.97120	.25516	.96690	13
48	.18738	.98229	.20450	.97887	.22155	.97515	.23853	.97113	.25545	.96682	12
49	.18767	.98223	.20478	.97881	.22183	.97508	.23882	.97106	.25573	.96675	11
50	.18795	.98218	.20507	.97875	.22212	.97502	.23910	.97100	.25601	.96667	10
51	.18824	.98212	.20535	.97869	.22240	.97496	.23938	.97093	.25629	.96660	9
52	.18852	.98207	.20563	.97863	.22268	.97489	.23966	.97086	.25657	.96653	8
53	.18881	.98201	.20592	.97857	.22297	.97483	.23995	.97079	.25685	.96645	7
54	.18910	.98196	.20620	.97851	.22325	.97476	.24023	.97072	.25713	.96638	6
55	.18938	.98190	.20649	.97845	.22353	.97470	.24051	.97065	.25741	.96630	5
56	.18967	.98185	.20677	.97839	.22382	.97463	.24079	.97058	.25769	.96623	4
57	.18995	.98179	.20706	.97833	.22410	.97457	.24108	.97051	.25798	.96615	3
58	.19024	.98174	.20734	.97827	.22438	.97450	.24136	.97044	.25826	.96608	2
59	.19052	.98168	.20763	.97821	.22467	.97444	.24164	.97037	.25854	.96600	1
60	.19081	.98163	.20791	.97815	.22495	.97437	.24192	.97030	.25882	.96593	0
′	Cosine	Sine	Cosine	Sine	Cosine	Sine	Cosine	Sine	Cosine	Sine	′
	79°		78°		77°		76°		75°		

′	15°		16°		17°		18°		19°		′
	Sine	Cosine	Sine	Cosine	Sine	Cosine	Sine	Cosine	Sine	Cosine	
0	.25882	.96593	.27564	.96126	.29237	.95630	.30902	.95106	.32557	.94552	60
1	.25910	.96585	.27592	.96118	.29265	.95622	.30929	.95097	.32584	.94542	59
2	.25938	.96578	.27620	.96110	.29293	.95613	.30957	.95088	.32612	.94533	58
3	.25966	.96570	.27648	.96102	.29321	.95605	.30985	.95079	.32639	.94523	57
4	.25994	.96562	.27676	.96094	.29348	.95596	.31012	.95070	.32667	.94514	56
5	.26022	.96555	.27704	.96086	.29376	.95588	.31040	.95061	.32694	.94504	55
6	.26050	.96547	.27731	.96078	.29404	.95579	.31068	.95052	.32722	.94495	54
7	.26079	.96540	.27759	.96070	.29432	.95571	.31095	.95043	.32749	.94485	53
8	.26107	.96532	.27787	.96062	.29460	.95562	.31123	.95033	.32777	.94476	52
9	.26135	.96524	.27815	.96054	.29487	.95554	.31151	.95024	.32804	.94466	51
10	.26163	.96517	.27843	.96046	.29515	.95545	.31178	.95015	.32832	.94457	50
11	.26191	.96509	.27871	.96037	.29543	.95536	.31206	.95006	.32859	.94447	49
12	.26219	.96502	.27899	.96029	.29571	.95528	.31233	.94997	.32887	.94438	48
13	.26247	.96494	.27927	.96021	.29599	.95519	.31261	.94988	.32914	.94428	47
14	.26275	.96486	.27955	.96013	.29626	.95511	.31289	.94979	.32942	.94418	46
15	.26303	.96479	.27983	.96005	.29654	.95502	.31316	.94970	.32969	.94409	45
16	.26331	.96471	.28011	.95997	.29682	.95493	.31344	.94961	.32997	.94399	44
17	.26359	.96463	.28039	.95989	.29710	.95485	.31372	.94952	.33024	.94390	43
18	.26387	.96456	.28067	.95981	.29737	.95476	.31399	.94943	.33051	.94380	42
19	.26415	.96448	.28095	.95972	.29765	.95467	.31427	.94933	.33079	.94370	41
20	.26443	.96440	.28123	.95964	.29793	.95459	.31454	.94924	.33106	.94361	40
21	.26471	.96433	.28150	.95956	.29821	.95450	.31482	.94915	.33134	.94351	39
22	.26500	.96425	.28178	.95948	.29849	.95441	.31510	.94906	.33161	.94342	38
23	.26528	.96417	.28206	.95940	.29876	.95433	.31537	.94897	.33189	.94332	37
24	.26556	.96410	.28234	.95931	.29904	.95424	.31565	.94888	.33216	.94322	36
25	.26584	.96402	.28262	.95923	.29932	.95415	.31593	.94878	.33244	.94313	35
26	.26612	.96394	.28290	.95915	.29960	.95407	.31620	.94869	.33271	.94303	34
27	.26640	.96386	.28318	.95907	.29987	.95398	.31648	.94860	.33298	.94293	33
28	.26668	.96379	.28346	.95898	.30015	.95389	.31675	.94851	.33326	.94284	32
29	.26696	.96371	.28374	.95890	.30043	.95380	.31703	.94842	.33353	.94274	31
30	.26724	.96363	.28402	.95882	.30071	.95372	.31730	.94832	.33381	.94264	30
31	.26752	.96355	.28429	.95874	.30098	.95363	.31758	.94823	.33408	.94254	29
32	.26780	.96347	.28457	.95865	.30126	.95354	.31786	.94814	.33436	.94245	28
33	.26808	.96340	.28485	.95857	.30154	.95345	.31813	.94805	.33463	.94235	27
34	.26836	.96332	.28513	.95849	.30182	.95337	.31841	.94795	.33490	.94225	26
35	.26864	.96324	.28541	.95841	.30209	.95328	.31868	.94786	.33518	.94215	25
36	.26892	.96316	.28569	.95832	.30237	.95319	.31896	.94777	.33545	.94206	24
37	.26920	.96308	.28597	.95824	.30265	.95310	.31923	.94768	.33573	.94196	23
38	.26948	.96301	.28625	.95816	.30292	.95301	.31951	.94758	.33600	.94186	22
39	.26976	.96293	.28652	.95807	.30320	.95293	.31979	.94749	.33627	.94176	21
40	.27004	.96285	.28680	.95799	.30348	.95284	.32006	.94740	.33655	.94167	20
41	.27032	.96277	.28708	.95791	.30376	.95275	.32034	.94730	.33682	.94157	19
42	.27060	.96269	.28736	.95782	.30403	.95266	.32061	.94721	.33710	.94147	18
43	.27088	.96261	.28764	.95774	.30431	.95257	.32089	.94712	.33737	.94137	17
44	.27116	.96253	.28792	.95766	.30459	.95248	.32116	.94702	.33764	.94127	16
45	.27144	.96246	.28820	.95757	.30486	.95240	.32144	.94693	.33792	.94118	15
46	.27172	.96238	.28847	.95749	.30514	.95231	.32171	.94684	.33819	.94108	14
47	.27200	.96230	.28875	.95740	.30542	.95222	.32199	.94674	.33846	.94098	13
48	.27228	.96222	.28903	.95732	.30570	.95213	.32227	.94665	.33874	.94088	12
49	.27256	.96214	.28931	.95724	.30597	.95204	.32254	.94656	.33901	.94078	11
50	.27284	.96206	.28959	.95715	.30625	.95195	.32282	.94646	.33929	.94068	10
51	.27312	.96198	.28987	.95707	.30653	.95186	.32309	.94637	.33956	.94058	9
52	.27340	.96190	.29015	.95698	.30680	.95177	.32337	.94627	.33983	.94049	8
53	.27368	.96182	.29042	.95690	.30708	.95168	.32364	.94618	.34011	.94039	7
54	.27396	.96174	.29070	.95681	.30736	.95159	.32392	.94609	.34038	.94029	6
55	.27424	.96166	.29098	.95673	.30763	.95150	.32419	.94599	.34065	.94019	5
56	.27452	.96158	.29126	.95664	.30791	.95142	.32447	.94590	.34093	.94009	4
57	.27480	.96150	.29154	.95656	.30819	.95133	.32474	.94580	.34120	.93999	3
58	.27508	.96142	.29182	.95647	.30846	.95124	.32502	.94571	.34147	.93989	2
59	.27536	.96134	.29209	.95639	.30874	.95115	.32529	.94561	.34175	.93979	1
60	.27564	.96126	.29237	.95630	.30902	.95106	.32557	.94552	.34202	.93969	0
′	Cosine	Sine	Cosine	Sine	Cosine	Sine	Cosine	Sine	Cosine	Sine	′
	74°		73°		72°		71°		70°		

′	20° Sine	Cosine	21° Sine	Cosine	22° Sine	Cosine	23° Sine	Cosine	24° Sine	Cosine	′
0	.34202	.93969	.35837	.93358	.37461	.92718	.39073	.92050	.40674	.91355	60
1	.34229	.93959	.35864	.93348	.37488	.92707	.39100	.92039	.40700	.91343	59
2	.34257	.93949	.35891	.93337	.37515	.92697	.39127	.92028	.40727	.91331	58
3	.34284	.93939	.35918	.93327	.37542	.92686	.39153	.92016	.40753	.91319	57
4	.34311	.93929	.35945	.93316	.37569	.92675	.39180	.92005	.40780	.91307	56
5	.34339	.93919	.35973	.93306	.37595	.92664	.39207	.91994	.40806	.91295	55
6	.34366	.93909	.36000	.93295	.37622	.92653	.39234	.91982	.40833	.91283	54
7	.34393	.93899	.36027	.93285	.37649	.92642	.39260	.91971	.40860	.91272	53
8	.34421	.93889	.36054	.93274	.37676	.92631	.39287	.91959	.40886	.91260	52
9	.34448	.93879	.36081	.93264	.37703	.92620	.39314	.91948	.40913	.91248	51
10	.34475	.93869	.36108	.93253	.37730	.92609	.39341	.91936	.40939	.91236	50
11	.34503	.93859	.36135	.93243	.37757	.92598	.39367	.91925	.40966	.91224	49
12	.34530	.93849	.36162	.93232	.37784	.92587	.39394	.91914	.40992	.91212	48
13	.34557	.93839	.36190	.93222	.37811	.92576	.39421	.91902	.41019	.91200	47
14	.34584	.93829	.36217	.93211	.37838	.92565	.39448	.91891	.41045	.91188	46
15	.34612	.93819	.36244	.93201	.37865	.92554	.39474	.91879	.41072	.91176	45
16	.34639	.93809	.36271	.93190	.37892	.92543	.39501	.91868	.41098	.91164	44
17	.34666	.93799	.36298	.93180	.37919	.92532	.39528	.91856	.41125	.91152	43
18	.34694	.93789	.36325	.93169	.37946	.92521	.39555	.91845	.41151	.91140	42
19	.34721	.93779	.36352	.93159	.37973	.92510	.39581	.91833	.41178	.91128	41
20	.34748	.93769	.36379	.93148	.37999	.92499	.39608	.91822	.41204	.91116	40
21	.34775	.93759	.36406	.93137	.38026	.92488	.39635	.91810	.41231	.91104	39
22	.34803	.93748	.36434	.93127	.38053	.92477	.39661	.91799	.41257	.91092	38
23	.34830	.93738	.36461	.93116	.38080	.92466	.39688	.91787	.41284	.91080	37
24	.34857	.93728	.36488	.93106	.38107	.92455	.39715	.91775	.41310	.91068	36
25	.34884	.93718	.36515	.93095	.38134	.92444	.39741	.91764	.41337	.91056	35
26	.34912	.93708	.36542	.93084	.38161	.92432	.39768	.91752	.41363	.91044	34
27	.34939	.93698	.36569	.93074	.38188	.92421	.39795	.91741	.41390	.91032	33
28	.34966	.93688	.36596	.93063	.38215	.92410	.39822	.91729	.41416	.91020	32
29	.34993	.93677	.36623	.93052	.38241	.92399	.39848	.91718	.41443	.91008	31
30	.35021	.93667	.36650	.93042	.38268	.92388	.39875	.91706	.41469	.90996	30
31	.35048	.93657	.36677	.93031	.38295	.92377	.39902	.91694	.41496	.90984	29
32	.35075	.93647	.36704	.93020	.38322	.92366	.39928	.91683	.41522	.90972	28
33	.35102	.93637	.36731	.93010	.38349	.92355	.39955	.91671	.41549	.90960	27
34	.35130	.93626	.36758	.92999	.38376	.92343	.39982	.91660	.41575	.90948	26
35	.35157	.93616	.36785	.92988	.38403	.92332	.40008	.91648	.41602	.90936	25
36	.35184	.93606	.36812	.92978	.38430	.92321	.40035	.91636	.41628	.90924	24
37	.35211	.93596	.36839	.92967	.38456	.92310	.40062	.91625	.41655	.90911	23
38	.35239	.93585	.36867	.92956	.38483	.92299	.40088	.91613	.41681	.90899	22
39	.35266	.93575	.36894	.92945	.38510	.92287	.40115	.91601	.41707	.90887	21
40	.35293	.93565	.36921	.92935	.38537	.92276	.40141	.91590	.41734	.90875	20
41	.35320	.93555	.36948	.92924	.38564	.92265	.40168	.91578	.41760	.90863	19
42	.35347	.93544	.36975	.92913	.38591	.92254	.40195	.91566	.41787	.90851	18
43	.35375	.93534	.37002	.92902	.38617	.92243	.40221	.91555	.41813	.90839	17
44	.35402	.93524	.37029	.92892	.38644	.92231	.40248	.91543	.41840	.90826	16
45	.35429	.93514	.37056	.92881	.38671	.92220	.40275	.91531	.41866	.90814	15
46	.35456	.93503	.37083	.92870	.38698	.92209	.40301	.91519	.41892	.90802	14
47	.35484	.93493	.37110	.92859	.38725	.92198	.40328	.91508	.41919	.90790	13
48	.35511	.93483	.37137	.92849	.38752	.92186	.40355	.91496	.41945	.90778	12
49	.35538	.93472	.37164	.92838	.38778	.92175	.40381	.91484	.41972	.90766	11
50	.35565	.93462	.37191	.92827	.38805	.92164	.40408	.91472	.41998	.90753	10
51	.35592	.93452	.37218	.92816	.38832	.92152	.40434	.91461	.42024	.90741	9
52	.35619	.93441	.37245	.92805	.38859	.92141	.40461	.91449	.42051	.90729	8
53	.35647	.93431	.37272	.92794	.38886	.92130	.40488	.91437	.42077	.90717	7
54	.35674	.93420	.37299	.92784	.38912	.92119	.40514	.91425	.42104	.90704	6
55	.35701	.93410	.37326	.92773	.38939	.92107	.40541	.91414	.42130	.90692	5
56	.35728	.93400	.37353	.92762	.38966	.92096	.40567	.91402	.42156	.90680	4
57	.35755	.93389	.37380	.92751	.38993	.92085	.40594	.91390	.42183	.90668	3
58	.35782	.93379	.37407	.92740	.39020	.92073	.40621	.91378	.42209	.90655	2
59	.35810	.93368	.37434	.92729	.39046	.92062	.40647	.91366	.42235	.90643	1
60	.35837	.93358	.37461	.92718	.39073	.92050	.40674	.91355	.42262	.90631	0
′	Cosine	Sine	Cosine	Sine	Cosine	Sine	Cosine	Sine	Cosine	Sine	′
	69°		68°		67°		66°		65°		

′	25° Sine	25° Cosine	26° Sine	26° Cosine	27° Sine	27° Cosine	28° Sine	28° Cosine	29° Sine	29° Cosine	′
0	.42262	.90631	.43837	.89879	.45399	.89101	.46947	.88295	.48481	.87462	60
1	.42288	.90618	.43863	.89867	.45425	.89087	.46973	.88281	.48506	.87448	59
2	.42315	.90606	.43889	.89854	.45451	.89074	.46999	.88267	.48532	.87434	58
3	.42341	.90594	.43916	.89841	.45477	.89061	.47024	.88254	.48557	.87420	57
4	.42367	.90582	.43942	.89828	.45503	.89048	.47050	.88240	.48583	.87406	56
5	.42394	.90569	.43968	.89816	.45529	.89035	.47076	.88226	.48608	.87391	55
6	.42420	.90557	.43994	.89803	.45554	.89021	.47101	.88213	.48634	.87377	54
7	.42446	.90545	.44020	.89790	.45580	.89008	.47127	.88199	.48659	.87363	53
8	.42473	.90532	.44046	.89777	.45606	.88995	.47153	.88185	.48684	.87349	52
9	.42499	.90520	.44072	.89764	.45632	.88981	.47178	.88172	.48710	.87335	51
10	.42525	.90507	.44098	.89752	.45658	.88968	.47204	.88158	.48735	.87321	50
11	.42552	.90495	.44124	.89739	.45684	.88955	.47229	.88144	.48761	.87306	49
12	.42578	.90483	.44151	.89726	.45710	.88942	.47255	.88130	.48786	.87292	48
13	.42604	.90470	.44177	.89713	.45736	.88928	.47281	.88117	.48811	.87278	47
14	.42631	.90458	.44203	.89700	.45762	.88915	.47306	.88103	.48837	.87264	46
15	.42657	.90446	.44229	.89687	.45787	.88902	.47332	.88089	.48862	.87250	45
16	.42683	.90433	.44255	.89674	.45813	.88888	.47358	.88075	.48888	.87235	44
17	.42709	.90421	.44281	.89662	.45839	.88875	.47383	.88062	.48913	.87221	43
18	.42736	.90408	.44307	.89649	.45865	.88862	.47409	.88048	.48938	.87207	42
19	.42762	.90396	.44333	.89636	.45891	.88848	.47434	.88034	.48964	.87193	
20	.42788	.90383	.44359	.89623	.45917	.88835	.47460	.88020	.48989	.87178	
21	.42815	.90371	.44385	.89610	.45942	.88822	.47486	.88006	.49014	.87164	
22	.42841	.90358	.44411	.89597	.45968	.88808	.47511	.87993	.49040	.87150	
23	.42867	.90346	.44437	.89584	.45994	.88795	.47537	.87979	.49065	.87136	
24	.42894	.90334	.44464	.89571	.46020	.88782	.47562	.87965	.49090	.87121	
25	.42920	.90321	.44490	.89558	.46046	.88768	.47588	.87951	.49116	.87107	
26	.42946	.90309	.44516	.89545	.46072	.88755	.47614	.87937	.49141	.87093	
27	.42972	.90296	.44542	.89532	.46097	.88741	.47639	.87923	.49166	.87079	
28	.42999	.90284	.44568	.89519	.46123	.88728	.47665	.87909	.49192	.87064	
29	.43025	.90271	.44594	.89506	.46149	.88715	.47690	.87896	.49217	.87050	
30	.43051	.90259	.44620	.89493	.46175	.88701	.47716	.87882	.49242	.87036	
31	.43077	.90246	.44646	.89480	.46201	.88688	.47741	.87868	.49268	.87021	
32	.43104	.90233	.44672	.89467	.46226	.88674	.47767	.87854	.49293	.87007	
33	.43130	.90221	.44698	.89454	.46252	.88661	.47793	.87840	.49318	.86993	
34	.43156	.90208	.44724	.89441	.46278	.88647	.47818	.87826	.49344	.86978	
35	.43182	.90196	.44750	.89428	.46304	.88634	.47844	.87812	.49369	.86964	
36	.43209	.90183	.44776	.89415	.46330	.88620	.47869	.87798	.49394	.86949	
37	.43235	.90171	.44802	.89402	.46355	.88607	.47895	.87784	.49419	.86935	
38	.43261	.90158	.44828	.89389	.46381	.88593	.47920	.87770	.49445	.86921	
39	.43287	.90146	.44854	.89376	.46407	.88580	.47946	.87756	.49470	.86906	
40	.43313	.90133	.44880	.89363	.46433	.88566	.47971	.87743	.49495	.86892	
41	.43340	.90120	.44906	.89350	.46458	.88553	.47997	.87729	.49521	.86878	19
42	.43366	.90108	.44932	.89337	.46484	.88539	.48022	.87715	.49546	.86863	18
43	.43392	.90095	.44958	.89324	.46510	.88526	.48048	.87701	.49571	.86849	
44	.43418	.90082	.44984	.89311	.46536	.88512	.48073	.87687	.49596	.86834	
45	.43445	.90070	.45010	.89298	.46561	.88499	.48099	.87673	.49622	.86820	
46	.43471	.90057	.45036	.89285	.46587	.88485	.48124	.87659	.49647	.86805	
47	.43497	.90045	.45062	.89272	.46613	.88472	.48150	.87645	.49672	.86791	
48	.43523	.90032	.45088	.89259	.46639	.88458	.48175	.87631	.49697	.86777	12
49	.43549	.90019	.45114	.89245	.46664	.88445	.48201	.87617	.49723	.86762	11
50	.43575	.90007	.45140	.89232	.46690	.88431	.48226	.87603	.49748	.86748	
51	.43602	.89994	.45166	.89219	.46716	.88417	.48252	.87589	.49773	.86733	
52	.43628	.89981	.45192	.89206	.46742	.88404	.48277	.87575	.49798	.86719	
53	.43654	.89968	.45218	.89193	.46767	.88390	.48303	.87561	.49824	.86704	
54	.43680	.89956	.45243	.89180	.46793	.88377	.48328	.87546	.49849	.86690	
55	.43706	.89943	.45269	.89167	.46819	.88363	.48354	.87532	.49874	.86675	
56	.43733	.89930	.45295	.89153	.46844	.88349	.48379	.87518	.49899	.86661	
57	.43759	.89918	.45321	.89140	.46870	.88336	.48405	.87504	.49924	.86646	
58	.43785	.89905	.45347	.89127	.46896	.88322	.48430	.87490	.49950	.86632	2
59	.43811	.89892	.45373	.89114	.46921	.88308	.48456	.87476	.49975	.86617	1
60	.43837	.89879	.45399	.89101	.46947	.88295	.48481	.87462	.50000	.86603	0
′	Cosine	Sine	Cosine	Sine	Cosine	Sine	Cosine	Sine	Cosine	Sine	′
	64°		63°		62°		61°		60°		

′	30° Sine	30° Cosine	31° Sine	31° Cosine	32° Sine	32° Cosine	33° Sine	33° Cosine	34° Sine	34° Cosine	′
0	.50000	.86603	.51504	.85717	.52992	.84805	.54464	.83867	.55919	.82904	60
1	.50025	.86588	.51529	.85702	.53017	.84789	.54488	.83851	.55943	.82887	59
2	.50050	.86573	.51554	.85687	.53041	.84774	.54513	.83835	.55968	.82871	58
3	.50076	.86559	.51579	.85672	.53066	.84759	.54537	.83819	.55992	.82855	57
4	.50101	.86544	.51604	.85657	.53091	.84743	.54561	.83804	.56016	.82839	56
5	.50126	.86530	.51628	.85642	.53115	.84728	.54586	.83788	.56040	.82822	55
6	.50151	.86515	.51653	.85627	.53140	.84712	.54610	.83772	.56064	.82806	54
7	.50176	.86501	.51678	.85612	.53164	.84697	.54635	.83756	.56088	.82790	53
8	.50201	.86486	.51703	.85597	.53189	.84681	.54659	.83740	.56112	.82773	52
9	.50227	.86471	.51728	.85582	.53214	.84666	.54683	.83724	.56136	.82757	51
10	.50252	.86457	.51753	.85567	.53238	.84650	.54708	.83708	.56160	.82741	50
11	.50277	.86442	.51778	.85551	.53263	.84635	.54732	.83692	.56184	.82724	49
12	.50302	.86427	.51803	.85536	.53288	.84619	.54756	.83676	.56208	.82708	48
13	.50327	.86413	.51828	.85521	.53312	.84604	.54781	.83660	.56232	.82692	47
14	.50352	.86398	.51852	.85506	.53337	.84588	.54805	.83645	.56256	.82675	46
15	.50377	.86384	.51877	.85491	.53361	.84573	.54829	.83629	.56280	.82659	45
16	.50403	.86369	.51902	.85476	.53386	.84557	.54854	.83613	.56305	.82643	44
17	.50428	.86354	.51927	.85461	.53411	.84542	.54878	.83597	.56329	.82626	43
18	.50453	.86340	.51952	.85446	.53435	.84526	.54902	.83581	.56353	.82610	42
19	.50478	.86325	.51977	.85431	.53460	.84511	.54927	.83565	.56377	.82593	41
20	.50503	.86310	.52002	.85416	.53484	.84495	.54951	.83549	.56401	.82577	40
21	.50528	.86295	.52026	.85401	.53509	.84480	.54975	.83533	.56425	.82561	39
22	.50553	.86281	.52051	.85385	.53534	.84464	.54999	.83517	.56449	.82544	38
23	.50578	.86266	.52076	.85370	.53558	.84448	.55024	.83501	.56473	.82528	37
24	.50603	.86251	.52101	.85355	.53583	.84433	.55048	.83485	.56497	.82511	36
25	.50628	.86237	.52126	.85340	.53607	.84417	.55072	.83469	.56521	.82495	35
26	.50654	.86222	.52151	.85325	.53632	.84402	.55097	.83453	.56545	.82478	34
27	.50679	.86207	.52175	.85310	.53656	.84386	.55121	.83437	.56569	.82462	33
28	.50704	.86192	.52200	.85294	.53681	.84370	.55145	.83421	.56593	.82446	32
29	.50729	.86178	.52225	.85279	.53705	.84355	.55169	.83405	.56617	.82429	31
30	.50754	.86163	.52250	.85264	.53730	.84339	.55194	.83389	.56641	.82413	30
31	.50779	.86148	.52275	.85249	.53754	.84324	.55218	.83373	.56665	.82396	29
32	.50804	.86133	.52299	.85234	.53779	.84308	.55242	.83356	.56689	.82380	28
33	.50829	.86119	.52324	.85218	.53804	.84292	.55266	.83340	.56713	.82363	27
34	.50854	.86104	.52349	.85203	.53828	.84277	.55291	.83324	.56736	.82347	26
35	.50879	.86089	.52374	.85188	.53853	.84261	.55315	.83308	.56760	.82330	25
36	.50904	.86074	.52399	.85173	.53877	.84245	.55339	.83292	.56784	.82314	24
37	.50929	.86059	.52423	.85157	.53902	.84230	.55363	.83276	.56808	.82297	23
38	.50954	.86045	.52448	.85142	.53926	.84214	.55388	.83260	.56832	.82281	22
39	.50979	.86030	.52473	.85127	.53951	.84198	.55412	.83244	.56856	.82264	21
40	.51004	.86015	.52498	.85112	.53975	.84182	.55436	.83228	.56880	.82248	20
41	.51029	.86000	.52522	.85096	.54000	.84167	.55460	.83212	.56904	.82231	19
42	.51054	.85985	.52547	.85081	.54024	.84151	.55484	.83195	.56928	.82214	18
43	.51079	.85970	.52572	.85066	.54049	.84135	.55509	.83179	56952	.82198	17
44	.51104	.85956	.52597	.85051	.54073	.84120	.55533	.83163	.56976	.82181	16
45	.51129	.85941	.52621	.85035	.54097	.84104	.55557	.83147	.57000	.82165	15
46	.51154	.85926	.52646	.85020	.54122	.84088	.55581	.83131	.57024	.82148	14
47	.51179	.85911	.52671	.85005	.54146	.84072	.55605	.83115	.57047	.82132	13
48	.51204	.85896	.52696	.84989	.54171	.84057	.55630	.83098	.57071	.82115	12
49	.51229	.85881	.52720	.84974	.54195	.84041	.55654	.83082	.57095	.82098	11
50	.51254	.85866	.52745	.84959	.54220	.84025	.55678	.83066	.57119	.82082	10
51	.51279	.85851	.52770	.84943	.54244	.84009	.55702	.83050	.57143	.82065	9
52	.51304	.85836	.52794	.84928	.54269	.83994	.55726	.83034	.57167	.82048	8
53	.51329	.85821	.52819	.84913	.54293	.83978	.55750	.83017	.57191	.82032	7
54	.51354	.85806	.52844	.84897	.54317	.83962	.55775	.83001	.57215	.82015	6
55	.51379	.85792	.52869	.84882	.54342	.83946	.55799	.82985	.57238	.81999	5
56	.51404	.85777	.52893	.84866	.54366	.83930	.55823	.82969	.57262	.81982	4
57	.51429	.85762	.52918	.84851	.54391	.83915	.55847	.82953	.57286	.81965	3
58	.51454	.85747	.52943	.84836	.54415	.83899	.55871	.82936	.57310	.81949	2
59	.51479	.85732	.52967	.84820	.54440	.83883	.55895	.82920	.57334	.81932	1
60	.51504	.85717	.52992	.84805	.54464	.83867	.55919	.82904	.57358	.81915	0
′	Cosine	Sine	Cosine	Sine	Cosine	Sine	Cosine	Sine	Cosine	Sine	′
	59°		58°		57°		56°		55°		

′	35° Sine	35° Cosine	36° Sine	36° Cosine	37° Sine	37° Cosine	38° Sine	38° Cosine	39° Sine	39° Cosine	′
0	.57358	.81915	.58779	.80902	.60182	.79864	.61566	.78801	.62932	.77715	60
1	.57381	.81899	.58802	.80885	.60205	.79846	.61589	.78783	.62955	.77696	59
2	.57405	.81882	.58826	.80867	.60228	.79829	.61612	.78765	.62977	.77678	58
3	.57429	.81865	.58849	.80850	.60251	.79811	.61635	.78747	.63000	.77660	57
4	.57453	.81848	.58873	.80833	.60274	.79793	.61658	.78729	.63022	.77641	56
5	.57477	.81832	.58896	.80816	.60298	.79776	.61681	.78711	.63045	.77623	55
6	.57501	.81815	.58920	.80799	.60321	.79758	.61704	.78694	.63068	.77605	54
7	.57524	.81798	.58943	.80782	.60344	.79741	.61726	.78676	.63090	.77586	53
8	.57548	.81782	.58967	.80765	.60367	.79723	.61749	.78658	.63113	.77568	52
9	.57572	.81765	.58990	.80748	.60390	.79706	.61772	.78640	.63135	.77550	51
10	.57596	.81748	.59014	.80730	.60414	.79688	.61795	.78622	.63158	.77531	50
11	.57619	.81731	.59037	.80713	.60437	.79671	.61818	.78604	.63180	.77513	49
12	.57643	.81714	.59061	.80696	.60460	.79653	.61841	.78586	.63203	.77494	48
13	.57667	.81698	.59084	.80679	.60483	.79635	.61864	.78568	.63225	.77476	47
14	.57691	.81681	.59108	.80662	.60506	.79618	.61887	.78550	.63248	.77458	46
15	.57715	.81664	.59131	.80644	.60529	.79600	.61909	.78532	.63271	.77439	45
16	.57738	.81647	.59154	.80627	.60553	.79583	.61932	.78514	.63293	.77421	44
17	.57762	.81631	.59178	.80610	.60576	.79565	.61955	.78496	.63316	.77402	43
18	.57786	.81614	.59201	.80593	.60599	.79547	.61978	.78478	.63338	.77384	42
19	.57810	.81597	.59225	.80576	.60622	.79530	.62001	.78460	.63361	.77366	41
20	.57833	.81580	.59248	.80558	.60645	.79512	.62024	.78442	.63383	.77347	40
21	.57857	.81563	.59272	.80541	.60668	.79494	.62046	.78424	.63406	.77329	39
22	.57881	.81546	.59295	.80524	.60691	.79477	.62069	.78405	.63428	.77310	38
23	.57904	.81530	.59318	.80507	.60714	.79459	.62092	.78387	.63451	.77292	37
24	.57928	.81513	.59342	.80489	.60738	.79441	.62115	.78369	.63473	.77273	36
25	.57952	.81496	.59365	.80472	.60761	.79424	.62138	.78351	.63496	.77255	35
26	.57976	.81479	.59389	.80455	.60784	.79406	.62160	.78333	.63518	.77236	34
27	.57999	.81462	.59412	.80438	.60807	.79388	.62183	.78315	.63540	.77218	33
28	.58023	.81445	.59436	.80420	.60830	.79371	.62206	.78297	.63563	.77199	32
29	.58047	.81428	.59459	.80403	.60853	.79353	.62229	.78279	.63585	.77181	31
30	.58070	.81412	.59482	.80386	.60876	.79335	.62251	.78261	.63608	.77162	30
31	.58094	.81395	.59506	.80368	.60899	.79318	.62274	.78243	.63630	.77144	29
32	.58118	.81378	.59529	.80351	.60922	.79300	.62297	.78225	.63653	.77125	28
33	.58141	.81361	.59552	.80334	.60945	.79282	.62320	.78206	.63675	.77107	27
34	.58165	.81344	.59576	.80316	.60968	.79264	.62342	.78188	.63698	.77088	26
35	.58189	.81327	.59599	.80299	.60991	.79247	.62365	.78170	.63720	.77070	25
36	.58212	.81310	.59622	.80282	.61015	.79229	.62388	.78152	.63742	.77051	24
37	.58236	.81293	.59646	.80264	.61038	.79211	.62411	.78134	.63765	.77033	23
38	.58260	.81276	.59669	.80247	.61061	.79193	.62433	.78116	.63787	.77014	22
39	.58283	.81259	.59693	.80230	.61084	.79176	.62456	.78098	.63810	.76996	21
40	.58307	.81242	.59716	.80212	.61107	.79158	.62479	.78079	.63832	.76977	20
41	.58330	.81225	.59739	.80195	.61130	.79140	.62502	.78061	.63854	.76959	19
42	.58354	.81208	.59763	.80178	.61153	.79122	.62524	.78043	.63877	.76940	18
43	.58378	.81191	.59786	.80160	.61176	.79105	.62547	.78025	.63899	.76921	17
44	.58401	.81174	.59809	.80143	.61199	.79087	.62570	.78007	.63922	.76903	16
45	.58425	.81157	.59832	.80125	.61222	.79069	.62592	.77988	.63944	.76884	15
46	.58449	.81140	.59856	.80108	.61245	.79051	.62615	.77970	.63966	.76866	14
47	.58472	.81123	.59879	.80091	.61268	.79033	.62638	.77952	.63989	.76847	13
48	.58496	.81106	.59902	.80073	.61291	.79016	.62660	.77934	.64011	.76828	12
49	.58519	.81089	.59926	.80056	.61314	.78998	.62683	.77916	.64033	.76810	11
50	.58543	.81072	.59949	.80038	.61337	.78980	.62706	.77897	.64056	.76791	10
51	.58567	.81055	.59972	.80021	.61360	.78962	.62728	.77879	.64078	.76772	9
52	.58590	.81038	.59995	.80003	.61383	.78944	.62751	.77861	.64100	.76754	8
53	.58614	.81021	.60019	.79986	.61406	.78926	.62774	.77843	.64123	.76735	7
54	.58637	.81004	.60042	.79968	.61429	.78908	.62796	.77824	.64145	.76717	6
55	.58661	.80987	.60065	.79951	.61451	.78891	.62819	.77806	.64167	.76698	5
56	.58684	.80970	.60089	.79934	.61474	.78873	.62842	.77788	.64190	.76679	4
57	.58708	.80953	.60112	.79916	.61497	.78855	.62864	.77769	.64212	.76661	3
58	.58731	.80936	.60135	.79899	.61520	.78837	.62887	.77751	.64234	.76642	2
59	.58755	.80919	.60158	.79881	.61543	.78819	.62909	.77733	.64256	.76623	1
60	.58779	.80902	.60182	.79864	.61566	.78801	.62932	.77715	.64279	.76604	0
′	Cosine	Sine	Cosine	Sine	Cosine	Sine	Cosine	Sine	Cosine	Sine	′
	54°		53°		52°		51°		50°		

| ′ | 40° | | 41° | | 42° | | 43° | | 44° | | ′ |
	Sine	Cosine	Sine	Cosine	Sine	Cosine	Sine	Cosine	Sine	Cosine	
0	.64279	.76604	.65606	.75471	.66913	.74314	.68200	.73135	.69466	.71934	60
1	.64301	.76586	.65628	.75452	.66935	.74295	.68221	.73116	.69487	.71914	59
2	.64323	.76567	.65650	.75433	.66956	.74276	.68242	.73096	.69508	.71894	58
3	.64346	.76548	.65672	.75414	.66978	.74256	.68264	.73076	.69529	.71873	57
4	.64368	.76530	.65694	.75395	.66999	.74237	.68285	.73056	.69549	.71853	56
5	.64390	.76511	.65716	.75375	.67021	.74217	.68306	.73036	.69570	.71833	55
6	.64412	.76492	.65738	.75356	.67043	.74198	.68327	.73016	.69591	.71813	54
7	.64435	.76473	.65759	.75337	.67064	.74178	.68349	.72996	.69612	.71792	53
8	.64457	.76455	.65781	.75318	.67086	.74159	.68370	.72976	.69633	.71772	52
9	.64479	.76436	.65803	.75299	.67107	.74139	.68391	.72957	.69654	.71752	51
10	.64501	.76417	.65825	.75280	.67129	.74120	.68412	.72937	.69675	.71732	50
11	.64524	.76398	.65847	.75261	.67151	.74100	.68434	.72917	.69696	.71711	49
12	.64546	.76380	.65869	.75241	.67172	.74080	.68455	.72897	.69717	.71691	48
13	.64568	.76361	.65891	.75222	.67194	.74061	.68476	.72877	.69737	.71671	47
14	.64590	.76342	.65913	.75203	.67215	.74041	.68497	.72857	.69758	.71650	46
15	.64612	.76323	.65935	.75184	.67237	.74022	.68518	.72837	.69779	.71630	45
16	.64635	.76304	.65956	.75165	.67258	.74002	.68539	.72817	.69800	.71610	44
17	.64657	.76286	.65978	.75146	.67280	.73983	.68561	.72797	.69821	.71590	43
18	.64679	.76267	.66000	.75126	.67301	.73963	.68582	.72777	.69842	.71569	42
19	.64701	.76248	.66022	.75107	.67323	.73944	.68603	.72757	.69862	.71549	41
20	.64723	.76229	.66044	.75088	.67344	.73924	.68624	.72737	.69883	.71529	40
21	.64746	.76210	.66066	.75069	.67366	.73904	.68645	.72717	.69904	.71508	39
22	.64768	.76192	.66088	.75050	.67387	.73885	.68666	.72697	.69925	.71488	38
23	.64790	.76173	.66109	.75030	.67409	.73865	.68688	.72677	.69946	.71468	37
24	.64812	.76154	.66131	.75011	.67430	.73846	.68709	.72657	.69966	.71447	36
25	.64834	.76135	.66153	.74992	.67452	.73826	.68730	.72637	.69987	.71427	35
26	.64856	.76116	.66175	.74973	.67473	.73806	.68751	.72617	.70008	.71407	34
27	.64878	.76097	.66197	.74953	.67495	.73787	.68772	.72597	.70029	.71386	33
28	.64901	.76078	.66218	.74934	.67516	.73767	.68793	.72577	.70049	.71366	32
29	.64923	.76059	.66240	.74915	.67538	.73747	.68814	.72557	.70070	.71345	31
30	.64945	.76041	.66262	.74896	.67559	.73728	.68835	.72537	.70091	.71325	30
31	.64967	.76022	.66284	.74876	.67580	.73708	.68857	.72517	.70112	.71305	29
32	.64989	.76003	.66306	.74857	.67602	.73688	.68878	.72497	.70132	.71284	28
33	.65011	.75984	.66327	.74838	.67623	.73669	.68899	.72477	.70153	.71264	27
34	.65033	.75965	.66349	.74818	.67645	.73649	.68920	.72457	.70174	.71243	26
35	.65055	.75946	.66371	.74799	.67666	.73629	.68941	.72437	.70195	.71223	25
36	.65077	.75927	.66393	.74780	.67688	.73610	.68962	.72417	.70215	.71203	24
37	.65100	.75908	.66414	.74760	.67709	.73590	.68983	.72397	.70236	.71182	23
38	.65122	.75889	.66436	.74741	.67730	.73570	.69004	.72377	.70257	.71162	22
39	.65144	.75870	.66458	.74722	.67752	.73551	.69025	.72357	.70277	.71141	21
40	.65166	.75851	.66480	.74703	.67773	.73531	.69046	.72337	.70298	.71121	20
41	.65188	.75832	.66501	.74683	.67795	.73511	.69067	.72317	.70319	.71100	19
42	.65210	.75813	.66523	.74664	.67816	.73491	.69088	.72297	.70339	.71080	18
43	.65232	.75794	.66545	.74644	.67837	.73472	.69109	.72277	.70360	.71059	17
44	.65254	.75775	.66566	.74625	.67859	.73452	.69130	.72257	.70381	.71039	16
45	.65276	.75756	.66588	.74606	.67880	.73432	.69151	.72236	.70401	.71019	15
46	.65298	.75738	.66610	.74586	.67901	.73413	.69172	.72216	.70422	.70998	14
47	.65320	.75719	.66632	.74567	.67923	.73393	.69193	.72196	.70443	.70978	13
48	.65342	.75700	.66653	.74548	.67944	.73373	.69214	.72176	.70463	.70957	12
49	.65364	.75680	.66675	.74528	.67965	.73353	.69235	.72156	.70484	.70937	11
50	.65386	.75661	.66697	.74509	.67987	.73333	.69256	.72136	.70505	.70916	10
51	.65408	.75642	.66718	.74489	.68008	.73314	.69277	.72116	.70525	.70896	9
52	.65430	.75623	.66740	.74470	.68029	.73294	.69298	.72095	.70546	.70875	8
53	.65452	.75604	.66762	.74451	.68051	.73274	.69319	.72075	.70567	.70855	7
54	.65474	.75585	.66783	.74431	.68072	.73254	.69340	.72055	.70587	.70834	6
55	.65496	.75566	.66805	.74412	.68093	.73234	.69361	.72035	.70608	.70813	5
56	.65518	.75547	.66827	.74392	.68115	.73215	.69382	.72015	.70628	.70793	4
57	.65540	.75528	.66848	.74373	.68136	.73195	.69403	.71995	.70649	.70772	3
58	.65562	.75509	.66870	.74353	.68157	.73175	.69424	.71974	.70670	.70752	2
59	.65584	.75490	.66891	.74334	.68179	.73155	.69445	.71954	.70690	.70731	1
60	.65606	.75471	.66913	.74314	.68200	.73135	.69466	.71934	.70711	.70711	0
′	Cosine	Sine	Cosine	Sine	Cosine	Sine	Cosine	Sine	Cosine	Sine	′
	49°		48°		47°		46°		45°		

′	0°		1°		2°		3°		4°		′
	Tang	Cotang	Tang	Cotang	Tang	Cotang	Tang	Cotang	Tang	Cotang	
0	.00000	Infinite	.01746	57.2900	.03492	28.6363	.05241	19.0811	.06993	14.3007	60
1	.00029	3437.75	.01775	56.3506	.03521	28.3994	.05270	18.9755	.07022	14.2411	59
2	.00058	1718.87	.01804	55.4415	.03550	28.1664	.05299	18.8711	.07051	14.1821	58
3	.00087	1145.92	.01833	54.5613	.03579	27.9372	.05328	18.7678	.07080	14.1235	57
4	.00116	859.436	.01862	53.7086	.03609	27.7117	.05357	18.6656	.07110	14.0655	56
5	.00145	687.549	.01891	52.8821	.03638	27.4899	.05387	18.5645	.07139	14.0079	55
6	.00175	572.957	.01920	52.0807	.03667	27.2715	.05416	18.4645	.07168	13.9507	54
7	.00204	491.106	.01949	51.3032	.03696	27.0566	.05445	18.3655	.07197	13.8940	53
8	.00233	429.718	.01978	50.5485	.03725	26.8450	.05474	18.2677	.07227	13.8378	52
9	.00262	381.971	.02007	49.8157	.03754	26.6367	.05503	18.1708	.07256	13.7821	51
10	.00291	343.774	.02036	49.1039	.03783	26.4316	.05533	18.0750	.07285	13.7267	50
11	.00320	312.521	.02066	48.4121	.03812	26.2296	.05562	17.9802	.07314	13.6719	49
12	.00349	286.478	.02095	47.7395	.03842	26.0307	.05591	17.8863	.07344	13.6174	48
13	.00378	264.441	.02124	47.0853	.03871	25.8348	.05620	17.7934	.07373	13.5634	47
14	.00407	245.552	.02153	46.4489	.03900	25.6418	.05649	17.7015	.07402	13.5098	46
15	.00436	229.182	.02182	45.8294	.03929	25.4517	.05678	17.6106	.07431	13.4566	45
16	.00465	214.858	.02211	45.2261	.03958	25.2644	.05708	17.5205	.07461	13.4039	44
17	.00495	202.219	.02240	44.6386	.03987	25.0798	.05737	17.4314	.07490	13.3515	43
18	.00524	190.984	.02269	44.0661	.04016	24.8978	.05766	17.3432	.07519	13.2996	42
19	.00553	180.932	.02298	43.5081	.04046	24.7185	.05795	17.2558	.07548	13.2480	41
20	.00582	171.885	.02328	42.9641	.04075	24.5418	.05824	17.1693	.07578	13.1969	40
21	.00611	163.700	.02357	42.4335	.04104	24.3675	.05854	17.0837	.07607	13.1461	39
22	.00640	156.259	.02386	41.9158	.04133	24.1957	.05883	16.9950	.07636	13.0958	38
23	.00669	149.465	.02415	41.4106	.04162	24.0263	.05912	16.9150	.07665	13.0458	37
24	.00698	143.237	.02444	40.9174	.04191	23.8593	.05941	16.8319	.07695	12.9962	36
25	.00727	137.507	.02473	40.4358	.04220	23.6945	.05970	16.7496	.07724	12.9469	35
26	.00756	132.219	.02502	39.9655	.04250	23.5321	.05999	16.6681	.07753	12.8981	34
27	.00785	127.321	.02531	39.5059	.04279	23.3718	.06029	16.5874	.07782	12.8496	33
28	.00815	122.774	.02560	39.0568	.04308	23.2137	.06058	16.5075	.07812	12.8014	32
29	.00844	118.540	.02589	38.6177	.04337	23.0577	.06087	16.4283	.07841	12.7536	31
30	.00873	114.589	.02619	38.1885	.04366	22.9038	.06116	16.3499	.07870	12.7062	30
31	.00902	110.892	.02648	37.7686	.04395	22.7519	.06145	16.2722	.07899	12.6591	29
32	.00931	107.426	.02677	37.3579	.04424	22.6020	.06175	16.1952	.07929	12.6124	28
33	.00960	104.171	.02706	36.9560	.04454	22.4541	.06204	16.1190	.07958	12.5660	27
34	.00989	101.107	.02735	36.5627	.04483	22.3081	.06233	16.0435	.07987	12.5199	26
35	.01018	98.2179	.02764	36.1776	.04512	22.1640	.06262	15.9687	.08017	12.4742	25
36	.01047	95.4895	.02793	35.8006	.04541	22.0217	.06291	15.8945	.08046	12.4288	24
37	.01076	92.9085	.02822	35.4313	.04570	21.8813	.06321	15.8211	.08075	12.3838	23
38	.01105	90.4633	.02851	35.0695	.04599	21.7426	.06350	15.7483	.08104	12.3390	22
39	.01135	88.1436	.02881	34.7151	.04628	21.6056	.06379	15.6762	.08134	12.2946	21
40	.01164	85.9398	.02910	34.3678	.04658	21.4704	.06408	15.6048	.08163	12.2505	20
41	.01193	83.8435	.02939	34.0273	.04687	21.3369	.06437	15.5340	.08192	12.2067	19
42	.01222	81.8470	.02968	33.6935	.04716	21.2049	.06467	15.4638	.08221	12.1632	18
43	.01251	79.9434	.02997	33.3662	.04745	21.0747	.06496	15.3943	.08251	12.1201	17
44	.01280	78.1263	.03026	33.0452	.04774	20.9460	.06525	15.3254	.08280	12.0772	16
45	.01309	76.3900	.03055	32.7303	.04803	20.8188	.06554	15.2571	.08309	12.0346	15
46	.01338	74.7292	.03084	32.4213	.04833	20.6932	.06584	15.1893	.08339	11.9923	14
47	.01367	73.1390	.03114	32.1181	.04862	20.5691	.06613	15.1222	.08368	11.9504	13
48	.01396	71.6151	.03143	31.8205	.04891	20.4465	.06642	15.0557	.08397	11.9087	12
49	.01425	70.1533	.03172	31.5284	.04920	20.3253	.06671	14.9898	.08427	11.8673	11
50	.01455	68.7501	.03201	31.2416	.04949	20.2056	.06700	14.9244	.08456	11.8262	10
51	.01484	67.4019	.03230	30.9599	.04978	20.0872	.06730	14.8596	.08485	11.7853	9
52	.01513	66.1055	.03259	30.6833	.05007	19.9702	.06759	14.7954	.08514	11.7448	8
53	.01542	64.8580	.03288	30.4116	.05037	19.8546	.06788	14.7317	.08544	11.7045	7
54	.01571	63.6567	.03317	30.1446	.05066	19.7403	.06817	14.6685	.08573	11.6645	6
55	.01600	62.4992	.03346	29.8823	.05095	19.6273	.06847	14.6059	.08602	11.6248	5
56	.01629	61.3829	.03376	29.6245	.05124	19.5156	.06876	14.5438	.08632	11.5853	4
57	.01658	60.3058	.03405	29.3711	.05153	19.4051	.06905	14.4823	.08661	11.5461	3
58	.01687	59.2659	.03434	29.1220	.05182	19.2959	.06934	14.4212	.08690	11.5072	2
59	.01716	58.2612	.03463	28.8771	.05212	19.1879	.06963	14.3607	.08720	11.4685	1
60	.01746	57.2900	.03492	28.6363	.05241	19.0811	.06993	14.3007	.08749	11.4301	0
′	Cotang	Tang	Cotang	Tang	Cotang	Tang	Cotang	Tang	Cotang	Tang	′
	89°		88°		87°		86°		85°		

′	5° Tang	Cotang	6° Tang	Cotang	7° Tang	Cotang	8° Tang	Cotang	9° Tang	Cotang	′
0	.08749	11.4301	.10510	9.51436	.12278	8.14435	.14054	7.11537	.15838	6.31375	60
1	.08778	11.3919	.10540	9.48781	.12308	8.12481	.14084	7.10038	.15868	6.30189	59
2	.08807	11.3540	.10569	9.46141	.12338	8.10536	.14113	7.08546	.15898	6.29007	58
3	.08837	11.3163	.10599	9.43515	.12367	8.08600	.14143	7.07059	.15928	6.27829	57
4	.08866	11.2789	.10628	9.40904	.12397	8.06674	.14173	7.05579	.15958	6.26655	56
5	.08895	11.2417	.10657	9.38307	.12426	8.04756	.14202	7.04105	.15988	6.25486	55
6	.08925	11.2048	.10687	9.35724	.12456	8.02848	.14232	7.02637	.16017	6.24321	54
7	.08954	11.1681	.10716	9.33155	.12485	8.00948	.14262	7.01174	.16047	6.23160	53
8	.08983	11.1316	.10746	9.30599	.12515	7.99058	.14291	6.99718	.16077	6.22003	52
9	.09013	11.0954	.10775	9.28058	.12544	7.97176	.14321	6.98268	.16107	6.20851	51
10	.09042	11.0594	.10805	9.25530	.12574	7.95302	.14351	6.96823	.16137	6.19703	50
11	.09071	11.0237	.10834	9.23016	.12603	7.93438	.14381	6.95385	.16167	6.18559	49
12	.09101	10.9882	.10863	9.20516	.12633	7.91582	.14410	6.93952	.16196	6.17419	48
13	.09130	10.9529	.10893	9.18028	.12662	7.89734	.14440	6.92525	.16226	6.16283	47
14	.09159	10.9178	.10922	9.15554	.12692	7.87895	.14470	6.91104	.16256	6.15151	46
15	.09189	10.8829	.10952	9.13093	.12722	7.86064	.14499	6.89688	.16286	6.14023	45
16	.09218	10.8483	.10981	9.10646	.12751	7.84242	.14529	6.88278	.16316	6.12899	44
17	.09247	10.8139	.11011	9.08211	.12781	7.82428	.14559	6.86874	.16346	6.11779	43
18	.09277	10.7797	.11040	9.05789	.12810	7.80622	.14588	6.85475	.16376	6.10664	42
19	.09306	10.7457	.11070	9.03379	.12840	7.78825	.14618	6.84082	.16405	6.09552	41
20	.09335	10.7119	.11099	9.00983	.12869	7.77035	.14648	6.82694	.16435	6.08444	40
21	.09365	10.6783	.11128	8.98598	.12899	7.75254	.14678	6.81312	.16465	6.07340	39
22	.09394	10.6450	.11158	8.96227	.12929	7.73480	.14707	6.79936	.16495	6.06240	38
23	.09423	10.6118	.11187	8.93867	.12958	7.71715	.14737	6.78564	.16525	6.05143	37
24	.09453	10.5789	.11217	8.91520	.12988	7.69957	.14767	6.77199	.16555	6.04051	36
25	.09482	10.5462	.11246	8.89185	.13017	7.68208	.14796	6.75838	.16585	6.02962	35
26	.09511	10.5136	.11276	8.86862	.13047	7.66466	.14826	6.74483	.16615	6.01878	34
27	.09541	10.4813	.11305	8.84551	.13076	7.64732	.14856	6.73133	.16645	6.00797	33
28	.09570	10.4491	.11335	8.82252	.13106	7.63005	.14886	6.71789	.16674	5.99720	32
29	.09600	10.4172	.11364	8.79964	.13136	7.61287	.14915	6.70450	.16704	5.98646	31
30	.09629	10.3854	.11394	8.77689	.13165	7.59575	.14945	6.69116	.16734	5.97576	30
31	.09658	10.3538	.11423	8.75425	.13195	7.57872	.14975	6.67787	.16764	5.96510	29
32	.09688	10.3224	.11452	8.73172	.13224	7.56176	.15005	6.66463	.16794	5.95448	28
33	.09717	10.2913	.11482	8.70931	.13254	7.54487	.15034	6.65144	.16824	5.94390	27
34	.09746	10.2602	.11511	8.68701	.13284	7.52806	.15064	6.63831	.16854	5.93335	26
35	.09776	10.2294	.11541	8.66482	.13313	7.51132	.15094	6.62523	.16884	5.92263	25
36	.09805	10.1988	.11570	8.64275	.13343	7.49465	.15124	6.61219	.16914	5.91236	24
37	.09834	10.1683	.11600	8.62078	.13372	7.47806	.15153	6.59921	.16944	5.90191	23
38	.09864	10.1381	.11629	8.59893	.13402	7.46154	.15183	6.58627	.16974	5.89151	22
39	.09893	10.1080	.11659	8.57718	.13432	7.44509	.15213	6.57339	.17004	5.88114	21
40	.09923	10.0780	.11688	8.55555	.13461	7.42871	.15243	6.56055	.17033	5.87080	20
41	.09952	10.0483	.11718	8.53402	.13491	7.41240	.15272	6.54777	.17063	5.86051	19
42	.09981	10.0187	.11747	8.51259	.13521	7.39616	.15302	6.53503	.17093	5.85024	18
43	.10011	9.98931	.11777	8.49128	.13550	7.37999	.15332	6.52234	.17123	5.84001	17
44	.10040	9.96007	.11806	8.47007	.13580	7.36389	.15362	6.50970	.17153	5.82982	16
45	.10069	9.93101	.11836	8.44896	.13609	7.34786	.15391	6.49710	.17183	5.81966	15
46	.10099	9.90211	.11865	8.42795	.13639	7.33190	.15421	6.48456	.17213	5.80953	14
47	.10128	9.87338	.11895	8.40705	.13669	7.31600	.15451	6.47206	.17243	5.79944	13
48	.10158	9.84482	.11924	8.38625	.13698	7.30018	.15481	6.45961	.17273	5.78938	12
49	.10187	9.81641	.11954	8.36555	.13728	7.28442	.15511	6.44720	.17303	5.77936	11
50	.10216	9.78817	.11983	8.34496	.13758	7.26873	.15540	6.43484	.17333	5.76937	10
51	.10246	9.76009	.12013	8.32446	.13787	7.25310	.15570	6.42253	.17363	5.75941	9
52	.10275	9.73217	.12042	8.30406	.13817	7.23754	.15600	6.41026	.17393	5.74949	8
53	.10305	9.70441	.12072	8.28376	.13846	7.22204	.15630	6.39804	.17423	5.73960	7
54	.10334	9.67680	.12101	8.26355	.13876	7.20661	.15660	6.38587	.17453	5.72974	6
55	.10363	9.64935	.12131	8.24345	.13906	7.19125	.15689	6.37374	.17483	5.71992	5
56	.10393	9.62205	.12160	8.22344	.13935	7.17594	.15719	6.36165	.17513	5.71013	4
57	.10422	9.59490	.12190	8.20352	.13965	7.16071	.15749	6.34961	.17543	5.70037	3
58	.10452	9.56791	.12219	8.18370	.13995	7.14553	.15779	6.33761	.17573	5.69064	2
59	.10481	9.54106	.12249	8.16398	.14024	7.13042	.15809	6.32566	.17603	5.68094	1
60	.10510	9.51436	.12278	8.14435	.14054	7.11537	.15838	6.31375	.17633	5.67128	0
′	Cotang	Tang	Cotang	Tang	Cotang	Tang	Cotang	Tang	Cotang	Tang	′
	84°		83°		82°		81°		80°		

′	10°		11°		12°		13°		14°		′
	Tang	Cotang	Tang	Cotang	Tang	Cotang	Tang	Cotang	Tang	Cotang	
0	.17633	5.67128	.19438	5.14455	.21256	4.70463	.23087	4.33148	.24933	4.01078	60
1	.17663	5.66165	.19468	5.13658	.21286	4.69791	.23117	4.32573	.24964	4.00582	59
2	.17693	5.65205	.19498	5.12862	.21316	4.69121	.23148	4.32001	.24995	4.00086	58
3	.17723	5.64248	.19529	5.12069	.21347	4.68452	.23179	4.31430	.25026	3.99592	57
4	.17753	5.63295	.19559	5.11279	.21377	4.67786	.23209	4.30860	.25056	3.99099	56
5	.17783	5.62344	.19589	5.10490	.21408	4.67121	.23240	4.30291	.25087	3.98607	55
6	.17813	5.61397	.19619	5.09704	.21438	4.66458	.23271	4.29724	.25118	3.98117	54
7	.17843	5.60452	.19649	5.08921	.21469	4.65797	.23301	4.29159	.25149	3.97627	53
8	.17873	5.59511	.19680	5.08139	.21499	4.65138	.23332	4.28595	.25180	3.97139	52
9	.17903	5.58573	.19710	5.07360	.21529	4.64480	.23363	4.28032	.25211	3.96651	51
10	.17933	5.57638	.19740	5.06584	.21560	4.63825	.23393	4.27471	.25242	3.96165	50
11	.17963	5.56706	.19770	5.05809	.21590	4.63171	.23424	4.26911	.25273	3.95680	49
12	.17993	5.55777	.19801	5.05037	.21621	4.62518	.23455	4.26352	.25304	3.95196	48
13	.18023	5.54851	.19831	5.04267	.21651	4.61868	.23485	4.25795	.25335	3.94713	47
14	.18053	5.53927	.19861	5.03499	.21682	4.61219	.23516	4.25239	.25366	3.94232	46
15	.18083	5.53007	.19891	5.02734	.21712	4.60572	.23547	4.24685	.25397	3.93751	45
16	.18113	5.52090	.19921	5.01971	.21743	4.59927	.23578	4.24132	.25428	3.93271	44
17	.18143	5.51176	.19952	5.01210	.21773	4.59283	.23608	4.23580	.25459	3.92793	43
18	.18173	5.50264	.19982	5.00451	.21804	4.58641	.23639	4.23030	.25490	3.92316	42
19	.18203	5.49356	.20012	4.99695	.21834	4.58001	.23670	4.22481	.25521	3.91839	41
20	.18233	5.48451	.20042	4.98940	.21864	4.57363	.23700	4.21933	.25552	3.91364	40
21	.18263	5.47548	.20073	4.98188	.21895	4.56726	.23731	4.21387	.25583	3.90890	39
22	.18293	5.46648	.20103	4.97438	.21925	4.56091	.23762	4.20842	.25614	3.90417	38
23	.18323	5.45751	.20133	4.96690	.21956	4.55458	.23793	4.20298	.25645	3.89945	37
24	.18353	5.44857	.20164	4.95945	.21986	4.54826	.23823	4.19756	.25676	3.89474	36
25	.18384	5.43966	.20194	4.95201	.22017	4.54196	.23854	4.19215	.25707	3.89004	35
26	.18414	5.43077	.20224	4.94460	.22047	4.53568	.23885	4.18675	.25738	3.88536	34
27	.18444	5.42192	.20254	4.93721	.22078	4.52941	.23916	4.18137	.25769	3.88068	33
28	.18474	5.41309	.20285	4.92984	.22108	4.52316	.23946	4.17600	.25800	3.87601	32
29	.18504	5.40429	.20315	4.92249	.22139	4.51693	.23977	4.17064	.25831	3.87136	31
30	.18534	5.39552	.20345	4.91516	.22169	4.51071	.24008	4.16530	.25862	3.86671	30
31	.18564	5.38677	.20376	4.90785	.22200	4.50451	.24039	4.15997	.25893	3.86208	29
32	.18594	5.37805	.20406	4.90056	.22231	4.49832	.24069	4.15465	.25924	3.85745	28
33	.18624	5.36936	.20436	4.89330	.22261	4.49215	.24100	4.14934	.25955	3.85284	27
34	.18654	5.36070	.20466	4.88605	.22292	4.48600	.24131	4.14405	.25986	3.84824	26
35	.18684	5.35206	.20497	4.87882	.22322	4.47986	.24162	4.13877	.26017	3.84364	25
36	.18714	5.34345	.20527	4.87162	.22353	4.47374	.24193	4.13350	.26048	3.83906	24
37	.18745	5.33487	.20557	4.86444	.22383	4.46764	.24223	4.12825	.26079	3.83449	23
38	.18775	5.32631	.20588	4.85727	.22414	4.46155	.24254	4.12301	.26110	3.82992	22
39	.18805	5.31778	.20618	4.85013	.22444	4.45548	.24285	4.11778	.26141	3.82537	21
40	.18835	5.30928	.20648	4.84300	.22475	4.44942	.24316	4.11256	.26172	3.82083	20
41	.18865	5.30080	.20679	4.83590	.22505	4.44338	.24347	4.10736	.26203	3.81630	19
42	.18895	5.29235	.20709	4.82882	.22536	4.43735	.24377	4.10216	.26235	3.81177	18
43	.18925	5.28393	.20739	4.82175	.22567	4.43134	.24408	4.09699	.26266	3.80726	17
44	.18955	5.27553	.20770	4.81471	.22597	4.42534	.24439	4.09182	.26297	3.80276	16
45	.18986	5.26715	.20800	4.80769	.22628	4.41936	.24470	4.08666	.26328	3.79827	15
46	.19016	5.25880	.20830	4.80068	.22658	4.41340	.24501	4.08152	.26359	3.79378	14
47	.19046	5.25048	.20861	4.79370	.22689	4.40745	.24532	4.07639	.26390	3.78931	13
48	.19076	5.24218	.20891	4.78673	.22719	4.40152	.24562	4.07127	.26421	3.78485	12
49	.19106	5.23391	.20921	4.77978	.22750	4.39560	.24593	4.06616	.26452	3.78040	11
50	.19136	5.22566	.20952	4.77286	.22781	4.38969	.24624	4.06107	.26483	3.77595	10
51	.19166	5.21744	.20982	4.76595	.22811	4.38381	.24655	4.05599	.26515	3.77152	9
52	.19197	5.20925	.21013	4.75906	.22842	4.37793	.24686	4.05092	.26546	3.76709	8
53	.19227	5.20107	.21043	4.75219	.22872	4.37207	.24717	4.04586	.26577	3.76268	7
54	.19257	5.19293	.21073	4.74534	.22903	4.36623	.24747	4.04081	.26608	3.75828	6
55	.19287	5.18480	.21104	4.73851	.22934	4.36040	.24778	4.03578	.26639	3.75388	5
56	.19317	5.17671	.21134	4.73170	.22964	4.35459	.24809	4.03076	.26670	3.74950	4
57	.19347	5.16863	.21164	4.72490	.22995	4.34879	.24840	4.02574	.26701	3.74512	3
58	.19378	5.16058	.21195	4.71813	.23026	4.34300	.24871	4.02074	.26733	3.74075	2
59	.19408	5.15256	.21225	4.71137	.23056	4.33723	.24902	4.01576	.26764	3.73640	1
60	.19438	5.14455	.21256	4.70463	.23087	4.33148	.24933	4.01078	.26795	3.73205	0
	Cotang	Tang	Cotang	Tang	Cotang	Tang	Cotang	Tang	Cotang	Tang	
′	79°		78°		77°		76°		75°		′

′	15° Tang	15° Cotang	16° Tang	16° Cotang	17° Tang	17° Cotang	18° Tang	18° Cotang	19° Tang	19° Cotang	′
0	.26795	3.73205	.28675	3.48741	.30573	3.27085	.32492	3.07768	.34433	2.90421	60
1	.26826	3.72771	.28706	3.48359	.30605	3.26745	.32524	3.07464	.34465	2.90147	59
2	.26857	3.72338	.28738	3.47977	.30637	3.26406	.32556	3.07160	.34498	2.89873	58
3	.26888	3.71907	.28769	3.47596	.30669	3.26067	.32588	3.06857	.34530	2.89600	57
4	.26920	3.71476	.28800	3.47216	.30700	3.25729	.32621	3.06554	.34563	2.89327	56
5	.26951	3.71046	.28832	3.46837	.30732	3.25392	.32653	3.06252	.34596	2.89055	55
6	.26982	3.70616	.28864	3.46458	.30764	3.25055	.32685	3.05950	.34628	2.88783	54
7	.27013	3.70188	.28895	3.46080	.30796	3.24719	.32717	3.05649	.34661	2.88511	53
8	.27044	3.69761	.28927	3.45703	.30828	3.24383	.32749	3.05349	.34693	2.88240	52
9	.27076	3.69335	.28958	3.45327	.30860	3.24049	.32782	3.05049	.34726	2.87970	51
10	.27107	3.68909	.28990	3.44951	.30891	3.23714	.32814	3.04749	.34758	2.87700	50
11	.27138	3.68485	.29021	3.44576	.30923	3.23381	.32846	3.04450	.34791	2.87430	49
12	.27169	3.68061	.29053	3.44202	.30955	3.23048	.32878	3.04152	.34824	2.87161	48
13	.27201	3.67638	.29084	3.43829	.30987	3.22715	.32911	3.03854	.34856	2.86892	47
14	.27232	3.67217	.29116	3.43456	.31019	3.22384	.32943	3.03556	.34889	2.86624	46
15	.27263	3.66796	.29147	3.43084	.31051	3.22053	.32975	3.03260	.34922	2.86356	45
16	.27294	3.66376	.29179	3.42713	.31083	3.21722	.33007	3.02963	.34954	2.86089	44
17	.27326	3.65957	.29210	3.42343	.31115	3.21392	.33040	3.02667	.34987	2.85822	43
18	.27357	3.65538	.29242	3.41973	.31147	3.21063	.33072	3.02372	.35020	2.85555	42
19	.27388	3.65121	.29274	3.41604	.31178	3.20734	.33104	3.02077	.35052	2.85289	41
20	.27419	3.64705	.29305	3.41236	.31210	3.20406	.33136	3.01783	.35085	2.85023	40
21	.27451	3.64289	.29337	3.40869	.31242	3.20079	.33169	3.01489	.35118	2.84758	39
22	.27482	3.63874	.29368	3.40502	.31274	3.19752	.33201	3.01196	.35150	2.84494	38
23	.27513	3.63461	.29400	3.40136	.31306	3.19426	.33233	3.00903	.35183	2.84229	37
24	.27545	3.63048	.29432	3.39771	.31338	3.19100	.33266	3.00611	.35216	2.83965	36
25	.27576	3.62636	.29463	3.39406	.31370	3.18775	.33298	3.00319	.35248	2.83702	35
26	.27607	3.62224	.29495	3.39042	.31402	3.18451	.33330	3.00028	.35281	2.83439	34
27	.27638	3.61814	.29526	3.38679	.31434	3.18127	.33363	2.99738	.35314	2.83176	33
28	.27670	3.61405	.29558	3.38317	.31466	3.17804	.33395	2.99447	.35346	2.82914	32
29	.27701	3.60996	.29590	3.37955	.31498	3.17481	.33427	2.99158	.35379	2.82653	31
30	.27732	3.60588	.29621	3.37594	.31530	3.17159	.33460	2.98868	.35412	2.82391	30
31	.27764	3.60181	.29653	3.37234	.31562	3.16838	.33492	2.98580	.35445	2.82130	29
32	.27795	3.59775	.29685	3.36875	.31594	3.16517	.33524	2.98292	.35477	2.81870	28
33	.27826	3.59370	.29716	3.36516	.31626	3.16197	.33557	2.98004	.35510	2.81610	27
34	.27858	3.58966	.29748	3.36158	.31658	3.15877	.33589	2.97717	.35543	2.81350	26
35	.27889	3.58562	.29780	3.35800	.31690	3.15558	.33621	2.97430	.35576	2.81091	25
36	.27921	3.58160	.29811	3.35443	.31722	3.15240	.33654	2.97144	.35608	2.80833	24
37	.27952	3.57758	.29843	3.35087	.31754	3.14922	.33686	2.96858	.35641	2.80574	23
38	.27983	3.57357	.29875	3.34732	.31786	3.14605	.33718	2.96573	.35674	2.80316	22
39	.28015	3.56957	.29906	3.34377	.31818	3.14288	.33751	2.96288	.35707	2.80059	21
40	.28046	3.56557	.29938	3.34023	.31850	3.13972	.33783	2.96004	.35740	2.79802	20
41	.28077	3.56159	.29970	3.33670	.31882	3.13656	.33816	2.95721	.35772	2.79545	19
42	.28109	3.55761	.30001	3.33317	.31914	3.13341	.33848	2.95437	.35805	2.79289	18
43	.28140	3.55364	.30033	3.32965	.31946	3.13027	.33881	2.95155	.35838	2.79033	17
44	.28172	3.54968	.30065	3.32614	.31978	3.12713	.33913	2.94872	.35871	2.78778	16
45	.28203	3.54573	.30097	3.32264	.32010	3.12400	.33945	2.94591	.35904	2.78523	15
46	.28234	3.54179	.30128	3.31914	.32042	3.12087	.33978	2.94309	.35937	2.78269	14
47	.28266	3.53785	.30160	3.31565	.32074	3.11775	.34010	2.94028	.35969	2.78014	13
48	.28297	3.53393	.30192	3.31216	.32106	3.11464	.34043	2.93748	.36002	2.77761	12
49	.28329	3.53001	.30224	3.30868	.32139	3.11153	.34075	2.93468	.36035	2.77507	11
50	.28360	3.52609	.30255	3.30521	.32171	3.10842	.34108	2.93189	.36068	2.77254	10
51	.28391	3.52219	.30287	3.30174	.32203	3.10532	.34140	2.92910	.36101	2.77002	9
52	.28423	3.51829	.30319	3.29829	.32235	3.10223	.34173	2.92632	.36134	2.76750	8
53	.28454	3.51441	.30351	3.29483	.32267	3.09914	.34205	2.92354	.36167	2.76498	7
54	.28486	3.51053	.30382	3.29139	.32299	3.09606	.34238	2.92076	.36199	2.76247	6
55	.28517	3.50666	.30414	3.28795	.32331	3.09298	.34270	2.91799	.36232	2.75996	5
56	.28549	3.50279	.30446	3.28452	.32363	3.08991	.34303	2.91523	.36265	2.75746	4
57	.28580	3.49894	.30478	3.28109	.32396	3.08685	.34335	2.91246	.36298	2.75496	3
58	.28612	3.49509	.30509	3.27767	.32428	3.08379	.34368	2.90971	.36331	2.75246	2
59	.28643	3.49125	.30541	3.27426	.32460	3.08073	.34400	2.90696	.36364	2.74997	1
60	.28675	3.48741	.30573	3.27085	.32492	3.07768	.34433	2.90421	.36397	2.74748	0
′	Cotang	Tang	Cotang	Tang	Cotang	Tang	Cotang	Tang	Cotang	Tang	′
	74°		73°		72°		71°		70°		

′	20° Tang	20° Cotang	21° Tang	21° Cotang	22° Tang	22° Cotang	23° Tang	23° Cotang	24° Tang	24° Cotang	′
0	.36397	2.74748	.38386	2.60509	.40403	2.47509	.42447	2.35585	.44523	2.24604	60
1	.36430	2.74499	.38420	2.60283	.40436	2.47302	.42482	2.35395	.44558	2.24428	59
2	.36463	2.74251	.38453	2.60057	.40470	2.47095	.42516	2.35205	.44593	2.24252	58
3	.36496	2.74004	.38487	2.59831	.40504	2.46888	.42551	2.35015	.44627	2.24077	57
4	.36529	2.73756	.38520	2.59606	.40538	2.46682	.42585	2.34825	.44662	2.23902	56
5	.36562	2.73509	.38553	2.59381	.40572	2.46476	.42619	2.34636	.44697	2.23727	55
6	.36595	2.73263	.38587	2.59156	.40606	2.46270	.42654	2.34447	.44732	2.23553	54
7	.36628	2.73017	.38620	2.58932	.40640	2.46065	.42688	2.34258	.44767	2.23378	53
8	.36661	2.72771	.38654	2.58708	.40674	2.45860	.42722	2.34069	.44802	2.23204	52
9	.36694	2.72526	.38687	2.58484	.40707	2.45655	.42757	2.33881	.44837	2.23030	51
10	.36727	2.72281	.38721	2.58261	.40741	2.45451	.42791	2.33693	.44872	2.22857	50
11	.36760	2.72036	.38754	2.58038	.40775	2.45246	.42826	2.33505	.44907	2.22683	49
12	.36793	2.71792	.38787	2.57815	.40809	2.45043	.42860	2.33317	.44942	2.22510	48
13	.36826	2.71548	.38821	2.57593	.40843	2.44839	.42894	2:33130	.44977	2.22337	47
14	.36859	2.71305	.38854	2.57371	.40877	2.44636	.42929	2.32943	.45012	2.22164	46
15	.36892	2.71062	.38888	2.57150	.40911	2.44433	.42963	2.32756	.45047	2.21992	45
16	.36925	2.70819	.38921	2.56928	.40945	2.44230	.42998	2.32570	.45082	2.21819	44
17	.36958	2.70577	.38955	2.56707	.40979	2.44027	.43032	2.32383	.45117	2.21647	43
18	.36991	2.70335	.38988	2.56487	.41013	2.43825	.43067	2.32197	.45152	2.21475	42
19	.37024	2.70094	.39022	2.56266	.41047	2.43623	.43101	2.32012	.45187	2.21304	41
20	.37057	2.69853	.39055	2.56046	.41081	2.43422	.43136	2.31826	.45222	2.21132	40
21	.37090	2.69612	.39089	2.55827	.41115	2.43220	.43170	2.31641	.45257	2.20961	39
22	.37123	2.69371	.39122	2.55608	.41149	2.43019	.43205	2.31456	.45292	2.20790	38
23	.37157	2.69131	.39156	2.55389	.41183	2.42819	.43230	2.31271	.45327	2.20619	37
24	.37190	2.68892	.39190	2.55170	.41217	2.42618	.43274	2.31086	.45362	2.20449	36
25	.37223	2.68653	.39223	2.54952	.41251	2.42418	.43308	2.30902	.45397	2.20278	35
26	.37256	2.68414	.39257	2.54734	.41285	2.42218	.43343	2.30718	.45432	2.20108	34
27	.37289	2.68175	.39290	2.54516	.41319	2.42019	.43378	2.30534	.45467	2.19938	33
28	.37322	2.67937	.39324	2.54299	.41353	2.41819	.43412	2.30351	.45502	2.19769	32
29	.37355	2.67700	.39357	2.54082	.41387	2.41620	.43447	2.30167	.45538	2.19599	31
30	.37388	2.67462	.39391	2.53865	.41421	2.41421	.43481	2.29984	.45573	2.19430	30
31	.37422	2.67225	.39425	2.53648	.41455	2.41223	.43516	2.29801	.45608	2.19261	29
32	.37455	2.66989	.39458	2.53432	.41490	2.41025	.43550	2.29619	.45643	2.19093	28
33	.37488	2.66752	.39492	2.53217	.41524	2.40827	.43585	2.29437	.45678	2.18923	27
34	.37521	2.66516	.39526	2.53001	.41558	2.40629	.43620	2.29254	.45713	2.18755	26
35	.37554	2.66281	.39559	2.52786	.41592	2.40432	.43654	2.29073	.45748	2.18587	25
36	.37588	2.66046	.39593	2.52571	.41626	2.40235	.43689	2.28891	.45784	2.18419	24
37	.37621	2.65811	.39626	2.52357	.41660	2.40038	.43724	2.28710	.45819	2.18251	23
38	.37654	2.65576	.39660	2.52142	.41694	2.39841	.43758	2.28528	.45854	2.18084	22
39	.37687	2.65342	.39694	2.51929	.41728	2.39645	.43793	2.28348	.45889	2.17916	21
40	.37720	2.65109	.39727	2.51715	.41763	2.39449	.43828	2.28167	.45924	2.17749	20
41	.37754	2.64875	.39761	2.51502	.41797	2.39253	.43862	2.27987	.45960	2.17582	19
42	.37787	2.64642	.39795	2.51289	.41831	2.39058	.43897	2.27806	.45995	2.17416	18
43	.37820	2.64410	.39829	2.51076	.41865	2.38863	.43932	2.27626	.46030	2.17249	17
44	.37853	2.64177	.39862	2.50864	.41899	2.38668	.43966	2.27447	.46065	2.17083	16
45	.37887	2.63945	.39896	2.50652	.41933	2.38473	.44001	2.27267	.46101	2.16917	15
46	.37920	2.63714	.39930	2.50440	.41968	2.38279	.44036	2.27088	.46136	2.16751	14
47	.37953	2.63483	.39963	2.50229	.42002	2.38084	.44071	2.26909	.46171	2.16585	13
48	.37986	2.63252	.39997	2.50018	.42036	2.37891	.44105	2.26730	.46206	2.16420	12
49	.38020	2.63021	.40031	2.49807	.42070	2.37697	.44140	2.26552	.46242	2.16255	11
50	.38053	2.62791	.40065	2.49597	.42105	2.37504	.44175	2.26374	.46277	2.16090	10
51	.38086	2.62561	.40098	2.49386	.42139	2.37311	.44210	2.26196	.46312	2.15925	9
52	.38120	2.62332	.40132	2.49177	.42173	2.37118	.44244	2.26018	.46348	2.15760	8
53	.38153	2.62103	.40166	2.48967	.42207	2.36925	.44279	2.25840	.46383	2.15596	7
54	.38186	2.61874	.40200	2.48758	.42242	2.36733	.44314	2.25663	.46418	2.15432	6
55	.38220	2.61646	.40234	2.48549	.42276	2.36541	.44349	2.25486	.46454	2.15268	5
56	.38253	2.61418	.40267	2.48340	.42310	2.36349	.44384	2.25309	.46489	2.15104	4
57	.38286	2.61190	.40301	2.48132	.42345	2.36158	.44418	2.25132	.46525	2.14940	3
58	.38320	2.60963	.40335	2.47924	.42379	2.35967	.44453	2.24956	.46560	2.14777	2
59	.38353	2.60736	.40369	2.47716	.42413	2.35776	.44488	2.24780	.46595	2.14614	1
60	.38386	2.60509	.40403	2.47509	.42447	2.35585	.44523	2.24604	.46631	2.14451	0
′	69° Cotang	69° Tang	68° Cotang	68° Tang	67° Cotang	67° Tang	66° Cotang	66° Tang	65° Cotang	65° Tang	′

′	25° Tang	25° Cotang	26° Tang	26° Cotang	27° Tang	27° Cotang	28° Tang	28° Cotang	29° Tang	29° Cotang	′
0	.46631	2.14451	.48773	2.05030	.50953	1.96261	.53171	1.88073	.55431	1.80405	60
1	.46666	2.14288	.48809	2.04879	.50989	1.96120	.53208	1.87941	.55469	1.80281	59
2	.46702	2.14125	.48845	2.04728	.51026	1.95979	.53246	1.87809	.55507	1.80158	58
3	.46737	2.13963	.48881	2.04577	.51063	1.95838	.53283	1.87677	.55545	1.80034	57
4	.46772	2.13801	.48917	2.04426	.51099	1.95698	.53320	1.87546	.55583	1.79911	56
5	.46808	2.13639	.48953	2.04276	.51136	1.95557	.53358	1.87415	.55621	1.79788	55
6	.46843	2.13477	.48989	2.04125	.51173	1.95417	.53395	1.87283	.55659	1.79665	54
7	.46879	2.13316	.49026	2.03975	.51209	1.95277	.53432	1.87152	55697	1.79542	53
8	.46914	2.13154	.49062	2.03825	.51246	1.95137	.53470	1.87021	.55736	1.79419	52
9	.46950	2.12993	.49098	2.03675	.51283	1.94997	.53507	1.86891	.55774	1.79296	51
10	.46985	2.12832	.49134	2.03526	.51319	1.94858	.53545	1.86760	.55812	1.79174	50
11	.47021	2.12671	.49170	2.03376	.51356	1.94718	.53582	1.86630	.55850	1.79051	49
12	.47056	2.12511	.49206	2.03227	.51393	1.94579	.53620	1.86499	.55888	1.78929	48
13	.47092	2.12350	.49242	2.03078	.51430	1.94440	.53657	1.86369	.55926	1.78807	47
14	.47128	2.12190	.49278	2.02929	.51467	1.94301	.53694	1.86239	.55964	1.78685	46
15	.47163	2.12030	.49315	2.02780	.51503	1.94162	.53732	1.86109	.56003	1.78563	45
16	.47199	2.11871	.49351	2.02631	.51540	1.94023	.53769	1.85979	.56041	1.78441	44
17	.47234	2.11711	.49387	2.02483	.51577	1.93885	.53807	1.85850	.56079	1.78319	43
18	.47270	2.11552	.49423	2.02335	.51614	1.93746	.53844	1.85720	.56117	1.78198	42
19	.47305	2.11392	.49459	2.02187	.51651	1.93608	.53882	1.85591	.56156	1.78077	41
20	.47341	2.11233	.49495	2.02039	.51688	1.93470	.53920	1.85462	.56194	1.77955	40
21	.47377	2.11075	.49532	2.01891	.51724	1.93332	.53957	1.85333	.56232	1.77834	39
22	.47412	2.10916	.49568	2.01743	.51761	1.93195	.53995	1.85204	.56270	1.77713	38
23	.47448	2.10758	.49604	2.01596	.51798	1.93057	.54032	1.85075	.56309	1.77592	37
24	.47483	2.10600	.49640	2.01449	.51835	1.92920	.54070	1.84946	.56347	1.77471	36
25	.47519	2.10442	.49677	2.01302	.51872	1.92782	.54107	1.84818	.56385	1.77351	35
26	.47555	2.10284	.49713	2.01155	.51909	1.92645	.54145	1.84689	.56424	1.77230	34
27	.47590	2.10126	.49749	2.01008	.51946	1.92508	.54183	1.84561	.56462	1.77110	33
28	.47626	2.09969	.49786	2.00862	.51983	1.92371	.54220	1.84433	.56501	1.76990	32
29	.47662	2.09811	.49822	2.00715	.52020	1.92235	.54258	1.84305	.56539	1.76869	31
30	.47698	2.09654	.49858	2.00569	.52057	1.92098	.54296	1.84177	.56577	1.76749	30
31	.47733	2.09498	.49894	2.00423	.52094	1.91962	.54333	1.84049	.56616	1.76629	29
32	.47769	2.09341	.49931	2.00277	.52131	1.91826	.54371	1.83922	.56654	1.76510	28
33	.47805	2.09184	.49967	2.00131	.52168	1.91690	.54409	1.83794	.56693	1.76390	27
34	.47840	2.09028	.50004	1.99986	.52205	1.91554	.54446	1.83667	.56731	1.76271	26
35	.47876	2.08872	.50040	1.99841	.52242	1.91418	.54484	1.83540	.56769	1.76151	25
36	.47912	2.08716	.50076	1.99695	.52279	1.91282	.54522	1.83413	.56808	1.76032	24
37	.47948	2.08560	.50113	1.99550	.52316	1.91147	.54560	1.83286	.56846	1.75913	23
38	.47984	2.08405	.50149	1.99406	.52353	1.91012	.54597	1.83159	.56885	1.75794	22
39	.48019	2.08250	.50185	1.99261	.52390	1.90876	.54635	1.83033	.56923	1.75675	21
40	.48055	2.08094	.50222	1.99116	.52427	1.90741	.54673	1.82906	.56962	1.75556	20
41	.48091	2.07939	.50258	1.98972	.52464	1.90607	.54711	1.82780	.57000	1.75437	19
42	.48127	2.07785	.50295	1.98828	.52501	1.90472	.54748	1.82654	.57039	1.75319	18
43	.48163	2.07630	.50331	1.98684	.52538	1.90337	.54786	1.82528	.57078	1.75200	17
44	.48198	2.07476	.50368	1.98540	.52575	1.90203	.54824	1.82402	.57116	1.75082	16
45	.48234	2.07321	.50404	1.98396	.52613	1.90069	.54862	1.82276	.57155	1.74964	15
46	.48270	2.07167	.50441	1.98253	.52650	1.89935	.54900	1.82150	.57193	1.74846	14
47	.48306	2.07014	.50477	1.98110	.52687	1.89801	.54938	1.82025	.57232	1.74728	13
48	.48342	2.06860	.50514	1.97966	.52724	1.89667	.54975	1.81899	.57271	1.74610	12
49	.48378	2.06706	.50550	1.97823	.52761	1.89533	.55013	1.81774	.57309	1.74492	11
50	.48414	2.06553	.50587	1.97681	.52798	1.89400	.55051	1.81649	.57348	1.74375	10
51	.48450	2.06400	.50623	1.97538	.52836	1.89266	.55089	1.81524	.57386	1.74257	9
52	.48486	2.06247	.50660	1.97395	.52873	1.89133	.55127	1.81399	.57425	1.74140	8
53	.48521	2.06094	.50696	1.97253	.52910	1.89000	.55165	1.81274	.57464	1.74022	7
54	.48557	2.05942	.50733	1.97111	.52947	1.88867	.55203	1.81150	.57503	1.73905	6
55	.48593	2.05790	.50769	1.96969	.52985	1.88734	.55241	1.81025	.57541	1.73788	5
56	.48629	2.05637	.50806	1.96827	.53022	1.88602	.55279	1.80901	.57580	1.73671	4
57	.48665	2.05485	.50843	1.96685	.53059	1.88469	.55317	1.80777	.57619	1.73555	3
58	.48701	2.05333	.50879	1.96544	.53096	1.88337	.55355	1.80653	.57657	1.73438	2
59	.48737	2.05182	.50916	1.96402	.53134	1.88205	.55393	1.80529	.57696	1.73321	1
60	.48773	2.05030	.50953	1.96261	.53171	1.88073	.55431	1.80405	.57735	1.73205	0
′	Cotang	Tang	Cotang	Tang	Cotang	Tang	Cotang	Tang	Cotang	Tang	′
	64°		63°		62°		61°		60°		

′	30° Tang	30° Cotang	31° Tang	31° Cotang	32° Tang	32° Cotang	33° Tang	33° Cotang	34° Tang	34° Cotang	′
0	.57735	1.73205	.60086	1.66428	.62487	1.60033	.64941	1.53986	.67451	1.48256	60
1	.57774	1.73089	.60126	1.66318	.62527	1.59930	.64982	1.53888	.67493	1.48163	59
2	.57813	1.72973	.60165	1.66209	.62568	1.59826	.65024	1.53791	.67536	1.48070	58
3	.57851	1.72857	.60205	1.66099	.62608	1.59723	.65065	1.53693	.67578	1.47977	57
4	.57890	1.72741	.60245	1.65990	.62649	1.59620	.65106	1.53595	.67620	1.47885	56
5	.57929	1.72625	.60284	1.65881	.62689	1.59517	.65148	1.53497	.67663	1.47792	55
6	.57968	1.72509	.60324	1.65772	.62730	1.59414	.65189	1.53400	.67705	1.47699	54
7	.58007	1.72393	.60364	1.65663	.62770	1.59311	.65231	1.53302	.67748	1.47607	53
8	.58046	1.72278	.60403	1.65554	.62811	1.59208	.65272	1.53205	.67790	1.47514	52
9	.58085	1.72163	.60443	1.65445	.62852	1.59105	.65314	1.53107	.67832	1.47422	51
10	.58124	1.72047	.60483	1.65337	.62892	1.59002	.65355	1.53010	.67875	1.47330	50
11	.58162	1.71932	.60522	1.65228	.62933	1.58900	.65397	1.52913	.67917	1.47238	49
12	.58201	1.71817	.60562	1.65120	.62973	1.58797	.65438	1.52816	.67960	1.47146	48
13	.58240	1.71702	.60602	1.65011	.63014	1.58695	.65480	1.52719	.68002	1.47053	47
14	.58279	1.71588	.60642	1.64903	.63055	1.58593	.65521	1.52622	.68045	1.46962	46
15	.58318	1.71473	.60681	1.64795	.63095	1.58490	.65563	1.52525	.68088	1.46870	45
16	.58357	1.71358	.60721	1.64687	.63136	1.58388	.65604	1.52429	.68130	1.46778	44
17	.58396	1.71244	.60761	1.64579	.63177	1.58286	.65646	1.52332	.68173	1.46686	43
18	.58435	1.71129	.60801	1.64471	.63217	1.58184	.65688	1.52235	.68215	1.46595	42
19	.58474	1.71015	.60841	1.64363	.63258	1.58083	.65729	1.52139	.68258	1.46503	41
20	.58513	1.70901	.60881	1.64256	.63299	1.57981	.65771	1.52043	.68301	1.46411	40
21	.58552	1.70787	.60921	1.64148	.63340	1.57879	.65813	1.51946	.68343	1.46320	39
22	.58591	1.70673	.60960	1.64041	.63380	1.57778	.65854	1.51850	.68386	1.46229	38
23	.58631	1.70560	.61000	1.63934	.63421	1.57676	.65896	1.51754	.68429	1.46137	37
24	.58670	1.70446	.61040	1.63826	.63462	1.57575	.65938	1.51658	.68471	1.46046	36
25	.58709	1.70332	.61080	1.63719	.63503	1.57474	.65980	1.51562	.68514	1.45955	35
26	.58748	1.70219	.61120	1.63612	.63544	1.57372	.66021	1.51466	.68557	1.45864	34
27	.58787	1.70106	.61160	1.63505	.63584	1.57271	.66063	1.51370	.68600	1.45773	33
28	.58826	1.69992	.61200	1.63398	.63625	1.57170	.66105	1.51275	.68642	1.45682	32
29	.58865	1.69879	.61240	1.63292	.63666	1.57069	.66147	1.51179	.68685	1.45592	31
30	.58905	1.69766	.61280	1.63185	.63707	1.56969	.66189	1.51084	.68728	1.45501	30
31	.58944	1.69653	.61320	1.63079	.63748	1.56868	.66230	1.50988	.68771	1.45410	29
32	.58983	1.69541	.61360	1.62972	.63789	1.56767	.66272	1.50893	.68814	1.45320	28
33	.59022	1.69428	.61400	1.62866	.63830	1.56667	.66314	1.50797	.68857	1.45229	27
34	.59061	1.69316	.61440	1.62760	.63871	1.56566	.66356	1.50702	.68900	1.45139	26
35	.59101	1.69203	.61480	1.62654	.63912	1.56466	.66398	1.50607	.68942	1.45049	25
36	.59140	1.69091	.61520	1.62548	.63953	1.56366	.66440	1.50512	.68985	1.44958	24
37	.59179	1.68979	.61561	1.62442	.63994	1.56265	.66482	1.50417	.69028	1.44868	23
38	.59218	1.68866	.61601	1.62336	.64035	1.56165	.66524	1.50322	.69071	1.44778	22
39	.59258	1.68754	.61641	1.62230	.64076	1.56065	.66566	1.50228	.69114	1.44688	21
40	.59297	1.68643	.61681	1.62125	.64117	1.55966	.66608	1.50133	.69157	1.44598	20
41	.59336	1.68531	.61721	1.62019	.64158	1.55866	.66650	1.50038	.69200	1.44508	19
42	.59376	1.68419	.61761	1.61914	.64199	1.55766	.66692	1.49944	.69243	1.44418	18
43	.59415	1.68308	.61801	1.61808	.64240	1.55666	.66734	1.49849	.69286	1.44329	17
44	.59454	1.68196	.61842	1.61703	.64281	1.55567	.66776	1.49755	.69329	1.44239	16
45	.59494	1.68085	.61882	1.61598	.64322	1.55467	.66818	1.49661	.69372	1.44149	15
46	.59533	1.67974	.61922	1.61493	.64363	1.55368	.66860	1.49566	.69416	1.44060	14
47	.59573	1.67863	.61962	1.61388	.64404	1.55269	.66902	1.49472	.69459	1.43970	13
48	.59612	1.67752	.62003	1.61283	.64446	1.55170	.66944	1.49378	.69502	1.43881	12
49	.59651	1.67641	.62043	1.61179	.64487	1.55071	.66986	1.49284	.69545	1.43792	11
50	.59691	1.67530	.62083	1.61074	.64528	1.54972	.67028	1.49190	.69588	1.43703	10
51	.59730	1.67419	.62124	1.60970	.64569	1.54873	.67071	1.49097	.69631	1.43614	9
52	.59770	1.67309	.62164	1.60865	.64610	1.54774	.67113	1.49003	.69675	1.43525	8
53	.59809	1.67198	.62204	1.60761	.64652	1.54675	.67155	1.48909	.69718	1.43436	7
54	.59849	1.67088	.62245	1.60657	.64693	1.54576	.67197	1.48816	.69761	1.43347	6
55	.59888	1.66978	.62285	1.60553	.64734	1.54478	.67239	1.48722	.69804	1.43258	5
56	.59928	1.66867	.62325	1.60449	.64775	1.54379	.67282	1.48629	.69847	1.43169	4
57	.59967	1.66757	.62366	1.60345	.64817	1.54281	.67324	1.48536	.69891	1.43080	3
58	.60007	1.66647	.62406	1.60241	.64858	1.54183	.67366	1.48442	.69934	1.42992	2
59	.60046	1.66538	.62446	1.60137	.64899	1.54085	.67409	1.48349	.69977	1.42903	1
60	.60086	1.66428	.62487	1.60033	.64941	1.53986	.67451	1.48256	.70021	1.42815	0
′	Cotang	Tang	Cotang	Tang	Cotang	Tang	Cotang	Tang	Cotang	Tang	′
	59°		58°		57°		56°		55°		

′	35° Tang	35° Cotang	36° Tang	36° Cotang	37° Tang	37° Cotang	38° Tang	38° Cotang	39° Tang	39° Cotang	′
0	.70021	1.42815	.72654	1.37638	.75355	1.32704	.78129	1.27994	.80978	1.23490	60
1	.70064	1.42726	.72699	1.37554	.75401	1.32624	.78175	1.27917	.81027	1.23416	59
2	.70107	1.42638	.72743	1.37470	.75447	1.32544	.78222	1.27841	.81075	1.23343	58
3	.70151	1.42550	.72788	1.37386	.75492	1.32464	.78269	1.27764	.81123	1.23270	57
4	.70194	1.42462	.72832	1.37302	.75538	1.32384	.78316	1.27688	.81171	1.23196	56
5	.70238	1.42374	.72877	1.37218	.75584	1.32304	.78363	1.27611	.81220	1.23123	55
6	.70281	1.42286	.72921	1.37134	.75629	1.32224	.78410	1.27535	.81268	1.23050	54
7	.70325	1.42198	.72966	1.37050	.75675	1.32144	.78457	1.27458	.81316	1.22977	53
8	.70368	1.42110	.73010	1.36967	.75721	1.32064	.78504	1.27382	.81364	1.22904	52
9	.70412	1.42022	.73055	1.36883	.75767	1.31984	.78551	1.27306	.81413	1.22831	51
10	.70455	1.41934	.73100	1.36800	.75812	1.31904	.78598	1.27230	.81461	1.22758	50
11	.70499	1.41847	.73144	1.36716	.75858	1.31825	.78645	1.27153	.81510	1.22685	49
12	.70542	1.41759	.73189	1.36633	.75904	1.31745	.78692	1.27077	.81558	1.22612	48
13	.70586	1.41672	.73234	1.36549	.75950	1.31666	.78739	1.27001	.81606	1.22539	47
14	.70629	1.41584	.73278	1.36466	.75996	1.31586	.78786	1.26925	.81655	1.22467	46
15	.70673	1.41497	.73323	1.36383	.76042	1.31507	.78834	1.26849	.81703	1.22394	45
16	.70717	1.41409	.73368	1.36300	.76088	1.31427	.78881	1.26774	.81752	1.22321	44
17	.70760	1.41322	.73413	1.36217	.76134	1.31348	.78928	1.26698	.81800	1.22249	43
18	.70804	1.41235	.73457	1.36134	.76180	1.31269	.78975	1.26622	.81849	1.22176	42
19	.70848	1.41148	.73502	1.36051	.76226	1.31190	.79022	1.26546	.81898	1.22104	41
20	.70891	1.41061	.73547	1.35968	.76272	1.31110	.79070	1.26471	.81946	1.22031	40
21	.70935	1.40974	.73592	1.35885	.76318	1.31031	.79117	1.26395	.81995	1.21959	39
22	.70979	1.40887	.73637	1.35802	.76364	1.30952	.79164	1.26319	.82044	1.21886	38
23	.71023	1.40800	.73681	1.35719	.76410	1.30873	.79212	1.26244	.82092	1.21814	37
24	.71066	1.40714	.73726	1.35637	.76456	1.30795	.79259	1.26169	.82141	1.21742	36
25	.71110	1.40627	.73771	1.35554	.76502	1.30716	.79306	1.26093	.82190	1.21670	35
26	.71154	1.40540	.73816	1.35472	.76548	1.30637	.79354	1.26018	.82238	1.21598	34
27	.71198	1.40454	.73861	1.35389	.76594	1.30558	.79401	1.25943	.82287	1.21526	33
28	.71242	1.40367	.73906	1.35307	.76640	1.30480	.79449	1.25867	.82336	1.21454	32
29	.71285	1.40281	.73951	1.35224	.76686	1.30401	.79496	1.25792	.82385	1.21382	31
30	.71329	1.40195	.73996	1.35142	.76733	1.30323	.79544	1.25717	.82434	1.21310	30
31	.71373	1.40109	.74041	1.35060	.76779	1.30244	.79591	1.25642	.82483	1.21238	29
32	.71417	1.40022	.74086	1.34978	.76825	1.30166	.79639	1.25567	.82531	1.21166	28
33	.71461	1.39936	.74131	1.34896	.76871	1.30087	.79686	1.25492	.82580	1.21094	27
34	.71505	1.39850	.74176	1.34814	.76918	1.30009	.79734	1.25417	.82629	1.21023	26
35	.71549	1.39764	.74221	1.34732	.76964	1.29931	.79781	1.25343	.82678	1.20951	25
36	.71593	1.39679	.74267	1.34650	.77010	1.29853	.79829	1.25268	.82727	1.20879	24
37	.71637	1.39593	.74312	1.34568	.77057	1.29775	.79877	1.25193	.82776	1.20808	23
38	.71681	1.39507	.74357	1.34487	.77103	1.29696	.79924	1.25118	.82825	1.20736	22
39	.71725	1.39421	.74402	1.34405	.77149	1.29618	.79972	1.25044	.82874	1.20665	21
40	.71769	1.39336	.74447	1.34323	.77196	1.29541	.80020	1.24969	.82923	1.20593	20
41	.71813	1.39250	.74492	1.34242	.77242	1.29463	.80067	1.24895	.82972	1.20522	19
42	.71857	1.39165	.74538	1.34160	.77289	1.29385	.80115	1.24820	.83022	1.20451	18
43	.71901	1.39079	.74583	1.34079	.77335	1.29307	.80163	1.24746	.83071	1.20379	17
44	.71946	1.38994	.74628	1.33998	.77382	1.29229	.80211	1.24672	.83120	1.20308	16
45	.71990	1.38909	.74674	1.33916	.77428	1.29152	.80258	1.24597	.83169	1.20237	15
46	.72034	1.38824	.74719	1.33835	.77475	1.29074	.80306	1.24523	.83218	1.20166	14
47	.72078	1.38738	.74764	1.33754	.77521	1.28997	.80354	1.24449	.83268	1.20095	13
48	.72122	1.38653	.74810	1.33673	.77568	1.28919	.80402	1.24375	.83317	1.20024	12
49	.72167	1.38568	.74855	1.33592	.77615	1.28842	.80450	1.24301	.83366	1.19953	11
50	.72211	1.38484	.74900	1.33511	.77661	1.28764	.80498	1.24227	.83415	1.19882	10
51	.72255	1.38399	.74946	1.33430	.77708	1.28687	.80546	1.24153	.83465	1.19811	9
52	.72299	1.38314	.74991	1.33349	.77754	1.28610	.80594	1.24079	.83514	1.19740	8
53	.72344	1.38229	.75037	1.33268	.77801	1.28533	.80642	1.24005	.83564	1.19669	7
54	.72388	1.38145	.75082	1.33187	.77848	1.28456	.80690	1.23931	.83613	1.19599	6
55	.72432	1.38060	.75128	1.33107	.77895	1.28379	.80738	1.23858	.83662	1.19528	5
56	.72477	1.37976	.75173	1.33026	.77941	1.28302	.80786	1.23784	.83712	1.19457	4
57	.72521	1.37891	.75219	1.32946	.77988	1.28225	.80834	1.23710	.83761	1.19387	3
58	.72565	1.37807	.75264	1.32865	.78035	1.28148	.80882	1.23637	.83811	1.19316	2
59	.72610	1.37722	.75310	1.32785	.78082	1.28071	.80930	1.23563	.83860	1.19246	1
60	.72654	1.37638	.75355	1.32704	.78129	1.27994	.80978	1.23490	.83910	1.19175	0
′	Cotang	Tang	Cotang	Tang	Cotang	Tang	Cotang	Tang	Cotang	Tang	′
	54°		53°		52°		51°		50°		

′	40°		41°		42°		43°		44°		′
	Tang	Cotang	Tang	Cotang	Tang	Cotang	Tang	Cotang	Tang	Cotang	
0	.83910	1.19175	.86929	1.15037	.90040	1.11061	.93252	1.07237	.96569	1.03553	60
1	.83960	1.19105	.86980	1.14969	.90093	1.10996	.93306	1.07174	.96625	1.03493	59
2	.84009	1.19035	.87031	1.14902	.90146	1.10931	.93360	1.07112	.96681	1.03433	58
3	.84059	1.18964	.87082	1.14834	.90199	1.10867	.93415	1.07049	.96738	1.03372	57
4	.84108	1.18894	.87133	1.14767	.90251	1.10802	.93469	1.06987	.96794	1.03312	56
5	.84158	1.18824	.87184	1.14699	.90304	1.10737	.93524	1.06925	.96850	1.03252	55
6	.84208	1.18754	.87236	1.14632	.90357	1.10672	.93578	1.06862	.96907	1.03192	54
7	.84258	1.18684	.87287	1.14565	.90410	1.10607	.93633	1.06800	.96963	1.03132	53
8	.84307	1.18614	.87338	1.14498	.90463	1.10543	.93688	1.06738	.97020	1.03072	52
9	.84357	1.18544	.87389	1.14430	.90516	1.10478	.93742	1.06676	.97076	1.03012	51
10	.84407	1.18474	.87441	1.14363	.90569	1.10414	.93797	1.06613	.97133	1.02952	50
11	.84457	1.18404	.87492	1.14296	.90621	1.10349	.93852	1.06551	.97189	1.02892	49
12	.84507	1.18334	.87543	1.14229	.90674	1.10285	.93906	1.06489	.97246	1.02832	48
13	.84556	1.18264	.87595	1.14162	.90727	1.10220	.93961	1.06427	.97302	1.02772	47
14	.84606	1.18194	.87646	1.14095	.90781	1.10156	.94016	1.06365	.97359	1.02713	46
15	.84656	1.18125	.87698	1.14028	.90834	1.10091	.94071	1.06303	.97416	1.02653	45
16	.84706	1.18055	.87749	1.13961	.90887	1.10027	.94125	1.06241	.97472	1.02593	44
17	.84756	1.17986	.87801	1.13894	.90940	1.09963	.94180	1.06179	.97529	1.02533	43
18	.84806	1.17916	.87852	1.13828	.90993	1.09899	.94235	1.06117	.97586	1.02474	42
19	.84856	1.17846	.87904	1.13761	.91046	1.09834	.94290	1.06056	.97643	1.02414	41
20	.84906	1.17777	.87955	1.13694	.91099	1.09770	.94345	1.05994	.97700	1.02355	40
21	.84956	1.17708	.88007	1.13627	.91153	1.09706	.94400	1.05932	.97756	1.02295	39
22	.85006	1.17638	.88059	1.13561	.91206	1.09642	.94455	1.05870	.97813	1.02236	38
23	.85057	1.17569	.88110	1.13494	.91259	1.09578	.94510	1.05809	.97870	1.02176	37
24	.85107	1.17500	.88162	1.13428	.91313	1.09514	.94565	1.05747	.97927	1.02117	36
25	.85157	1.17430	.88214	1.13361	.91366	1.09450	.94620	1.05685	.97984	1.02057	35
26	.85207	1.17361	.88265	1.13295	.91419	1.09386	.94676	1.05624	.98041	1.01998	34
27	.85257	1.17292	.88317	1.13228	.91473	1.09322	.94731	1.05562	.98098	1.01939	33
28	.85308	1.17223	.88369	1.13162	.91526	1.09258	.94786	1.05501	.98155	1.01879	32
29	.85358	1.17154	.88421	1.13096	.91580	1.09195	.94841	1.05439	.98213	1.01820	31
30	.85408	1.17085	.88473	1.13029	.91633	1.09131	.94896	1.05378	.98270	1.01761	30
31	.85458	1.17016	.88524	1.12963	.91687	1.09067	.94952	1.05317	.98327	1.01702	29
32	.85509	1.16947	.88576	1.12897	.91740	1.09003	.95007	1.05255	.98384	1.01642	28
33	.85559	1.16878	.88628	1.12831	.91794	1.08940	.95062	1.05194	.98441	1.01583	27
34	.85609	1.16809	.88680	1.12765	.91847	1.08876	.95118	1.05133	.98499	1.01524	26
35	.85660	1.16741	.88732	1.12699	.91901	1.08813	.95173	1.05072	.98556	1.01465	25
36	.85710	1.16672	.88784	1.12633	.91955	1.08749	.95229	1.05010	.98613	1.01406	24
37	.85761	1.16603	.88836	1.12567	.92008	1.08686	.95284	1.04949	.98671	1.01347	23
38	.85811	1.16535	.88888	1.12501	.92062	1.08622	.95340	1.04888	.98728	1.01288	22
39	.85862	1.16466	.88940	1.12435	.92116	1.08559	.95395	1.04827	.98786	1.01229	21
40	.85912	1.16398	.88992	1.12369	.92170	1.08496	.95451	1.04766	.98843	1.01170	20
41	.85963	1.16329	.89045	1.12303	.92224	1.08432	.95506	1.04705	.98901	1.01112	19
42	.86014	1.16261	.89097	1.12238	.92277	1.08369	.95562	1.04644	.98958	1.01053	18
43	.86064	1.16192	.89149	1.12172	.92331	1.08306	.95618	1.04583	.99016	1.00994	17
44	.86115	1.16124	.89201	1.12106	.92385	1.08243	.95673	1.04522	.99073	1.00935	16
45	.86166	1.16056	.89253	1.12041	.92439	1.08179	.95729	1.04461	.99131	1.00876	15
46	.86216	1.15987	.89306	1.11975	.92493	1.08116	.95785	1.04401	.99189	1.00818	14
47	.86267	1.15919	.89358	1.11909	.92547	1.08053	.95841	1.04340	.99247	1.00759	13
48	.86318	1.15851	.89410	1.11844	.92601	1.07990	.95897	1.04279	.99304	1.00701	12
49	.86368	1.15783	.89463	1.11778	.92655	1.07927	.95952	1.04218	.99362	1.00642	11
50	.86419	1.15715	.89515	1.11713	.92709	1.07864	.96008	1.04158	.99420	1.00583	10
51	.86470	1.15647	.89567	1.11648	.02763	1.07801	.96064	1.04097	.99478	1.00525	9
52	.86521	1.15579	.89620	1.11582	.92817	1.07738	.96120	1.04036	.99536	1.00467	8
53	.86572	1.15511	.89672	1.11517	.92872	1.07676	.96176	1.03976	.99594	1.00408	7
54	.86623	1.15443	.89725	1.11452	.92926	1.07613	.96232	1.03915	.99652	1.00350	6
55	.86674	1.15375	.89777	1.11387	.92980	1.07550	.96288	1.03855	.99710	1.00291	5
56	.86725	1.15308	.89830	1.11321	.93034	1.07487	.96344	1.03794	.99768	1.00233	4
57	.86776	1.15240	.89883	1.11256	.93088	1.07425	.96400	1.03734	.99826	1.00175	3
58	.86827	1.15172	.89935	1.11191	.93143	1.07362	.96457	1.03674	.99884	1.00116	2
59	.86878	1.15104	.89988	1.11126	.93197	1.07299	.96513	1.03613	.99942	1.00058	1
60	.86929	1.15037	.90040	1.11061	.93252	1.07237	.96569	1.03553	1.00000	1.00000	0
′	Cotang	Tang	Cotang	Tang	Cotang	Tang	Cotang	Tang	Cotang	Tang	′
	49°		48°		47°		46°		45°		

LOGARITHMIC TRIGONOMETRIC FUNCTIONS

0°

″	′	log sin	d	S	T	log tan	c. d.	log cot	C	log cos	
				$\bar{6}$.685					5.314		
0	0									0.00000	60
60	1	$\bar{4}$.46373		57	57	$\bar{4}$.46373		3.53627	43	0.00000	59
120	2	$\bar{4}$.76476	30103	57	57	$\bar{4}$.76476	30103	3.23524	43	0.00000	58
180	3	$\bar{4}$.94085	17609	57	58	$\bar{4}$.94085	17609	3.05915	43	0.00000	57
240	4	$\bar{3}$.06579	12494	57	58	$\bar{3}$.06579	12494	2.93421	42	0.00000	56
300	5	$\bar{3}$.16270	9691	57	58	$\bar{3}$.16270	9691	2.83730	42	0.00000	55
360	6	$\bar{3}$.24188	7918	57	58	$\bar{3}$.24188	7918	2.75812	42	0.00000	54
420	7	$\bar{3}$.30882	6694	57	58	$\bar{3}$.30882	6694	2.69118	42	0.00000	53
480	8	$\bar{3}$.36682	5800	57	58	$\bar{3}$.36682	5800	2.63318	42	0.00000	52
540	9	$\bar{3}$.41797	5115	57	58	$\bar{3}$.41797	5115	2.58203	42	0.00000	51
600	10	$\bar{3}$.46373	4576	57	58	$\bar{3}$.46373	4576	2.53627	42	0.00000	50
660	11	$\bar{3}$.50512	4139	57	58	$\bar{3}$.50512	4139	2.49488	42	0.00000	49
720	12	$\bar{3}$.54291	3779	57	58	$\bar{3}$.54291	3779	2.45709	42	0.00000	48
780	13	$\bar{3}$.57767	3476	57	58	$\bar{3}$.57767	3476	2.42233	42	0.00000	47
840	14	$\bar{3}$.60985	3218	57	58	$\bar{3}$.60986	3219	2.39014	42	0.00000	46
900	15	$\bar{3}$.63982	2997	57	58	$\bar{3}$.63982	2996	2.36018	42	0.00000	45
960	16	$\bar{3}$.66784	2802	57	58	$\bar{3}$.66785	2803	2.33215	42	0.00000	44
1020	17	$\bar{3}$.69417	2633	57	58	$\bar{3}$.69418	2633	2.30582	42	$\bar{1}$.99999	43
1080	18	$\bar{3}$.71900	2483	57	58	$\bar{3}$.71900	2482	2.28100	42	$\bar{1}$.99999	42
1140	19	$\bar{3}$.74248	2348	57	58	$\bar{3}$.74248	2348	2.25752	42	$\bar{1}$.99999	41
1200	20	$\bar{3}$.76475	2227	57	58	$\bar{3}$.76476	2228	2.23524	42	$\bar{1}$.99999	40
1260	21	$\bar{3}$.78594	2119	57	58	$\bar{3}$.78595	2119	2.21405	42	$\bar{1}$.99999	39
1320	22	$\bar{3}$.80615	2021	57	58	$\bar{3}$.80615	2020	2.19385	42	$\bar{1}$.99999	38
1380	23	$\bar{3}$.82545	1930	57	58	$\bar{3}$.82546	1931	2.17454	42	$\bar{1}$.99999	37
1440	24	$\bar{3}$.84393	1848	57	58	$\bar{3}$.84394	1848	2.15606	42	$\bar{1}$.99999	36
1500	25	$\bar{3}$.86166	1773	57	58	$\bar{3}$.86167	1773	2.13833	42	$\bar{1}$.99999	35
1560	26	$\bar{3}$.87870	1704	57	58	$\bar{3}$.87871	1704	2.12129	42	$\bar{1}$.99999	34
1620	27	$\bar{3}$.89509	1639	57	58	$\bar{3}$.89510	1639	2.10490	42	$\bar{1}$.99999	33
1680	28	$\bar{3}$.91088	1579	57	58	$\bar{3}$.91089	1579	2.08911	42	$\bar{1}$.99999	32
1740	29	$\bar{3}$.92612	1524	57	59	$\bar{3}$.92613	1524	2.07387	41	$\bar{1}$.99998	31
1800	30	$\bar{3}$.94084	1472	57	59	$\bar{3}$.94086	1473	2.05914	41	$\bar{1}$.99998	30
1860	31	$\bar{3}$.95508	1424	57	59	$\bar{3}$.95510	1424	2.04490	41	$\bar{1}$.99998	29
1920	32	$\bar{3}$.96887	1379	57	59	$\bar{3}$.96889	1379	2.03111	41	$\bar{1}$.99998	28
1980	33	$\bar{3}$.98223	1336	57	59	$\bar{3}$.98225	1336	2.01775	41	$\bar{1}$.99998	27
2040	34	$\bar{3}$.99520	1297	57	59	$\bar{3}$.99522	1297	2.00478	41	$\bar{1}$.99998	26
2100	35	$\bar{2}$.00779	1259	57	59	$\bar{2}$.00781	1259	1.99219	41	$\bar{1}$.99998	25
2160	36	$\bar{2}$.02002	1223	57	59	$\bar{2}$.02004	1223	1.97996	41	$\bar{1}$.99998	24
2220	37	$\bar{2}$.03192	1190	57	59	$\bar{2}$.03194	1190	1.96806	41	$\bar{1}$.99997	23
2280	38	$\bar{2}$.04350	1158	57	59	$\bar{2}$.04353	1159	1.95647	41	$\bar{1}$.99997	22
2340	39	$\bar{2}$.05478	1128	57	59	$\bar{2}$.05481	1128	1.94519	41	$\bar{1}$.99997	21
2400	40	$\bar{2}$.06578	1100	57	59	$\bar{2}$.06581	1100	1.93419	41	$\bar{1}$.99997	20
2460	41	$\bar{2}$.07650	1072	56	60	$\bar{2}$.07653	1072	1.92347	40	$\bar{1}$.99997	19
2520	42	$\bar{2}$.08696	1046	56	60	$\bar{2}$.08700	1047	1.91300	40	$\bar{1}$.99997	18
2580	43	$\bar{2}$.09718	1022	56	60	$\bar{2}$.09722	1022	1.90278	40	$\bar{1}$.99997	17
2640	44	$\bar{2}$.10717	999	56	60	$\bar{2}$.10720	998	1.89280	40	$\bar{1}$.99996	16
2700	45	$\bar{2}$.11693	976	56	60	$\bar{2}$.11696	976	1.88304	40	$\bar{1}$.99996	15
2760	46	$\bar{2}$.12647	954	56	60	$\bar{2}$.12651	955	1.87349	40	$\bar{1}$.99996	14
2820	47	$\bar{2}$.13581	934	56	60	$\bar{2}$.13585	934	1.86415	40	$\bar{1}$.99996	13
2880	48	$\bar{2}$.14495	914	56	60	$\bar{2}$.14500	915	1.85500	40	$\bar{1}$.99996	12
2940	49	$\bar{2}$.15391	896	56	60	$\bar{2}$.15395	895	1.84605	40	$\bar{1}$.99996	11
3000	50	$\bar{2}$.16268	877	56	61	$\bar{2}$.16273	878	1.83727	39	$\bar{1}$.99995	10
3060	51	$\bar{2}$.17128	860	56	61	$\bar{2}$.17133	860	1.82867	39	$\bar{1}$.99995	9
3120	52	$\bar{2}$.17971	843	56	61	$\bar{2}$.17976	843	1.82024	39	$\bar{1}$.99995	8
3180	53	$\bar{2}$.18798	827	56	61	$\bar{2}$.18804	828	1.81196	39	$\bar{1}$.99995	7
3240	54	$\bar{2}$.19610	812	56	61	$\bar{2}$.19616	812	1.80384	39	$\bar{1}$.99995	6
3300	55	$\bar{2}$.20407	797	56	61	$\bar{2}$.20413	797	1.79587	39	$\bar{1}$.99994	5
3360	56	$\bar{2}$.21189	782	56	61	$\bar{2}$.21195	782	1.78805	39	$\bar{1}$.99994	4
3420	57	$\bar{2}$.21958	769	55	61	$\bar{2}$.21964	769	1.78036	39	$\bar{1}$.99994	3
3480	58	$\bar{2}$.22713	755	55	62	$\bar{2}$.22720	756	1.77280	38	$\bar{1}$.99994	2
3540	59	$\bar{2}$.23456	743	55	62	$\bar{2}$.23462	742	1.76538	38	$\bar{1}$.99994	1
3600	60	$\bar{2}$.24186	730	55	62	$\bar{2}$.24192	730	1.75808	38	$\bar{1}$.99993	0
				$\bar{6}$.685					5.314		
		log cos	d	S	T	log cot	c. d.	log tan	C	log sin	′

89°

"	'	log sin	d	S	T	log tan	c. d.	log cot	C	log cos	'
				6̄.685					5.314		
3600	0	3̄.24186	717	55	62	3̄.24192	718	1.75808	38	1̄.99993	60
3660	1	3̄.24903	706	55	62	3̄.24910	706	1.75090	38	1̄.99993	59
3720	2	3̄.25609	695	55	62	3̄.25616	696	1.74384	38	1̄.99993	58
3780	3	3̄.26304	684	55	62	3̄.26312	684	1.73688	38	1̄.99993	57
3840	4	3̄.26988	673	55	63	3̄.26996	673	1.73004	37	1̄.99992	56
3900	5	3̄.27661	663	55	63	3̄.27669	663	1.72331	37	1̄.99992	55
3960	6	3̄.28324	653	55	63	3̄.28332	654	1.71668	37	1̄.99992	54
4020	7	3̄.28977	644	55	63	3̄.28986	643	1.71014	37	1̄.99992	53
4080	8	3̄.29621	634	55	63	3̄.29629	634	1.70371	37	1̄.99992	52
4140	9	3̄.30255	624	55	63	3̄.30263	625	1.69737	37	1̄.99991	51
4200	10	3̄.30879	616	54	63	3̄.30888	617	1.69112	37	1̄.99991	50
4260	11	3̄.31495	608	54	64	3̄.31505	607	1.68495	36	1̄.99991	49
4320	12	3̄.32103	599	54	64	3̄.32112	599	1.67888	36	1̄.99990	48
4380	13	3̄.32702	590	54	64	3̄.32711	591	1.67289	36	1̄.99990	47
4440	14	3̄.33292	583	54	64	3̄.33302	584	1.66698	36	1̄.99990	46
4500	15	3̄.33875	575	54	64	3̄.33886	575	1.66114	36	1̄.99990	45
4560	16	3̄.34450	568	54	65	3̄.34461	568	1.65539	35	1̄.99989	44
4620	17	3̄.35018	560	54	65	3̄.35029	561	1.64971	35	1̄.99989	43
4680	18	3̄.35578	553	54	65	3̄.35590	553	1.64410	35	1̄.99989	42
4740	19	3̄.36131	547	54	65	3̄.36143	546	1.63857	35	1̄.99989	41
4800	20	3̄.36678	539	54	65	3̄.36689	540	1.63311	35	1̄.99988	40
4860	21	3̄.37217	533	53	66	3̄.37229	533	1.62771	34	1̄.99988	39
4920	22	3̄.37750	526	53	66	3̄.37762	527	1.62238	34	1̄.99988	38
4980	23	3̄.38276	520	53	66	3̄.38289	520	1.61711	34	1̄.99987	37
5040	24	3̄.38796	514	53	66	3̄.38809	514	1.61191	34	1̄.99987	36
5100	25	3̄.39310	508	53	66	3̄.39323	509	1.60677	34	1̄.99987	35
5160	26	3̄.39818	502	53	67	3̄.39832	502	1.60168	33	1̄.99986	34
5220	27	3̄.40320	496	53	67	3̄.40334	496	1.59666	33	1̄.99986	33
5280	28	3̄.40816	491	53	67	3̄.40830	491	1.59170	33	1̄.99986	32
5340	29	3̄.41307	485	53	67	3̄.41321	486	1.58679	33	1̄.99985	31
5400	30	3̄.41792	480	53	67	3̄.41807	480	1.58193	33	1̄.99985	30
5460	31	3̄.42272	474	52	68	3̄.42287	475	1.57713	32	1̄.99985	29
5520	32	3̄.42746	470	52	68	3̄.42762	470	1.57238	32	1̄.99984	28
5580	33	3̄.43216	464	52	68	3̄.43232	464	1.56768	32	1̄.99984	27
5640	34	3̄.43680	459	52	68	3̄.43696	460	1.56304	32	1̄.99984	26
5700	35	3̄.44139	455	52	69	3̄.44156	455	1.55844	31	1̄.99983	25
5760	36	3̄.44594	450	52	69	3̄.44611	450	1.55389	31	1̄.99983	24
5820	37	3̄.45044	445	52	69	3̄.45061	446	1.54939	31	1̄.99983	23
5880	38	3̄.45489	441	52	69	3̄.45507	441	1.54493	31	1̄.99982	22
5940	39	3̄.45930	436	51	69	3̄.45948	437	1.54052	31	1̄.99982	21
6000	40	3̄.46366	433	51	70	3̄.46385	432	1.53615	30	1̄.99982	20
6060	41	3̄.46799	427	51	70	3̄.46817	428	1.53183	30	1̄.99981	19
6120	42	3̄.47226	424	51	70	3̄.47245	424	1.52755	30	1̄.99981	18
6180	43	3̄.47650	419	51	70	3̄.47669	420	1.52331	30	1̄.99981	17
6240	44	3̄.48069	416	51	71	3̄.48089	416	1.51911	29	1̄.99980	16
6300	45	3̄.48485	411	51	71	3̄.48505	412	1.51495	29	1̄.99980	15
6360	46	3̄.48896	408	51	71	3̄.48917	408	1.51083	29	1̄.99979	14
6420	47	3̄.49304	404	50	72	3̄.49325	404	1.50675	28	1̄.99979	13
6480	48	3̄.49708	400	50	72	3̄.49729	401	1.50271	28	1̄.99979	12
6540	49	3̄.50108	396	50	72	3̄.50130	397	1.49870	28	1̄.99978	11
6600	50	3̄.50504	393	50	72	3̄.50527	393	1.49473	28	1̄.99978	10
6660	51	3̄.50897	390	50	73	3̄.50920	390	1.49080	27	1̄.99977	9
6720	52	3̄.51287	386	50	73	3̄.51310	386	1.48690	27	1̄.99977	8
6780	53	3̄.51673	382	50	73	3̄.51696	383	1.48304	27	1̄.99977	7
6840	54	3̄.52055	379	50	73	3̄.52079	380	1.47921	27	1̄.99976	6
6900	55	3̄.52434	.76	49	74	3̄.52459	376	1.47541	26	1̄.99976	5
6960	56	3̄.52810	373	49	74	3̄.52835	373	1.47165	26	1̄.99975	4
7020	57	3̄.53183	369	49	74	3̄.53208	370	1.46792	26	1̄.99975	3
7080	58	3̄.53552	367	49	75	3̄.53578	367	1.46422	25	1̄.99974	2
7140	59	3̄.53919	363	49	75	3̄.53945	363	1.46055	25	1̄.99974	1
7200	60	3̄.54282		49	75	3̄.54308		1.45692	25	1̄.99974	0
				6̄.685					5.314		
		log cos	d	S	T	log cot	c. d.	log tan	C	log sin	'

88°

"	'	log sin	d	S	T	log tan	c. d.	log cot	C	log cos	
				6̄.685					5.314		
7200	0	2̄.54282	360	49	75	2̄.54308	361	1.45692	25	1̄.99974	60
7260	1	2̄.54642	357	49	75	2̄.54669	358	1.45331	25	1̄.99973	59
7320	2	2̄.54999	355	48	76	2̄.55027	355	1.44973	24	1̄.99973	58
7380	3	2̄.55354	351	48	76	2̄.55382	352	1.44618	24	1̄.99972	57
7440	4	2̄.55705	349	48	76	2̄.55734	349	1.44266	24	1̄.99972	56
7500	5	2̄.56054	346	48	77	2̄.56083	346	1.43917	23	1̄.99971	55
7560	6	2̄.56400	343	48	77	2̄.56429	344	1.43571	23	1̄.99971	54
7620	7	2̄.56743	341	48	77	2̄.56773	341	1.43227	23	1̄.99970	53
7680	8	2̄.57084	337	47	78	2̄.57114	338	1.42886	22	1̄.99970	52
7740	9	2̄.57421	336	47	78	2̄.57452	336	1.42548	22	1̄.99969	51
7800	10	2̄.57757	332	47	78	2̄.57788	333	1.42212	22	1̄.99969	50
7860	11	2̄.58089	330	47	79	2̄.58121	330	1.41879	21	1̄.99968	49
7920	12	2̄.58419	328	47	79	2̄.58451	328	1.41549	21	1̄.99968	48
7980	13	2̄.58747	325	47	79	2̄.58779	326	1.41221	21	1̄.99967	47
8040	14	2̄.59072	323	46	79	2̄.59105	323	1.40895	21	1̄.99967	46
8100	15	2̄.59395	320	46	80	2̄.59428	321	1.40572	20	1̄.99967	45
8160	16	2̄.59715	318	46	80	2̄.59749	319	1.40251	20	1̄.99966	44
8220	17	2̄.60033	316	46	80	2̄.60068	316	1.39932	20	1̄.99966	43
8280	18	2̄.60349	313	46	81	2̄.60384	314	1.39616	19	1̄.99965	42
8340	19	2̄.60662	311	46	81	2̄.60698	311	1.39302	19	1̄.99964	41
8400	20	2̄.60973	309	45	82	2̄.61009	310	1.38991	18	1̄.99964	40
8460	21	2̄.61282	307	45	82	2̄.61319	307	1.38681	18	1̄.99963	39
8520	22	2̄.61589	305	45	82	2̄.61626	305	1.38374	18	1̄.99963	38
8580	23	2̄.61894	302	45	83	2̄.61931	303	1.38069	17	1̄.99962	37
8640	24	2̄.62196	301	45	83	2̄.62234	301	1.37766	17	1̄.99962	36
8700	25	2̄.62497	298	45	83	2̄.62535	299	1.37465	17	1̄.99961	35
8760	26	2̄.62795	296	44	84	2̄.62834	297	1.37166	16	1̄.99961	34
8820	27	2̄.63091	294	44	84	2̄.63131	295	1.36869	16	1̄.99960	33
8880	28	2̄.63385	293	44	84	2̄.63426	292	1.36574	16	1̄.99960	32
8940	29	2̄.63678	290	44	85	2̄.63718	291	1.36282	15	1̄.99959	31
9000	30	2̄.63968	288	44	85	2̄.64009	289	1.35991	15	1̄.99959	30
9060	31	2̄.64256	287	44	85	2̄.64298	287	1.35702	15	1̄.99958	29
9120	32	2̄.64543	284	43	86	2̄.64585	285	1.35415	14.	1̄.99958	28
9180	33	2̄.64827	283	43	86	2̄.64870	284	1.35130	14	1̄.99957	27
9240	34	2̄.65110	281	43	87	2̄.65154	281	1.34846	13	1̄.99956	26
9300	35	2̄.65391	279	43	87	2̄.65435	280	1.34565	13	1̄.99956	25
9360	36	2̄.65670	277	43	87	2̄.65715	278	1.34285	13	1̄.99955	24
9420	37	2̄.65947	276	42	88	2̄.65993	276	1.34007	12	1̄.99955	23
9480	38	2̄.66223	274	42	88	2̄.66269	274	1.33731	12	1̄.99954	22
9540	39	2̄.66497	272	42	88	2̄.66543	273	1.33457	12	1̄.99954	21
9600	40	2̄.66769	270	42	89	2̄.66816	271	1.33184	11	1̄.99953	20
9660	41	2̄.67039	269	42	89	2̄.67087	269	1.32913	11	1̄.99952	19
9720	42	2̄.67308	267	41	90	2̄.67356	268	1.32644	10	1̄.99952	18
9780	43	2̄.67575	266	41	90	2̄.67624	266	1.32376	10	1̄.99951	17
9840	44	2̄.67841	263	41	90	2̄.67890	264	1.32110	10	1̄.99951	16
9900	45	2̄.68104	263	41	91	2̄.68154	263	1.31846	09	1̄.99950	15
9960	46	2̄.68367	260	41	91	2̄.68417	261	1.31583	09	1̄.99949	14
10020	47	2̄.68627	259	40	92	2̄.68678	260	1.31322	08	1̄.99949	13
10080	48	2̄.68886	258	40	92	2̄.68938	258	1.31062	08	1̄.99948	12
10141	49	2̄.69144	256	40	92	2̄.69196	257	1.30804	08	1̄.99948	11
10200	50	2̄.69400	254	40	93	2̄.69453	255	1.30547	07	1̄.99947	10
10260	51	2̄.69654	253	40	93	2̄.69708	254	1.30292	07	1̄.99946	9
10320	52	2̄.69907	252	39	94	2̄.69962	252	1.30038	06	1̄.99946	8
10380	53	2̄.70159	250	39	94	2̄.70214	251	1.29786	06	1̄.99945	7
10440	54	2̄.70409	249	39	95	2̄.70465	249	1.29535	05	1̄.99944	6
10500	55	2̄.70658	247	39	95	2̄.70714	248	1.29286	05	1̄.99944	5
10560	56	2̄.70905	246	39	95	2̄.70962	246	1.29038	05	1̄.99943	4
10620	57	2̄.71151	244	38	96	2̄.71208	245	1.28792	04	1̄.99942	3
10680	58	2̄.71395	243	38	96	2̄.71453	244	1.28547	04	1̄.99942	2
10740	59	2̄.71638	242	38	97	2̄.71697	243	1.28303	03	1̄.99941	1
10800	60	2̄.71880		38	97	2̄.71940		1.28060	03	1̄.99940	0
				6̄.685					5.314		
		log cos	d	S	T	log cot	c. d.	log tan	C	log sin	'

′	log sin	d	log tan	c. d.	log cot	log cos		p. p.
0	2.71880		2.71940		1.28060	1.99940	60	
1	2.72120	240	2.72181	241	1.27819	1.99940	59	
2	2.72359	239	2.72420	239	1.27580	1.99939	58	
3	2.72597	238	2.72659	239	1.27341	1.99938	57	
4	2.72834	237	2.72896	237	1.27104	1.99938	56	
5	2.73069	235	2.73132	236	1.26868	1.99937	55	
6	2.73303	234	2.73366	234	1.26634	1.99936	54	
7	2.73535	232	2.73600	234	1.26400	1.99936	53	
8	2.73767	232	2.73832	232	1.26168	1.99935	52	
9	2.73997	230	2.74063	231	1.25937	1.99934	51	
10	2.74226	229	2.74292	229	1.25708	1.99934	50	
11	2.74454	228	2.74521	229	1.25479	1.99933	49	
12	2.74680	226	2.74748	227	1.25252	1.99932	48	
13	2.74906	226	2.74974	226	1.25026	1.99932	47	
14	2.75130	224	2.75199	225	1.24801	1.99931	46	
15	2.75353	223	2.75423	224	1.24577	1.99930	45	
16	2.75575	222	2.75645	222	1.24355	1.99929	44	
17	2.75795	220	2.75867	222	1.24133	1.99929	43	
18	2.76015	220	2.76087	220	1.23913	1.99928	42	
19	2.76234	219	2.76306	219	1.23694	1.99927	41	
20	2.76451	217	2.76525	219	1.23475	1.99926	40	
21	2.76667	216	2.76742	217	1.23258	1.99926	39	
22	2.76883	216	2.76958	216	1.23042	1.99925	38	
23	2.77097	214	2.77173	215	1.22827	1.99924	37	
24	2.77310	213	2.77387	214	1.22613	1.99923	36	
25	2.77522	212	2.77600	213	1.22400	1.99923	35	
26	2.77733	211	2.77811	211	1.22189	1.99922	34	
27	2.77943	210	2.78022	211	1.21978	1.99921	33	
28	2.78152	209	2.78232	210	1.21768	1.99920	32	
29	2.78360	208	2.78441	209	1.21559	1.99920	31	
30	2.78568	208	2.78649	208	1.21351	1.99919	30	
31	2.78774	206	2.78855	206	1.21145	1.99918	29	
32	2.78979	205	2.79061	206	1.20939	1.99917	28	
33	2.79183	204	2.79266	205	1.20734	1.99917	27	
34	2.79386	203	2.79470	204	1.20530	1.99916	26	
35	2.79588	202	2.79673	203	1.20327	1.99915	25	
36	2.79789	201	2.79875	202	1.20125	1.99914	24	
37	2.79990	201	2.80076	201	1.19924	1.99913	23	
38	2.80189	199	2.80277	201	1.19723	1.99913	22	
39	2.80388	199	2.80476	199	1.19524	1.99912	21	
40	2.80585	197	2.80674	198	1.19326	1.99911	20	
41	2.80782	197	2.80872	198	1.19128	1.99910	19	
42	2.80978	196	2.81068	196	1.18932	1.99909	18	
43	2.81173	195	2.81264	196	1.18736	1.99909	17	
44	2.81367	194	2.81459	195	1.18541	1.99908	16	
45	2.81560	193	2.81653	194	1.18347	1.99907	15	
46	2.81752	192	2.81846	193	1.18154	1.99906	14	
47	2.81944	192	2.82038	192	1.17962	1.99905	13	
48	2.82134	190	2.82220	192	1.17770	1.99904	12	
49	2.82324	190	2.82420	190	1.17580	1.99904	11	
50	2.82513	189	2.82610	190	1.17390	1.99903	10	
51	2.82701	188	2.82799	189	1.17201	1.99902	9	
52	2.82888	187	2.82987	188	1.17013	1.99901	8	
53	2.83075	187	2.83175	188	1.16825	1.99900	7	
54	2.83261	186	2.83361	186	1.16639	1.99899	6	
55	2.83446	185	2.83547	186	1.16453	1.99898	5	
56	2.83630	184	2.83732	185	1.16268	1.99898	4	
57	2.83813	183	2.83916	184	1.16084	1.99897	3	
58	2.83996	183	2.84100	184	1.15900	1.99896	2	
59	2.84177	181	2.84282	182	1.15718	1.99895	1	
60	2.84358	181	2.84464	182	1.15536	1.99894	0	

log cos	d	log cot	c. d.	log tan	log sin	′	p. p.

p. p.

	238	234	229
6	23.8	23.4	22.9
7	27.8	27.3	26.7
8	31.7	31.2	30.5
9	35.7	35.1	34.4
10	39.7	39.0	38.2
20	79.3	78.0	76.3
30	119.0	117.0	114.5
40	158.7	156.0	152.7
50	198.3	195.0	190.8

	225	220	216
6	22.5	22.0	21.6
7	26.3	25.7	25.2
8	30.0	29.3	28.8
9	33.8	33.0	32.4
10	37.5	36.7	36.0
20	75.0	73.3	72.0
30	112.5	110.0	108.0
40	150.0	146.7	144.0
50	187.5	183.3	180.0

	212	208	204
6	21.2	20.8	20.4
7	24.7	24.3	23.8
8	28.3	27.7	27.2
9	31.8	31.2	30.6
10	35.3	34.7	34.0
20	70.7	69.3	68.0
30	106.0	104.0	102.0
40	141.3	138.7	136.0
50	176.7	173.3	170.0

	201	197	193
6	20.1	19.7	19.3
7	23.5	23.0	22.5
8	26.8	26.3	25.7
9	30.2	29.6	29.0
10	33.5	32.8	32.2
20	67.0	65.7	64.3
30	100.5	98.5	96.5
40	134.0	131.3	128.7
50	167.5	164.2	160.8

	189	185	181
6	18.9	18.5	18.1
7	22.1	21.6	21.1
8	25.2	24.7	24.1
9	28.4	27.8	27.2
10	31.5	30.8	30.2
20	63.0	61.7	60.3
30	94.5	92.5	90.5
40	126.0	123.3	120.7
50	157.5	154.2	150.8

	4	3	2	1
6	0.4	0.3	0.2	0.1
7	0.5	0.4	0.2	0.1
8	0.5	0.4	0.3	0.1
9	0.6	0.5	0.3	0.2
10	0.7	0.5	0.3	0.2
20	1.3	1.0	0.7	0.3
30	2.0	1.5	1.0	0.5
40	2.7	2.0	1.3	0.7
50	3.3	2.5	1.7	0.8

′	log sin	d	log tan	c. d.	log cot	log cos	
0	$\bar{2}$.84358		$\bar{2}$.84464		1.15536	$\bar{1}$.99894	60
1	$\bar{2}$.84539	181	$\bar{2}$.84646	182	1.15354	$\bar{1}$.99893	59
2	$\bar{2}$.84718	179	$\bar{2}$.84826	180	1.15174	$\bar{1}$.99892	58
3	$\bar{2}$.84897	179	$\bar{2}$.85006	180	1.14994	$\bar{1}$.99891	57
4	$\bar{2}$.85075	178	$\bar{2}$.85185	179	1.14815	$\bar{1}$.99891	56
5	$\bar{2}$.85252	177	$\bar{2}$.85363	178	1.14637	$\bar{1}$.99890	55
6	$\bar{2}$.85429	177	$\bar{2}$.85540	177	1.14460	$\bar{1}$.99889	54
7	$\bar{2}$.85605	176	$\bar{2}$.85717	177	1.14283	$\bar{1}$.99888	53
8	$\bar{2}$.85780	175	$\bar{2}$.85893	176	1.14107	$\bar{1}$.99887	52
9	$\bar{2}$.85955	175	$\bar{2}$.86069	176	1.13931	$\bar{1}$.99886	51
10	$\bar{2}$.86128	173	$\bar{2}$.86243	174	1.13757	$\bar{1}$.99885	50
11	$\bar{2}$.86301		$\bar{2}$.86417		1.13582	$\bar{1}$.99884	49
12	$\bar{2}$.86474	173	$\bar{2}$.86591	174	1.13409	$\bar{1}$.99883	48
13	$\bar{2}$.86645	171	$\bar{2}$.86763	172	1.13237	$\bar{1}$.99882	47
14	$\bar{2}$.86816	171	$\bar{2}$.86935	172	1.13065	$\bar{1}$.99881	46
15	$\bar{2}$.86987	171	$\bar{2}$.87106	171	1.12894	$\bar{1}$.99880	45
16	$\bar{2}$.87156	169	$\bar{2}$.87277	171	1.12723	$\bar{1}$.99879	44
17	$\bar{2}$.87325	169	$\bar{2}$.87447	170	1.12553	$\bar{1}$.99879	43
18	$\bar{2}$.87494	169	$\bar{2}$.87616	169	1.12384	$\bar{1}$.99878	42
19	$\bar{2}$.87661	167	$\bar{2}$.87785	169	1.12215	$\bar{1}$.99877	41
20	$\bar{2}$.87829	168	$\bar{2}$.87953	168	1.12047	$\bar{1}$.99876	40
21	$\bar{2}$.87995		$\bar{2}$.88120		1.11880	$\bar{1}$.99875	39
22	$\bar{2}$.88161	166	$\bar{2}$.88287	167	1.11713	$\bar{1}$.99874	38
23	$\bar{2}$.88326	165	$\bar{2}$.88453	166	1.11547	$\bar{1}$.99873	37
24	$\bar{2}$.88490	164	$\bar{2}$.88618	165	1.11382	$\bar{1}$.99872	36
25	$\bar{2}$.88654	164	$\bar{2}$.88783	165	1.11217	$\bar{1}$.99871	35
26	$\bar{2}$.88817	163	$\bar{2}$.88948	165	1.11052	$\bar{1}$.99870	34
27	$\bar{2}$.88980	163	$\bar{2}$.89111	163	1.10889	$\bar{1}$.99869	33
28	$\bar{2}$.89142	162	$\bar{2}$.89274	163	1.10726	$\bar{1}$.99868	32
29	$\bar{2}$.89304	162	$\bar{2}$.89437	163	1.10563	$\bar{1}$.99867	31
30	$\bar{2}$.89464	160	$\bar{2}$.89598	161	1.10402	$\bar{1}$.99866	30
31	$\bar{2}$.89625		$\bar{2}$.89760		1.10240	$\bar{1}$.99865	29
32	$\bar{2}$.89784	159	$\bar{2}$.89920	160	1.10080	$\bar{1}$.99864	28
33	$\bar{2}$.89943	159	$\bar{2}$.90080	160	1.09920	$\bar{1}$.99863	27
34	$\bar{2}$.90102	159	$\bar{2}$.90240	160	1.09760	$\bar{1}$.99862	26
35	$\bar{2}$.90260	158	$\bar{2}$.90399	159	1.09601	$\bar{1}$.99861	25
36	$\bar{2}$.90417	157	$\bar{2}$.90557	158	1.09443	$\bar{1}$.99860	24
37	$\bar{2}$.90574	157	$\bar{2}$.90715	158	1.09285	$\bar{1}$.99859	23
38	$\bar{2}$.90730	156	$\bar{2}$.90872	157	1.09128	$\bar{1}$.99858	22
39	$\bar{2}$.90885	155	$\bar{2}$.91029	157	1.08971	$\bar{1}$.99857	21
40	$\bar{2}$.91040	155	$\bar{2}$.91185	156	1.08815	$\bar{1}$.99856	20
41	$\bar{2}$.91195		$\bar{2}$.91340		1.08660	$\bar{1}$.99855	19
42	$\bar{2}$.91349	154	$\bar{2}$.91495	155	1.08505	$\bar{1}$.99854	18
43	$\bar{2}$.91502	153	$\bar{2}$.91650	155	1.08350	$\bar{1}$.99853	17
44	$\bar{2}$.91655	153	$\bar{2}$.91803	153	1.08197	$\bar{1}$.99852	16
45	$\bar{2}$.91807	152	$\bar{2}$.91957	154	1.08043	$\bar{1}$.99851	15
46	$\bar{2}$.91959	152	$\bar{2}$.92110	153	1.07890	$\bar{1}$.99850	14
47	$\bar{2}$.92110	151	$\bar{2}$.92262	152	1.07738	$\bar{1}$.99848	13
48	$\bar{2}$.92261	151	$\bar{2}$.92414	152	1.07586	$\bar{1}$.99847	12
49	$\bar{2}$.92411	150	$\bar{2}$.92565	151	1.07435	$\bar{1}$.99846	11
50	$\bar{2}$.92561	150	$\bar{2}$.92716	151	1.07284	$\bar{1}$.99845	10
51	$\bar{2}$.92710		$\bar{2}$.92866		1.07134	$\bar{1}$.99844	9
52	$\bar{2}$.92859	149	$\bar{2}$.93016	150	1.06984	$\bar{1}$.99843	8
53	$\bar{2}$.93007	148	$\bar{2}$.93165	149	1.06835	$\bar{1}$.99842	7
54	$\bar{2}$.93154	147	$\bar{2}$.93313	148	1.06687	$\bar{1}$.99841	6
55	$\bar{2}$.93301	147	$\bar{2}$.93462	149	1.06538	$\bar{1}$.99840	5
56	$\bar{2}$.93448	147	$\bar{2}$.93609	147	1.06391	$\bar{1}$.99839	4
57	$\bar{2}$.93594	146	$\bar{2}$.93756	147	1.06244	$\bar{1}$.99838	3
58	$\bar{2}$.93740	146	$\bar{2}$.93903	147	1.06097	$\bar{1}$.99837	2
59	$\bar{2}$.93885	145	$\bar{2}$.94049	146	1.05951	$\bar{1}$.99836	1
60	$\bar{2}$.94030	145	$\bar{2}$.94195	146	1.05805	$\bar{1}$.99834	0
	log cos	d	log cot	c. d.	log tan	log sin	′

p. p.

	181	179	177
6	18.1	17.9	17.7
7	21.1	20.9	20.7
8	24.1	23.9	23.6
9	27.2	26.9	26.6
10	30.2	29.8	29.5
20	60.3	59.7	59.0
30	90.5	89.5	88.5
40	120.7	119.3	118.0
50	150.8	149.2	147.5

	175	173	171
6	17.5	17.3	17.1
7	20.4	20.2	20.0
8	23.3	23.1	22.8
9	26.3	26.0	25.7
10	29.2	28.8	28.5
20	58.3	57.7	57.0
30	87.5	86.5	85.5
40	116.7	115.3	114.0
50	145.8	144.2	142.5

	168	166	164
6	16.8	16.6	16.4
7	19.6	19.4	19.1
8	22.4	22.1	21.9
9	25.2	24.9	24.6
10	28.0	27.7	27.3
20	56.0	55.3	54.7
30	84.0	83.0	82.0
40	112.0	110.7	109.3
50	140.0	138.3	136.7

	162	159	157
6	16.2	15.9	15.7
7	18.9	18.6	18.3
8	21.6	21.2	20.9
9	24.3	23.9	23.6
10	27.0	26.5	26.2
20	54.0	53.0	52.3
30	81.0	79.5	78.5
40	108.0	106.0	104.7
50	135.0	132.5	130.8

	155	153	151
6	15.5	15.3	15.1
7	18.1	17.9	17.6
8	20.7	20.4	20.1
9	23.3	23.0	22.7
10	25.8	25.5	25.2
20	51.7	51.0	50.3
30	77.5	76.5	75.5
40	103.3	102.0	100.7
50	129.2	127.5	125.8

	149	147	λ
6	14.9	14.7	0.1
7	17.4	17.2	0.1
8	19.9	19.6	0.1
9	22.4	22.1	0.2
10	24.8	24.5	0.2
20	49.7	49.0	0.3
30	74.5	73.5	0.5
40	99.3	98.0	0.7
50	124.2	122.5	0.8

′	log sin	d	log tan	c. d.	log cot	log cos	′
0	2̄.94030	144	2̄.94195	145	1.05805	1̄.99834	60
1	2̄.94174	143	2̄.94340	145	1.05660	1̄.99833	59
2	2̄.94317	144	2̄.94485	145	1.05515	1̄.99832	58
3	2̄.94461	142	2̄.94630	143	1.05370	1̄.99831	57
4	2̄.94603	143	2̄.94773	144	1.05227	1̄.99830	56
5	2̄.94746	141	2̄.94917	143	1.05083	1̄.99829	55
6	2̄.94887	142	2̄.95060	142	1.04940	1̄.99828	54
7	2̄.95029	141	2̄.95202	142	1.04798	1̄.99827	53
8	2̄.95170	140	2̄.95344	142	1.04656	1̄.99825	52
9	2̄.95310	140	2̄.95486	141	1.04514	1̄.99824	51
10	2̄.95450	139	2̄.95627	140	1.04373	1̄.99823	50
11	2̄.95589	139	2̄.95767	141	1.04233	1̄.99822	49
12	2̄.95728	139	2̄.95908	139	1.04092	1̄.99821	48
13	2̄.95867	138	2̄.96047	140	1.03953	1̄.99820	47
14	2̄.96005	138	2̄.96187	138	1.03813	1̄.99819	46
15	2̄.96143	137	2̄.96325	139	1.03675	1̄.99817	45
16	2̄.96280	137	2̄.96464	138	1.03536	1̄.99816	44
17	2̄.96417	136	2̄.96602	137	1.03398	1̄.99815	43
18	2̄.96553	136	2̄.96739	138	1.03261	1̄.99814	42
19	2̄.96689	136	2̄.96877	136	1.03123	1̄.99813	41
20	2̄.96825	135	2̄.97013	137	1.02987	1̄.99812	40
21	2̄.96960	135	2̄.97150	135	1.02850	1̄.99810	39
22	2̄.97095	134	2̄.97285	136	1.02715	1̄.99809	38
23	2̄.97229	134	2̄.97421	135	1.02579	1̄.99808	37
24	2̄.97363	133	2̄.97556	135	1.02444	1̄.99807	36
25	2̄.97496	133	2̄.97691	134	1.02309	1̄.99806	35
26	2̄.97629	133	2̄.97825	134	1.02175	1̄.99804	34
27	2̄.97762	132	2̄.97959	133	1.02041	1̄.99803	33
28	2̄.97894	132	2̄.98092	133	1.01908	1̄.99802	32
29	2̄.98026	131	2̄.98225	133	1.01775	1̄.99801	31
30	2̄.98157	131	2̄.98358	132	1.01642	1̄.99800	30
31	2̄.98288	131	2̄.98490	132	1.01510	1̄.99798	29
32	2̄.98419	130	2̄.98622	131	1.01378	1̄.99797	28
33	2̄.98549	130	2̄.98753	131	1.01247	1̄.99796	27
34	2̄.98679	129	2̄.98884	131	1.01116	1̄.99795	26
35	2̄.98808	129	2̄.99015	130	1.00985	1̄.99793	25
36	2̄.98937	129	2̄.99145	130	1.00855	1̄.99792	24
37	2̄.99066	128	2̄.99275	130	1.00725	1̄.99791	23
38	2̄.99194	128	2̄.99405	129	1.00595	1̄.99790	22
39	2̄.99322	128	2̄.99534	128	1.00466	1̄.99788	21
40	2̄.99450	127	2̄.99662	129	1.00338	1̄.99787	20
41	2̄.99577	127	2̄.99791	128	1.00209	1̄.99786	19
42	2̄.99704	126	2̄.99919	127	1.00081	1̄.99785	18
43	2̄.99830	126	1̄.00046	128	0.99954	1̄.99783	17
44	2̄.99956	126	1̄.00174	127	0.99826	1̄.99782	16
45	1̄.00082	125	1̄.00301	126	0.99699	1̄.99781	15
46	1̄.00207	125	1̄.00427	126	0.99573	1̄.99780	14
47	1̄.00332	124	1̄.00553	126	0.99447	1̄.99778	13
48	1̄.00456	125	1̄.00679	126	0.99321	1̄.99777	12
49	1̄.00581	123	1̄.00805	125	0.99195	1̄.99776	11
50	1̄.00704	124	1̄.00930	125	0.99070	1̄.99775	10
51	1̄.00828	123	1̄.01055	124	0.98945	1̄.99773	9
52	1̄.00951	123	1̄.01179	124	0.98821	1̄.99772	8
53	1̄.01074	122	1̄.01303	124	0.98697	1̄.99771	7
54	1̄.01196	122	1̄.01427	123	0.98573	1̄.99769	6
55	1̄.01318	122	1̄.01550	123	0.98450	1̄.99768	5
56	1̄.01440	121	1̄.01673	123	0.98327	1̄.99767	4
57	1̄.01561	121	1̄.01796	122	0.98204	1̄.99765	3
58	1̄.01682	121	1̄.01918	122	0.98082	1̄.99764	2
59	1̄.01803	120	1̄.02040	122	0.97960	1̄.99763	1
60	1̄.01923		1̄.02162		0.97838	1̄.99761	0
	log cos	d	log cot	c. d.	log tan	log sin	′

p. p.

	145	143	141
6	14.5	14.3	14.1
7	16.9	16.7	16.5
8	19.3	19.1	18.8
9	21.8	21.5	21.2
10	24.2	23.8	23.5
20	48.3	47.7	47.0
30	72.5	71.5	70.5
40	96.7	95.3	94.0
50	120.8	119.2	117.5

	139	138	136
6	13.9	13.8	13.6
7	16.2	16.1	15.9
8	18.5	18.4	18.1
9	20.9	20.7	20.4
10	23.2	23.0	22.7
20	46.3	46.0	45.3
30	69.5	69.0	68.0
40	92.7	92.0	90.7
50	115.8	115.0	113.3

	135	133	131
6	13.5	13.3	13.1
7	15.8	15.5	15.3
8	18.0	17.7	17.5
9	20.3	20.0	19.7
10	22.5	22.2	21.8
20	45.0	44.3	43.7
30	67.5	66.5	65.5
40	90.0	88.7	87.3
50	112.5	110.8	109.2

	129	128	126
6	12.9	12.8	12.6
7	15.1	14.9	14.7
8	17.2	17.1	16.8
9	19.4	19.2	18.9
10	21.5	21.3	21.0
20	43.0	42.7	42.0
30	64.5	64.0	63.0
40	86.0	85.3	84.0
50	107.5	106.7	105.0

	125	123	122
6	12.5	12.3	12.2
7	14.6	14.4	14.2
8	16.7	16.4	16.3
9	18.8	18.5	18.3
10	20.8	20.5	20.3
20	41.7	41.0	40.7
30	62.5	61.5	61.0
40	83.3	82.0	81.3
50	104.2	102.5	101.7

	121	120	1
6	12.1	12.0	0.1
7	14.1	14.0	0.1
8	16.1	16.0	0.1
9	18.2	18.0	0.2
10	20.2	20.0	0.2
20	40.3	40.0	0.3
30	60.5	60.0	0.5
40	80.7	80.0	0.7
50	100.8	100.0	0.8

p. p.

'	log sin	d	log tan	c. d.	log cot	log cos	'
0	1̄.01923		1̄.02162		0.97838	1̄.99761	60
1	1̄.02043	120	1̄.02283	121	0.97717	1̄.99760	59
2	1̄.02163	120	1̄.02404	121	0.97596	1̄.99759	58
3	1̄.02283	120	1̄.02525	121	0.97475	1̄.99757	57
4	1̄.02402	119	1̄.02645	120	0.97355	1̄.99756	56
5	1̄.02520	118	1̄.02766	121	0.97234	1̄.99755	55
6	1̄.02639	119	1̄.02885	119	0.97115	1̄.99753	54
7	1̄.02757	118	1̄.03005	120	0.96995	1̄.99752	53
8	1̄.02874	117	1̄.03124	119	0.96876	1̄.99751	52
9	1̄.02992	118	1̄.03242	118	0.96758	1̄.99749	51
10	1̄.03109	117	1̄.03361	119	0.96639	1̄.99748	50
11	1̄.03226	117	1̄.03479	118	0.96521	1̄.99747	49
12	1̄.03342	116	1̄.03597	118	0.96403	1̄.99745	48
13	1̄.03458	116	1̄.03714	117	0.96286	1̄.99744	47
14	1̄.03574	116	1̄.03832	118	0.96168	1̄.99742	46
15	1̄.03690	116	1̄.03948	116	0.96052	1̄.99741	45
16	1̄.03805	115	1̄.04065	117	0.95935	1̄.99740	44
17	1̄.03920	115	1̄.04181	116	0.95819	1̄.99738	43
18	1̄.04034	114	1̄.04297	116	0.95703	1̄.99737	42
19	1̄.04149	115	1̄.04413	116	0.95587	1̄.99736	41
20	1̄.04262	113	1̄.04528	115	0.95472	1̄.99734	40
21	1̄.04376	114	1̄.04643	115	0.95357	1̄.99733	39
22	1̄.04490	114	1̄.04758	115	0.95242	1̄.99731	38
23	1̄.04603	113	1̄.04873	115	0.95127	1̄.99730	37
24	1̄.04715	112	1̄.04987	114	0.95013	1̄.99728	36
25	1̄.04828	113	1̄.05101	114	0.94899	1̄.99727	35
26	1̄.04940	112	1̄.05214	113	0.94786	1̄.99726	34
27	1̄.05052	112	1̄.05328	114	0.94672	1̄.99724	33
28	1̄.05164	112	1̄.05441	113	0.94559	1̄.99723	32
29	1̄.05275	111	1̄.05553	112	0.94447	1̄.99721	31
30	1̄.05386	111	1̄.05666	113	0.94334	1̄.99720	30
31	1̄.05497	111	1̄.05778	112	0.94222	1̄.99718	29
32	1̄.05607	110	1̄.05890	112	0.94110	1̄.99717	28
33	1̄.05717	110	1̄.06002	112	0.93998	1̄.99716	27
34	1̄.05827	110	1̄.06113	111	0.93887	1̄.99714	26
35	1̄.05937	110	1̄.06224	111	0.93776	1̄.99713	25
36	1̄.06046	109	1̄.06335	111	0.93665	1̄.99711	24
37	1̄.06155	109	1̄.06445	110	0.93555	1̄.99710	23
38	1̄.06264	109	1̄.06556	111	0.93444	1̄.99708	22
39	1̄.06372	108	1̄.06666	110	0.93334	1̄.99707	21
40	1̄.06481	109	1̄.06775	109	0.93225	1̄.99705	20
41	1̄.06589	108	1̄.06885	110	0.93115	1̄.99704	19
42	1̄.06696	107	1̄.06994	109	0.93006	1̄.99702	18
43	1̄.06804	108	1̄.07103	109	0.92897	1̄.99701	17
44	1̄.06911	107	1̄.07211	108	0.92789	1̄.99699	16
45	1̄.07018	107	1̄.07320	109	0.92680	1̄.99698	15
46	1̄.07124	106	1̄.07428	108	0.92572	1̄.99696	14
47	1̄.07231	107	1̄.07536	108	0.92464	1̄.99695	13
48	1̄.07337	106	1̄.07643	107	0.92357	1̄.99693	12
49	1̄.07442	105	1̄.07751	108	0.92249	1̄.99692	11
50	1̄.07548	106	1̄.07858	107	0.92142	1̄.99690	10
51	1̄.07653	105	1̄.07964	106	0.92036	1̄.99689	9
52	1̄.07758	105	1̄.08071	107	0.91929	1̄.99687	8
53	1̄.07863	105	1̄.08177	106	0.91823	1̄.99686	7
54	1̄.07968	105	1̄.08283	106	0.91717	1̄.99684	6
55	1̄.08072	104	1̄.08389	106	0.91611	1̄.99683	5
56	1̄.08176	104	1̄.08495	106	0.91505	1̄.99681	4
57	1̄.08280	104	1̄.08600	105	0.91400	1̄.99680	3
58	1̄.08383	103	1̄.08705	105	0.91295	1̄.99678	2
59	1̄.08486	103	1̄.08810	105	0.91190	1̄.99677	1
60	1̄.08589	103	1̄.08914	104	0.91086	1̄.99675	0

'	log cos	d	log cot	c. d.	log tan	log sin	'

p. p.

	121	120	119
6	12.1	12.0	11.9
7	14.1	14.0	13.9
8	16.1	16.0	15.9
9	18.2	18.0	17.9
10	20.2	20.0	19.8
20	40.3	40.0	39.7
30	60.5	60.0	59.5
40	80.7	80.0	79.3
50	100.8	100.0	99.2

	118	117	116
6	11.8	11.7	11.6
7	13.8	13.7	13.5
8	15.7	15.6	15.5
9	17.7	17.6	17.4
10	19.7	19.5	19.3
20	39.3	39.0	38.7
30	59.0	58.5	58.0
40	78.7	78.0	77.3
50	98.3	97.5	96.7

	115	114	113
6	11.5	11.4	11.3
7	13.4	13.3	13.2
8	15.3	15.2	15.1
9	17.3	17.1	17.0
10	19.2	19.0	18.8
20	38.3	38.0	37.7
30	57.5	57.0	56.5
40	76.7	76.0	75.3
50	95.8	95.0	94.2

	112	111	110
6	11.2	11.1	11.0
7	13.1	13.0	12.8
8	14.9	14.8	14.7
9	16.8	16.7	16.5
10	18.7	18.5	18.3
20	37.3	37.0	36.7
30	56.0	55.5	55.0
40	74.7	74.0	73.3
50	93.3	92.5	91.7

	109	108	107
6	10.9	10.8	10.7
7	12.7	12.6	12.5
8	14.5	14.4	14.3
9	16.4	16.2	16.1
10	18.2	18.0	17.8
20	36.3	36.0	35.7
30	54.5	54.0	53.5
40	72.7	72.0	71.3
50	90.8	90.0	89.2

	106	105	104
6	10.6	10.5	10.4
7	12.4	12.3	12.1
8	14.1	14.0	13.9
9	15.9	15.8	15.6
10	17.7	17.5	17.3
20	35.3	35.0	34.7
30	53.0	52.5	52.0
40	70.7	70.0	69.3
50	88.3	87.5	86.7

′	log sin	d	log tan	c. d.	log cot	log cos	
0	1̄.08589	103	1̄.08914	105	0.91086	1̄.99675	60
1	1̄.08692	103	1̄.09019	104	0.90981	1̄.99674	59
2	1̄.08795	102	1̄.09123	104	0.90877	1̄.99672	58
3	1̄.08897	102	1̄.09227	103	0.90773	1̄.99670	57
4	1̄.08990	102	1̄.09330	104	0.90670	1̄.99669	56
5	1̄.09101	101	1̄.09434	103	0.90566	1̄.99667	55
6	1̄.09202	102	1̄.09537	103	0.90463	1̄.99666	54
7	1̄.09304	101	1̄.09640	102	0.90360	1̄.99664	53
8	1̄.09405	101	1̄.09742	103	0.90258	1̄.99663	52
9	1̄.09506	100	1̄.09845	102	0.90155	1̄.99661	51
10	1̄.09606	101	1̄.09947	102	0.90053	1̄.99659	50
11	1̄.09707	100	1̄.10049	101	0.89951	1̄.99658	49
12	1̄.09807	100	1̄.10150	102	0.89850	1̄.99656	48
13	1̄.09907	99	1̄.10252	101	0.89748	1̄.99655	47
14	1̄.10006	100	1̄.10353	101	0.89647	1̄.99653	46
15	1̄.10106	99	1̄.10454	101	0.89546	1̄.99651	45
16	1̄.10205	99	1̄.10555	101	0.89445	1̄.99650	44
17	1̄.10304	98	1̄.10656	100	0.89344	1̄.99648	43
18	1̄.10402	99	1̄.10756	100	0.89244	1̄.99647	42
19	1̄.10501	98	1̄.10856	100	0.89144	1̄.99645	41
20	1̄.10599	98	1̄.10956	100	0.89044	1̄.99643	40
21	1̄.10697	98	1̄.11056	99	0.88944	1̄.99642	39
22	1̄.10795	98	1̄.11155	99	0.88845	1̄.99640	38
23	1̄.10893	97	1̄.11254	99	0.88746	1̄.99638	37
24	1̄.10990	97	1̄.11353	99	0.88647	1̄.99637	36
25	1̄.11087	97	1̄.11452	99	0.88548	1̄.99635	35
26	1̄.11184	97	1̄.11551	98	0.88449	1̄.99633	34
27	1̄.11281	96	1̄.11649	98	0.88351	1̄.99632	33
28	1̄.11377	97	1̄.11747	98	0.88253	1̄.99630	32
29	1̄.11474	96	1̄.11845	98	0.88155	1̄.99629	31
30	1̄.11570	96	1̄.11943	97	0.88057	1̄.99627	30
31	1̄.11666	95	1̄.12040	98	0.87960	1̄.99625	29
32	1̄.11761	96	1̄.12138	97	0.87862	1̄.99624	28
33	1̄.11857	95	1̄.12235	97	0.87765	1̄.99622	27
34	1̄.11952	95	1̄.12332	96	0.87668	1̄.99620	26
35	1̄.12047	95	1̄.12428	97	0.87572	1̄.99618	25
36	1̄.12142	94	1̄.12525	96	0.87475	1̄.99617	24
37	1̄.12236	95	1̄.12621	96	0.87379	1̄.99615	23
38	1̄.12331	94	1̄.12717	96	0.87283	1̄.99613	22
39	1̄.12425	94	1̄.12813	96	0.87187	1̄.99612	21
40	1̄.12519	93	1̄.12909	95	0.87091	1̄.99610	20
41	1̄.12612	94	1̄.13004	95	0.86996	1̄.99608	19
42	1̄.12706	93	1̄.13099	95	0.86901	1̄.99607	18
43	1̄.12799	93	1̄.13194	95	0.86806	1̄.99605	17
44	1̄.12892	93	1̄.13289	95	0.86711	1̄.99603	16
45	1̄.12985	93	1̄.13384	94	0.86616	1̄.99601	15
46	1̄.13078	93	1̄.13478	95	0.86522	1̄.99600	14
47	1̄.13171	92	1̄.13573	94	0.86427	1̄.99598	13
48	1̄.13263	92	1̄.13667	94	0.86333	1̄.99596	12
49	1̄.13355	92	1̄.13761	93	0.86239	1̄.99595	11
50	1̄.13447	92	1̄.13854	94	0.86146	1̄.99593	10
51	1̄.13539	91	1̄.13948	93	0.86052	1̄.99591	9
52	1̄.13630	92	1̄.14041	93	0.85959	1̄.99589	8
53	1̄.13722	91	1̄.14134	93	0.85866	1̄.99588	7
54	1̄.13813	91	1̄.14227	93	0.85773	1̄.99586	6
55	1̄.13904	90	1̄.14320	92	0.85680	1̄.99584	5
56	1̄.13994	91	1̄.14412	92	0.85588	1̄.99582	4
57	1̄.14085	90	1̄.14504	93	0.85496	1̄.99581	3
58	1̄.14175	91	1̄.14597	91	0.85403	1̄.99579	2
59	1̄.14266	90	1̄.14688	92	0.85312	1̄.99577	1
60	1̄.14356		1̄.14780		0.85220	1̄.99575	0
	log cos	d	log cot	c. d.	log tan	log sin	′

p. p.

	105	104	103
6	10.5	10.4	10.3
7	12.3	12.1	12.0
8	14.0	13.9	13.7
9	15.8	15.6	15.5
10	17.5	17.3	17.2
20	35.0	34.7	34.3
30	52.5	52.0	51.5
40	70.0	69.3	68.7
50	87.5	86.7	85.8

	102	101	100
6	10.2	10.1	10.0
7	11.9	11.8	11.7
8	13.6	13.5	13.3
9	15.3	15.2	15.0
10	17.0	16.8	16.7
20	34.0	33.7	33.3
30	51.0	50.5	50.0
40	68.0	67.3	66.7
50	85.0	84.2	83.3

	99	98
6	9.9	9.8
7	11.6	11.4
8	13.2	13.1
9	14.9	14.7
10	16.5	16.3
20	33.0	32.7
30	49.5	49.0
40	66.0	65.3
50	82.5	81.7

	97	96	95
6	9.7	9.6	9.5
7	11.3	11.2	11.1
8	12.9	12.8	12.7
9	14.6	14.4	14.3
10	16.2	16.0	15.8
20	32.3	32.0	31.7
30	48.5	48.0	47.5
40	64.7	64.0	63.3
50	80.8	80.0	79.2

	94	93	92
6	9.4	9.3	9.2
7	11.0	10.9	10.7
8	12.5	12.4	12.3
9	14.1	14.0	13.8
10	15.7	15.5	15.3
20	31.3	31.0	30.7
30	47.0	46.5	46.0
40	62.7	62.0	61.3
50	78.3	77.5	76.7

	91	90	2
6	9.1	9.0	0.2
7	10.6	10.5	0.2
8	12.1	12.0	0.3
9	13.7	13.5	0.3
10	15.2	15.0	0.3
20	30.3	30.0	0.7
30	45.5	45.0	1.0
40	60.7	60.0	1.3
50	75.8	75.0	1.7

8°

′	log sin	d	log tan	c. d.	log cot	log cos	′
0	1̄.14356	89	1̄.14780	92	0.85220	1̄.99575	60
1	1̄.14445	90	1̄.14872	91	0.85128	1̄.99574	59
2	1̄.14535	89	1̄.14963	91	0.85037	1̄.99572	58
3	1̄.14624	90	1̄.15054	91	0.84946	1̄.99570	57
4	1̄.14714	89	1̄.15145	91	0.84855	1̄.99568	56
5	1̄.14803	88	1̄.15236	91	0.84764	1̄.99566	55
6	1̄.14891	89	1̄.15327	91	0.84673	1̄.99565	54
7	1̄.14980	89	1̄.15417	90	0.84583	1̄.99563	53
8	1̄.15069	88	1̄.15508	91	0.84492	1̄.99561	52
9	1̄.15157	88	1̄.15598	90	0.84402	1̄.99559	51
10	1̄.15245	88	1̄.15688	90	0.84312	1̄.99557	50
11	1̄.15333	88	1̄.15777	90	0.84223	1̄.99556	49
12	1̄.15421	87	1̄.15867	89	0.84133	1̄.99554	48
13	1̄.15508	88	1̄.15956	90	0.84044	1̄.99552	47
14	1̄.15596	87	1̄.16046	89	0.83954	1̄.99550	46
15	1̄.15683	87	1̄.16135	89	0.83865	1̄.99548	45
16	1̄.15770	87	1̄.16224	88	0.83776	1̄.99546	44
17	1̄.15857	87	1̄.16312	89	0.83688	1̄.99545	43
18	1̄.15944	86	1̄.16401	88	0.83599	1̄.99543	42
19	1̄.16030	86	1̄.16489	88	0.83511	1̄.99541	41
20	1̄.16116	87	1̄.16577	88	0.83423	1̄.99539	40
21	1̄.16203	86	1̄.16665	88	0.83335	1̄.99537	39
22	1̄.16289	85	1̄.16753	88	0.83247	1̄.99535	38
23	1̄.16374	86	1̄.16841	87	0.83159	1̄.99533	37
24	1̄.16460	85	1̄.16928	88	0.83072	1̄.99532	36
25	1̄.16545	86	1̄.17016	87	0.82984	1̄.99530	35
26	1̄.16631	85	1̄.17103	87	0.82897	1̄.99528	34
27	1̄.16716	85	1̄.17190	87	0.82810	1̄.99526	33
28	1̄.16801	85	1̄.17277	86	0.82723	1̄.99524	32
29	1̄.16886	84	1̄.17363	87	0.82637	1̄.99522	31
30	1̄.16970	85	1̄.17450	86	0.82550	1̄.99520	30
31	1̄.17055	84	1̄.17536	86	0.82464	1̄.99518	29
32	1̄.17139	84	1̄.17622	86	0.82378	1̄.99517	28
33	1̄.17223	84	1̄.17708	86	0.82292	1̄.99515	27
34	1̄.17307	84	1̄.17794	86	0.82206	1̄.99513	26
35	1̄.17391	83	1̄.17880	85	0.82120	1̄.99511	25
36	1̄.17474	84	1̄.17965	86	0.82035	1̄.99509	24
37	1̄.17558	83	1̄.18051	85	0.81949	1̄.99507	23
38	1̄.17641	83	1̄.18136	85	0.81864	1̄.99505	22
39	1̄.17724	83	1̄.18221	85	0.81779	1̄.99503	21
40	1̄.17807	83	1̄.18306	85	0.81694	1̄.99501	20
41	1̄.17890	83	1̄.18391	84	0.81609	1̄.99499	19
42	1̄.17973	82	1̄.18475	85	0.81525	1̄.99497	18
43	1̄.18055	82	1̄.18560	84	0.81440	1̄.99495	17
44	1̄.18137	83	1̄.18644	84	0.81356	1̄.99494	16
45	1̄.18220	82	1̄.18728	84	0.81272	1̄.99492	15
46	1̄.18302	81	1̄.18812	84	0.81188	1̄.99490	14
47	1̄.18383	82	1̄.18896	83	0.81104	1̄.99488	13
48	1̄.18465	82	1̄.18979	84	0.81021	1̄.99486	12
49	1̄.18547	81	1̄.19063	83	0.80937	1̄.99484	11
50	1̄.18628	81	1̄.19146	83	0.80854	1̄.99482	10
51	1̄.18709	81	1̄.19229	83	0.80771	1̄.99480	9
52	1̄.18790	81	1̄.19312	83	0.80688	1̄.99478	8
53	1̄.18871	81	1̄.19395	83	0.80605	1̄.99476	7
54	1̄.18952	81	1̄.19478	83	0.80522	1̄.99474	6
55	1̄.19033	80	1̄.19561	82	0.80439	1̄.99472	5
56	1̄.19113	80	1̄.19643	82	0.80357	1̄.99470	4
57	1̄.19193	80	1̄.19725	82	0.80275	1̄.99468	3
58	1̄.19273	80	1̄.19807	82	0.80193	1̄.99466	2
59	1̄.19353	80	1̄.19889	82	0.80111	1̄.99464	1
60	1̄.19433		1̄.19971		0.80029	1̄.99462	0

	log cos	d	log cot	c. d.	log tan	log sin	′

p. p.

	92	91	90
6	9.2	9.1	9.0
7	10.7	10.6	10.5
8	12.3	12.1	12.0
9	13.8	13.7	13.5
10	15.3	15.2	15.0
20	30.7	30.3	30.0
30	46.0	45.5	45.0
40	61.3	60.7	60.0
50	76.7	75.8	75.0

	89	88
6	8.9	8.8
7	10.4	10.3
8	11.9	11.7
9	13.4	13.2
10	14.8	14.7
20	29.7	29.3
30	44.5	44.0
40	59.3	58.7
50	74.2	73.3

	87	86
6	8.7	8.6
7	10.2	10.0
8	11.6	11.5
9	13.1	12.9
10	14.5	14.3
20	29.0	28.7
30	43.5	43.0
40	58.0	57.3
50	72.5	71.7

	85	84
6	8.5	8.4
7	9.9	9.8
8	11.3	11.2
9	12.8	12.6
10	14.2	14.0
20	28.3	28.0
30	42.5	42.0
40	56.7	56.0
50	70.8	70.0

	83	82
6	8.3	8.2
7	9.7	9.6
8	11.1	10.9
9	12.5	12.3
10	13.8	13.7
20	27.7	27.3
30	41.5	41.0
40	55.3	54.7
50	69.2	68.3

	81	80	2
6	8.1	8.0	0.2
7	9.5	9.3	0.2
8	10.8	10.7	0.3
9	12.2	12.0	0.3
10	13.5	13.3	0.3
20	27.0	26.7	0.7
30	40.5	40.0	1.0
40	54.0	53.3	1.3
50	67.5	66.7	1.7

′	log sin	d	log tan	c. d.	log cot	log cos	′
0	1̄.19433	80	1̄.19971	82	0.80029	1̄.99462	60
1	1̄.19513	79	1̄.20053	81	0.79947	1̄.99460	59
2	1̄.19592	80	1̄.20134	82	0.79866	1̄.99458	58
3	1̄.19672	79	1̄.20216	81	0.79784	1̄.99456	57
4	1̄.19751	79	1̄.20297	81	0.79703	1̄.99454	56
5	1̄.19830	79	1̄.20378	81	0.79622	1̄.99452	55
6	1̄.19909	79	1̄.20459	81	0.79541	1̄.99450	54
7	1̄.19988	79	1̄.20540	81	0.79460	1̄.99448	53
8	1̄.20067	79	1̄.20621	80	0.79379	1̄.99446	52
9	1̄.20145	78	1̄.20701	81	0.79299	1̄.99444	51
10	1̄.20223	78	1̄.20782	80	0.79218	1̄.99442	50
		79		80			
11	1̄.20302	78	1̄.20862	80	0.79138	1̄.99440	49
12	1̄.20380	78	1̄.20942	80	0.79058	1̄.99438	48
13	1̄.20458	77	1̄.21022	80	0.78978	1̄.99436	47
14	1̄.20535	78	1̄.21102	80	0.78898	1̄.99434	46
15	1̄.20613	78	1̄.21182	79	0.78818	1̄.99432	45
16	1̄.20691	77	1̄.21261	80	0.78739	1̄.99429	44
17	1̄.20768	77	1̄.21341	79	0.78659	1̄.99427	43
18	1̄.20845	77	1̄.21420	79	0.78580	1̄.99425	42
19	1̄.20922	77	1̄.21499	79	0.78501	1̄.99423	41
20	1̄.20999	77	1̄.21578	79	0.78422	1̄.99421	40
		77		79			
21	1̄.21076	77	1̄.21657	79	0.78343	1̄.99419	39
22	1̄.21153	76	1̄.21736	78	0.78264	1̄.99417	38
23	1̄.21229	77	1̄.21814	79	0.78186	1̄.99415	37
24	1̄.21306	76	1̄.21893	78	0.78107	1̄.99413	36
25	1̄.21382	76	1̄.21971	78	0.78029	1̄.99411	35
26	1̄.21458	76	1̄.22049	78	0.77951	1̄.99409	34
27	1̄.21534	76	1̄.22127	78	0.77873	1̄.99407	33
28	1̄.21610	75	1̄.22205	78	0.77795	1̄.99404	32
29	1̄.21685	76	1̄.22283	78	0.77717	1̄.99402	31
30	1̄.21761	76	1̄.22361	78	0.77639	1̄.99400	30
		75		77			
31	1̄.21836	76	1̄.22438	78	0.77562	1̄.99398	29
32	1̄.21912	75	1̄.22516	77	0.77484	1̄.99396	28
33	1̄.21987	75	1̄.22593	77	0.77407	1̄.99394	27
34	1̄.22062	75	1̄.22670	77	0.77330	1̄.99392	26
35	1̄.22137	74	1̄.22747	77	0.77253	1̄.99390	25
36	1̄.22211	75	1̄.22824	77	0.77176	1̄.99388	24
37	1̄.22286	75	1̄.22901	76	0.77099	1̄.99385	23
38	1̄.22361	74	1̄.22977	77	0.77023	1̄.99383	22
39	1̄.22435	74	1̄.23054	76	0.76946	1̄.99381	21
40	1̄.22509	74	1̄.23130	76	0.76870	1̄.99379	20
		74		76			
41	1̄.22583	74	1̄.23206	77	0.76794	1̄.99377	19
42	1̄.22657	74	1̄.23283	76	0.76717	1̄.99375	18
43	1̄.22731	74	1̄.23359	76	0.76641	1̄.99372	17
44	1̄.22805	73	1̄.23435	75	0.76565	1̄.99370	16
45	1̄.22878	74	1̄.23510	76	0.76490	1̄.99368	15
46	1̄.22952	73	1̄.23586	75	0.76411	1̄.99366	14
47	1̄.23025	73	1̄.23661	76	0.76339	1̄.99364	13
48	1̄.23098	73	1̄.23737	75	0.76263	1̄.99362	12
49	1̄.23171	73	1̄.23812	75	0.76188	1̄.99359	11
50	1̄.23244	73	1̄.23887	75	0.76113	1̄.99357	10
		73		75			
51	1̄.23317	73	1̄.23962	75	0.76038	1̄.99355	9
52	1̄.23390	72	1̄.24037	75	0.75963	1̄.99353	8
53	1̄.23462	73	.24112	75	0.75888	1̄.99351	7
54	1̄.23535	72	1̄.24186	74	0.75814	1̄.99348	6
55	1̄.23607	72	1̄.24261	75	0.75739	1̄.99346	5
56	1̄.23679	73	1̄.24335	74	0.75665	1̄.99344	4
57	1̄.23752	71	1̄.24410	75	0.75590	1̄.99342	3
58	1̄.23823	72	1̄.24484	74	0.75516	1̄.99340	2
59	1̄.23895	72	1̄.24558	74	0.75442	1̄.99337	1
60	1̄.23967	72	1̄.24632	74	0.75368	1̄.99335	0
	log cos	d	log cot	c. d.	log tan	log sin	′

p. p.

	82	81	80
6	8.2	8.1	8.0
7	9.6	9.5	9.3
8	10.9	10.8	10.7
9	12.3	12.2	12.0
10	13.7	13.5	13.3
20	27.3	27.0	26.7
30	41.0	40.5	40.0
40	54.7	54.0	53.3
50	68.3	67.5	66.7

	79	78
6	7.9	7.8
7	9.2	9.1
8	10.5	10.4
9	11.9	11.7
10	13.2	13.0
20	26.3	26.0
30	39.5	39.0
40	52.7	52.0
50	65.8	65.0

	77	76
6	7.7	7.6
7	9.0	8.9
8	10.3	10.1
9	11.6	11.4
10	12.8	12.7
20	25.7	25.3
30	38.5	38.0
40	51.3	50.7
50	64.2	63.3

	75	74
6	7.5	7.4
7	8.8	8.6
8	10.0	9.9
9	11.3	11.1
10	12.5	12.3
20	25.0	24.7
30	37.5	37.0
40	50.0	49.3
50	62.5	61.7

	73	72
6	7.3	7.2
7	8.5	8.4
8	9.7	9.6
9	11.0	10.8
10	12.2	12.0
20	24.3	24.0
30	36.5	36.0
40	48.7	48.0
50	60.8	60.0

	71	3	2
6	7.1	0.3	0.2
7	8.3	0.4	0.2
8	9.5	0.4	0.3
9	10.7	0.5	0.3
10	11.8	0.5	0.3
20	23.7	1.0	0.7
30	35.5	1.5	1.0
40	47.3	2.0	1.3
50	59.2	2.5	1.7

p. p.

'	log sin	d	log tan	ç. d.	log cot	log cos	
0	1̄.23967	72	1̄.24632	74	0.75368	1̄.99335	60
1	1̄.24039	71	1̄.24706	73	0.75294	1̄.99333	59
2	1̄.24110	71	1̄.24779	74	0.75221	1̄.99331	58
3	1̄.24181	72	1̄.24853	73	0.75147	1̄.99328	57
4	1̄.24253	71	1̄.24926	74	0.75074	1̄.99326	56
5	1̄.24324	71	1̄.25000	73	0.75000	1̄.99324	55
6	1̄.24395	71	1̄.25073	73	0.74927	1̄.99322	54
7	1̄.24466	70	1̄.25146	73	0.74854	1̄.99319	53
8	1̄.24536	71	1̄.25219	73	0.74781	1̄.99317	52
9	1̄.24607	70	1̄.25292	73	0.74708	1̄.99315	51
10	1̄.24677	71	1̄.25365	72	0.74635	1̄.99313	50
11	1̄.24748	70	1̄.25437	73	0.74563	1̄.99310	49
12	1̄.24818	70	1̄.25510	72	0.74490	1̄.99308	48
13	1̄.24888	70	1̄.25582	73	0.74418	1̄.99306	47
14	1̄.24958	70	1̄.25655	72	0.74345	1̄.99304	46
15	1̄.25028	70	1̄.25727	72	0.74273	1̄.99301	45
16	1̄.25098	70	1̄.25799	72	0.74201	1̄.99299	44
17	1̄.25168	69	1̄.25871	72	0.74129	1̄.99297	43
18	1̄.25237	70	1̄.25943	72	0.74057	1̄.99294	42
19	1̄.25307	69	1̄.26015	71	0.73985	1̄.99292	41
20	1̄.25376	69	1̄.26086	72	0.73914	1̄.99290	40
21	1̄.25445	69	1̄.26158	71	0.73842	1̄.99288	39
22	1̄.25514	69	1̄.26229	72	0.73771	1̄.99285	38
23	1̄.25583	69	1̄.26301	71	0.73699	1̄.99283	37
24	1̄.25652	69	1̄.26372	71	0.73628	1̄.99281	36
25	1̄.25721	69	1̄.26443	71	0.73557	1̄.99278	35
26	1̄.25790	68	1̄.26514	71	0.73486	1̄.99276	34
27	1̄.25858	69	1̄.26585	70	0.73415	1̄.99274	33
28	1̄.25927	68	1̄.26655	71	0.73345	1̄.99271	32
29	1̄.25995	68	1̄.26726	71	0.73274	1̄.99269	31
30	1̄.26063	68	1̄.26797	70	0.73203	1̄.99267	30
31	1̄.26131	68	1̄.26867	70	0.73133	1̄.99264	29
32	1̄.26199	68	1̄.26937	71	0.73063	1̄.99262	28
33	1̄.26267	68	1̄.27008	70	0.72992	1̄.99260	27
34	1̄.26335	68	1̄.27078	70	0.72922	1̄.99257	26
35	1̄.26403	67	1̄.27148	70	0.72852	1̄.99255	25
36	1̄.26470	68	1̄.27218	70	0.72782	1̄.99252	24
37	1̄.26538	67	1̄.27288	69	0.72712	1̄.99250	23
38	1̄.26605	67	1̄.27357	70	0.72643	1̄.99248	22
39	1̄.26672	67	1̄.27427	69	0.72573	1̄.99245	21
40	1̄.26739	67	1̄.27496	70	0.72504	1̄.99243	20
41	1̄.26806	67	1̄.27566	69	0.72434	1̄.99241	19
42	1̄.26873	67	1̄.27635	69	0.72365	1̄.99238	18
43	1̄.26940	67	1̄.27704	69	0.72296	1̄.99236	17
44	1̄.27007	66	1̄.27773	69	0.72227	1̄.99233	16
45	1̄.27073	67	1̄.27842	69	0.72158	1̄.99231	15
46	1̄.27140	66	1̄.27911	69	0.72089	1̄.99229	14
47	1̄.27206	67	1̄.27980	69	0.72020	1̄.99226	13
48	1̄.27273	66	1̄.28049	68	0.71951	1̄.99224	12
49	1̄.27339	66	1̄.28117	69	0.71883	1̄.99221	11
50	1̄.27405	66	1̄.28186	68	0.71814	1̄.99219	10
51	1̄.27471	66	1̄.28254	69	0.71746	1̄.99217	9
52	1̄.27537	65	1̄.28323	68	0.71677	1̄.99214	8
53	1̄.27602	66	1̄.28391	68	0.71609	1̄.99212	7
54	1̄.27668	66	1̄.28459	68	0.71541	1̄.99209	6
55	1̄.27734	65	1̄.28527	68	0.71473	1̄.99207	5
56	1̄.27799	65	1̄.28595	67	0.71405	1̄.99204	4
57	1̄.27864	66	1̄.28662	68	0.71338	1̄.99202	3
58	1̄.27930	65	1̄.28730	68	0.71270	1̄.99200	2
59	1̄.27995	65	1̄.28798	67	0.71202	1̄.99197	1
60	1̄.28060		1̄.28865		0.71135	1̄.99195	0
	log cos	d	log cot	c. d.	log tan	log sin	'

p. p.

	74	73
6	7.4	7.3
7	8.6	8.5
8	9.9	9.7
9	11.1	11.0
10	12.3	12.2
20	24.7	24.3
30	37.0	36.5
40	49.3	48.7
50	61.7	60.8

	72	71
6	7.2	7.1
7	8.4	8.3
8	9.6	9.5
9	10.8	10.7
10	12.0	11.8
20	24.0	23.7
30	36.0	35.5
40	48.0	47.3
50	60.0	59.2

	70	69
6	7.0	6.9
7	8.2	8.1
8	9.3	9.2
9	10.5	10.4
10	11.7	11.5
20	23.3	23.0
30	35.0	34.5
40	46.7	46.0
50	58.3	57.5

	68	67
6	6.8	6.7
7	7.9	7.8
8	9.1	8.9
9	10.2	10.1
10	11.3	11.2
20	22.7	22.3
30	34.0	33.5
40	45.3	44.7
50	56.7	55.8

	66	65
6	6.6	6.5
7	7.7	7.6
8	8.8	8.7
9	9.9	9.8
10	11.0	10.8
20	22.0	21.7
30	33.0	32.5
40	44.0	43.3
50	55.0	54.2

	3	2
6	0.3	0.2
7	0.4	0.2
8	0.4	0.3
9	0.5	0.3
10	0.5	0.3
20	1.0	0.7
30	1.5	1.0
40	2.0	1.3
50	2.5	1.7

′	log sin	d	log tan	c. d.	log cot	log cos	
0	$\bar{1}$.28060	65	$\bar{1}$.28865	68	0.71135	$\bar{1}$.99195	60
1	$\bar{1}$.28125	65	$\bar{1}$.28933	67	0.71067	$\bar{1}$.99192	59
2	$\bar{1}$.28190	64	$\bar{1}$.29000	67	0.71000	$\bar{1}$.99190	58
3	$\bar{1}$.28254	65	$\bar{1}$.29067	67	0.70933	$\bar{1}$.99187	57
4	$\bar{1}$.28319	65	$\bar{1}$.29134	67	0.70866	$\bar{1}$.99185	56
5	$\bar{1}$.28384	64	$\bar{1}$.29201	67	0.70799	$\bar{1}$.99182	55
6	$\bar{1}$.28448	64	$\bar{1}$.29268	67	0.70732	$\bar{1}$.99180	54
7	$\bar{1}$.28512	65	$\bar{1}$.29335	67	0.70665	$\bar{1}$.99177	53
8	$\bar{1}$.28577	64	$\bar{1}$.29402	66	0.70598	$\bar{1}$.99175	52
9	$\bar{1}$.28641	64	$\bar{1}$.29468	67	0.70532	$\bar{1}$.99172	51
10	$\bar{1}$.28705	64	$\bar{1}$.29535	66	0.70465	$\bar{1}$.99170	50
11	$\bar{1}$.28769	64	$\bar{1}$.29601	67	0.70399	$\bar{1}$.99167	49
12	$\bar{1}$.28833	63	$\bar{1}$.29668	66	0.70332	$\bar{1}$.99165	48
13	$\bar{1}$.28896	64	$\bar{1}$.29734	66	0.70266	$\bar{1}$.99162	47
14	$\bar{1}$.28960	64	$\bar{1}$.29800	66	0.70200	$\bar{1}$.99160	46
15	$\bar{1}$.29024	63	$\bar{1}$.29866	66	0.70134	$\bar{1}$.99157	45
16	$\bar{1}$.29087	63	$\bar{1}$.29932	66	0.70068	$\bar{1}$.99155	44
17	$\bar{1}$.29150	64	$\bar{1}$.29998	66	0.70002	$\bar{1}$.99152	43
18	$\bar{1}$.29214	63	$\bar{1}$.30064	66	0.69936	$\bar{1}$.99150	42
19	$\bar{1}$.29277	63	$\bar{1}$.30130	65	0.69870	$\bar{1}$.99147	41
20	$\bar{1}$.29340	63	$\bar{1}$.30195	66	0.69805	$\bar{1}$.99145	40
21	$\bar{1}$.29403	63	$\bar{1}$.30261	65	0.69739	$\bar{1}$.99142	39
22	$\bar{1}$.29466	63	$\bar{1}$.30326	65	0.69674	$\bar{1}$.99140	38
23	$\bar{1}$.29529	62	$\bar{1}$.30391	66	0.69609	$\bar{1}$.99137	37
24	$\bar{1}$.29591	63	$\bar{1}$.30457	65	0.69543	$\bar{1}$.99135	36
25	$\bar{1}$.29654	62	$\bar{1}$.30522	65	0.69478	$\bar{1}$.99132	35
26	$\bar{1}$.29716	63	$\bar{1}$.30587	65	0.69413	$\bar{1}$.99130	34
27	$\bar{1}$.29779	62	$\bar{1}$.30652	65	0.69348	$\bar{1}$.99127	33
28	$\bar{1}$.29841	62	$\bar{1}$.30717	65	0.69283	$\bar{1}$.99124	32
29	$\bar{1}$.29903	63	$\bar{1}$.30782	64	0.69218	$\bar{1}$.99122	31
30	$\bar{1}$.29966	62	$\bar{1}$.30846	65	0.69154	$\bar{1}$.99119	30
31	$\bar{1}$.30028	62	$\bar{1}$.30911	64	0.69089	$\bar{1}$.99117	29
32	$\bar{1}$.30090	61	$\bar{1}$.30975	65	0.69025	$\bar{1}$.99114	28
33	$\bar{1}$.30151	62	$\bar{1}$.31040	64	0.68960	$\bar{1}$.99112	27
34	$\bar{1}$.30213	62	$\bar{1}$.31104	64	0.68896	$\bar{1}$.99109	26
35	$\bar{1}$.30275	61	$\bar{1}$.31168	65	0.68832	$\bar{1}$.99106	25
36	$\bar{1}$.30336	62	$\bar{1}$.31233	64	0.68767	$\bar{1}$.99104	24
37	$\bar{1}$.30398	61	$\bar{1}$.31297	64	0.68703	$\bar{1}$.99101	23
38	$\bar{1}$.30459	62	$\bar{1}$.31361	64	0.68639	$\bar{1}$.99099	22
39	$\bar{1}$.30521	61	$\bar{1}$.31425	64	0.68575	$\bar{1}$.99096	21
40	$\bar{1}$.30582	61	$\bar{1}$.31489	64	0.68511	$\bar{1}$.99093	20
41	$\bar{1}$.30643	61	$\bar{1}$.31552	64	0.68448	$\bar{1}$.99091	19
42	$\bar{1}$.30704	61	$\bar{1}$.31616	63	0.68384	$\bar{1}$.99088	18
43	$\bar{1}$.30765	61	$\bar{1}$.31679	64	0.68321	$\bar{1}$.99086	17
44	$\bar{1}$.30826	61	$\bar{1}$.31743	63	0.68257	$\bar{1}$.99083	16
45	$\bar{1}$.30887	60	$\bar{1}$.31806	64	0.68194	$\bar{1}$.99080	15
46	$\bar{1}$.30947	61	$\bar{1}$.31870	63	0.68130	$\bar{1}$.99078	14
47	$\bar{1}$.31008	60	$\bar{1}$.31933	63	0.68067	$\bar{1}$.99075	13
48	$\bar{1}$.31068	61	$\bar{1}$.31996	63	0.68004	$\bar{1}$.99072	12
49	$\bar{1}$.31129	60	$\bar{1}$.32059	63	0.67941	$\bar{1}$.99070	11
50	$\bar{1}$.31189	61	$\bar{1}$.32122	63	0.67878	$\bar{1}$.99067	10
51	$\bar{1}$.31250	60	$\bar{1}$.32185	63	0.67815	$\bar{1}$.99064	9
52	$\bar{1}$.31310	60	$\bar{1}$.32248	63	0.67752	$\bar{1}$.99062	8
53	$\bar{1}$.31370	60	$\bar{1}$.32311	62	0.67689	$\bar{1}$.99059	7
54	$\bar{1}$.31430	60	$\bar{1}$.32373	63	0.67627	$\bar{1}$.99056	6
55	$\bar{1}$.31490	59	$\bar{1}$.32436	62	0.67564	$\bar{1}$.99054	5
56	$\bar{1}$.31549	60	$\bar{1}$.32498	63	0.67502	$\bar{1}$.99051	4
57	$\bar{1}$.31609	60	$\bar{1}$.32561	62	0.67439	$\bar{1}$.99048	3
58	$\bar{1}$.31669	60	$\bar{1}$.32623	62	0.67377	$\bar{1}$.99046	2
59	$\bar{1}$.31728	59	$\bar{1}$.32685	62	0.67315	$\bar{1}$.99043	1
60	$\bar{1}$.31788	60	$\bar{1}$.32747		0.67253	$\bar{1}$.99040	0
	log cos	d	log cot	c. d.	log tan	log sin	′

p. p.

	68	67
6	6.8	6.7
7	7.9	7.8
8	9.1	8.9
9	10.2	10.1
10	11.3	11.2
20	22.7	22.3
30	34.0	33.5
40	45.3	44.7
50	56.7	55.8

	66	65
6	6.6	6.5
7	7.7	7.6
8	8.8	8.7
9	9.9	9.8
10	11.0	10.8
20	22.0	21.7
30	33.0	32.5
40	44.0	43.3
50	55.0	54.2

	64	63
6	6.4	6.3
7	7.5	7.4
8	8.5	8.4
9	9.6	9.5
10	10.7	10.5
20	21.3	21.0
30	32.0	31.5
40	42.7	42.0
50	53.3	52.5

	62	61
6	6.2	6.1
7	7.2	7.1
8	8.3	8.1
9	9.3	9.2
10	10.3	10.2
20	20.7	20.3
30	31.0	30.5
40	41.3	40.7
50	51.7	50.8

	60	59
6	6.0	5.9
7	7.0	6.9
8	8.0	7.9
9	9.0	8.9
10	10.0	9.8
20	20.0	19.7
30	30.0	29.5
40	40.0	39.3
50	50.0	49.2

	3	2
6	0.3	0.2
7	0.4	0.2
8	0.4	0.3
9	0.5	0.3
10	0.5	0.3
20	1.0	0.7
30	1.5	1.0
40	2.0	1.3
50	2.5	1.7

′	log sin	d	log tan	c. d.	log cot	log cos		p. p.
0	1̄.31788	59	1̄.32747	63	0.67253	1̄.99040	60	
1	1̄.31847	60	1̄.32810	62	0.67190	1̄.99038	59	
2	1̄.31907	59	1̄.32872	61	0.67128	1̄.99035	58	
3	1̄.31966	59	1̄.32933	62	0.67067	1̄.99032	57	
4	1̄.32025	59	1̄.32995	62	0.67005	1̄.99030	56	
5	1̄.32084	59	1̄.33057	62	0.66943	1̄.99027	55	
6	1̄.32143	59	1̄.33119	61	0.66881	1̄.99024	54	
7	1̄.32202	59	1̄.33180	62	0.66820	1̄.99022	53	
8	1̄.32261	58	1̄.33242	61	0.66758	1̄.99019	52	
9	1̄.32319	59	1̄.33303	62	0.66697	1̄.99016	51	
10	1̄.32378	59	1̄.33365	61	0.66635	1̄.99013	50	
11	1̄.32437	58	1̄.33426	61	0.66574	1̄.99011	49	
12	1̄.32495	58	1̄.33487	61	0.66513	1̄.99008	48	
13	1̄.32553	59	1̄.33548	61	0.66452	1̄.99005	47	
14	1̄.32612	58	1̄.33609	61	0.66391	1̄.99002	46	
15	1̄.32670	58	1̄.33670	61	0.66330	1̄.99000	45	
16	1̄.32728	58	1̄.33731	61	0.66269	1̄.98997	44	
17	1̄.32786	58	1̄.33792	61	0.66208	1̄.98994	43	
18	1̄.32844	58	1̄.33853	60	0.66147	1̄.98991	42	
19	1̄.32902	58	1̄.33913	61	0.66087	1̄.98989	41	
20	1̄.32960	58	1̄.33974	60	0.66026	1̄.98986	40	
21	1̄.33018	57	1̄.34034	61	0.65966	1̄.98983	39	
22	1̄.33075	58	1̄.34095	60	0.65905	1̄.98980	38	
23	1̄.33133	57	1̄.34155	60	0.65845	1̄.98978	37	
24	1̄.33190	58	1̄.34215	61	0.65785	1̄.98975	36	
25	1̄.33248	57	1̄.34276	60	0.65724	1̄.98972	35	
26	1̄.33305	57	1̄.34336	60	0.65664	1̄.98969	34	
27	1̄.33362	58	1̄.34396	60	0.65604	1̄.98967	33	
28	1̄.33420	57	1̄.34456	60	0.65544	1̄.98964	32	
29	1̄.33477	57	1̄.34516	60	0.65484	1̄.98961	31	
30	1̄.33534	57	1̄.34576	59	0.65424	1̄.98958	30	
31	1̄.33591	56	1̄.34635	60	0.65365	1̄.98955	29	
32	1̄.33647	57	1̄.34695	60	0.65305	1̄.98953	28	
33	1̄.33704	57	1̄.34755	59	0.65245	1̄.98950	27	
34	1̄.33761	57	1̄.34814	60	0.65186	1̄.98947	26	
35	1̄.33818	56	1̄.34874	59	0.65126	1̄.98944	25	
36	1̄.33874	57	1̄.34933	59	0.65067	1̄.98941	24	
37	1̄.33931	56	1̄.34992	59	0.65008	1̄.98938	23	
38	1̄.33987	56	1̄.35051	60	0.64949	1̄.98936	22	
39	1̄.34043	57	1̄.35111	59	0.64889	1̄.98933	21	
40	1̄.34100	56	1̄.35170	59	0.64830	1̄.98930	20	
41	1̄.34156	56	1̄.35229	59	0.64771	1̄.98927	19	
42	1̄.34212	56	1̄.35288	59	0.64712	1̄.98924	18	
43	1̄.34268	56	1̄.35347	58	0.64653	1̄.98921	17	
44	1̄.34324	56	1̄.35405	59	0.64595	1̄.98919	16	
45	1̄.34380	56	1̄.35464	59	0.64536	1̄.98916	15	
46	1̄.34436	55	1̄.35523	58	0.64477	1̄.98913	14	
47	1̄.34491	56	1̄.35581	59	0.64419	1̄.98910	13	
48	1̄.34547	55	1̄.35640	58	0.64360	1̄.98907	12	
49	1̄.34602	56	1̄.35698	59	0.64302	1̄.98904	11	
50	1̄.34658	55	1̄.35757	58	0.64243	1̄.98901	10	
51	1̄.34713	56	1̄.35815	58	0.64185	1̄.98898	9	
52	1̄.34769	55	1̄.35873	58	0.64127	1̄.98896	8	
53	1̄.34824	55	1̄.35931	58	0.64069	1̄.98893	7	
54	1̄.34879	55	1̄.35989	58	0.64011	1̄.98890	6	
55	1̄.34934	55	1̄.36047	58	0.63953	1̄.98887	5	
56	1̄.34989	55	1̄.36105	58	0.63895	1̄.98884	4	
57	1̄.35044	55	1̄.36163	58	0.63837	1̄.98881	3	
58	1̄.35099	55	1̄.36221	58	0.63779	1̄.98878	2	
59	1̄.35154	55	1̄.36279	57	0.63721	1̄.98875	1	
60	1̄.35209		1̄.36336		0.63664	1̄.98872	0	
	log cos	d	log cot	c. d.	log tan	log sin	′	p. p.

p. p.

	63	62
6	6.3	6.2
7	7.4	7.2
8	8.4	8.3
9	9.5	9.3
10	10.5	10.3
20	21.0	20.7
30	31.5	31.0
40	42.0	41.3
50	52.5	51.7

	61	60
6	6.1	6.0
7	7.1	7.0
8	8.1	8.0
9	9.2	9.0
10	10.2	10.0
20	20.3	20.0
30	30.5	30.0
40	40.7	40.0
50	50.8	50.0

	59
6	5.9
7	6.9
8	7.9
9	8.9
10	9.8
20	19.7
30	29.5
40	39.3
59	49.2

	58	57
6	5.8	5.7
7	6.8	6.7
8	7.7	7.6
9	8.7	8.6
10	9.7	9.5
20	19.3	19.0
30	29.0	28.5
40	38.7	38.0
50	48.3	47.5

	56	55
6	5.6	5.5
7	6.5	6.4
8	7.5	7.3
9	8.4	8.3
10	9.3	9.2
20	18.7	18.3
30	28.0	27.5
40	37.3	36.7
50	46.7	45.8

	3	2
6	0.3	0.2
7	0.4	0.2
8	0.4	0.3
9	0.5	0.3
10	0.5	0.3
20	1.0	0.7
30	1.5	1.0
40	2.0	1.3
50	2.5	1.7

′	log sin	d	log tan	c. d.	log cot	ʹlog cos		p. p.		
0	Ī.35209		Ī.36336		0.63664	Ī.98872	60		**58**	**57**
1	Ī.35263	54	Ī.36394	58	0.63606	Ī.98869	59	6	5.8	5.7
2	Ī.35318	55	Ī.36452	58	0.63548	Ī.98867	58	7	6.8	6.7
3	Ī.35373	55	Ī.36509	57	0.63491	Ī.98864	57	8	7.7	7.6
4	Ī.35427	54	Ī.36566	57	0.63434	Ī.98861	56	9	8.7	8.6
5	Ī.35481	54	Ī.36624	58	0.63376	Ī.98858	55	10	9.7	9.5
6	Ī.35536	55	Ī.36681	57	0.63319	Ī.98855	54	20	19.3	19.0
7	Ī.35590	54	Ī.36738	57	0.63262	Ī.98852	53	30	29.0	28.5
8	Ī.35644	54	Ī.36795	57	0.63205	Ī.98849	52	40	38.7	38.3
9	Ī.35698	54	Ī.36852	57	0.63148	Ī.98846	51	50	48.3	47.5
10	Ī.35752	54	Ī.36909	57	0.63091	Ī.98843	50			
		54		57					**58**	**55**
11	Ī.35806	54	Ī.36966	57	0.63034	Ī.98840	49	6	5.6	5.5
12	Ī.35860	54	Ī.37023	57	0.62977	Ī.98837	48	7	6.5	6.4
13	Ī.35914	54	Ī.37080	57	0.62920	Ī.98834	47	8	7.5	7.3
14	Ī.35968	54	Ī.37137	56	0.62863	Ī.98831	46	9	8.4	8.3
15	Ī.36022	54	Ī.37193	57	0.62807	Ī.98828	45	10	9.3	9.2
16	Ī.36075	53	Ī.37250	56	0.62750	Ī.98825	44	20	18.7	18.3
17	Ī.36129	54	Ī.37306	57	0.62694	Ī.98822	43	30	28.0	27.5
18	Ī.36182	53	Ī.37363	56	0.62637	Ī.98819	42	40	37.3	36.7
19	Ī.36236	54	Ī.37419	57	0.62581	Ī.98816	41	50	46.7	45.8
20	Ī.36289	53	Ī.37476	56	0.62524	Ī.98813	40			
		53		56						**54**
21	Ī.36342	53	Ī.37532	56	0.62468	Ī.98810	39	6		5.4
22	Ī.36395	54	Ī.37588	56	0.62412	Ī.98807	38	7		6.3
23	Ī.36449	53	Ī.37644	56	0.62356	Ī.98804	37	8		7.2
24	Ī.36502	53	Ī.37700	56	0.62300	Ī.98801	36	9		8.1
25	Ī.36555	53	Ī.37756	56	0.62244	Ī.98798	35	10		9.0
26	Ī.36608	52	Ī.37812	56	0.62188	Ī.98795	34	20		18.0
27	Ī.36660	53	Ī.37868	56	0.62132	Ī.98792	33	30		27.0
28	Ī.36713	53	Ī.37924	56	0.62076	Ī.98789	32	40		36.0
29	Ī.36766	53	Ī.37980	55	0.62020	Ī.98786	31	50		45.0
30	Ī.36819	52	Ī.38035	56	0.61965	Ī.98783	30			
		52		56					**53**	**52**
31	Ī.36871	53	Ī.38091	56	0.61909	Ī.98780	29	6	5.3	5.2
32	Ī.36924	52	Ī.38147	55	0.61853	Ī.98777	28	7	6.2	6.1
33	Ī.36976	52	Ī.38202	55	0.61798	Ī.98774	27	8	7.1	6.9
34	Ī.37028	53	Ī.38257	56	0.61743	Ī.98771	26	9	8.0	7.8
35	Ī.37081	52	Ī.38313	55	0.61687	Ī.98768	25	10	8.8	8.7
36	Ī.37133	52	Ī.38368	55	0.61632	Ī.98765	24	20	17.7	17.3
37	Ī.37185	52	Ī.38423	56	0.61577	Ī.98762	23	30	26.5	26.0
38	Ī.37237	52	Ī.38479	55	0.61521	Ī.98759	22	40	35.3	34.7
39	Ī.37289	52	Ī.38534	55	0.61466	Ī.98756	21	50	44.2	43.3
40	Ī.37341	52	Ī.38589	55	0.61411	Ī.98753	20			
		52		55					**51**	**4**
41	Ī.37393	52	Ī.38644	55	0.61356	Ī.98750	19	6	5.1	0.4
42	Ī.37445	52	Ī.38699	55	0.61301	Ī.98746	18	7	6.0	0.5
43	Ī.37497	52	Ī.38754	54	0.61246	Ī.98743	17	8	6.8	0.5
44	Ī.37549	51	Ī.38808	55	0.61192	Ī.98740	16	9	7.7	0.6
45	Ī.37600	52	Ī.38863	55	0.61137	Ī.98737	15	10	8.5	0.7
46	Ī.37652	51	Ī.38918	54	0.61082	Ī.98734	14	20	17.0	1.3
47	Ī.37703	52	Ī.38972	55	0.61028	Ī.98731	13	30	25.5	2.0
48	Ī.37755	51	Ī.39027	55	0.60973	Ī.98728	12	40	34.0	2.7
49	Ī.37806	52	Ī.39082	54	0.60918	Ī.98725	11	50	42.5	3.3
50	Ī.37858	51	Ī.39136	54	0.60864	Ī.98722	10			
		51		54					**3**	**2**
51	Ī.37909	51	Ī.39190	55	0.60810	Ī.98719	9	6	0.3	0.2
52	Ī.37960	51	Ī.39245	54	0.60755	Ī.98715	8	7	0.4	0.2
53	Ī.38011	51	Ī.39299	54	0.60701	Ī.98712	7	8	0.4	0.3
54	Ī.38062	51	Ī.39353	54	0.60647	Ī.98709	6	9	0.5	0.3
55	Ī.38113	51	Ī.39407	54	0.60593	Ī.98706	5	10	0.5	0.3
56	Ī.38164	51	Ī.39461	54	0.60539	Ī.98703	4	20	1.0	0.7
57	Ī.38215	51	Ī.39515	54	0.60485	Ī.98700	3	30	1.5	1.0
58	Ī.38266	51	Ī.39569	54	0.60431	Ī.98697	2	40	2.0	1.3
59	Ī.38317	51	Ī.39623	54	0.60377	Ī.98694	1	50	2.5	1.7
60	Ī.38368		Ī.39677	54	0.60323	Ī.98690	0			

log cos	d	log cot	c. d.	log tan	log sin	′	p. p.

′	log sin	d	log tan	c. d.	log cot	log cos	d	′
0	1̄.38368	50	1̄.3967?	54	0.60323	1̄.98690	3	60
1	1̄.38418	51	1̄.39731	54	0.60269	1̄.98687	3	59
2	1̄.38469	50	1̄.39785	53	0.60215	1̄.98684	3	58
3	1̄.38519	51	1̄.39838	54	0.60162	1̄.98681	3	57
4	1̄.38570	50	1̄.39892	53	0.60108	1̄.98678	3	56
5	1̄.38620	50	1̄.39945	54	0.60055	1̄.98675	3	55
6	1̄.38670	51	1̄.39999	53	0.60001	1̄.98671	4	54
7	1̄.38721	50	1̄.40052	54	0.59948	1̄.98668	3	53
8	1̄.38771	50	1̄.40106	53	0.59894	1̄.98665	3	52
9	1̄.38821	50	1̄.40159	53	0.59841	1̄.98662	3	51
10	1̄.38871	50	1̄.40212	53	0.59788	1̄.98659	3	50
11	1̄.38921	50	1̄.40266	54	0.59734	1̄.98656	3	49
12	1̄.38971	50	1̄.40319	53	0.59681	1̄.98652	4	48
13	1̄.39021	50	1̄.40372	53	0.59628	1̄.98649	3	47
14	1̄.39071	50	1̄.40425	53	0.59575	1̄.98646	3	46
15	1̄.39121	50	1̄.40478	53	0.59522	1̄.98643	3	45
16	1̄.39170	49	1̄.40531	53	0.59469	1̄.98640	3	44
17	1̄.39220	50	1̄.40584	53	0.59416	1̄.98636	4	43
18	1̄.39270	50	1̄.40636	52	0.59364	1̄.98633	3	42
19	1̄.39319	49	1̄.40689	53	0.59311	1̄.98630	3	41
20	1̄.39369	50	1̄.40742	53	0.59258	1̄.98627	3	40
21	1̄.39418	49	1̄.40795	53	0.59205	1̄.98623	4	39
22	1̄.39467	49	1̄.40847	52	0.59153	1̄.98620	3	38
23	1̄.39517	50	1̄.40900	53	0.59100	1̄.98617	3	37
24	1̄.39566	49	1̄.40952	52	0.59048	1̄.98614	3	36
25	1̄.39615	49	1̄.41005	53	0.58995	1̄.98610	4	35
26	1̄.39664	49	1̄.41057	52	0.58943	1̄.98607	3	34
27	1̄.39713	49	1̄.41109	52	0.58891	1̄.98604	3	33
28	1̄.39762	49	1̄.41161	52	0.58839	1̄.98601	3	32
29	1̄.39811	49	1̄.41214	53	0.58786	1̄.98597	4	31
30	1̄.39860	49	1̄.41266	52	0.58734	1̄.98594	3	30
31	1̄.39909	49	1̄.41318	52	0.58682	1̄.98591	3	29
32	1̄.39958	49	1̄.41370	52	0.58630	1̄.98588	3	28
33	1̄.40006	48	1̄.41422	52	0.58578	1̄.98584	4	27
34	1̄.40055	49	1̄.41474	52	0.58526	1̄.98581	3	26
35	1̄.40103	48	1̄.41526	52	0.58474	1̄.98578	3	25
36	1̄.40152	49	1̄.41578	52	0.58422	1̄.98574	4	24
37	1̄.40200	48	1̄.41629	51	0.58371	1̄.98571	3	23
38	1̄.40249	49	1̄.41681	52	0.58319	1̄.98568	3	22
39	1̄.40297	48	1̄.41733	52	0.58267	1̄.98565	3	21
40	1̄.40346	49	1̄.41784	51	0.58216	1̄.98561	4	20
41	1̄.40394	48	1̄.41836	52	0.58164	1̄.98558	3	19
42	1̄.40442	48	1̄.41887	51	0.58113	1̄.98555	3	18
43	1̄.40490	48	1̄.41939	52	0.58061	1̄.98551	4	17
44	1̄.40538	48	1̄.41990	51	0.58010	1̄.98548	3	16
45	1̄.40586	48	1̄.42041	51	0.57959	1̄.98545	3	15
46	1̄.40634	48	1̄.42093	52	0.57907	1̄.98541	4	14
47	1̄.40682	48	1̄.42144	51	0.57856	1̄.98538	3	13
48	1̄.40730	48	1̄.42195	51	0.57805	1̄.98535	3	12
49	1̄.40778	48	1̄.42246	51	0.57754	1̄.98531	4	11
50	1̄.40825	47	1̄.42297	51	0.57703	1̄.98528	3	10
51	1̄.40873	48	1̄.42348	51	0.57652	1̄.98525	3	9
52	1̄.40921	48	1̄.42399	51	0.57601	1̄.98521	4	8
53	1̄.40968	47	1̄.42450	51	0.57550	1̄.98518	3	7
54	1̄.41016	48	1̄.42501	51	0.57499	1̄.98515	3	6
55	1̄.41063	47	1̄.42552	51	0.57448	1̄.98511	4	5
56	1̄.41111	48	1̄.42603	51	0.57397	1̄.98508	3	4
57	1̄.41158	47	1̄.42653	50	0.57347	1̄.98505	3	3
58	1̄.41205	47	1̄.42704	51	0.57296	1̄.98501	4	2
59	1̄.41252	47	1̄.42755	51	0.57245	1̄.98498	3	1
60	1̄.41300	48	1̄.42805	50	0.57195	1̄.98494	4	0
	log cos	d	log cot	c. d.	log tan	log sin	d	′

p. p.

	54	53
6	5.4	5.3
7	6.3	6.2
8	7.2	7.1
9	8.1	8.0
10	9.0	8.8
20	18.0	17.7
30	27.0	26.5
40	36.0	35.3
50	45.0	44.2

	52	51
6	5.2	5.1
7	6.1	6.0
8	6.9	6.8
9	7.8	7.7
10	8.7	8.5
20	17.3	17.0
30	26.0	25.5
40	34.7	34.0
50	43.3	42.5

	50	49
6	5.0	4.9
7	5.8	5.7
8	6.7	6.5
9	7.5	7.4
10	8.3	8.2
20	16.7	16.3
30	25.0	24.5
40	33.3	32.7
50	41.7	40.8

	48	47
6	4.8	4.7
7	5.6	5.5
8	6.4	6.3
9	7.2	7.1
10	8.0	7.8
20	16.0	15.7
30	24.0	23.5
40	32.0	31.3
50	40.0	39.2

	4	3
6	0.4	0.3
7	0.5	0.4
8	0.5	0.4
9	0.6	0.5
10	0.7	0.5
20	1.3	1.0
30	2.0	1.5
40	2.7	2.0
50	3.3	2.5

′	log sin	d	log tan	c. d.	log cot	log cos	d	′
0	1̄.41300	47	1̄.42805	51	0.57195	1̄.98494	3	60
1	1̄.41347	47	1̄.42856	50	0.57144	1̄.98491	3	59
2	1̄.41394	47	1̄.42906	51	0.57094	1̄.98488	4	58
3	1̄.41441	47	1̄.42957	50	0.57043	1̄.98484	3	57
4	1̄.41488	47	1̄.43007	50	0.56993	1̄.98481	4	56
5	1̄.41535	47	1̄.43057	51	0.56943	1̄.98477	3	55
6	1̄.41582	46	1̄.43108	50	0.56892	1̄.98474	3	54
7	1̄.41628	47	1̄.43158	50	0.56842	1̄.98471	4	53
8	1̄.41675	47	1̄.43208	50	0.56792	1̄.98467	3	52
9	1̄.41722	46	1̄.43258	50	0.56742	1̄.98464	4	51
10	1̄.41768	47	1̄.43308	50	0.56692	1̄.98460	3	50
11	1̄.41815	46	1̄.43358	50	0.56642	1̄.98457	4	49
12	1̄.41861	47	1̄.43408	50	0.56592	1̄.98453	3	48
13	1̄.41908	46	1̄.43458	50	0.56542	1̄.98450	3	47
14	1̄.41954	47	1̄.43508	50	0.56492	1̄.98447	4	46
15	1̄.42001	46	1̄.43558	49	0.56442	1̄.98443	3	45
16	1̄.42047	46	1̄.43607	50	0.56393	1̄.98440	4	44
17	1̄.42093	47	1̄.43657	50	0.56343	1̄.98436	3	43
18	1̄.42140	46	1̄.43707	49	0.56293	1̄.98433	4	42
19	1̄.42186	46	1̄.43756	50	0.56244	1̄.98429	3	41
20	1̄.42232	46	1̄.43806	49	0.56194	1̄.98426	4	40
21	1̄.42278	46	1̄.43855	50	0.56145	1̄.98422	3	39
22	1̄.42324	46	1̄.43905	49	0.56095	1̄.98419	4	38
23	1̄.42370	46	1̄.43954	50	0.56046	1̄.98415	3	37
24	1̄.42416	45	1̄.44004	49	0.55996	1̄.98412	3	36
25	1̄.42461	46	1̄.44053	49	0.55947	1̄.98409	4	35
26	1̄.42507	46	1̄.44102	49	0.55898	1̄.98405	3	34
27	1̄.42553	46	1̄.44151	50	0.55849	1̄.98402	4	33
28	1̄.42599	45	1̄.44201	49	0.55799	1̄.98398	3	32
29	1̄.42644	46	1̄.44250	49	0.55750	1̄.98395	4	31
30	1̄.42690	45	1̄.44299	49	0.55701	1̄.98391	3	30
31	1̄.42735	46	1̄.44348	49	0.55652	1̄.98388	4	29
32	1̄.42781	45	1̄.44397	49	0.55603	1̄.98384	3	28
33	1̄.42826	46	1̄.44446	49	0.55554	1̄.98381	4	27
34	1̄.42872	45	1̄.44495	49	0.55505	1̄.98377	4	26
35	1̄.42917	45	1̄.44544	48	0.55456	1̄.98373	3	25
36	1̄.42962	46	1̄.44592	49	0.55408	1̄.98370	4	24
37	1̄.43008	45	1̄.44641	49	0.55359	1̄.98366	3	23
38	1̄.43053	45	1̄.44690	48	0.55310	1̄.98363	4	22
39	1̄.43098	45	1̄.44738	49	0.55262	1̄.98359	3	21
40	1̄.43143	45	1̄.44787	49	0.55213	1̄.98356	4	20
41	1̄.43188	45	1̄.44836	48	0.55164	1̄.98352	3	19
42	1̄.43233	45	1̄.44884	49	0.55116	1̄.98349	4	18
43	1̄.43278	45	1̄.44933	48	0.55067	1̄.98345	3	17
44	1̄.43323	44	1̄.44981	48	0.55019	1̄.98342	4	16
45	1̄.43367	45	1̄.45029	49	0.54971	1̄.98338	4	15
46	1̄.43412	45	1̄.45078	48	0.54922	1̄.98334	3	14
47	1̄.43457	45	1̄.45126	48	0.54874	1̄.98331	4	13
48	1̄.43502	44	1̄.45174	48	0.54826	1̄.98327	3	12
49	1̄.43546	45	1̄.45222	49	0.54778	1̄.98324	4	11
50	1̄.43591	44	1̄.45271	48	0.54729	1̄.98320	3	10
51	1̄.43635	45	1̄.45319	48	0.54681	1̄.98317	4	9
52	1̄.43680	44	1̄.45367	48	0.54633	1̄.98313	4	8
53	1̄.43724	45	1̄.45415	48	0.54585	1̄.98309	3	7
54	1̄.43769	44	1̄.45463	48	0.54537	1̄.98306	4	6
55	1̄.43813	44	1̄.45511	48	0.54489	1̄.98302	3	5
56	1̄.43857	44	1̄.45559	47	0.54441	1̄.98299	4	4
57	1̄.43901	45	1̄.45606	48	0.54394	1̄.98295	4	3
58	1̄.43946	44	1̄.45654	48	0.54346	1̄.98291	3	2
59	1̄.43990	44	1̄.45702	48	0.54298	1̄.98288	4	1
60	1̄.44034		1̄.45750		0.54250	1̄.98284		0

	log cos	d	log cot	c. d.	log tan	log sin	d	′

p. p.

	51	50
6	5.1	5.0
7	6.0	5.8
8	6.8	6.7
9	7.7	7.5
10	8.5	8.3
20	17.0	16.7
30	25.5	25.0
40	34.0	33.3
50	42.5	41.7

	48	48
6	4.9	4.8
7	5.7	5.6
8	6.5	6.4
9	7.4	7.2
10	8.2	8.0
20	16.3	16.0
30	24.5	24.0
40	32.7	32.0
50	40.8	40.0

	47	46
6	4.7	4.5
7	5.5	5.4
8	6.3	6.1
9	7.1	6.9
10	7.8	7.7
20	15.7	15.3
30	23.5	23.0
40	31.3	30.7
50	39.2	38.3

	45	44
6	4.5	4.4
7	5.3	5.1
8	6.0	5.9
9	6.8	6.6
10	7.5	7.3
20	15.0	14.7
30	22.5	22.0
40	30.0	29.3
50	37.5	36.7

	4	3
6	0.4	0.3
7	0.5	0.4
8	0.5	0.4
9	0.6	0.5
10	0.7	0.5
20	1.3	1.0
30	2.0	1.5
40	2.7	2.0
50	3.3	2.5

'	log sin	d	log tan	c. d.	log cot	log cos	d	
0	1̄.44034	44	1̄.45750	47	0.54250	1̄.98284	3	60
1	1̄.44078	44	1̄.45797	48	0.54203	1̄.98281	4	59
2	1̄.44122	44	1̄.45845	47	0.54155	1̄.98277	4	58
3	1̄.44166	44	1̄.45892	48	0.54108	1̄.98273	3	57
4	1̄.44210	43	1̄.45940	47	0.54060	1̄.98270	4	56
5	1̄.44253	44	1̄.45987	48	0.54013	1̄.98266	4	55
6	1̄.44297	44	1̄.46035	47	0.53965	1̄.98262	3	54
7	1̄.44341	44	1̄.46082	48	0.53918	1̄.98259	4	53
8	1̄.44385	43	1̄.46130	47	0.53870	1̄.98255	4	52
9	1̄.44428	44	1̄.46177	47	0.53823	1̄.98251	3	51
10	1̄.44472	44	1̄.46224	47	0.53776	1̄.98248	4	50
11	1̄.44516	43	1̄.46271	48	0.53729	1̄.98244	4	49
12	1̄.44559	43	1̄.46319	47	0.53681	1̄.98240	3	48
13	1̄.44602	44	1̄.46366	47	0.53634	1̄.98237	4	47
14	1̄.44646	43	1̄.46413	47	0.53587	1̄.98233	4	46
15	1̄.44689	44	1̄.46460	47	0.53540	1̄.98229	3	45
16	1̄.44733	43	1̄.46507	47	0.53493	1̄.98226	4	44
17	1̄.44776	43	1̄.46554	47	0.53446	1̄.98222	4	43
18	1̄.44819	43	1̄.46601	47	0.53399	1̄.98218	3	42
19	1̄.44862	43	1̄.46648	46	0.53352	1̄.98215	4	41
20	1̄.44905	43	1̄.46694	47	0.53306	1̄.98211	4	40
21	1̄.44948	44	1̄.46741	47	0.53259	1̄.98207	3	39
22	1̄.44991	43	1̄.46788	47	0.53212	1̄.98204	4	38
23	1̄.45035	42	1̄.46835	46	0.53165	1̄.98200	4	37
24	1̄.45077	43	1̄.46881	47	0.53119	1̄.98196	4	36
25	1̄.45120	43	1̄.46928	47	0.53072	1̄.98192	3	35
26	1̄.45163	43	1̄.46975	46	0.53025	1̄.98189	4	34
27	1̄.45206	43	1̄.47021	47	0.52979	1̄.98185	4	33
28	1̄.45249	43	1̄.47068	46	0.52932	1̄.98181	4	32
29	1̄.45292	42	1̄.47114	46	0.52886	1̄.98177	3	31
30	1̄.45334	43	1̄.47160	47	0.52840	1̄.98174	4	30
31	1̄.45377	42	1̄.47207	46	0.52793	1̄.98170	4	29
32	1̄.45419	43	1̄.47253	46	0.52747	1̄.98166	4	28
33	1̄.45462	42	1̄.47299	47	0.52701	1̄.98162	3	27
34	1̄.45504	43	1̄.47346	46	0.52654	1̄.98159	4	26
35	1̄.45547	42	1̄.47392	46	0.52608	1̄.98155	4	25
36	1̄.45589	43	1̄.47438	46	0.52562	1̄.98151	4	24
37	1̄.45632	42	1̄.47484	46	0.52516	1̄.98147	3	23
38	1̄.45674	42	1̄.47530	46	0.52470	1̄.98144	4	22
39	1̄.45716	42	1̄.47576	46	0.52424	1̄.98140	4	21
40	1̄.45758	43	1̄.47622	46	0.52378	1̄.98136	4	20
41	1̄.45801	42	1̄.47668	46	0.52332	1̄.98132	3	19
42	1̄.45843	42	1̄.47714	46	0.52286	1̄.98129	4	18
43	1̄.45885	42	1̄.47760	46	0.52240	1̄.98125	4	17
44	1̄.45927	42	1̄.47806	46	0.52194	1̄.98121	4	16
45	1̄.45969	42	1̄.47852	45	0.52148	1̄.98117	4	15
46	1̄.46011	42	1̄.47897	46	0.52103	1̄.98113	3	14
47	1̄.46053	42	1̄.47943	46	0.52057	1̄.98110	4	13
48	1̄.46095	41	1̄.47989	46	0.52011	1̄.98106	4	12
49	1̄.46136	42	1̄.48035	46	0.51965	1̄.98102	4	11
50	1̄.46178	42	1̄.48080	45	0.51920	1̄.98098	4	10
51	1̄.46220	42	1̄.48126	46	0.51874	1̄.98094	4	9
52	1̄.46262	41	1̄.48171	45	0.51829	1̄.98090	3	8
53	1̄.46303	42	1̄.48217	46	0.51783	1̄.98087	4	7
54	1̄.46345	41	1̄.48262	45	0.51738	1̄.98083	4	6
55	1̄.46386	42	1̄.48307	45	0.51693	1̄.98079	4	5
56	1̄.46428	41	1̄.48353	46	0.51647	1̄.98075	4	4
57	1̄.46469	42	1̄.48398	45	0.51602	1̄.98071	4	3
58	1̄.46511	41	1̄.48443	45	0.51557	1̄.98067	4	2
59	1̄.46552	42	1̄.48489	46	0.51511	1̄.98063	3	1
60	1̄.46594		1̄.48534		0.51406	1̄.98060		0

	log cos	d	log cot	c. d.	log tan	log sin	d	'

p. p.

	48	47
6	4.8	4.7
7	5.6	5.5
8	6.4	6.3
9	7.2	7.1
10	8.0	7.8
20	16.0	15.7
30	24.0	23.5
40	32.0	31.3
50	40.0	39.2

	46	45
6	4.6	4.5
7	5.4	5.3
8	6.1	6.0
9	6.9	6.8
10	7.7	7.5
20	15.3	15.0
30	23.0	22.5
40	30.7	30.0
50	38.3	37.5

	44	43
6	4.4	4.3
7	5.1	5.0
8	5.9	5.7
9	6.6	6.5
10	7.3	7.2
20	14.7	14.3
30	22.0	21.5
40	29.3	28.7
50	36.7	35.8

	42	41
6	4.2	4.1
7	4.9	4.8
8	5.6	5.5
9	6.3	6.2
10	7.0	6.8
20	14.0	13.7
30	21.0	20.5
40	28.0	27.3
50	35.0	34.2

	4	3
6	0.4	0.3
7	0.5	0.4
8	0.5	0.4
9	0.6	0.5
10	0.7	0.5
20	1.3	1.0
30	2.0	1.5
40	2.7	2.0
50	3.3	2.5

′	log sin	d	log tan	c. d.	log cot	log cos	d	
0	1̄.46594		1̄.48534		0.51466	1̄.98060		60
1	1̄.46635	41	1̄.48579	45	0.51421	1̄.98056	4	59
2	1̄.46676	41	1̄.48624	45	0.51376	1̄.98052	4	58
3	1̄.46717	41	1̄.48669	45	0.51331	1̄.98048	4	57
4	1̄.46758	41	1̄.48714	45	0.51286	1̄.98044	4	56
5	1̄.46800	42	1̄.48759	45	0.51241	1̄.98040	4	55
6	1̄.46841	41	1̄.48804	45	0.51196	1̄.98036	4	54
7	1̄.46882	41	1̄.48849	45	0.51151	1̄.98032	4	53
8	1̄.46923	41	1̄.48894	45	0.51106	1̄.98029	3	52
9	1̄.46964	41	1̄.48939	45	0.51061	1̄.98025	4	51
10	1̄.47005	41	1̄.48984	45	0.51016	1̄.98021	4	50
		40		45			4	
11	1̄.47045	41	1̄.49029	45	0.50971	1̄.98017	4	49
12	1̄.47086	41	1̄.49073	44	0.50927	1̄.98013	4	48
13	1̄.47127	41	1̄.49118	45	0.50882	1̄.98009	4	47
14	1̄.47168	41	1̄.49163	45	0.50837	1̄.98005	4	46
15	1̄.47209	41	1̄.49207	44	0.50793	1̄.98001	4	45
16	1̄.47249	40	1̄.49252	45	0.50748	1̄.97997	4	44
17	1̄.47290	41	1̄.49296	44	0.50704	1̄.97993	4	43
18	1̄.47330	40	1̄.49341	45	0.50659	1̄.97989	4	42
19	1̄.47371	41	1̄.49385	44	0.50615	1̄.97986	3	41
20	1̄.47411	40	1̄.49430	45	0.50570	1̄.97982	4	40
		41		44			4	
21	1̄.47452	40	1̄.49474	45	0.50526	1̄.97978	4	39
22	1̄.47492	41	1̄.49519	44	0.50481	1̄.97974	4	38
23	1̄.47533	40	1̄.49563	44	0.50437	1̄.97970	4	37
24	1̄.47573	40	1̄.49607	45	0.50393	1̄.97966	4	36
25	1̄.47613	41	1̄.49652	44	0.50348	1̄.97962	4	35
26	1̄.47654	40	1̄.49696	44	0.50304	1̄.97958	4	34
27	1̄.47694	40	1̄.49740	44	0.50260	1̄.97954	4	33
28	1̄.47734	40	1̄.49784	44	0.50216	1̄.97950	4	32
29	1̄.47774	40	1̄.49828	44	0.50172	1̄.97946	4	31
30	1̄.47814	40	1̄.49872	44	0.50128	1̄.97942	4	30
		40		44			4	
31	1̄.47854	40	1̄.49916	44	0.50084	1̄.97938	4	29
32	1̄.47894	40	1̄.49960	44	0.50040	1̄.97934	4	28
33	1̄.47934	40	1̄.50004	44	0.49996	1̄.97930	4	27
34	1̄.47974	40	1̄.50048	44	0.49952	1̄.97926	4	26
35	1̄.48014	40	1̄.50092	44	0.49908	1̄.97922	4	25
36	1̄.48054	40	1̄.50136	44	0.49864	1̄.97918	4	24
37	1̄.48094	39	1̄.50180	43	0.49820	1̄.97914	4	23
38	1̄.48133	40	1̄.50223	44	0.49777	1̄.97910	4	22
39	1̄.48173	40	1̄.50267	44	0.49733	1̄.97906	4	21
40	1̄.48213	39	1̄.50311	44	0.49689	1̄.97902	4	20
		39		44			4	
41	1̄.48252	40	1̄.50355	43	0.49645	1̄.97898	4	19
42	1̄.48292	40	1̄.50398	44	0.49602	1̄.97894	4	18
43	1̄.48332	39	1̄.50442	43	0.49558	1̄.97890	4	17
44	1̄.48371	40	1̄.50485	44	0.49515	1̄.97886	4	16
45	1̄.48411	39	1̄.50529	43	0.49471	1̄.97882	4	15
46	1̄.48450	40	1̄.50572	44	0.49428	1̄.97878	4	14
47	1̄.48490	39	1̄.50616	43	0.49384	1̄.97874	4	13
48	1̄.48529	39	1̄.50659	44	0.49341	1̄.97870	4	12
49	1̄.48568	39	1̄.50703	43	0.49297	1̄.97866	4	11
50	1̄.48607	39	1̄.50746	43	0.49254	1̄.97861	5	10
		40		43			4	
51	1̄.48647	39	1̄.50789	44	0.49211	1̄.97857	4	9
52	1̄.48686	39	1̄.50833	43	0.49167	1̄.97853	4	8
53	1̄.48725	39	1̄.50876	43	0.49124	1̄.97849	4	7
54	1̄.48754	39	1̄.50919	43	0.49081	1̄.97845	4	6
55	1̄.48803	39	1̄.50962	43	0.49038	1̄.97841	4	5
56	1̄.48842	39	1̄.51005	43	0.48995	1̄.97837	4	4
57	1̄.48881	39	1̄.51048	43	0.48952	1̄.97833	4	3
58	1̄.48920	39	1̄.51092	44	0.48908	1̄.97829	4	2
59	1̄.48959	39	1̄.51135	43	0.48865	1̄.97825	4	1
60	1̄.48998	39	1̄.51178	43	0.48822	1̄.97821	4	0
	log cos	d	log cot	c. d.	log tan	log sin	d	′

p. p.

	45	44
6	4.5	4.4
7	5.3	5.1
8	6.0	5.9
9	6.8	6.6
10	7.5	7.3
20	15.0	14.7
30	22.5	22.0
40	30.0	29.3
50	37.5	36.7

	43
6	4.3
7	5.0
8	5.7
9	6.5
10	7.2
20	14.3
30	21.5
40	28.7
50	35.8

	42	41
6	4.2	4.1
7	4.9	4.8
8	5.6	5.5
9	6.3	6.2
10	7.0	6.8
20	14.0	13.7
30	21.0	20.5
40	28.0	27.3
50	35.0	34.2

	40	39
6	4.0	3.9
7	4.7	4.6
8	5.3	5.2
9	6.0	5.9
10	6.7	6.5
20	13.3	13.0
30	20.0	19.5
40	26.7	26.0
50	33.3	32.5

	5	4	3
6	0.5	0.4	0.3
7	0.6	0.5	0.4
8	0.7	0.5	0.4
9	0.8	0.6	0.5
10	0.8	0.7	0.5
20	1.7	1.3	1.0
30	2.5	2.0	1.5
40	3.3	2.7	2.0
50	4.2	3.3	2.5

18°

′	log sin	d	log tan	c. d.	log cot	log cos	d	′
0	1.48998		1.51178		0.48822	1.97821		60
1	1.49037	39	1.51221	43	0.48779	1.97817	4	59
2	1.49076	39	1.51264	43	0.48736	1.97812	5	58
3	1.49115	39	1.51306	42	0.48694	1.97808	4	57
4	1.49153	38	1.51349	43	0.48651	1.97804	4	56
5	1.49192	39	1.51392	43	0.48608	1.97800	4	55
6	1.49231	39	1.51435	43	0.48565	1.97796	4	54
7	1.49269	38	1.51478	43	0.48522	1.97792	4	53
8	1.49308	39	1.51520	43	0.48480	1.97788	4	52
9	1.49347	39	1.51563	43	0.48437	1.97784	4	51
10	1.49385	38	1.51606	43	0.48394	1.97779	5	50
11	1.49424	39	1.51648	42	0.48352	1.97775	4	49
12	1.49462	38	1.51691	43	0.48309	1.97771	4	48
13	1.49500	38	1.51734	43	0.48266	1.97767	4	47
14	1.49539	39	1.51776	42	0.48224	1.97763	4	46
15	1.49577	38	1.51819	43	0.48181	1.97759	4	45
16	1.49615	38	1.51861	42	0.48139	1.97754	5	44
17	1.49654	39	1.51903	42	0.48097	1.97750	4	43
18	1.49692	38	1.51946	43	0.48054	1.97746	4	42
19	1.49730	38	1.51988	42	0.48012	1.97742	4	41
20	1.49768	38	1.52031	43	0.47969	1.97738	4	40
21	1.49806	38	1.52073	42	0.47927	1.97734	4	39
22	1.49844	38	1.52115	42	0.47885	1.97729	5	38
23	1.49882	38	1.52157	42	0.47843	1.97725	4	37
24	1.49920	38	1.52200	43	0.47800	1.97721	4	36
25	1.49958	38	1.52242	42	0.47758	1.97717	4	35
26	1.49996	38	1.52284	42	0.47716	1.97713	4	34
27	1.50034	38	1.52326	42	0.47674	1.97708	5	33
28	1.50072	38	1.52368	42	0.47632	1.97704	4	32
29	1.50110	38	1.52410	42	0.47590	1.97700	4	31
30	1.50148	38	1.52452	42	0.47548	1.97696	4	30
31	1.50185	37	1.52494	42	0.47506	1.97691	5	29
32	1.50223	38	1.52536	42	0.47464	1.97687	4	28
33	1.50261	38	1.52578	42	0.47422	1.97683	4	27
34	1.50298	37	1.52620	42	0.47380	1.97679	4	26
35	1.50336	38	1.52661	41	0.47339	1.97674	5	25
36	1.50374	38	1.52703	42	0.47297	1.97670	4	24
37	1.50411	37	1.52745	42	0.47255	1.97666	4	23
38	1.50449	38	1.52787	42	0.47213	1.97662	4	22
39	1.50486	37	1.52829	42	0.47171	1.97657	5	21
40	1.50523	37	1.52870	41	0.47130	1.97653	4	20
41	1.50561	38	1.52912	42	0.47088	1.97649	4	19
42	1.50598	37	1.52953	41	0.47047	1.97645	4	18
43	1.50635	37	1.52995	42	0.47005	1.97640	5	17
44	1.50673	38	1.53037	42	0.46963	1.97636	4	16
45	1.50710	37	1.53078	41	0.46922	1.97632	4	15
46	1.50747	37	1.53120	42	0.46880	1.97628	4	14
47	1.50784	37	1.53161	41	0.46839	1.97623	5	13
48	1.50821	37	1.53202	42	0.46798	1.97619	4	12
49	1.50858	37	1.53244	41	0.46756	1.97615	4	11
50	1.50896	38	1.53285	41	0.46715	1.97610	5	10
51	1.50933	37	1.53327	42	0.46673	1.97606	4	9
52	1.50970	37	1.53368	41	0.46632	1.97602	4	8
53	1.51007	37	1.53409	41	0.46591	1.97597	5	7
54	1.51043	36	1.53450	41	0.46550	1.97593	4	6
55	1.51080	37	1.53492	42	0.46508	1.97589	4	5
56	1.51117	37	1.53533	41	0.46467	1.97584	5	4
57	1.51154	37	1.53574	41	0.46426	1.97580	4	3
58	1.51191	37	1.53615	41	0.46385	1.97576	4	2
59	1.51227	36	1.53656	41	0.46344	1.97571	5	1
60	1.51264	37	1.53697	41	0.46303	1.97567	4	0

| | log cos | d | log cot | c. d. | log tan | log sin | d | ′ |

p. p.

	43	42
6	4.3	4.2
7	5.0	4.9
8	5.7	5.6
9	6.5	6.3
10	7.2	7.0
20	14.3	14.0
30	21.5	21.0
40	28.7	28.0
50	35.8	35.0

	41
6	4.1
7	4.8
8	5.5
9	6.2
10	6.8
20	13.7
30	20.5
40	27.3
50	34.1

	39	38
6	3.9	3.8
7	4.6	4.4
8	5.2	5.1
9	5.9	5.7
10	6.5	6.3
20	13.0	12.7
30	19.5	19.0
40	26.0	25.3
50	32.5	31.7

	37	36
6	3.7	3.6
7	4.3	4.2
8	4.9	4.8
9	5.6	5.4
10	6.2	6.0
20	12.3	12.0
30	18.5	18.0
40	24.7	24.0
50	30.8	30.0

	5	4
6	0.5	0.4
7	0.6	0.5
8	0.7	0.5
9	0.8	0.6
10	0.8	0.7
20	1.7	1.3
30	2.5	2.0
40	3.3	2.7
50	4.2	3.3

71°

′	log sin	d	log tan	c. d.	log cot	log cos	d	′
0	$\bar{1}$.51264	37	$\bar{1}$.53697	41	0.46303	$\bar{1}$.97567	4	60
1	$\bar{1}$.51301	37	$\bar{1}$.53738	41	0.46262	$\bar{1}$.97563	5	59
2	$\bar{1}$.51338	36	$\bar{1}$.53779	41	0.46221	$\bar{1}$.97558	4	58
3	$\bar{1}$.51374	37	$\bar{1}$.53820	41	0.46180	$\bar{1}$.97554	4	57
4	$\bar{1}$.51411	36	$\bar{1}$.53861	41	0.46139	$\bar{1}$.97550	5	56
5	$\bar{1}$.51447	37	$\bar{1}$.53902	41	0.46098	$\bar{1}$.97545	4	55
6	$\bar{1}$.51484	36	$\bar{1}$.53943	41	0.46057	$\bar{1}$.97541	5	54
7	$\bar{1}$.51520	37	$\bar{1}$.53984	41	0.46016	$\bar{1}$.97536	4	53
8	$\bar{1}$.51557	36	$\bar{1}$.54025	40	0.45975	$\bar{1}$.97532	4	52
9	$\bar{1}$.51593	36	$\bar{1}$.54065	41	0.45935	$\bar{1}$.97528	5	51
10	$\bar{1}$.51629	37	$\bar{1}$.54106	41	0.45894	$\bar{1}$.97523	4	50
11	$\bar{1}$.51666	36	$\bar{1}$.54147	40	0.45853	$\bar{1}$.97519	4	49
12	$\bar{1}$.51702	36	$\bar{1}$.54187	41	0.45813	$\bar{1}$.97515	5	48
13	$\bar{1}$.51738	36	$\bar{1}$.54228	41	0.45772	$\bar{1}$.97510	4	47
14	$\bar{1}$.51774	37	$\bar{1}$.54269	40	0.45731	$\bar{1}$.97506	5	46
15	$\bar{1}$.51811	36	$\bar{1}$.54309	41	0.45691	$\bar{1}$.97501	4	45
16	$\bar{1}$.51847	36	$\bar{1}$.54350	40	0.45650	$\bar{1}$.97497	5	44
17	$\bar{1}$.51883	36	$\bar{1}$.54390	41	0.45610	$\bar{1}$.97492	4	43
18	$\bar{1}$.51919	36	$\bar{1}$.54431	40	0.45569	$\bar{1}$.97488	4	42
19	$\bar{1}$.51955	36	$\bar{1}$.54471	41	0.45529	$\bar{1}$.97484	5	41
20	$\bar{1}$.51991	36	$\bar{1}$.54512	40	0.45488	$\bar{1}$.97479	4	40
21	$\bar{1}$.52027	36	$\bar{1}$.54552	41	0.45448	$\bar{1}$.97475	5	39
22	$\bar{1}$.52063	36	$\bar{1}$.54593	40	0.45407	$\bar{1}$.97470	4	38
23	$\bar{1}$.52099	36	$\bar{1}$.54633	40	0.45367	$\bar{1}$.97466	5	37
24	$\bar{1}$.52135	36	$\bar{1}$.54673	41	0.45327	$\bar{1}$.97461	4	36
25	$\bar{1}$.52171	36	$\bar{1}$.54714	40	0.45286	$\bar{1}$.97457	4	35
26	$\bar{1}$.52207	35	$\bar{1}$.54754	40	0.45246	$\bar{1}$.97453	5	34
27	$\bar{1}$.52242	36	$\bar{1}$.54794	41	0.45206	$\bar{1}$.97448	4	33
28	$\bar{1}$.52278	36	$\bar{1}$.54835	40	0.45165	$\bar{1}$.97444	5	32
29	$\bar{1}$.52314	36	$\bar{1}$.54875	40	0.45125	$\bar{1}$.97439	4	31
30	$\bar{1}$.52350	35	$\bar{1}$.54915	40	0.45085	$\bar{1}$.97435	5	30
31	$\bar{1}$.52385	36	$\bar{1}$.54955	40	0.45045	$\bar{1}$.97430	4	29
32	$\bar{1}$.52421	35	$\bar{1}$.54995	40	0.45005	$\bar{1}$.97426	5	28
33	$\bar{1}$.52456	36	$\bar{1}$.55035	40	0.44965	$\bar{1}$.97421	4	27
34	$\bar{1}$.52492	35	$\bar{1}$.55075	40	0.44925	$\bar{1}$.97417	5	26
35	$\bar{1}$.52527	36	$\bar{1}$.55115	40	0.44885	$\bar{1}$.97412	4	25
36	$\bar{1}$.52563	35	$\bar{1}$.55155	40	0.44845	$\bar{1}$.97408	5	24
37	$\bar{1}$.52598	36	$\bar{1}$.55195	40	0.44805	$\bar{1}$.97403	4	23
38	$\bar{1}$.52634	35	$\bar{1}$.55235	40	0.44765	$\bar{1}$.97399	5	22
39	$\bar{1}$.52669	36	$\bar{1}$.55275	40	0.44725	$\bar{1}$.97394	4	21
40	$\bar{1}$.52705	35	$\bar{1}$.55315	40	0.44685	$\bar{1}$.97390	5	20
41	$\bar{1}$.52740	35	$\bar{1}$.55355	40	0.44645	$\bar{1}$.97385	4	19
42	$\bar{1}$.52775	36	$\bar{1}$.55395	39	0.44605	$\bar{1}$.97381	5	18
43	$\bar{1}$.52811	35	$\bar{1}$.55434	40	0.44566	$\bar{1}$.97376	4	17
44	$\bar{1}$.52846	35	$\bar{1}$.55474	40	0.44526	$\bar{1}$.97372	5	16
45	$\bar{1}$.52881	35	$\bar{1}$.55514	40	0.44486	$\bar{1}$.97367	4	15
46	$\bar{1}$.52916	35	$\bar{1}$.55554	39	0.44446	$\bar{1}$.97363	5	14
47	$\bar{1}$.52951	35	$\bar{1}$.55593	40	0.44407	$\bar{1}$.97358	5	13
48	$\bar{1}$.52986	35	$\bar{1}$.55633	40	0.44367	$\bar{1}$.97353	4	12
49	$\bar{1}$.53021	35	$\bar{1}$.55673	39	0.44327	$\bar{1}$.97349	5	11
50	$\bar{1}$.53056	36	$\bar{1}$.55712	40	0.44288	$\bar{1}$.97344	4	10
51	$\bar{1}$.53092	34	$\bar{1}$.55752	39	0.44248	$\bar{1}$.97340	5	9
52	$\bar{1}$.53126	35	$\bar{1}$.55791	40	0.44209	$\bar{1}$.97335	4	8
53	$\bar{1}$.53161	35	$\bar{1}$.55831	39	0.44169	$\bar{1}$.97331	5	7
54	$\bar{1}$.53196	35	$\bar{1}$.55870	40	0.44130	$\bar{1}$.97326	4	6
55	$\bar{1}$.53231	35	$\bar{1}$.55910	39	0.44090	$\bar{1}$.97322	5	5
56	$\bar{1}$.53266	35	$\bar{1}$.55949	40	0.44051	$\bar{1}$.97317	5	4
57	$\bar{1}$.53301	35	$\bar{1}$.55989	39	0.44011	$\bar{1}$.97312	4	3
58	$\bar{1}$.53336	35	$\bar{1}$.56028	39	0.43972	$\bar{1}$.97308	5	2
59	$\bar{1}$.53370	34	$\bar{1}$.56067	40	0.43933	$\bar{1}$.97303	4	1
60	$\bar{1}$.53405	35	$\bar{1}$.56107		0.43893	$\bar{1}$.97299		0
	log cos	d	log cot	c. d.	log tan	log sin	d	′

p. p.

	41	40
6	4.1	4.0
7	4.8	4.7
8	5.5	5.3
9	6.2	6.0
10	6.8	6.7
20	13.7	13.3
30	20.5	20.0
40	27.3	26.7
50	34.2	33.3

	39
6	3.9
7	4.6
8	5.2
9	5.9
10	6.5
20	13.0
30	19.5
40	26.0
50	32.5

	37	36
6	3.7	3.6
7	4.3	4.2
8	4.9	4.8
9	5.6	5.4
10	6.2	6.0
20	12.3	12.0
30	18.5	18.0
40	24.7	24.0
50	30.8	30.0

	35	34
6	3.5	3.4
7	4.1	4.0
8	4.7	4.5
9	5.3	5.1
10	5.8	5.7
20	11.7	11.3
30	17.5	17.0
40	23.3	22.7
50	29.2	28.3

	5	4
6	0.5	0.4
7	0.6	0.5
8	0.7	0.5
9	0.8	0.6
10	0.8	0.7
20	1.7	1.3
30	2.5	2.0
40	3.3	2.7
50	4.2	3.3

′	log sin	d	log tan	c. d.	log cot	log cos	d	
0	$\bar{1}$.53405	35	$\bar{1}$.56107	39	0.43893	$\bar{1}$.97299	5	60
1	$\bar{1}$.53440	35	$\bar{1}$.56146	39	0.43854	$\bar{1}$.97294	5	59
2	$\bar{1}$.53475	34	$\bar{1}$.56185	39	0.43815	$\bar{1}$.97289	4	58
3	$\bar{1}$.53509	35	$\bar{1}$.56224	40	0.43776	$\bar{1}$.97285	5	57
4	$\bar{1}$.53544	34	$\bar{1}$.56264	39	0.43736	$\bar{1}$.97280	4	56
5	$\bar{1}$.53578	35	$\bar{1}$.56303	39	0.43697	$\bar{1}$.97276	5	55
6	$\bar{1}$.53613	34	$\bar{1}$.56342	39	0.43658	$\bar{1}$.97271	5	54
7	$\bar{1}$.53647	35	$\bar{1}$.56381	39	0.43619	$\bar{1}$.97266	4	53
8	$\bar{1}$.53682	34	$\bar{1}$.56420	39	0.43580	$\bar{1}$.97262	5	52
9	$\bar{1}$.53716	35	$\bar{1}$.56459	39	0.43541	$\bar{1}$.97257	5	51
10	$\bar{1}$.53751	34	$\bar{1}$.56498	39	0.43502	$\bar{1}$.97252	4	50
11	$\bar{1}$.53785	34	$\bar{1}$.56537	39	0.43463	$\bar{1}$.97248	5	49
12	$\bar{1}$.53819	35	$\bar{1}$.56576	39	0.43424	$\bar{1}$.97243	5	48
13	$\bar{1}$.53854	34	$\bar{1}$.56615	39	0.43385	$\bar{1}$.97238	4	47
14	$\bar{1}$.53888	34	$\bar{1}$.56654	39	0.43346	$\bar{1}$.97234	5	46
15	$\bar{1}$.53922	35	$\bar{1}$.56693	39	0.43307	$\bar{1}$.97229	5	45
16	$\bar{1}$.53957	34	$\bar{1}$.56732	39	0.43268	$\bar{1}$.97224	4	44
17	$\bar{1}$.53991	34	$\bar{1}$.56771	39	0.43229	$\bar{1}$.97220	5	43
18	$\bar{1}$.54025	34	$\bar{1}$.56810	39	0.43190	$\bar{1}$.97215	5	42
19	$\bar{1}$.54059	34	$\bar{1}$.56849	38	0.43151	$\bar{1}$.97210	4	41
20	$\bar{1}$.54093	34	$\bar{1}$.56887	39	0.43113	$\bar{1}$.97206	5	40
21	$\bar{1}$.54127	34	$\bar{1}$.56926	39	0.43074	$\bar{1}$.97201	5	39
22	$\bar{1}$.54161	34	$\bar{1}$.56965	39	0.43035	$\bar{1}$.97196	4	38
23	$\bar{1}$.54195	34	$\bar{1}$.57004	38	0.42996	$\bar{1}$.97192	5	37
24	$\bar{1}$.54229	34	$\bar{1}$.57042	39	0.42958	$\bar{1}$.97187	5	36
25	$\bar{1}$.54263	34	$\bar{1}$.57081	39	0.42919	$\bar{1}$.97182	4	35
26	$\bar{1}$.54297	34	$\bar{1}$.57120	38	0.42880	$\bar{1}$.97178	5	34
27	$\bar{1}$.54331	34	$\bar{1}$.57158	39	0.42842	$\bar{1}$.97173	5	33
28	$\bar{1}$.54365	34	$\bar{1}$.57197	38	0.42803	$\bar{1}$.97168	5	32
29	$\bar{1}$.54399	34	$\bar{1}$.57235	39	0.42765	$\bar{1}$.97163	4	31
30	$\bar{1}$.54433	33	$\bar{1}$.57274	38	0.42726	$\bar{1}$.97159	5	30
31	$\bar{1}$.54466	34	$\bar{1}$.57312	39	0.42688	$\bar{1}$.97154	5	29
32	$\bar{1}$.54500	34	$\bar{1}$.57351	38	0.42649	$\bar{1}$.97149	4	28
33	$\bar{1}$.54534	33	$\bar{1}$.57389	39	0.42611	$\bar{1}$.97145	5	27
34	$\bar{1}$.54567	34	$\bar{1}$.57428	38	0.42572	$\bar{1}$.97140	5	26
35	$\bar{1}$.54601	34	$\bar{1}$.57466	38	0.42534	$\bar{1}$.97135	5	25
36	$\bar{1}$.54635	33	$\bar{1}$.57504	39	0.42496	$\bar{1}$.97130	4	24
37	$\bar{1}$.54668	34	$\bar{1}$.57543	38	0.42457	$\bar{1}$.97126	5	23
38	$\bar{1}$.54702	33	$\bar{1}$.57581	38	0.42419	$\bar{1}$.97121	5	22
39	$\bar{1}$.54735	34	$\bar{1}$.57619	39	0.42381	$\bar{1}$.97116	5	21
40	$\bar{1}$.54769	33	$\bar{1}$.57658	38	0.42342	$\bar{1}$.97111	4	20
41	$\bar{1}$.54802	34	$\bar{1}$.57696	38	0.42304	$\bar{1}$.97107	5	19
42	$\bar{1}$.54836	33	$\bar{1}$.57734	38	0.42266	$\bar{1}$.97102	5	18
43	$\bar{1}$.54869	34	$\bar{1}$.57772	38	0.42228	$\bar{1}$.97097	5	17
44	$\bar{1}$.54903	33	$\bar{1}$.57810	39	0.42190	$\bar{1}$.97092	5	16
45	$\bar{1}$.54936	33	$\bar{1}$.57849	38	0.42151	$\bar{1}$.97087	4	15
46	$\bar{1}$.54969	34	$\bar{1}$.57887	38	0.42113	$\bar{1}$.97083	5	14
47	$\bar{1}$.55003	33	$\bar{1}$.57925	38	0.42075	$\bar{1}$.97078	5	13
48	$\bar{1}$.55036	33	$\bar{1}$.57963	38	0.42037	$\bar{1}$.97073	5	12
49	$\bar{1}$.55069	33	$\bar{1}$.58001	38	0.41999	$\bar{1}$.97068	5	11
50	$\bar{1}$.55102	34	$\bar{1}$.58039	38	0.41961	$\bar{1}$.97063	4	10
51	$\bar{1}$.55136	33	$\bar{1}$.58077	38	0.41923	$\bar{1}$.97059	5	9
52	$\bar{1}$.55169	33	$\bar{1}$.58115	38	0.41885	$\bar{1}$.97054	5	8
53	$\bar{1}$.55202	33	$\bar{1}$.58153	38	0.41847	$\bar{1}$.97049	5	7
54	$\bar{1}$.55235	33	$\bar{1}$.58191	38	0.41809	$\bar{1}$.97044	5	6
55	$\bar{1}$.55268	33	$\bar{1}$.58229	38	0.41771	$\bar{1}$.97039	4	5
56	$\bar{1}$.55301	33	$\bar{1}$.58267	37	0.41733	$\bar{1}$.97035	5	4
57	$\bar{1}$.55334	33	$\bar{1}$.58304	38	0.41696	$\bar{1}$.97030	5	3
58	$\bar{1}$.55367	33	$\bar{1}$.58342	38	0.41658	$\bar{1}$.97025	5	2
59	$\bar{1}$.55400	33	$\bar{1}$.58380	38	0.41620	$\bar{1}$.97020	5	1
60	$\bar{1}$.55433		$\bar{1}$.58418		0.41582	$\bar{1}$.97015		0

	log cos	d	log cot	c. d.	log tan	log sin	d	′

p. p.

	40	39
6	4.0	3.9
7	4.7	4.6
8	5.3	5.2
9	6.0	5.9
10	6.7	6.5
20	13.3	13.0
30	20.0	19.5
40	26.7	26.0
50	33.3	32.5

	38	37
6	3.8	3.7
7	4.4	4.3
8	5.1	4.9
9	5.7	5.6
10	6.3	6.2
20	12.7	12.3
30	19.0	18.5
40	25.3	24.7
50	31.7	30.8

	35
6	3.5
7	4.1
8	4.7
9	5.3
10	5.8
20	11.7
30	17.5
40	23.3
50	29.2

	34	33
6	3.4	3.3
7	4.0	3.9
8	4.5	4.4
9	5.1	5.0
10	5.7	5.5
20	11.3	11.0
30	17.0	16.5
40	22.7	22.0
50	28.3	27.5

	5	4
6	0.5	0.4
7	0.6	0.5
8	0.7	0.5
9	0.8	0.6
10	0.8	0.7
20	1.7	1.3
30	2.5	2.0
40	3.3	2.7
50	4.2	3.3

'	log sin	d	log tan	c. d.	log cot	log cos	d	'
0	1̄.55433	33	1̄.58418	37	0.41582	1̄.97015	5	60
1	1̄.55466	33	1̄.58455	38	0.41545	1̄.97010	5	59
2	1̄.55499	33	1̄.58493	38	0.41507	1̄.97005	4	58
3	1̄.55532	32	1̄.58531	38	0.41469	1̄.97001	5	57
4	1̄.55564	33	1̄.58569	37	0.41431	1̄.96996	5	56
5	1̄.55597	33	1̄.58606	38	0.41394	1̄.96991	5	55
6	1̄.55630	33	1̄.58644	37	0.41356	1̄.96986	5	54
7	1̄.55663	32	1̄.58681	38	0.41319	1̄.96981	5	53
8	1̄.55695	33	1̄.58719	38	0.41281	1̄.96976	5	52
9	1̄.55728	33	1̄.58757	37	0.41243	1̄.96971	5	51
10	1̄.55761	32	1̄.58794	38	0.41206	1̄.96966	5	50
11	1̄.55793	33	1̄.58832	37	0.41168		5	49
12	1̄.55826	32	1̄.58869	38	0.41131	1̄.96957	5	48
13	1̄.55858	33	1̄.58907	37	0.41093	1̄.96952	5	47
14	1̄.55891	32	1̄.58944	37	0.41056	1̄.96947	5	46
15	1̄.55923	33	1̄.58981	38	0.41019	1̄.96942	5	45
16	1̄.55956	32	1̄.59019	37	0.40981	1̄.96937	5	44
17	1̄.55988	33	1̄.59056	38	0.40944	1̄.96932	5	43
18	1̄.56021	32	1̄.59094	37	0.40906	1̄.96927	5	42
19	1̄.56053	32	1̄.59131	37	0.40869	1̄.96922	5	41
20	1̄.56085	33	1̄.59168	37	0.40832	1̄.96917	5	40
21	1̄.56118	32	1̄.59205	38	0.40795	1̄.96912	5	39
22	1̄.56150	32	1̄.59243	37	0.40757		5	38
23	1̄.56182	33	1̄.59280	37	0.40720	1̄.96903	4	37
24	1̄.56215	32	1̄.59317	37	0.40683	1̄.96898	5	36
25	1̄.56247	32	1̄.59354	37	0.40646	1̄.96893	5	35
26	1̄.56279	32	1̄.59391	38	0.40609	1̄.96888	5	34
27	1̄.56311	32	1̄.59429	37	0.40571	1̄.96883	5	33
28	1̄.56343	32	1̄.59466	37	0.40534	1̄.96878	5	32
29	1̄.56375	33	1̄.59503	37	0.40497	1̄.96873	5	31
30	1̄.56408	32	1̄.59540	37	0.40460	1̄.96868	5	30
31	1̄.56440	32	1̄.59577	37	0.40423	1̄.96863	5	29
32	1̄.56472	32	1̄.59614	37	0.40386	1̄.96858	5	28
33	1̄.56504	32	1̄.59651	37	0.40349	1̄.96853	5	27
34	1̄.56536	32	1̄.59688	37	0.40312	1̄.96848	5	26
35	1̄.56568	31	1̄.59725	37	0.40275	1̄.96843	5	25
36	1̄.56599	32	1̄.59762	37	0.40238	1̄.96838	5	24
37	1̄.56631	32	1̄.59799	37	0.40201	1̄.96833	5	23
38	1̄.56663	32	1̄.59835	36	0.40165	1̄.96828	5	22
39	1̄.56695	32	1̄.59872	37	0.40128	1̄.96823	5	21
40	1̄.56727	32	1̄.59909	37	0.40091	1̄.96818	5	20
41	1̄.56759	31	1̄.59946	37	0.40054	1̄.96813	5	19
42	1̄.56790	32	1̄.59983	36	0.40017	1̄.96808	5	18
43	1̄.56822	32	1̄.60019	37	0.39981	1̄.96803	5	17
44	1̄.56854	32	1̄.60056	37	0.39944	1̄.96798	5	16
45	1̄.56886	31		37	0.39907	1̄.96793	5	15
46	1̄.56917	32	1̄.60130	36	0.39870	1̄.96788	5	14
47	1̄.56949	31	1̄.60166	37	0.39834	1̄.96783	5	13
48	1̄.56980	32	1̄.60203	37	0.39797	1̄.96778	6	12
49	1̄.57012	32	1̄.60240	36	0.39760	1̄.96772	5	11
50	1̄.57044	31	1̄.60276	37	0.39724	1̄.96767	5	10
51	1̄.57075	32	1̄.60313	36	0.39687	1̄.96762	5	9
52	1̄.57107	31	1̄.60349	37	0.39651	1̄.96757	5	8
53	1̄.57138	31	1̄.60386	36	0.39614	1̄.96752	5	7
54	1̄.57169	32	1̄.60422	37	0.39578	1̄.96747	5	6
55	1̄.57201	31	1̄.60459	36	0.39541	1̄.96742	5	5
56	1̄.57232	32	1̄.60495	37	0.39505	1̄.96737	5	4
57	1̄.57264	31	1̄.60532	36	0.39468	1̄.96732	5	3
58	1̄.57295	31	1̄.60568	37	0.39432	1̄.96727	5	2
59	1̄.57326	32		36	0.39395	1̄.96722	5	1
60	1̄.57358		1̄.60641		0.39359	1̄.96717		0
	log cos	d	log cot	c. d.	log tan	log sin	d	'

p. p.

	38	37
6	3.8	3.7
7	4.4	4.3
8	5.1	4.9
9	5.7	5.6
10	6.3	6.2
20	12.7	12.3
30	19.0	18.5
40	25.3	24.7
50	31.7	30.8

	36	33
6	3.6	3.3
7	4.2	3.9
8	4.8	4.4
9	5.4	5.0
10	6.0	5.5
20	12.0	11.0
30	18.0	16.5
40	24.0	22.0
50	30.0	27.5

	32
6	3.2
7	3.7
8	4.3
9	4.8
10	5.3
20	10.7
30	16.0
40	21.3
50	26.7

	31	6
6	3.1	0.6
7	3.6	0.7
8	4.1	0.8
9	4.7	0.9
10	5.2	1.0
20	10.3	2.0
30	15.5	3.0
40	20.7	4.0
50	25.8	5.0

	5	4
6	0.5	0.4
7	0.6	0.5
8	0.7	0.5
9	0.8	0.6
10	0.8	0.7
20	1.7	1.3
30	2.5	2.0
40	3.3	2.7
50	4.2	3.3

′	log sin	d	log tan	c. d.	log cot	log cos	d	′
0	1̄.57358		1̄.60641		0.39359	1̄.96717		60
1	1̄.57389	31	1̄.60677	36	0.39323	1̄.96711	6	59
2	1̄.57420	31	1̄.60714	37	0.39286	1̄.96706	5	58
3	1̄.57451	31	1̄.60750	36	0.39250	1̄.96701	5	57
4	1̄.57482	31	1̄.60786	36	0.39214	1̄.96696	5	56
5	1̄.57514	32	1̄.60823	37	0.39177	1̄.96691	5	55
6	1̄.57545	31	1̄.60859	36	0.39141	1̄.96686	5	54
7	1̄.57576	31	1̄.60895	36	0.39105	1̄.96681	5	53
8	1̄.57607	31	1̄.60931	36	0.39069	1̄.96676	6	52
9	1̄.57638	31	1̄.60967	36	0.39033	1̄.96670	5	51
10	1̄.57669	31	1̄.61004	37	0.38996	1̄.96665	5	50
11	1̄.57700	31	1̄.61040	36	0.38960	1̄.96660	5	49
12	1̄.57731	31	1̄.61076	36	0.38924	1̄.96655	5	48
13	1̄.57762	31	1̄.61112	36	0.38888	1̄.96650	5	47
14	1̄.57793	31	1̄.61148	36	0.38852	1̄.96645	5	46
15	1̄.57824	31	1̄.61184	36	0.38816	1̄.96640	6	45
16	1̄.57855	31	1̄.61220	36	0.38780	1̄.96634	5	44
17	1̄.57885	30	1̄.61256	36	0.38744	1̄.96629	5	43
18	1̄.57916	31	1̄.61292	36	0.38708	1̄.96624	5	42
19	1̄.57947	31	1̄.61328	36	0.38672	1̄.96619	5	41
20	1̄.57978	31	1̄.61364	36	0.38636	1̄.96614	6	40
21	1̄.58008	30	1̄.61400	36	0.38600	1̄.96608	5	39
22	1̄.58039	31	1̄.61436	36	0.38564	1̄.96603	5	38
23	1̄.58070	31	1̄.61472	36	0.38528	1̄.96598	5	37
24	1̄.58101	31	1̄.61508	36	0.38492	1̄.96593	5	36
25	1̄.58131	30	1̄.61544	36	0.38456	1̄.96588	6	35
26	1̄.58162	31	1̄.61579	35	0.38421	1̄.96582	5	34
27	1̄.58192	30	1̄.61615	36	0.38385	1̄.96577	5	33
28	1̄.58223	31	1̄.61651	36	0.38349	1̄.96572	5	32
29	1̄.58253	30	1̄.61687	36	0.38313	1̄.96567	5	31
30	1̄.58284	31	1̄.61722	35	0.38278	1̄.96562	6	30
31	1̄.58314	30	1̄.61758	36	0.38242	1̄.96556	5	29
32	1̄.58345	31	1̄.61794	36	0.38206	1̄.96551	5	28
33	1̄.58375	30	1̄.61830	36	0.38170	1̄.96546	5	27
34	1̄.58406	31	1̄.61865	35	0.38135	1̄.96541	6	26
35	1̄.58436	30	1̄.61901	36	0.38099	1̄.96535	5	25
36	1̄.58467	31	1̄.61936	35	0.38064	1̄.96530	5	24
37	1̄.58497	30	1̄.61972	36	0.38028	1̄.96525	5	23
38	1̄.58527	30	1̄.62008	35	0.37992	1̄.96520	6	22
39	1̄.58557	30	1̄.62043	36	0.37957	1̄.96514	5	21
40	1̄.58588	31	1̄.62079	35	0.37921	1̄.96509	5	20
41	1̄.58618	30	1̄.62114	36	0.37886	1̄.96504	6	19
42	1̄.58648	30	1̄.62150	35	0.37850	1̄.96498	5	18
43	1̄.58678	30	1̄.62185	36	0.37815	1̄.96493	5	17
44	1̄.58709	31	1̄.62221	35	0.37779	1̄.96488	5	16
45	1̄.58739	30	1̄.62256	36	0.37744	1̄.96483	6	15
46	1̄.58769	30	1̄.62292	35	0.37708	1̄.96477	5	14
47	1̄.58799	30	1̄.62327	35	0.37673	1̄.96472	5	13
48	1̄.58829	30	1̄.62362	35	0.37638	1̄.96467	6	12
49	1̄.58859	30	1̄.62398	36	0.37602	1̄.96461	5	11
50	1̄.58889	30	1̄.62433	35	0.37567	1̄.96456	5	10
51	1̄.58919	30	1̄.62468	36	0.37532	1̄.96451	6	9
52	1̄.58949	30	1̄.62504	35	0.37496	1̄.96445	5	8
53	1̄.58979	30	1̄.62539	35	0.37461	1̄.96440	5	7
54	1̄.59009	30	1̄.62574	35	0.37426	1̄.96435	6	6
55	1̄.59039	30	1̄.62609	35	0.37391	1̄.96429	5	5
56	1̄.59069	30	1̄.62645	36	0.37355	1̄.96424	5	4
57	1̄.59098	29	1̄.62680	35	0.37320	1̄.96419	6	3
58	1̄.59128	30	1̄.62715	35	0.37285	1̄.96413	5	2
59	1̄.59158	30	1̄.62750	35	0.37250	1̄.96408	5	1
60	1̄.59188	30	1̄.62785	35	0.37215	1̄.96403	5	0
	log cos	d	log cot	c. d.	log tan	log sin	d	′

p. p.

	37	36
6	3.7	3.6
7	4.3	4.2
8	4.9	4.8
9	5.6	5.4
10	6.2	6.0
20	12.3	12.0
30	18.5	18.0
40	24.7	24.0
50	30.8	30.0

	35
6	3.5
7	4.1
8	4.7
9	5.3
10	5.8
20	11.7
30	17.5
40	23.3
50	29.2

	32	31
6	3.2	3.1
7	3.7	3.6
8	4.3	4.1
9	4.8	4.7
10	5.3	5.2
20	10.7	10.3
30	16.0	15.5
40	21.3	20.7
50	26.7	25.8

	30	29
6	3.0	2.9
7	3.5	3.4
8	4.0	3.9
9	4.5	4.4
10	5.0	4.8
20	10.0	9.7
30	15.0	14.5
40	20.0	19.3
50	25.0	24.2

	6	5
6	0.6	0.5
7	0.7	0.6
8	0.8	0.7
9	0.9	0.8
10	1.0	0.8
20	2.0	1.7
30	3.0	2.5
40	4.0	3.3
50	5.0	4.2

′	log sin	d	log tan	c. d.	log cot	log cos	d	′
0	1̄.59188	30	1̄.62785	35	0.37215	1̄.96403	6	60
1	1̄.59218	29	1̄.62820	35	0.37180	1̄.96397	5	59
2	1̄.59247	30	1̄.62855	35	0.37145	1̄.96392	5	58
3	1̄.59277	30	1̄.62890	36	0.37110	1̄.96387	6	57
4	1̄.59307	29	1̄.62926	35	0.37074	1̄.96381	5	56
5	1̄.59336	30	1̄.62961	35	0.37039	1̄.96376	6	55
6	1̄.59366	30	1̄.62996	35	0.37004	1̄.96370	5	54
7	1̄.59396	29	1̄.63031	35	0.36969	1̄.96365	5	53
8	1̄.59425	30	1̄.63066	35	0.36934	1̄.96360	6	52
9	1̄.59455	29	1̄.63101	34	0.36899	1̄.96354	5	51
10	1̄.59484	30	1̄.63135	35	0.36865	1̄.96349	6	50
11	1̄.59514	29	1̄.63170	35	0.36830	1̄.96343	5	49
12	1̄.59543	30	1̄.63205	35	0.36795	1̄.96338	5	48
13	1̄.59573	29	1̄.63240	35	0.36760	1̄.96333	6	47
14	1̄.59602	30	1̄.63275	35	0.36725	1̄.96327	5	46
15	1̄.59632	29	1̄.63310	35	0.36690	1̄.96322	6	45
16	1̄.59661	29	1̄.63345	34	0.36655	1̄.96316	5	44
17	1̄.59690	30	1̄.63379	35	0.36621	1̄.96311	6	43
18	1̄.59720	29	1̄.63414	35	0.36586	1̄.96305	5	42
19	1̄.59749	29	1̄.63449	35	0.36551	1̄.96300	6	41
20	1̄.59778	30	1̄.63484	35	0.36516	1̄.96294	5	40
21	1̄.59808	29	1̄.63519	34	0.36481	1̄.96289	5	39
22	1̄.59837	29	1̄.63553	35	0.36447	1̄.96284	6	38
23	1̄.59866	29	1̄.63588	35	0.36412	1̄.96278	5	37
24	1̄.59895	29	1̄.63623	34	0.36377	1̄.96273	6	36
25	1̄.59924	30	1̄.63657	35	0.36343	1̄.96267	5	35
26	1̄.59954	29	1̄.63692	34	0.36308	1̄.96262	6	34
27	1̄.59983	29	1̄.63726	35	0.36274	1̄.96256	5	33
28	1̄.60012	29	1̄.63761	35	0.36239	1̄.96251	6	32
29	1̄.60041	29	1̄.63796	34	0.36204	1̄.96245	5	31
30	1̄.60070	29	1̄.63830	35	0.36170	1̄.96240	6	30
31	1̄.60099	29	1̄.63865	34	0.36135	1̄.96234	5	29
32	1̄.60128	29	1̄.63899	35	0.36101	1̄.96229	6	28
33	1̄.60157	29	1̄.63934	34	0.36066	1̄.96223	5	27
34	1̄.60186	29	1̄.63968	35	0.36032	1̄.96218	6	26
35	1̄.60215	29	1̄.64003	34	0.35997	1̄.96212	5	25
36	1̄.60244	29	1̄.64037	35	0.35963	1̄.96207	6	24
37	1̄.60273	29	1̄.64072	34	0.35928	1̄.96201	5	23
38	1̄.60302	29	1̄.64106	34	0.35894	1̄.96196	6	22
39	1̄.60331	28	1̄.64140	35	0.35860	1̄.96190	6	21
40	1̄.60359	29	1̄.64175	34	0.35825	1̄.96185	5	20
41	1̄.60388	29	1̄.64209	34	0.35791	1̄.96179	6	19
42	1̄.60417	29	1̄.64243	35	0.35757	1̄.96174	5	18
43	1̄.60446	28	1̄.64278	34	0.35722	1̄.96168	6	17
44	1̄.60474	29	1̄.64312	34	0.35688	1̄.96162	6	16
45	1̄.60503	29	1̄.64346	35	0.35654	1̄.96157	5	15
46	1̄.60532	29	1̄.64381	34	0.35619	1̄.96151	6	14
47	1̄.60561	28	1̄.64415	34	0.35585	1̄.96146	5	13
48	1̄.60589	29	1̄.64449	34	0.35551	1̄.96140	6	12
49	1̄.60618	28	1̄.64483	34	0.35517	1̄.96135	5	11
50	1̄.60646	29	1̄.64517	35	0.35483	1̄.96129	6	10
51	1̄.60675	29	1̄.64552	34	0.35448	1̄.96123	5	9
52	1̄.60704	28	1̄.64586	34	0.35414	1̄.96118	6	8
53	1̄.60732	29	1̄.64620	34	0.35380	1̄.96112	5	7
54	1̄.60761	28	1̄.64654	34	0.35346	1̄.96107	6	6
55	1̄.60789	29	1̄.64688	34	0.35312	1̄.96101	6	5
56	1̄.60818	28	1̄.64722	34	0.35278	1̄.96095	5	4
57	1̄.60846	29	1̄.64756	34	0.35244	1̄.96090	6	3
58	1̄.60875	28	1̄.64790	34	0.35210	1̄.96084	5	2
59	1̄.60903	28	1̄.64824	34	0.35176	1̄.96079	6	1
60	1̄.60931		1̄.64858		0.35142	1̄.96073		0

| | log cos | d | log cot | c. d. | log tan | log sin | d | ′ |

p. p.

	36	35
6	3.6	3.5
7	4.2	4.1
8	4.8	4.7
9	5.4	5.3
10	6.0	5.8
20	12.0	11.7
30	18.0	17.5
40	24.0	23.3
50	30.0	29.2

	34
6	3.4
7	4.0
8	4.5
9	5.1
10	5.7
20	11.3
30	17.0
40	22.7
50	28.3

	30	29
6	3.0	2.9
7	3.5	3.4
8	4.0	3.9
9	4.5	4.4
10	5.0	4.8
20	10.0	9.7
30	15.0	14.5
40	20.0	19.3
50	25.0	24.2

	28
6	2.8
7	3.3
8	3.7
9	4.2
10	4.7
20	9.3
30	14.0
40	18.7
50	23.3

	6	5
6	0.6	0.5
7	0.7	0.6
8	0.8	0.7
9	0.9	0.8
10	1.0	0.8
20	2.0	1.7
30	3.0	2.5
40	4.0	3.3
50	5.0	4.2

′	log sin	d	log tan	c. d.	log cot	log cos	d	′
0	1̄.60931	29	1̄.64858	34	0.35142	1̄.96073	6	60
1	1̄.60960	28	1̄.64892	34	0.35108	1̄.96067	5	59
2	1̄.60988	28	1̄.64926	34	0.35074	1̄.96062	6	58
3	1̄.61016	29	1̄.64960	34	0.35040	1̄.96056	6	57
4	1̄.61045	28	1̄.64994	34	0.35006	1̄.96050	5	56
5	1̄.61073	28	1̄.65028	34	0.34972	1̄.96045	6	55
6	1̄.61101	28	1̄.65062	34	0.34938	1̄.96039	5	54
7	1̄.61129	29	1̄.65096	34	0.34904	1̄.96034	6	53
8	1̄.61158	28	1̄.65130	34	0.34870	1̄.96028	6	52
9	1̄.61186	28	1̄.65164	34	0.34836	1̄.96022	5	51
10	1̄.61214	28	1̄.65197	33	0.34803	1̄.96017	6	50
11	1̄.61242	28	1̄.65231	34	0.34769	1̄.96011	6	49
12	1̄.61270	28	1̄.65265	34	0.34735	1̄.96005	5	48
13	1̄.61298	28	1̄.65299	34	0.34701	1̄.96000	6	47
14	1̄.61326	28	1̄.65333	33	0.34667	1̄.95994	6	46
15	1̄.61354	28	1̄.65366	34	0.34634	1̄.95988	6	45
16	1̄.61382	29	1̄.65400	34	0.34600	1̄.95982	5	44
17	1̄.61411	27	1̄.65434	33	0.34566	1̄.95977	6	43
18	1̄.61438	28	1̄.65467	34	0.34533	1̄.95971	6	42
19	1̄.61466	28	1̄.65501	34	0.34499	1̄.95965	5	41
20	1̄.61494	28	1̄.65535	33	0.34465	1̄.95960	6	40
21	1̄.61522	28	1̄.65568	34	0.34432	1̄.95954	6	39
22	1̄.61550	28	1̄.65602	34	0.34398	1̄.95948	6	38
23	1̄.61578	28	1̄.65636	33	0.34364	1̄.95942	5	37
24	1̄.61606	28	1̄.65669	34	0.34331	1̄.95937	6	36
25	1̄.61634	28	1̄.65703	33	0.34297	1̄.95931	6	35
26	1̄.61662	27	1̄.65736	34	0.34264	1̄.95925	5	34
27	1̄.61689	28	1̄.65770	33	0.34230	1̄.95920	6	33
28	1̄.61717	28	1̄.65803	34	0.34197	1̄.95914	6	32
29	1̄.61745	28	1̄.65837	33	0.34163	1̄.95908	6	31
30	1̄.61773	27	1̄.65870	34	0.34130	1̄.95902	5	30
31	1̄.61800	28	1̄.65904	33	0.34096	1̄.95897	6	29
32	1̄.61828	28	1̄.65937	34	0.34063	1̄.95891	6	28
33	1̄.61856	27	1̄.65971	33	0.34029	1̄.95885	6	27
34	1̄.61883	28	1̄.66004	34	0.33996	1̄.95879	6	26
35	1̄.61911	28	1̄.66038	33	0.33962	1̄.95873	5	25
36	1̄.61939	27	1̄.66071	33	0.33929	1̄.95868	6	24
37	1̄.61966	28	1̄.66104	34	0.33896	1̄.95862	6	23
38	1̄.61994	27	1̄.66138	33	0.33862	1̄.95856	6	22
39	1̄.62021	28	1̄.66171	33	0.33829	1̄.95850	6	21
40	1̄.62049	27	1̄.66204	34	0.33796	1̄.95844	5	20
41	1̄.62076	28	1̄.66238	33	0.33762	1̄.95839	6	19
42	1̄.62104	27	1̄.66271	33	0.33729	1̄.95833	6	18
43	1̄.62131	28	1̄.66304	33	0.33696	1̄.95827	6	17
44	1̄.62159	27	1̄.66337	34	0.33663	1̄.95821	6	16
45	1̄.62186	28	1̄.66371	33	0.33629	1̄.95815	5	15
46	1̄.62214	27	1̄.66404	33	0.33596	1̄.95810	6	14
47	1̄.62241	27	1̄.66437	33	0.33563	1̄.95804	6	13
48	1̄.62268	28	1̄.66470	33	0.33530	1̄.95798	6	12
49	1̄.62296	27	1̄.66503	34	0.33497	1̄.95792	6	11
50	1̄.62323	27	1̄.66537	33	0.33463	1̄.95786	6	10
51	1̄.62350	27	1̄.66570	33	0.33430	1̄.95780	5	9
52	1̄.62377	28	1̄.66603	33	0.33397	1̄.95775	6	8
53	1̄.62405	27	1̄.66636	33	0.33364	1̄.95769	6	7
54	1̄.62432	27	1̄.66669	33	0.33331	1̄.95763	6	6
55	1̄.62459	27	1̄.66702	33	0.33298	1̄.95757	6	5
56	1̄.62486	27	1̄.66735	33	0.33265	1̄.95751	6	4
57	1̄.62513	28	1̄.66768	33	0.33232	1̄.95745	6	3
58	1̄.62541	27	1̄.66801	33	0.33199	1̄.95739	6	2
59	1̄.62568	27	1̄.66834	33	0.33166	1̄.95733	5	1
60	1̄.62595		1̄.66867		0.33133	1̄.95728		0
	log cos	d	log cot	c. d.	log tan	log sin	d	′

p. p.

	34	33
6	3.4	3.3
7	4.0	3.9
8	4.5	4.4
9	5.1	5.0
10	5.7	5.5
20	11.3	11.0
30	17.0	16.5
40	22.7	22.0
50	28.3	27.5

	29
6	2.9
7	3.4
8	3.9
9	4.4
10	4.8
20	9.7
30	14.5
40	19.3
50	24.2

	28
6	2.8
7	3.3
8	3.7
9	4.2
10	4.7
20	9.3
30	14.0
40	18.7
50	23.3

	27
6	2.7
7	3.2
8	3.6
9	4.1
10	4.5
20	9.0
30	13.5
40	18.0
50	22.5

	6	5
6	0.6	0.5
7	0.7	0.6
8	0.8	0.7
9	0.9	0.8
10	1.0	0.8
20	2.0	1.7
30	3.0	2.5
40	4.0	3.3
50	5.0	4.2

25°

′	log sin	d	log tan	c. d.	log cot	log cos	d	′
0	1̄.62595	27	1̄.66867	33	0.33133	1̄.95728	6	60
1	1̄.62622	27	1̄.66900	33	0.33100	1̄.95722	6	59
2	1̄.62649	27	1̄.66933	33	0.33067	1̄.95716	6	58
3	1̄.62676	27	1̄.66966	33	0.33034	1̄.95710	6	57
4	1̄.62703	27	1̄.66999	33	0.33001	1̄.95704	6	56
5	1̄.62730	27	1̄.67032	33	0.32968	1̄.95698	6	55
6	1̄.62757	27	1̄.67065	33	0.32935	1̄.95692	6	54
7	1̄.62784	27	1̄.67098	33	0.32902	1̄.95686	6	53
8	1̄.62811	27	1̄.67131	32	0.32869	1̄.95680	6	52
9	1̄.62838	27	1̄.67163	33	0.32837	1̄.95674	6	51
10	1̄.62865		1̄.67196		0.32804	1̄.95668		50
		27		33			5	
11	1̄.62892	26	1̄.67229	33	0.32771	1̄.95663	6	49
12	1̄.62918	27	1̄.67262	33	0.32738	1̄.95657	6	48
13	1̄.62945	27	1̄.67295	32	0.32705	1̄.95651	6	47
14	1̄.62972	27	1̄.67327	33	0.32673	1̄.95645	6	46
15	1̄.62999	27	1̄.67360	33	0.32640	1̄.95639	6	45
16	1̄.63026	26	1̄.67393	33	0.32607	1̄.95633	6	44
17	1̄.63052	27	1̄.67426	32	0.32574	1̄.95627	6	43
18	1̄.63079	27	1̄.67458	33	0.32542	1̄.95621	6	42
19	1̄.63106	27	1̄.67491	33	0.32509	1̄.95615	6	41
20	1̄.63133		1̄.67524		0.32476	1̄.95609		40
		26		32			6	
21	1̄.63159	27	1̄.67556	33	0.32444	1̄.95603	6	39
22	1̄.63186	27	1̄.67589	33	0.32411	1̄.95597	6	38
23	1̄.63213	26	1̄.67622	32	0.32378	1̄.95591	6	37
24	1̄.63239	27	1̄.67654	33	0.32346	1̄.95585	6	36
25	1̄.63266	26	1̄.67687	32	0.32313	1̄.95579	6	35
26	1̄.63292	27	1̄.67719	33	0.32281	1̄.95573	6	34
27	1̄.63319	26	1̄.67752	33	0.32248	1̄.95567	6	33
28	1̄.63345	27	1̄.67785	32	0.32215	1̄.95561	6	32
29	1̄.63372	26	1̄.67817	33	0.32183	1̄.95555	6	31
30	1̄.63398		1̄.67850		0.32150	1̄.95549		30
		27		32			6	
31	1̄.63425	26	1̄.67882	33	0.32118	1̄.95543	6	29
32	1̄.63451	27	1̄.67915	32	0.32085	1̄.95537	6	28
33	1̄.63478	26	1̄.67947	33	0.32053	1̄.95531	6	27
34	1̄.63504	27	1̄.67980	32	0.32020	1̄.95525	6	26
35	1̄.63531	26	1̄.68012	32	0.31988	1̄.95519	6	25
36	1̄.63557	26	1̄.68044	33	0.31956	1̄.95513	6	24
37	1̄.63583	27	1̄.68077	32	0.31923	1̄.95507	7	23
38	1̄.63610	26	1̄.68109	33	0.31891	1̄.95500	6	22
39	1̄.63636	26	1̄.68142	32	0.31858	1̄.95494	6	21
40	1̄.63662		1̄.68174		0.31826	1̄.95488		20
		27		32			6	
41	1̄.63689	26	1̄.68206	33	0.31794	1̄.95482	6	19
42	1̄.63715	26	1̄.68239	32	0.31761	1̄.95476	6	18
43	1̄.63741	26	1̄.68271	32	0.31729	1̄.95470	6	17
44	1̄.63767	27	1̄.68303	33	0.31697	1̄.95464	6	16
45	1̄.63794	26	1̄.68336	32	0.31664	1̄.95458	6	15
46	1̄.63820	26	1̄.68368	32	0.31632	1̄.95452	6	14
47	1̄.63846	26	1̄.68400	32	0.31600	1̄.95446	6	13
48	1̄.63872	26	1̄.68432	33	0.31568	1̄.95440	6	12
49	1̄.63898	26	1̄.68465	32	0.31535	1̄.95434	7	11
50	1̄.63924		1̄.68497		0.31503	1̄.95427		10
		26		32			6	
51	1̄.63950	26	1̄.68529	32	0.31471	1̄.95421	6	9
52	1̄.63976	26	1̄.68561	32	0.31439	1̄.95415	6	8
53	1̄.64002	26	1̄.68593	33	0.31407	1̄.95409	6	7
54	1̄.64028	26	1̄.68626	32	0.31374	1̄.95403	6	6
55	1̄.64054	26	1̄.68658	32	0.31342	1̄.95397	6	5
56	1̄.64080	26	1̄.68690	32	0.31310	1̄.95391	7	4
57	1̄.64106	26	1̄.68722	32	0.31278	1̄.95384	6	3
58	1̄.64132	26	1̄.68754	32	0.31246	1̄.95378	6	2
59	1̄.64158	26	1̄.68786	32	0.31214	1̄.95372	6	1
60	1̄.64184		1̄.68818		0.31182	1̄.95366		0
′	log cos	d	log cot	c. d.	log tan	log sin	d	′

p. p.

	33	32
6	3.3	3.2
7	3.9	3.7
8	4.4	4.3
9	5.0	4.8
10	5.5	5.3
20	11.0	10.7
30	16.5	16.0
40	22.0	21.3
50	27.5	26.7

	27
6	2.7
7	3.2
8	3.6
9	4.1
10	4.5
20	9.0
30	13.5
40	18.0
50	22.5

	26
6	2.6
7	3.0
8	3.5
9	3.9
10	4.3
20	8.7
30	13.0
40	17.3
50	21.7

	7
6	0.7
7	0.8
8	0.9
9	1.1
10	1.2
20	2.3
30	3.5
40	4.7
50	5.8

	8	5
6	0.6	0.5
7	0.7	0.6
8	0.8	0.7
9	0.9	0.8
10	1.0	0.8
20	2.0	1.7
30	3.0	2.5
40	4.0	3.3
50	5.0	4.2

64°

′	log sin	d	log tan	c. d.	log cot	log cos	d	′
0	1̄.64184	26	1̄.68818	32	0.31182	1̄.95366	6	60
1	1̄.64210	26	1̄.68850	32	0.31150	1̄.95360	6	59
2	1̄.64236	26	1̄.68882	32	0.31118	1̄.95354	6	58
3	1̄.64262	26	1̄.68914	32	0.31086	1̄.95348	7	57
4	1̄.64288	25	1̄.68946	32	0.31054	1̄.95341	6	56
5	1̄.64313	26	1̄.68978	32	0.31022	1̄.95335	6	55
6	1̄.64339	26	1̄.69010	32	0.30990	1̄.95329	6	54
7	1̄.64365	26	1̄.69042	32	0.30958	1̄.95323	6	53
8	1̄.64391	26	1̄.69074	32	0.30926	1̄.95317	7	52
9	1̄.64417	25	1̄.69106	32	0.30894	1̄.95310	6	51
10	1̄.64442	26	1̄.69138	32	0.30862	1̄.95304	6	50
11	1̄.64468	26	1̄.69170	32	0.30830	1̄.95298	6	49
12	1̄.64494	25	1̄.69202	32	0.30798	1̄.95292	6	48
13	1̄.64519	26	1̄.69234	32	0.30766	1̄.95286	7	47
14	1̄.64545	26	1̄.69266	32	0.30734	1̄.95279	6	46
15	1̄.64571	25	1̄.69298	31	0.30702	1̄.95273	6	45
16	1̄.64596	26	1̄.69329	32	0.30671	1̄.95267	6	44
17	1̄.64622	25	1̄.69361	32	0.30639	1̄.95261	7	43
18	1̄.64647	26	1̄.69393	32	0.30607	1̄.95254	6	42
19	1̄.64673	25	1̄.69425	32	0.30575	1̄.95248	6	41
20	1̄.64698	26	1̄.69457	31	0.30543	1̄.95242	6	40
21	1̄.64724	25	1̄.69488	32	0.30512	1̄.95236	7	39
22	1̄.64749	26	1̄.69520	32	0.30480	1̄.95229	6	38
23	1̄.64775	25	1̄.69552	32	0.30448	1̄.95223	6	37
24	1̄.64800	26	1̄.69584	31	0.30416	1̄.95217	6	36
25	1̄.64826	25	1̄.69615	32	0.30385	1̄.95211	7	35
26	1̄.64851	26	1̄.69647	32	0.30353	1̄.95204	6	34
27	1̄.64877	25	1̄.69679	31	0.30321	1̄.95198	6	33
28	1̄.64902	25	1̄.69710	32	0.30290	1̄.95192	7	32
29	1̄.64927	26	1̄.69742	32	0.30258	1̄.95185	6	31
30	1̄.64953	25	1̄.69774	31	0.30226	1̄.95179	6	30
31	1̄.64978	25	1̄.69805	32	0.30195	1̄.95173	6	29
32	1̄.65003	26	1̄.69837	31	0.30163	1̄.95167	7	28
33	1̄.65029	25	1̄.69868	32	0.30132	1̄.95160	6	27
34	1̄.65054	25	1̄.69900	32	0.30100	1̄.95154	6	26
35	1̄.65079	25	1̄.69932	31	0.30068	1̄.95148	7	25
36	1̄.65104	26	1̄.69963	32	0.30037	1̄.95141	6	24
37	1̄.65130	25	1̄.69995	31	0.30005	1̄.95135	6	23
38	1̄.65155	25	1̄.70026	32	0.29974	1̄.95129	7	22
39	1̄.65180	25	1̄.70058	31	0.29942	1̄.95122	6	21
40	1̄.65205	25	1̄.70089	32	0.29911	1̄.95116	6	20
41	1̄.65230	25	1̄.70121	31	0.29879	1̄.95110	7	19
42	1̄.65255	26	1̄.70152	32	0.29848	1̄.95103	6	18
43	1̄.65281	25	1̄.70184	31	0.29816	1̄.95097	7	17
44	1̄.65306	25	1̄.70215	32	0.29785	1̄.95090	6	16
45	1̄.65331	25	1̄.70247	31	0.29753	1̄.95084	6	15
46	1̄.65356	25	1̄.70278	31	0.29722	1̄.95078	7	14
47	1̄.65381	25	1̄.70309	32	0.29691	1̄.95071	6	13
48	1̄.65406	25	1̄.70341	31	0.29659	1̄.95065	6	12
49	1̄.65431	25	1̄.70372	32	0.29628	1̄.95059	7	11
50	1̄.65456	25	1̄.70404	31	0.29596	1̄.95052	6	10
51	1̄.65481	25	1̄.70435	31	0.29565	1̄.95046	7	9
52	1̄.65506	25	1̄.70466	32	0.29534	1̄.95039	6	8
53	1̄.65531	25	1̄.70498	31	0.29502	1̄.95033	6	7
54	1̄.65556	24	1̄.70529	31	0.29471	1̄.95027	7	6
55	1̄.65580	25	1̄.70560	32	0.29440	1̄.95020	6	5
56	1̄.65605	25	1̄.70592	31	0.29408	1̄.95014	7	4
57	1̄.65630	25	1̄.70623	31	0.29377	1̄.95007	6	3
58	1̄.65655	25	1̄.70654	31	0.29346	1̄.95001	6	2
59	1̄.65680	25	1̄.70685	32	0.29315	1̄.94995	7	1
60	1̄.65705		1̄.70717		0.29283	1̄.94988		0
	log cos	d	log cot	c. d.	log tan	log sin	d	′

p. p.

	32	31
6	3.2	3.1
7	3.7	3.6
8	4.3	4.1
9	4.8	4.7
10	5.3	5.2
20	10.7	10.3
30	16.0	15.5
40	21.3	20.7
50	26.7	25.8

	26
6	2.6
7	3.0
8	3.5
9	3.9
10	4.3
20	8.7
30	13.0
40	17.3
50	21.7

	25
6	2.5
7	2.9
8	3.3
9	3.8
10	4.2
20	8.3
30	12.5
40	16.7
50	20.8

	24
6	2.4
7	2.8
8	3.2
9	3.6
10	4.0
20	8.0
30	12.0
40	16.0
50	20.0

	7	6
6	0.7	0.6
7	0.8	0.7
8	0.9	0.8
9	1.1	0.9
10	1.2	1.0
20	2.3	2.0
30	3.5	3.0
40	4.7	4.0
50	5.8	5.0

'	log sin	d	log tan	c. d.	log cot	log cos	d	'
0	1̄.65705	24	1̄.70717	31	0.29283	1̄.94988	6	60
1	1̄.65729	25	1̄.70748	31	0.29252	1̄.94982	7	59
2	1̄.65754	25	1̄.70779	31	0.29221	1̄.94975	6	58
3	1̄.65779	25	1̄.70810	31	0.29190	1̄.94969	7	57
4	1̄.65804	24	1̄.70841	32	0.29159	1̄.94962	6	56
5	1̄.65828	25	1̄.70873	31	0.29127	1̄.94956	7	55
6	1̄.65853	25	1̄.70904	31	0.29096	1̄.94949	6	54
7	1̄.65878	24	1̄.70935	31	0.29065	1̄.94943	7	53
8	1̄.65902	25	1̄.70966	31	0.29034	1̄.94936	6	52
9	1̄.65927	25	1̄.70997	31	0.29003	1̄.94930	7	51
10	1̄.65952	24	1̄.71028	31	0.28972	1̄.94923	6	50
11	1̄.65976	25	1̄.71059	31	0.28941	1̄.94917	6	49
12	1̄.66001	24	1̄.71090	31	0.28910	1̄.94911	7	48
13	1̄.66025	25	1̄.71121	32	0.28879	1̄.94904	6	47
14	1̄.66050	25	1̄.71153	31	0.28847	1̄.94898	7	46
15	1̄.66075	24	1̄.71184	31	0.28816	1̄.94891	6	45
16	1̄.66099	25	1̄.71215	31	0.28785	1̄.94885	7	44
17	1̄.66124	24	1̄.71246	31	0.28754	1̄.94878	7	43
18	1̄.66148	25	1̄.71277	31	0.28723	1̄.94871	6	42
19	1̄.66173	24	1̄.71308	31	0.28692	1̄.94865	7	41
20	1̄.66197	24	1̄.71339	31	0.28661	1̄.94858	6	40
21	1̄.66221	25	1̄.71370	31	0.28630	1̄.94852	7	39
22	1̄.66246	24	1̄.71401	30	0.28599	1̄.94845	6	38
23	1̄.66270	25	1̄.71431	31	0.28569	1̄.94839	7	37
24	1̄.66295	24	1̄.71462	31	0.28538	1̄.94832	6	36
25	1̄.66319	24	1̄.71493	31	0.28507	1̄.94826	7	35
26	1̄.66343	25	1̄.71524	31	0.28476	1̄.94819	6	34
27	1̄.66368	24	1̄.71555	31	0.28445	1̄.94813	7	33
28	1̄.66392	24	1̄.71586	31	0.28414	1̄.94806	7	32
29	1̄.66416	25	1̄.71617	31	0.28383	1̄.94799	6	31
30	1̄.66441	24	1̄.71648	31	0.28352	1̄.94793	7	30
31	1̄.66465	24	1̄.71679	30	0.28321	1̄.94786	6	29
32	1̄.66489	24	1̄.71709	31	0.28291	1̄.94780	7	28
33	1̄.66513	24	1̄.71740	31	0.28260	1̄.94773	6	27
34	1̄.66537	25	1̄.71771	31	0.28229	1̄.94767	7	26
35	1̄.66562	24	1̄.71802	31	0.28198	1̄.94760	7	25
36	1̄.66586	24	1̄.71833	30	0.28167	1̄.94753	6	24
37	1̄.66610	24	1̄.71863	31	0.28137	1̄.94747	7	23
38	1̄.66634	24	1̄.71894	31	0.28106	1̄.94740	6	22
39	1̄.66658	24	1̄.71925	30	0.28075	1̄.94734	7	21
40	1̄.66682	24	1̄.71955	31	0.28045	1̄.94727	7	20
41	1̄.66706	25	1̄.71986	31	0.28014	1̄.94720	6	19
42	1̄.66731	24	1̄.72017	31	0.27983	1̄.94714	7	18
43	1̄.66755	24	1̄.72048	30	0.27952	1̄.94707	7	17
44	1̄.66779	24	1̄.72078	31	0.27922	1̄.94700	6	16
45	1̄.66803	24	1̄.72109	31	0.27891	1̄.94694	7	15
46	1̄.66827	24	1̄.72140	30	0.27860	1̄.94687	7	14
47	1̄.66851	24	1̄.72170	31	0.27830	1̄.94680	6	13
48	1̄.66875	24	1̄.72201	30	0.27799	1̄.94674	7	12
49	1̄.66899	23	1̄.72231	31	0.27769	1̄.94667	7	11
50	1̄.66922	24	1̄.72262	31	0.27738	1̄.94660	6	10
51	1̄.66946	24	1̄.72293	30	0.27707	1̄.94654	7	9
52	1̄.66970	24	1̄.72323	31	0.27677	1̄.94647	7	8
53	1̄.66994	24	1̄.72354	30	0.27646	1̄.94640	6	7
54	1̄.67018	24	1̄.72384	31	0.27616	1̄.94634	7	6
55	1̄.67042	24	1̄.72415	36	0.27585	1̄.94627	7	5
56	1̄.67066	24	1̄.72445	31	0.27555	1̄.94620	6	4
57	1̄.67090	23	1̄.72476	30	0.27524	1̄.94614	7	3
58	1̄.67113	24	1̄.72506	31	0.27494	1̄.94607	7	2
59	1̄.67137	24	1̄.72537	30	0.27463	1̄.94600	7	1
60	1̄.67161		1̄.72567		0.27433	1̄.94593		0

	log cos	d	log cot	c. d.	log tan	log sin	à	'	p. p.

p. p.

```
        32     31
  6  |  3.2  |  3.1
  7  |  3.7  |  3.6
  8  |  4.3  |  4.1
  9  |  4.8  |  4.7
 10  |  5.3  |  5.2
 20  | 10.7  | 10.3
 30  | 16.0  | 15.5
 40  | 21.3  | 20.7
 50  | 26.7  | 25.8

        30
  6  |  3.0
  7  |  3.5
  8  |  4.0
  9  |  4.5
 10  |  5.0
 20  | 10.0
 30  | 15.0
 40  | 20.0
 50  | 25.0

        25     24
  6  |  2.5  |  2.4
  7  |  2.9  |  2.8
  8  |  3.3  |  3.2
  9  |  3.8  |  3.6
 10  |  4.2  |  4.0
 20  |  8.3  |  8.0
 30  | 12.5  | 12.0
 40  | 16.7  | 16.0
 50  | 20.8  | 20.0

        23
  6  |  2.3
  7  |  2.7
  8  |  3.1
  9  |  3.5
 10  |  3.8
 20  |  7.7
 30  | 11.5
 40  | 15.3
 50  | 19.2

         7      6
  6  |  0.7  |  0.6
  7  |  0.8  |  0.7
  8  |  0.9  |  0.8
  9  |  1.1  |  0.9
 10  |  1.2  |  1.0
 20  |  2.3  |  2.0
 30  |  3.5  |  3.0
 40  |  4.7  |  4.0
 50  |  5.8  |  5.0
```

'	log sin	d	log tan	c. d.	log cot	log cos	d	'
0	1̄.67161	24	1̄.72567	31	0.27433	1̄.94593	6	60
1	1̄.67185	23	1̄.72598	30	0.27402	1̄.94587	7	59
2	1̄.67208	24	1̄.72628	31	0.27372	1̄.94580	7	58
3	1̄.67232	24	1̄.72659	30	0.27341	1̄.94573	6	57
4	1̄.67256	24	1̄.72689	31	0.27311	1̄.94567	7	56
5	1̄.67280	23	1̄.72720	30	0.27280	1̄.94560	7	55
6	1̄.67303	24	1̄.72750	30	0.27250	1̄.94553	7	54
7	1̄.67327	23	1̄.72780	31	0.27220	1̄.94546	6	53
8	1̄.67350	24	1̄.72811	30	0.27189	1̄.94540	7	52
9	1̄.67374	24	1̄.72841	31	0.27159	1̄.94533	7	51
10	1̄.67398	23	1̄.72872	30	0.27128	1̄.94526	7	50
11	1̄.67421	24	1̄.72902	30	0.27098	1̄.94519	6	49
12	1̄.67445	23	1̄.72932	31	0.27068	1̄.94513	7	48
13	1̄.67468	24	1̄.72963	30	0.27037	1̄.94506	7	47
14	1̄.67492	23	1̄.72993	30	0.27007	1̄.94499	7	46
15	1̄.67515	24	1̄.73023	31	0.26977	1̄.94492	7	45
16	1̄.67539	23	1̄.73054	30	0.26946	1̄.94485	6	44
17	1̄.67562	24	1̄.73084	30	0.26916	1̄.94479	7	43
18	1̄.67586	23	1̄.73114	30	0.26886	1̄.94472	7	42
19	1̄.67609	24	1̄.73144	31	0.26856	1̄.94465	7	41
20	1̄.67633	23	1̄.73175	30	0.26825	1̄.94458	7	40
21	1̄.67656	24	1̄.73205	30	0.26795	1̄.94451	6	39
22	1̄.67680	23	1̄.73235	30	0.26765	1̄.94445	7	38
23	1̄.67703	23	1̄.73265	30	0.26735	1̄.94438	7	37
24	1̄.67726	24	1̄.73295	31	0.26705	1̄.94431	7	36
25	1̄.67750	23	1̄.73326	30	0.26674	1̄.94424	7	35
26	1̄.67773	23	1̄.73356	30	0.26644	1̄.94417	7	34
27	1̄.67796	24	1̄.73386	30	0.26614	1̄.94410	6	33
28	1̄.67820	23	1̄.73416	30	0.26584	1̄.94404	7	32
29	1̄.67843	23	1̄.73446	30	0.26554	1̄.94397	7	31
30	1̄.67866	24	1̄.73476	31	0.26524	1̄.94390	7	30
31	1̄.67890	23	1̄.73507	30	0.26493	1̄.94383	7	29
32	1̄.67913	23	1̄.73537	30	0.26463	1̄.94376	7	28
33	1̄.67936	23	1̄.73567	30	0.26433	1̄.94369	7	27
34	1̄.67959	23	1̄.73597	30	0.26403	1̄.94362	6	26
35	1̄.67982	24	1̄.73627	30	0.26373	1̄.94355	7	25
36	1̄.68006	23	1̄.73657	30	0.36343	1̄.94349	7	24
37	1̄.68029	23	1̄.73687	30	0.26313	1̄.94342	7	23
38	1̄.68052	23	1̄.73717	30	0.26283	1̄.94335	7	22
39	1̄.68075	23	1̄.73747	30	0.26253	1̄.94328	7	21
40	1̄.68098	23	1̄.73777	30	0.26223	1̄.94321	7	20
41	1̄.68121	23	1̄.73807	30	0.26193	1̄.94314	7	19
42	1̄.68144	23	1̄.73837	30	0.26163	1̄ 94307	7	18
43	1̄.68167	23	1̄.73867	30	0.26133	1̄.94300	7	17
44	1̄.68190	23	1̄.73897	30	0.26103	1̄.94293	7	16
45	1̄.68213	24	1̄.73927	30	0.26073	1̄.94286	6	15
46	1̄.68237	23	1̄.73957	30	0.26043	1̄.94279	7	14
47	1̄.68260	23	1̄.73987	30	0.26013	1̄.94273	7	13
48	1̄.68283	22	1̄.74017	30	0.25983	1̄.94266	7	12
49	1̄.68305	23	1̄.74047	30	0.25953	1̄.94259	7	11
50	1̄.68328	23	1̄.74077	30	0.25923	1̄.94252	7	10
51	1̄.68251	23	1̄.74107	30	0.25893	1̄.94245	7	9
52	1̄.68374	23	1̄.74137	29	0.25863	1̄.94238	7	8
53	1̄.68397	23	1̄.74166	30	0.25834	1̄.94231	7	7
54	1̄.68420	23	1̄.74196	30	0.25804	1̄.94224	7	6
55	1̄.68443	23	1̄.74226	30	0.25774	1̄.94217	7	5
56	1̄.68466	23	1̄.74256	30	0.25744	1̄.94210	7	4
57	1̄.68489	23	1̄.74286	30	0.25714	1̄.94203	7	3
58	1̄.68512	22	1̄.74316	29	0.25684	1̄.94196	7	2
59	1̄.68534	23	1̄.74345	30	0.25655	1̄.94189	7	1
60	1̄.68557		1̄.74375		0.25625	1̄.94182		0
	log cos	d	log cot	c. d.	log tan	log sin	d	'

p. p.

	31	30
6	3.1	3.0
7	3.6	3.5
8	4.1	4.0
9	4.7	4.5
10	5.2	5.0
20	10.3	10.0
30	15.5	15.0
40	20.7	20.0
50	25.8	25.0

	29
6	2.9
7	3.4
8	3.9
9	4.4
10	4.8
20	9.7
30	14.5
40	19.3
50	24.2

	24	23
6	2.4	2.3
7	2.8	2.7
8	3.2	3.1
9	3.6	3.5
10	4.0	3.8
20	8.0	7.7
30	12.0	11.5
40	16.0	15.3
50	20.0	19.2

	22
6	2.2
7	2.6
8	2.9
9	3.3
10	3.7
20	7.3
30	11.0
40	14.7
50	18.3

	7	6
6	0.7	0.6
7	0.8	0.7
8	0.9	0.8
9	1.1	0.9
10	1.2	1.0
20	2.3	2.0
30	3.5	3.0
40	4.7	4.0
50	5.8	5.0

′	log sin	d	log tan	c. d.	log cot	log cos	d	′
0	1.68557	23	1.74375	30	0.25625	1.94182	7	60
1	1.68580	23	1.74405	30	0.25595	1.94175	7	59
2	1.68603	22	1.74435	30	0.25565	1.94168	7	58
3	1.68625	23	1.74465	29	0.25535	1.94161	7	57
4	1.68648	23	1.74494	30	0.25506	1.94154	7	56
5	1.68671	23	1.74524	30	0.25476	1.94147	7	55
6	1.68694	22	1.74554	29	0.25446	1.94140	7	54
7	1.68716	23	1.74583	30	0.25417	1.94133	7	53
8	1.68739	23	1.74613	30	0.25387	1.94126	7	52
9	1.68762	22	1.74643	30	0.25357	1.94119	7	51
10	1.68784	23	1.74673	29	0.25327	1.94112	7	50
11	1.68807	22	1.74702	30	0.25298	1.94105	7	49
12	1.68829	23	1.74732	30	0.25268	1.94098	8	48
13	1.68852	23	1.74762	29	0.25238	1.94090	7	47
14	1.68875	22	1.74791	30	0.25209	1.94083	7	46
15	1.68897	23	1.74821	30	0.25179	1.94076	7	45
16	1.68920	22	1.74851	29	0.25149	1.94069	7	44
17	1.68942	23	1.74880	30	0.25120	1.94062	7	43
18	1.68965	22	1.74910	29	0.25090	1.94055	7	42
19	1.68987	23	1.74939	30	0.25061	1.94048	7	41
20	1.69010	22	1.74969	29	0.25031	1.94041	7	40
21	1.69032	23	1.74998	30	0.25002	1.94034	7	39
22	1.69055	22	1.75028	30	0.24972	1.94027	7	38
23	1.69077	23	1.75058	29	0.24942	1.94020	8	37
24	1.69100	22	1.75087	30	0.24913	1.94012	7	36
25	1.69122	22	1.75117	29	0.24883	1.94005	7	35
26	1.69144	23	1.75146	30	0.24854	1.93998	7	34
27	1.69167	22	1.75176	30	0.24824	1.93991	7	33
28	1.69189	23	1.75205	29	0.24795	1.93984	7	32
29	1.69212	22	1.75235	30	0.24765	1.93977	7	31
30	1.69234	22	1.75264	29	0.24736	1.93970	7	30
31	1.69256	23	1.75294	29	0.24706	1.93963	8	29
32	1.69279	22	1.75323	30	0.24677	1.93955	7	28
33	1.69301	22	1.75353	29	0.24647	1.93948	7	27
34	1.69323	22	1.75382	29	0.24618	1.93941	7	26
35	1.69345	23	1.75411	30	0.24589	1.93934	7	25
36	1.69368	22	1.75441	29	0.24559	1.93927	7	24
37	1.69390	22	1.75470	30	0.24530	1.93920	8	23
38	1.69412	22	1.75500	29	0.24500	1.93912	7	22
39	1.69434	22	1.75529	29	0.24471	1.93905	7	21
40	1.69456	23	1.75558	30	0.24442	1.93898	7	20
41	1.69479	22	1.75588	29	0.24412	1.93891	7	19
42	1.69501	22	1.75617	30	0.24383	1.93884	8	18
43	1.69523	22	1.75647	29	0.24353	1.93876	7	17
44	1.69545	22	1.75676	29	0.24324	1.93869	7	16
45	1.69567	22	1.75705	30	0.24295	1.93862	7	15
46	1.69589	22	1.75735	29	0.24265	1.93855	8	14
47	1.69611	22	1.75764	29	0.24236	1.93847	7	13
48	1.69633	22	1.75793	29	0.24207	1.93840	7	12
49	1.69655	22	1.75822	30	0.24178	1.93833	7	11
50	1.69677	22	1.75852	29	0.24148	1.93826	7	10
51	1.69699	22	1.75881	29	0.24119	1.93819	8	9
52	1.69721	22	1.75910	29	0.24090	1.93811	7	8
53	1.69743	22	1.75939	30	0.24061	1.93804	7	7
54	1.69765	22	1.75969	29	0.24031	1.93797	8	6
55	1.69787	22	1.75998	29	0.24002	1.93789	7	5
56	1.69809	22	1.76027	29	0.23973	1.93782	7	4
57	1.69831	22	1.76056	30	0.23944	1.93775	7	3
58	1.69853	22	1.76086	29	0.23914	1.93768	8	2
59	1.69875	22	1.76115	29	0.23885	1.93760	7	1
60	1.69897		1.76144		0.23856	1.93753		0
	log cos	d	log cot	c. d.	log tan	log sin	d	′

p. p.

30
6	3.0
7	3.5
8	4.0
9	4.5
10	5.0
20	10.0
30	15.0
40	20.0
50	25.0

29
6	2.9
7	3.4
8	3.9
9	4.4
10	4.8
20	9.7
30	14.5
40	19.3
50	24.2

23
6	2.3
7	2.7
8	3.1
9	3.5
10	3.8
20	7.7
30	11.5
40	15.3
50	19.2

22
6	2.2
7	2.6
8	2.9
9	3.3
10	3.7
20	7.3
30	11.0
40	14.7
50	18.3

′	8	7
6	0.8	0.7
7	0.9	0.8
8	1.1	0.9
9	1.2	1.1
10	1.3	1.2
20	2.7	2.3
30	4.0	3.5
40	5.3	4.7
50	6.7	5.8

′	log sin	d	log tan	c. d.	log cot	log cos	d	
0	1̄.69897	22	1̄.76144	29	0.23856	1̄.93753	7	60
1	1̄.69919	22	1̄.76173	29	0.23827	1̄.93746	8	59
2	1̄.69941	22	1̄.76202	29	0.23798	1̄.93738	7	58
3	1̄.69963	21	1̄.76231	30	0.23769	1̄.93731	7	57
4	1̄.69984	22	1̄.76261	29	0.23739	1̄.93724	7	56
5	1̄.70006	22	1̄.76290	29	0.23710	1̄.93717	8	55
6	1̄.70028	22	1̄.76319	29	0.23681	1̄.93709	7	54
7	1̄.70050	22	1̄.76348	29	0.23652	1̄.93702	7	53
8	1̄.70072	21	1̄.76377	29	0.23623	1̄.93695	8	52
9	1̄.70093	22	1̄.76406	29	0.23594	1̄.93687	7	51
10	1̄.70115	22	1̄.76435	29	0.23565	1̄.93680	7	50
11	1̄.70137	22	1̄.76464	29	0.23536	1̄.93673	8	49
12	1̄.70159	21	1̄.76493	29	0.23507	1̄.93665	7	48
13	1̄.70180	22	1̄.76522	29	0.23478	1̄.93658	8	47
14	1̄.70202	22	1̄.76551	29	0.23449	1̄.93650	7	46
15	1̄.70224	21	1̄.76580	29	0.23420	1̄.93643	7	45
16	1̄.70245	22	1̄.76609	30	0.23391	1̄.93636	8	44
17	1̄.70267	21	1̄.76639	29	0.23361	1̄.93628	7	43
18	1̄.70288	22	1̄.76668	29	0.23332	1̄.93621	7	42
19	1̄.70310	22	1̄.76697	28	0.23303	1̄.93614	8	41
20	1̄.70332	21	1̄.76725	29	0.23275	1̄.93606	7	40
21	1̄.70353	22	1̄.76754	29	0.23246	1̄.93599	8	39
22	1̄.70375	21	1̄.76783	29	0.23217	1̄.93591	7	38
23	1̄.70396	22	1̄.76812	29	0.23188	1̄.93584	7	37
24	1̄.70418	21	1̄.76841	29	0.23159	1̄.93577	8	36
25	1̄.70439	22	1̄.76870	29	0.23130	1̄.93569	7	35
26	1̄.70461	21	1̄.76899	29	0.23101	1̄.93562	8	34
27	1̄.70482	22	1̄.76928	29	0.23072	1̄.93554	7	33
28	1̄.70504	21	1̄.76957	29	0.23043	1̄.93547	8	32
29	1̄.70525	22	1̄.76986	29	0.23014	1̄.93539	7	31
30	1̄.70547	21	1̄.77015	29	0.22985	1̄.93532	7	30
31	1̄.70568	22	1̄.77044	29	0.22956	1̄.93525	8	29
32	1̄.70590	21	1̄.77073	28	0.22927	1̄.93517	7	28
33	1̄.70611	22	1̄.77101	29	0.22899	1̄.93510	8	27
34	1̄.70633	21	1̄.77130	29	0.22870	1̄.93502	7	26
35	1̄.70654	21	1̄.77159	29	0.22841	1̄.93495	8	25
36	1̄.70675	22	1̄.77188	29	0.22812	1̄.93487	7	24
37	1̄.70697	21	1̄.77217	29	0.22783	1̄.93480	8	23
38	1̄.70718	21	1̄.77246	28	0.22754	1̄.93472	7	22
39	1̄.70739	22	1̄.77274	29	0.22726	1̄.93465	8	21
40	1̄.70761	21	1̄.77303	29	0.22697	1̄.93457	7	20
41	1̄.70782	21	1̄.77332	29	0.22668	1̄.93450	8	19
42	1̄.70803	21	1̄.77361	29	0.22639	1̄.93442	7	18
43	1̄.70824	22	1̄.77390	28	0.22610	1̄.93435	8	17
44	1̄.70846	21	1̄.77418	29	0.22582	1̄.93427	7	16
45	1̄.70867	21	1̄.77447	29	0.22553	1̄.93420	8	15
46	1̄.70888	21	1̄.77476	29	0.22524	1̄.93412	7	14
47	1̄.70909	22	1̄.77505	28	0.22495	1̄.93405	8	13
48	1̄.70931	21	1̄.77533	29	0.22467	1̄.93397	7	12
49	1̄.70952	21	1̄.77562	29	0.22438	1̄.93390	8	11
50	1̄.70973	21	1̄.77591	28	0.22409	1̄.93382	7	10
51	1̄.70994	21	1̄.77619	29	0.22381	1̄.93375	8	9
52	1̄.71015	21	1̄.77648	29	0.22352	1̄.93367	7	8
53	1̄.71036	22	1̄.77677	29	0.22323	1̄.93360	8	7
54	1̄.71058	21	1̄.77706	28	0.22294	1̄.93352	8	6
55	1̄.71079	21	1̄.77734	29	0.22266	1̄.93344	7	5
56	1̄.71100	21	1̄.77763	28	0.22237	1̄.93337	8	4
57	1̄.71121	21	1̄.77791	29	0.22209	1̄.93329	7	3
58	1̄.71142	21	1̄.77820	29	0.22180	1̄.93322	8	2
59	1̄.71163	21	1̄.77849	28	0.22151	1̄.93314	7	1
60	1̄.71184		1̄.77877		0.22123	1̄.93307		0
	log cos	d	log cot	c. d.	log tan	log sin	d	′

p. p.

	30	29
6	3.0	2.9
7	3.5	3.4
8	4.0	3.9
9	4.5	4.4
10	5.0	4.8
20	10.0	9.7
30	15.0	14.5
40	20.0	19.3
50	25.0	24.2

	28
6	2.8
7	3.3
8	3.7
9	4.2
10	4.7
20	9.3
30	14.0
40	18.7
50	23.3

	22
6	2.2
7	2.6
8	2.9
9	3.3
10	3.7
20	7.3
30	11.0
40	14.7
50	18.3

	21
6	2.1
7	2.5
8	2.8
9	3.2
10	3.5
20	7.0
30	10.5
40	14.0
50	17.5

	8	7
6	0.8	0.7
7	0.9	0.8
8	1.1	0.9
9	1.2	1.1
10	1.3	1.2
20	2.7	2.3
30	4.0	3.5
40	5.3	4.7
50	6.7	5.8

′	log sin	d	log tan	c. d.	log cot	log cos	d	′
0	1̄.71184	21	1̄.77877	29	0.22123	1̄.93307	8	60
1	1̄.71205	21	1̄.77906	29	0.22094	1̄.93299	8	59
2	1̄.71226	21	1̄.77935	28	0.22065	1̄.93291	7	58
3	1̄.71247	21	1̄.77963	29	0.22037	1̄.93284	8	57
4	1̄.71268	21	1̄.77992	28	0.22008	1̄.93276	7	56
5	1̄.71289	21	1̄.78020	29	0.21980	1̄.93269	8	55
6	1̄.71310	21	1̄.78049	28	0.21951	1̄.93261	8	54
7	1̄.71331	21	1̄.78077	29	0.21923	1̄.93253	7	53
8	1̄.71352	21	1̄.78106	29	0.21894	1̄.93246	8	52
9	1̄.71373	20	1̄.78135	28	0.21865	1̄.93238	8	51
10	1̄.71393	21	1̄.78163	29	0.21837	1̄.93230	7	50
11	1̄.71414	21	1̄.78192	28	0.21808	1̄.93223	8	49
12	1̄.71435	21	1̄.78220	29	0.21780	1̄.93215	8	48
13	1̄.71456	21	1̄.78249	28	0.21751	1̄.93207	7	47
14	1̄.71477	21	1̄.78277	29	0.21723	1̄.93200	8	46
15	1̄.71498	21	1̄.78306	28	0.21694	1̄.93192	8	45
16	1̄.71519	20	1̄.78334	29	0.21666	1̄.93184	7	44
17	1̄.71539	21	1̄.78363	28	0.21637	1̄.93177	8	43
18	1̄.71560	21	1̄.78391	28	0.21609	1̄.93169	8	42
19	1̄.71581	21	1̄.78419	29	0.21581	1̄.93161	7	41
20	1̄.71602	20	1̄.78448	28	0.21552	1̄.93154	8	40
21	1̄.71622	21	1̄.78476	29	0.21524	1̄.93146	8	39
22	1̄.71643	21	1̄.78505	28	0.21495	1̄.93138	7	38
23	1̄.71664	21	1̄.78533	29	0.21467	1̄.93131	8	37
24	1̄.71685	20	1̄.78562	28	0.21438	1̄.93123	8	36
25	1̄.71705	21	1̄.78590	28	0.21410	1̄.93115	7	35
26	1̄.71726	21	1̄.78618	29	0.21382	1̄.93108	8	34
27	1̄.71747	20	1̄.78647	28	0.21353	1̄.93100	8	33
28	1̄.71767	21	1̄.78675	29	0.21325	1̄.93092	8	32
29	1̄.71788	21	1̄.78704	28	0.21296	1̄.93084	7	31
30	1̄.71809	20	1̄.78732	28	0.21268	1̄.93077	8	30
31	1̄.71829	21	1̄.78760	29	0.21240	1̄.93069	8	29
32	1̄.71850	20	1̄.78789	28	0.21211	1̄.93061	8	28
33	1̄.71870	21	1̄.78817	28	0.21183	1̄.93053	7	27
34	1̄.71891	20	1̄.78845	29	0.21155	1̄.93046	8	26
35	1̄.71911	21	1̄.78874	28	0.21126	1̄.93038	8	25
36	1̄.71932	20	1̄.78902	28	0.21098	1̄.93030	8	24
37	1̄.71952	21	1̄.78930	29	0.21070	1̄.93022	8	23
38	1̄.71973	21	1̄.78959	28	0.21041	1̄.93014	7	22
39	1̄.71994	20	1̄.78987	28	0.21013	1̄.93007	8	21
40	1̄.72014	20	1̄.79015	28	0.20985	1̄.92999	8	20
41	1̄.72034	21	1̄.79043	29	0.20957	1̄.92991	8	19
42	1̄.72055	20	1̄.79072	28	0.20928	1̄.92983	7	18
43	1̄.72075	21	1̄.79100	28	0.20900	1̄.92976	8	17
44	1̄.72096	20	1̄.79128	28	0.20872	1̄.92968	8	16
45	1̄.72116	21	1̄.79156	29	0.20844	1̄.92960	8	15
46	1̄.72137	20	1̄.79185	28	0.20815	1̄.92952	8	14
47	1̄.72157	20	1̄.79213	28	0.20787	1̄.92944	8	13
48	1̄.72177	21	1̄.79241	28	0.20759	1̄.92936	7	12
49	1̄.72198	20	1̄.79269	28	0.20731	1̄.92929	8	11
50	1̄.72218	20	1̄.79297	29	0.20703	1̄.92921	8	10
51	1̄.72238	21	1̄.79326	28	0.20674	1̄.92913	8	9
52	1̄.72259	20	1̄.79354	28	0.20646	1̄.92905	8	8
53	1̄.72279	20	1̄.79382	28	0.20618	1̄.92897	8	7
54	1̄.72299	21	1̄.79410	28	0.20590	1̄.92889	8	6
55	1̄.72320	20	1̄.79438	28	0.20562	1̄.92881	7	5
56	1̄.72340	20	1̄.79466	29	0.20534	1̄.92874	8	4
57	1̄.72360	21	1̄.79495	28	0.20505	1̄.92866	8	3
58	1̄.72381	20	1̄.79523	28	0.20477	1̄.92858	8	2
59	1̄.72401	20	1̄.79551	28	0.20449	1̄.92850	8	1
60	1̄.72421		1̄.79579		0.20421	1̄.92842		0
	log cos	d	log cot	c. d.	log tan	log sin	d	′

p. p.

29		28		21		20	
6	2.9	6	2.8	6	2.1	6	2.0
7	3.4	7	3.3	7	2.5	7	2.3
8	3.9	8	3.7	8	2.8	8	2.7
9	4.4	9	4.2	9	3.2	9	3.0
10	4.8	10	4.7	10	3.5	10	3.3
20	9.7	20	9.3	20	7.0	20	6.7
30	14.5	30	14.0	30	10.5	30	10.0
40	19.3	40	18.7	40	14.0	40	13.3
50	24.2	50	23.3	50	17.5	50	16.7

	8	7
6	0.8	0.7
7	0.9	0.8
8	1.1	0.9
9	1.2	1.1
10	1.3	1.2
20	2.7	2.3
30	4.0	3.5
40	5.3	4.7
50	6.7	5.8

′	log sin	d	log tan	c. d.	log cot	log cos	d	′
0	1.72421	20	1.79579	28	0.20421	1.92842	8	60
1	1.72441	20	1.79607	28	0.20393	1.92834	8	59
2	1.72461	21	1.79635	28	0.20365	1.92826	8	58
3	1.72482	20	1.79663	28	0.20337	1.92818	8	57
4	1.72502	20	1.79691	28	0.20309	1.92810	7	56
5	1.72522	20	1.79719	28	0.20281	1.92803	8	55
6	1.72542	20	1.79747	29	0.20253	1.92795	8	54
7	1.72562	20	1.79776	28	0.20224	1.92787	8	53
8	1.72582	20	1.79804	28	0.20196	1.92779	8	52
9	1.72602	20	1.79832	28	0.20168	1.92771	8	51
10	1.72622	21	1.79860	28	0.20140	1.92763	8	50
11	1.72643	20	1.79888	28	0.20112	1.92755	8	49
12	1.72663	20	1.79916	28	0.20084	1.92747	8	48
13	1.72683	20	1.79944	28	0.20056	1.92739	8	47
14	1.72703	20	1.79972	28	0.20028	1.92731	8	46
15	1.72723	20	1.80000	28	0.20000	1.92723	8	45
16	1.72743	20	1.80028	28	0.19972	1.92715	8	44
17	1.72763	20	1.80056	28	0.19944	1.92707	8	43
18	1.72783	20	1.80084	28	0.19916	1.92699	8	42
19	1.72803	20	1.80112	28	0.19888	1.92691	8	41
20	1.72823	20	1.80140	28	0.19860	1.92683	8	40
21	1.72843	20	1.80168	27	0.19832	1.92675	8	39
22	1.72863	20	1.80195	28	0.19805	1.92667	8	38
23	1.72883	19	1.80223	28	0.19777	1.92659	8	37
24	1.72902	20	1.80251	28	0.19749	1.92651	8	36
25	1.72922	20	1.80279	28	0.19721	1.92643	8	35
26	1.72942	20	1.80307	28	0.19693	1.92635	8	34
27	1.72962	20	1.80335	28	0.19665	1.92627	8	33
28	1.72982	20	1.80363	28	0.19637	1.92619	8	32
29	1.73002	20	1.80391	28	0.19609	1.92611	8	31
30	1.73022	19	1.80419	28	0.19581	1.92603	8	30
31	1.73041	20	1.80447	27	0.19553	1.92595	8	29
32	1.73061	20	1.80474	28	0.19526	1.92587	8	28
33	1.73081	20	1.80502	28	0.19498	1.92579	8	27
34	1.73101	20	1.80530	28	0.19470	1.92571	8	26
35	1.73121	19	1.80558	28	0.19442	1.92563	8	25
36	1.73140	20	1.80586	28	0.19414	1.92555	9	24
37	1.73160	20	1.80614	28	0.19386	1.92546	8	23
38	1.73180	20	1.80642	27	0.19358	1.92538	8	22
39	1.73200	19	1.80669	28	0.19331	1.92530	8	21
40	1.73219	20	1.80697	28	0.19303	1.92522	8	20
41	1.73239	20	1.80725	28	0.19275	1.92514	8	19
42	1.73259	19	1.80753	28	0.19247	1.92506	8	18
43	1.73278	20	1.80781	27	0.19219	1.92498	8	17
44	1.73298	20	1.80808	28	0.19192	1.92490	8	16
45	1.73318	19	1.80836	28	0.19164	1.92482	9	15
46	1.73337	20	1.80864	28	0.19136	1.92473	8	14
47	1.73357	20	1.80892	27	0.19108	1.92465	8	13
48	1.73377	19	1.80919	28	0.19081	1.92457	8	12
49	1.73396	20	1.80947	28	0.19053	1.92449	8	11
50	1.73416	19	1.80975	28	0.19025	1.92441	8	10
51	1.73435	20	1.81003	27	0.18997	1.92433	8	9
52	1.73455	19	1.81030	28	0.18970	1.92425	9	8
53	1.73474	20	1.81058	28	0.18942	1.92416	8	7
54	1.73494	19	1.81086	27	0.18914	1.92408	8	6
55	1.73513	20	1.81113	28	0.18887	1.92400	8	5
56	1.73533	19	1.81141	28	0.18859	1.92392	8	4
57	1.73552	20	1.81169	28	0.18831	1.92384	8	3
58	1.73572	19	1.81196	27	0.18804	1.92376	9	2
59	1.73591	20	1.81224	28	0.18776	1.92367	8	1
60	1.73611		1.81252	28	0.18748	1.92359		0
	log cos	d	log cot	c. d.	log tan	log sin	d	′

p. p.

	29	28
6	2.9	2.8
7	3.4	3.3
8	3.9	3.7
9	4.4	4.2
10	4.8	4.7
20	9.7	9.3
30	14.5	14.0
40	19.3	18.7
50	24.2	23.3

	27
6	2.7
7	3.2
8	3.6
9	4.1
10	4.5
20	9.0
30	13.5
40	18.0
50	22.5

	21	20
6	2.1	2.0
7	2.5	2.3
8	2.8	2.7
9	3.2	3.0
10	3.5	3.3
20	7.0	6.7
30	10.5	10.0
40	14.0	13.3
50	17.5	16.7

	19
6	1.9
7	2.2
8	2.5
9	2.9
10	3.2
20	6.3
30	9.5
40	12.7
50	15.8

	9	8	7
6	0.9	0.8	0.7
7	1.1	0.9	0.8
8	1.2	1.1	0.9
9	1.4	1.2	1.1
10	1.5	1.3	1.2
20	3.0	2.7	2.3
30	4.5	4.0	3.5
40	6.0	5.3	4.7
50	7.5	6.7	5.8

log tan	c. d.	log cot	log cos	d	'
Ī.81252	27	0.18748	Ī.92359	8	
Ī.81279	28	0.18721	Ī.92351	8	
Ī.81307	28	0.18693	Ī.92343	8	
Ī.81335	27	0.18665	Ī.92335	9	
Ī.81362	28	0.18638	Ī.92326	8	
Ī.81390	28	0.18610	Ī.92318	8	
Ī.81418	27	0.18582	Ī.92310	8	
Ī.81445	28	0.18555	Ī.92302	9	
Ī.81473	27	0.18527	Ī.92293	8	
Ī.81500	28	0.18500	Ī.92285	8	
Ī.81528	28	0.18472	Ī.92277	8	
Ī.81556	27	0.18444	Ī.92269	9	
Ī.81583	28	0.18417	Ī.92260	8	
Ī.81611	27	0.18389	Ī.92252	8	
Ī.81638	28	0.18362	Ī.92244	9	
Ī.81666	27	0.18334	Ī.92235	8	
Ī.81693	28	0.18307	Ī.92227	8	
Ī.81721	27	0.18279	Ī.92219	8	
Ī.81748	28	0.18252	Ī.92211	9	
Ī.81776	27	0.18224	Ī.92202	8	
Ī.81803	28	0.18197	Ī.92194	8	
Ī.81831	27	0.18169	Ī.92186	9	
Ī.81858	28	0.18142	Ī.92177	8	
Ī.81886	27	0.18114	Ī.92169	8	
Ī.81913	28	0.18087	Ī.92161	9	
Ī.81941	27	0.18059	Ī.92152	8	
Ī.81968	28	0.18032	Ī.92144	8	
Ī.81996	27	0.18004	Ī.92136	9	
Ī.82023	28	0.17977	Ī.92127	8	
Ī.82051	27	0.17949	Ī.92119	8	
Ī.82078	28	0.17922	Ī.92111	9	
Ī.82106	27	0.17894	Ī.92102	8	
Ī.82133	28	0.17867	Ī.92094	8	
Ī.82161	27	0.17839	Ī.92086	9	
Ī.82188	27	0.17812	Ī.92077	8	
Ī.82215	28	0.17785	Ī.92069	9	
Ī.82243	27	0.17757	Ī.92060	8	
Ī.82270	28	0.17730	Ī.92052	8	
Ī.82298	27	0.17702	Ī.92044	9	
Ī.82325	27	0.17675	Ī.92035	8	
Ī.82352	28	0.17648	Ī.92027	9	
Ī.82380	27	0.17620	Ī.92018	8	
Ī.82407	28	0.17593	Ī.92010	8	
Ī.82435	27	0.17565	Ī.92002	9	
Ī.82462	27	0.17538	Ī.91993	8	
Ī.82489	28	0.17511	Ī.91985	9	
Ī.82517	27	0.17483	Ī.91976	8	
Ī.82544	27	0.17456	Ī.91968	9	
Ī.82571	28	0.17429	Ī.91959	8	
Ī.82599	27	0.17401	Ī.91951	9	
Ī.82626	27	0.17374	Ī.91942	8	
Ī.82653	28	0.17347	Ī.91934	9	9
Ī.82681	27	0.17319	Ī.91925	8	8
Ī.82708	27	0.17292	Ī.91917	9	7
Ī.82735	27	0.17265	Ī.91908	8	6
Ī.82762	28	0.17238	Ī.91900	9	5
Ī.82790	27	0.17210	Ī.91891	8	4
Ī.82817	27	0.17183	Ī.91883	9	3
Ī.82844	27	0.17156	Ī.91874	8	2
Ī.82871	28	0.17129	Ī.91866	9	1
Ī.82899		0.17101	Ī.91857	9	0

Far-left fragment (near the last rows): Ī.74756   19

p. p.

	28	27
6	2.8	2.7
7	3.3	3.2
8	3.7	3.6
9	4.2	4.1
10	4.7	4.5
20	9.3	9.0
30	14.0	13.5
40	18.7	18.0
50	23.3	22.5

	20
6	2.0
7	2.3
8	2.7
9	3.0
10	3.3
20	6.7
30	10.0
40	13.3
50	16.7

	19
6	1.9
7	2.2
8	2.5
9	2.9
10	3.2
20	6.3
30	9.5
40	12.7
50	15.8

	18
6	1.8
7	2.1
8	2.4
9	2.7
10	3.0
20	6.0
30	9.0
40	12.0
50	15.0

	9	8
6	0.9	0.8
7	1.1	0.9
8	1.2	1.1
9	1.4	1.2
10	1.5	1.3
20	3.0	2.7
30	4.5	4.0
40	6.0	5.3
50	7.5	6.7

log cos	d	log cot	c. d.	log tan	log sin	d	'	p. p.

′	log sin	d	log tan	c. d.	log cot	log cos	d		
0	1̄.74756	19	1̄.82899	27	0.17101	1̄.91857	8	60	
1	1̄.74775	19	1̄.82926	27	0.17074	1̄.91849	9	59	
2	1̄.74794	18	1̄.82953	27	0.17047	1̄.91840	8	58	
3	1̄.74812	19	1̄.82980	28	0.17020	1̄.91832	9	57	
4	1̄.74831	19	1̄.83008	27	0.16992	1̄.91823	8	56	
5	1̄.74850	18	1̄.83035	27	0.16965	1̄.91815	9	55	
6	1̄.74868	19	1̄.83062	27	0.16938	1̄.91806	8	54	
7	1̄.74887	19	1̄.83089	28	0.16911	1̄.91798	9	53	
8	1̄.74906	18	1̄.83117	27	0.16883	1̄.91789	8	52	
9	1̄.74924	19	1̄.83144	27	0.16856	1̄.91781	9	51	
10	1̄.74943	18	1̄.83171	27	0.16829	1̄.91772	9	50	
11	1̄.74961	19	1̄.83198	27	0.16802	1̄.91763	8	49	
12	1̄.74980	19	1̄.83225	27	0.16775	1̄.91755	9	48	
13	1̄.74999	18	1̄.83252	28	0.16748	1̄.91746	8	47	
14	1̄.75017	19	1̄.83280	27	0.16720	1̄.91738	9	46	
15	1̄.75036	18	1̄.83307	27	0.16693	1̄.91729	9	45	
16	1̄.75054	19	1̄.83334	27	0.16666	1̄.91720	8	44	
17	1̄.75073	18	1̄.83361	27	0.16639	1̄.91712	9	43	
18	1̄.75091	19	1̄.83388	27	0.16612	1̄.91703	8	42	
19	1̄.75110	18	1̄.83415	27	0.16585	1̄.91695	9	41	
20	1̄.75128	19	1̄.83442	28	0.16558	1̄.91686	9	40	
21	1̄.75147	18	1̄.83470	27	0.16530	1̄.91677	8	39	
22	1̄.75165	19	1̄.83497	27	0.16503	1̄.91669	9	38	
23	1̄.75184	18	1̄.83524	27	0.16476	1̄.91660	9	37	
24	1̄.75202	19	1̄.83551	27	0.16449	1̄.91651	8	36	
25	1̄.75221	18	1̄.83578	27	0.16422	1̄.91643	9	35	
26	1̄.75239	19	1̄.83605	27	0.16395	1̄.91634	9	34	
27	1̄.75258	18	1̄.83632	27	0.16368	1̄.91625	8	33	
28	1̄.75276	18	1̄.83659	27	0.16341	1̄.91617	9	32	
29	1̄.75294	19	1̄.83686	27	0.16314	1̄.91608	9	31	
30	1̄.75313	18	1̄.83713	27	0.16287	1̄.91599	8	30	
31	1̄.75331	19	1̄.83740	28	0.16260	1̄.91591	9	29	
32	1̄.75350	18	1̄.83768	27	0.16232	1̄.91582	9	28	
33	1̄.75368	18	1̄.83795	27	0.16205	1̄.91573	8	27	
34	1̄.75386	19	1̄.83822	27	0.16178	1̄.91565	9	26	
35	1̄.75405	18	1̄.83849	27	0.16151	1̄.91556	9	25	
36	1̄.75423	18	1̄.83876	27	0.16124	1̄.91547	9	24	
37	1̄.75441	18	1̄.83903	27	0.16097	1̄.91538	8	23	
38	1̄.75459	19	1̄.83930	27	0.16070	1̄.91530	9	22	
39	1̄.75478	18	1̄.83957	27	0.16043	1̄.91521	9	21	
40	1̄.75496	18	1̄.83984	27	0.16016	1̄.91512	8	20	
41	1̄.75514	19	1̄.84011	27	0.15989	1̄.91504	9	19	
42	1̄.75533	18	1̄.84038	27	0.15962	1̄.91495	9	18	
43	1̄.75551	18	1̄.84065	27	0.15935	1̄.91486	9	17	
44	1̄.75569	18	1̄.84092	27	0.15908	1̄.91477	8	16	
45	1̄.75587	18	1̄.84119	27	0.15881	1̄.91469	9	15	
46	1̄.75605	19	1̄.84146	27	0.15854	1̄.91460	9	14	
47	1̄.75624	18	1̄.84173	27	0.15827	1̄.91451	9	13	
48	1̄.75642	18	1̄.84200	27	0.15800	1̄.91442	9	12	
49	1̄.75660	18	1̄.84227	27	0.15773	1̄.91433	8	11	
50	1̄.75678	18	1̄.84254	26	0.15746	1̄.91425	9	10	
51	1̄.75696	18	1̄.84280	27	0.15720	1̄.91416	9	9	
52	1̄.75714	19	1̄.84307	27	0.15693	1̄.91407	9	8	
53	1̄.75733	18	1̄.84334	27	0.15666	1̄.91398	9	7	
54	1̄.75751	18	1̄.84361	27	0.15639	1̄.91389	8	6	
55	1̄.75769	18	1̄.84388	27	0.15612	1̄.91381	9	5	
56	1̄.75787	18	1̄.84415	27	0.15585	1̄.91372	9	4	
57	1̄.75805	18	1̄.84442	27	0.15558	1̄.91363	9	3	
58	1̄.75823	18	1̄.84469	27	0.15531	1̄.91354	9	2	
59	1̄.75841	18	1̄.84496	27	0.15504	1̄.91345	9	1	
60	1̄.75859		1̄.84523		0.15477	1̄.91336		0	

| | log cos | d | log cot | c. d. | log tan | log sin | d | ′ | p. p. |

p. p.

	28	27
6	2.8	2.7
7	3.3	3.2
8	3.7	3.6
9	4.2	4.1
10	4.7	4.5
20	9.3	9.0
30	14.0	13.5
40	18.7	18.0
50	23.3	22.5

	26
6	2.6
7	3.0
8	3.5
9	3.9
10	4.3
20	8.7
30	13.0
40	17.3
50	21.7

	19
6	1.9
7	2.2
8	2.5
9	2.9
10	3.2
20	6.3
30	9.5
40	12.7
50	15.8

	18
6	1.8
7	2.1
8	2.4
9	2.7
10	3.0
20	6.0
30	9.0
40	12.0
50	15.0

	9	8
6	0.9	0.8
7	1.1	0.9
8	1.2	1.1
9	1.4	1.2
10	1.5	1.3
20	3.0	2.7
30	4.5	4.0
40	6.0	5.3
50	7.5	6.7

′	log sin	d	log tan	c. d.	log cot	log cos	d		p. p.
0	1.75859	18	1.84523	27	0.15477	1.91336	8	60	
1	1.75877	18	1.84550	26	0.15450	1.91328	9	59	
2	1.75895	18	1.84576	27	0.15424	1.91319	9	58	**27** **28**
3	1.75913	18	1.84603	27	0.15397	1.91310	9	57	6 2.7 2.6
4	1.75931	18	1.84630	27	0.15370	1.91301	9	56	7 3.2 3.0
5	1.75949	18	1.84657	27	0.15343	1.91292	9	55	8 3.6 3.5
6	1.75967	18	1.84684	27	0.15316	1.91283	9	54	9 4.1 3.9
7	1.75985	18	1.84711	27	0.15289	1.91274	8	53	10 4.5 4.3
8	1.76003	18	1.84738	26	0.15262	1.91266	9	52	20 9.0 8.7
9	1.76021	18	1.84764	27	0.15236	1.91257	9	51	30 13.5 13.0
10	1.76039	18	1.84791	27	0.15209	1.91248	9	50	40 18.0 17.3
		18		27			9		50 22.5 21.7
11	1.76057	18	1.84818	27	0.15182	1.91239	9	49	
12	1.76075	18	1.84845	27	0.15155	1.91230	9	48	
13	1.76093	18	1.84872	27	0.15128	1.91221	9	47	
14	1.76111	18	1.84899	26	0.15101	1.91212	9	46	**18**
15	1.76129	17	1.84925	27	0.15075	1.91203	9	45	6 1.8
16	1.76146	18	1.84952	27	0.15048	1.91194	9	44	7 2.1
17	1.76164	18	1.84979	27	0.15021	1.91185	9	43	8 2.4
18	1.76182	18	1.85006	27	0.14994	1.91176	9	42	9 2.7
19	1.76200	18	1.85033	26	0.14967	1.91167	9	41	10 3.0
20	1.76218	18	1.85059	27	0.14941	1.91158	9	40	20 6.0
		18		27			9		30 9.0
21	1.76236	17	1.85086	27	0.14914	1.91149	8	39	40 12.0
22	1.76253	18	1.85113	27	0.14887	1.91141	9	38	50 15.0
23	1.76271	18	1.85140	26	0.14860	1.91132	9	37	
24	1.76289	18	1.85166	27	0.14834	1.91123	9	36	
25	1.76307	17	1.85193	27	0.14807	1.91114	9	35	
26	1.76324	18	1.85220	27	0.14780	1.91105	9	34	**17**
27	1.76342	18	1.85247	26	0.14753	1.91096	9	33	6 1.7
28	1.76360	18	1.85273	27	0.14727	1.91087	9	32	7 2.0
29	1.76378	17	1.85300	27	0.14700	1.91078	9	31	8 2.3
30	1.76395	18	1.85327	27	0.14673	1.91069	9	30	9 2.6
		18		27			9		10 2.8
31	1.76413	18	1.85354	26	0.14646	1.91060	9	29	20 5.7
32	1.76431	17	1.85380	27	0.14620	1.91051	9	28	30 8.5
33	1.76448	18	1.85407	27	0.14593	1.91042	9	27	40 11.3
34	1.76466	18	1.85434	26	0.14566	1.91033	10	26	50 14.2
35	1.76484	17	1.85460	27	0.14540	1.91023	9	25	
36	1.76501	18	1.85487	27	0.14513	1.91014	9	24	
37	1.76519	18	1.85514	26	0.14486	1.91005	9	23	
38	1.76537	17	1.85540	27	0.14460	1.90996	9	22	**10**
39	1.76554	18	1.85567	27	0.14433	1.90987	9	21	6 1.0
40	1.76572	18	1.85594	26	0.14406	1.90978	9	20	7 1.2
		18		26			9		8 1.3
41	1.76590	17	1.85620	27	0.14380	1.90969	9	19	9 1.5
42	1.76607	18	1.85647	27	0.14353	1.90960	9	18	10 1.7
43	1.76625	17	1.85674	26	0.14326	1.90951	9	17	20 3.3
44	1.76642	18	1.85700	27	0.14300	1.90942	9	16	30 5.0
45	1.76660	17	1.85727	27	0.14273	1.90933	9	15	40 6.7
46	1.76677	18	1.85754	26	0.14246	1.90924	9	14	50 8.3
47	1.76695	17	1.85780	27	0.14220	1.90915	9	13	
48	1.76712	18	1.85807	27	0.14193	1.90906	10	12	
49	1.76730	17	1.85834	26	0.14166	1.90896	9	11	
50	1.76747	18	1.85860	27	0.14140	1.90887	9	10	**9** **8**
		18		27			9		6 0.9 0.8
51	1.76765	17	1.85887	26	0.14113	1.90878	9	9	7 1.1 0.9
52	1.76782	18	1.85913	27	0.14087	1.90869	9	8	8 1.2 1.1
53	1.76800	17	1.85940	27	0.14060	1.90860	9	7	9 1.4 1.2
54	1.76817	18	1.85967	26	0.14033	1.90851	9	6	10 1.5 1.3
55	1.76835	17	1.85993	27	0.14007	1.90842	10	5	20 3.0 2.7
56	1.76852	18	1.86020	26	0.13980	1.90832	9	4	30 4.5 4.0
57	1.76870	17	1.86046	27	0.13954	1.90823	9	3	40 6.0 5.3
58	1.76887	17	1.86073	27	0.13927	1.90814	9	2	50 7.5 6.7
59	1.76904	17	1.86100	27	0.13900	1.90805	9	1	
60	1.76922	18	1.86126	26	0.13874	1.90796	9	0	

	log cos	d	log cot	c. d.	log tan	log sin	d	′	p. p.

′	log sin	d	log tan	c. d.	log cot	log cos	d	′
0	1̄.76922	17	1̄.86126	27	0.13874	1̄.90796	9	60
1	1̄.76939	18	1̄.86153	26	0.13847	1̄.90787	10	59
2	1̄.76957	17	1̄.86179	27	0.13821	1̄.90777	9	58
3	1̄.76974	17	1̄.86206	26	0.13794	1̄.90768	9	57
4	1̄.76991	18	1̄.86232	27	0.13768	1̄.90759	9	56
5	1̄.77009	17	1̄.86259	26	0.13741	1̄.90750	9	55
6	1̄.77026	17	1̄.86285	27	0.13715	1̄.90741	10	54
7	1̄.77043	18	1̄.86312	26	0.13688	1̄.90731	9	53
8	1̄.77061	17	1̄.86338	27	0.13662	1̄.90722	9	52
9	1̄.77078	17	1̄.86365	27	0.13635	1̄.90713	9	51
10	1̄.77095	17	1̄.86392	26	0.13608	1̄.90704	10	50
11	1̄.77112	18	1̄.86418	27	0.13582	1̄.90694	9	49
12	1̄.77130	17	1̄.86445	26	0.13555	1̄.90685	9	48
13	1̄.77147	17	1̄.86471	27	0.13529	1̄.90676	9	47
14	1̄.77164	17	1̄.86498	26	0.13502	1̄.90667	10	46
15	1̄.77181	18	1̄.86524	27	0.13476	1̄.90657	9	45
16	1̄.77199	17	1̄.86551	26	0.13449	1̄.90648	9	44
17	1̄.77216	17	1̄.86577	26	0.13423	1̄.90639	9	43
18	1̄.77233	17	1̄.86603	27	0.13397	1̄.90630	10	42
19	1̄.77250	18	1̄.86630	26	0.13370	1̄.90620	9	41
20	1̄.77268	17	1̄.86656	27	0.13344	1̄.90611	9	40
21	1̄.77285	17	1̄.86683	26	0.13317	1̄.90602	10	39
22	1̄.77302	17	1̄.86709	27	0.13291	1̄.90592	9	38
23	1̄.77319	17	1̄.86736	26	0.13264	1̄.90583	9	37
24	1̄.77336	17	1̄.86762	27	0.13238	1̄.90574	9	36
25	1̄.77353	17	1̄.86789	26	0.13211	1̄.90565	10	35
26	1̄.77370	17	1̄.86815	27	0.13185	1̄.90555	9	34
27	1̄.77387	18	1̄.86842	26	0.13158	1̄.90546	9	33
28	1̄.77405	17	1̄.86868	26	0.13132	1̄.90537	10	32
29	1̄.77422	17	1̄.86894	27	0.13106	1̄.90527	9	31
30	1̄.77439	17	1̄.86921	26	0.13079	1̄.90518	9	30
31	1̄.77456	17	1̄.86947	27	0.13053	1̄.90509	10	29
32	1̄.77473	17	1̄.86974	26	0.13026	1̄.90499	9	28
33	1̄.77490	17	1̄.87000	27	0.13000	1̄.90490	10	27
34	1̄.77507	17	1̄.87027	26	0.12973	1̄.90480	9	26
35	1̄.77524	17	1̄.87053	26	0.12947	1̄.90471	9	25
36	1̄.77541	17	1̄.87079	27	0.12921	1̄.90462	10	24
37	1̄.77558	17	1̄.87106	26	0.12894	1̄.90452	9	23
38	1̄.77575	17	1̄.87132	26	0.12868	1̄.90443	9	22
39	1̄.77592	17	1̄.87158	27	0.12842	1̄.90434	10	21
40	1̄.77609	17	1̄.87185	26	0.12815	1̄.90424	9	20
41	1̄.77626	17	1̄.87211	27	0.12789	1̄.90415	10	19
42	1̄.77643	17	1̄.87238	26	0.12762	1̄.90405	9	18
43	1̄.77660	17	1̄.87264	26	0.12736	1̄.90396	10	17
44	1̄.77677	17	1̄.87290	27	0.12710	1̄.90386	9	16
45	1̄.77694	17	1̄.87317	26	0.12683	1̄.90377	9	15
46	1̄.77711	17	1̄.87343	26	0.12657	1̄.90368	10	14
47	1̄.77728	16	1̄.87369	27	0.12631	1̄.90358	9	13
48	1̄.77744	17	1̄.87396	26	0.12604	1̄.90349	10	12
49	1̄.77761	17	1̄.87422	26	0.12578	1̄.90339	9	11
50	1̄.77778	17	1̄.87448	27	0.12552	1̄.90330	10	10
51	1̄.77795	17	1̄.87475	26	0.12525	1̄.90320	9	9
52	1̄.77812	17	1̄.87501	26	0.12499	1̄.90311	10	8
53	1̄.77829	17	1̄.87527	27	0.12473	1̄.90301	9	7
54	1̄.77846	16	1̄.87554	26	0.12446	1̄.90292	10	6
55	1̄.77862	17	1̄.87580	26	0.12420	1̄.90282	9	5
56	1̄.77879	17	1̄.87606	27	0.12394	1̄.90273	10	4
57	1̄.77896	17	1̄.87633	26	0.12367	1̄.90263	9	3
58	1̄.77913	17	1̄.87659	26	0.12341	1̄.90254	10	2
59	1̄.77930	16	1̄.87685	26	0.12315	1̄.90244	9	1
60	1̄.77946		1̄.87711		0.12289	1̄.90235		0

log cos	d	log cot	c. d.	log tan	log sin	d	′

p. p.

	27	28
6	2.7	2.6
7	3.2	3.0
8	3.6	3.5
9	4.1	3.9
10	4.5	4.3
20	9.0	8.7
30	13.5	13.0
40	18.0	17.3
50	22.5	21.7

	18
6	1.8
7	2.1
8	2.4
9	2.7
10	3.0
20	6.0
30	9.0
40	12.0
50	15.0

	17
6	1.7
7	2.0
8	2.3
9	2.6
10	2.8
20	5.7
30	8.5
40	11.3
50	14.2

	16
6	1.6
7	1.9
8	2.1
9	2.4
10	2.7
20	5.3
30	8.0
40	10.7
50	13.3

	10	9
6	1.0	0.9
7	1.2	1.1
8	1.3	1.2
9	1.5	1.4
10	1.7	1.5
20	3.3	3.0
30	5.0	4.5
40	6.7	6.0
50	8.3	7.5

′	log sin	d	log tan	c. d.	log cot	log cos	d	
0	1̅.77946	17	1̅.87711	27	0.12289	1̅.90235	10	60
1	1̅.77963	17	1̅.87738	26	0.12262	1̅.90225	9	59
2	1̅.77980	17	1̅.87764	26	0.12236	1̅.90216	10	58
3	1̅.77997	16	1̅.87790	27	0.12210	1̅.90206	9	57
4	1̅.78013	17	1̅.87817	26	0.12183	1̅.90197	10	56
5	1̅.78030	17	1̅.87843	26	0.12157	1̅.90187	9	55
6	1̅.78047	16	1̅.87869	26	0.12131	1̅.90178	10	54
7	1̅.78063	17	1̅.87895	27	0.12105	1̅.90168	9	53
8	1̅.78080	17	1̅.87922	26	0.12078	1̅.90159	10	52
9	1̅.78097	16	1̅.87948	26	0.12052	1̅.90149	10	51
10	1̅.78113	17	1̅.87974	26	0.12026	1̅.90139	9	50
11	1̅.78130	17	1̅.88000	27	0.12000	1̅.90130	10	49
12	1̅.78147	16	1̅.88027	26	0.11973	1̅.90120	9	48
13	1̅.78163	17	1̅.88053	26	0.11947	1̅.90111	10	47
14	1̅.78180	17	1̅.88079	26	0.11921	1̅.90101	10	46
15	1̅.78197	16	1̅.88105	26	0.11895	1̅.90091	9	45
16	1̅.78213	17	1̅.88131	27	0.11869	1̅.90082	10	44
17	1̅.78230	16	1̅.88158	26	0.11842	1̅.90072	9	43
18	1̅.78246	17	1̅.88184	26	0.11816	1̅.90063	10	42
19	1̅.78263	17	1̅.88210	26	0.11790	1̅.90053	10	41
20	1̅.78280	16	1̅.88236	26	0.11764	1̅.90043	9	40
21	1̅.78296	17	1̅.88262	27	0.11738	1̅.90034	10	39
22	1̅.78313	16	1̅.88289	26	0.11711	1̅.90024	10	38
23	1̅.78329	17	1̅.88315	26	0.11685	1̅.90014	9	37
24	1̅.78346	16	1̅.88341	26	0.11659	1̅.90005	10	36
25	1̅.78362	17	1̅.88367	26	0.11633	1̅.89995	10	35
26	1̅.78379	16	1̅.88393	27	0.11607	1̅.89985	9	34
27	1̅.78395	17	1̅.88420	26	0.11580	1̅.89976	10	33
28	1̅.78412	16	1̅.88446	26	0.11554	1̅.89966	10	32
29	1̅.78428	17	1̅.88472	26	0.11528	1̅.89956	9	31
30	1̅.78445	16	1̅.88498	26	0.11502	1̅.89947	10	30
31	1̅.78461	17	1̅.88524	26	0.11476	1̅.89937	10	29
32	1̅.78478	16	1̅.88550	27	0.11450	1̅.89927	9	28
33	1̅.78494	16	1̅.88577	26	0.11423	1̅.89918	10	27
34	1̅.78510	17	1̅.88603	26	0.11397	1̅.89908	10	26
35	1̅.78527	16	1̅.88629	26	0.11371	1̅.89898	10	25
36	1̅.78543	17	1̅.88655	26	0.11345	1̅.89888	9	24
37	1̅.78560	16	1̅.88681	26	0.11319	1̅.89879	10	23
38	1̅.78576	16	1̅.88707	26	0.11293	1̅.89869	10	22
39	1̅.78592	17	1̅.88733	26	0.11267	1̅.89859	10	21
40	1̅.78609	16	1̅.88759	27	0.11241	1̅.89849	9	20
41	1̅.78625	17	1̅.88786	26	0.11214	1̅.89840	10	19
42	1̅.78642	16	1̅.88812	26	0.11188	1̅.89830	10	18
43	1̅.78658	16	1̅.88838	26	0.11162	1̅.89820	10	17
44	1̅.78674	17	1̅.88864	26	0.11136	1̅.89810	9	16
45	1̅.78691	16	1̅.88890	26	0.11110	1̅.89801	10	15
46	1̅.78707	16	1̅.88916	26	0.11084	1̅.89791	10	14
47	1̅.78723	16	1̅.88942	26	0.11058	1̅.89781	10	13
48	1̅.78739	17	1̅.88968	26	0.11032	1̅.89771	10	12
49	1̅.78756	16	1̅.88994	26	0.11006	1̅.89761	9	11
50	1̅.78772	16	1̅.89020	26	0.10980	1̅.89752	10	10
51	1̅.78788	17	1̅.89046	27	0.10954	1̅.89742	10	9
52	1̅.78805	16	1̅.89073	26	0.10927	1̅.89732	10	8
53	1̅.78821	16	1̅.89099	26	0.10901	1̅.89722	10	7
54	1̅.78837	16	1̅.89125	26	0.10875	1̅.89712	10	6
55	1̅.78853	16	1̅.89151	26	0.10849	1̅.89702	9	5
56	1̅.78869	17	1̅.89177	26	0.10823	1̅.89693	10	4
57	1̅.78886	16	1̅.89203	26	0.10797	1̅.89683	10	3
58	1̅.78902	16	1̅.89229	26	0.10771	1̅.89673	10	2
59	1̅.78918	16	1̅.89255	26	0.10745	1̅.89663	10	1
60	1̅.78934		1̅.89281		0.10719	1̅.89653		0
	log cos	d	log cot	c. d.	log tan	log sin	d	′

p. p.

27
6	2.7
7	3.2
8	3.6
9	4.1
10	4.5
20	9.0
30	13.5
40	18.0
50	22.5

26
6	2.6
7	3.0
8	3.5
9	3.9
10	4.3
20	8.7
30	13.0
40	17.3
50	21.7

17
6	1.7
7	2.0
8	2.3
9	2.6
10	2.8
20	5.7
30	8.5
40	11.3
50	14.2

16
6	1.6
7	1.9
8	2.1
9	2.4
10	2.7
20	5.3
30	8.0
40	10.7
50	13.3

	10	**9**
6	1.0	0.9
7	1.2	1.1
8	1.3	1.2
9	1.5	1.4
10	1.7	1.5
20	3.3	3.0
30	5.0	4.5
40	6.7	6.0
50	8.3	7.5

′	log sin	d	log tan	c.d.	log cot	log cos	d	′
0	Ī.78934		Ī.89281		0.10719	Ī.89653		60
1	Ī.78950	16	Ī.89307	26	0.10693	Ī.89643	10	59
2	Ī.78967	17	Ī.89333	26	0.10667	Ī.89633	10	58
3	Ī.78983	16	Ī.89359	26	0.10641	Ī.89624	9	57
4	Ī.78999	16	Ī.89385	26	0.10615	Ī.89614	10	56
5	Ī.79015	16	Ī.89411	26	0.10589	Ī.89604	10	55
6	Ī.79031	16	Ī.89437	26	0.10563	Ī.89594	10	54
7	Ī.79047	16	Ī.89463	26	0.10537	Ī.89584	10	53
8	Ī.79063	16	Ī.89489	26	0.10511	Ī.89574	10	52
9	Ī.79079	16	Ī.89515	26	0.10485	Ī.89564	10	51
10	Ī.79095	16	Ī.89541	26	0.10459	Ī.89554	10	50
11	Ī.79111	17	Ī.89567	26	0.10433	Ī.89544	10	49
12	Ī.79128	16	Ī.89593	26	0.10407	Ī.89534	10	48
13	Ī.79144	16	Ī.89619	26	0.10381	Ī.89524	10	47
14	Ī.79160	16	Ī.89645	26	0.10355	Ī.89514	10	46
15	Ī.79176	16	Ī.89671	26	0.10329	Ī.89504	10	45
16	Ī.79192	16	Ī.89697	26	0.10303	Ī.89495	9	44
17	Ī.79208	16	Ī.89723	26	0.10277	Ī.89485	10	43
18	Ī.79224	16	Ī.89749	26	0.10251	Ī.89475	10	42
19	Ī.79240	16	Ī.89775	26	0.10225	Ī.89465	10	41
20	Ī.79256	16	Ī.89801	26	0.10199	Ī.89455	10	40
21	Ī.79272	16	Ī.89827	26	0.10173	Ī.89445	10	39
22	Ī.79288	16	Ī.89853	26	0.10147	Ī.89435	10	38
23	Ī.79304	15	Ī.89879	26	0.10121	Ī.89425	10	37
24	Ī.79319	16	Ī.89905	26	0.10095	Ī.89415	10	36
25	Ī.79335	16	Ī.89931	26	0.10069	Ī.89405	10	35
26	Ī.79351	16	Ī.89957	26	0.10043	Ī.89395	10	34
27	Ī.79367	16	Ī.89983	26	0.10017	Ī.89385	10	33
28	Ī.79383	16	Ī.90009	26	0.09991	Ī.89375	10	32
29	Ī.79399	16	Ī.90035	26	0.09965	Ī.89364	11	31
30	Ī.79415	16	Ī.90061	26	0.09939	Ī.89354	10	30
31	Ī.79431	16	Ī.90086	25	0.09914	Ī.89344	10	29
32	Ī.79447	16	Ī.90112	26	0.09888	Ī.89334	10	28
33	Ī.79463	15	Ī.90138	26	0.09862	Ī.89324	10	27
34	Ī.79478	16	Ī.90164	26	0.09836	Ī.89314	10	26
35	Ī.79494	16	Ī.90190	26	0.09810	Ī.89304	10	25
36	Ī.79510	16	Ī.90216	26	0.09784	Ī.89294	10	24
37	Ī.79526	16	Ī.90242	26	0.09758	Ī.89284	10	23
38	Ī.79542	16	Ī.90268	26	0.09732	Ī.89274	10	22
39	Ī.79558	15	Ī.90294	26	0.09706	Ī.89264	10	21
40	Ī.79573	16	Ī.90320	26	0.09680	Ī.89254	10	20
41	Ī.79589	16	Ī.90346	25	0.09654	Ī.89244	11	19
42	Ī.79605	16	Ī.90371	26	0.09629	Ī.89233	10	18
43	Ī.79621	15	Ī.90397	26	0.09603	Ī.89223	10	17
44	Ī.79636	16	Ī.90423	26	0.09577	Ī.89213	10	16
45	Ī.79652	16	Ī.90449	26	0.09551	Ī.89203	10	15
46	Ī.79668	16	Ī.90475	26	0.09525	Ī.89193	10	14
47	Ī.79684	15	Ī.90501	26	0.09499	Ī.89183	10	13
48	Ī.79699	16	Ī.90527	26	0.09473	Ī.89173	11	12
49	Ī.79715	16	Ī.90553	25	0.09447	Ī.89162	10	11
50	Ī.79731	15	Ī.90578	26	0.09422	Ī.89152	10	10
51	Ī.79746	16	Ī.90604	26	0.09396	Ī.89142	10	9
52	Ī.79762	16	Ī.90630	26	0.09370	Ī.89132	10	8
53	Ī.79778	15	Ī.90656	26	0.09344	Ī.89122	10	7
54	Ī.79793	16	Ī.90682	26	0.09318	Ī.89112	11	6
55	Ī.79809	16	Ī.90708	26	0.09292	Ī.89101	10	5
56	Ī.79825	15	Ī.90734	25	0.09266	Ī.89091	10	4
57	Ī.79840	16	Ī.90759	26	0.09241	Ī.89081	10	3
58	Ī.79856	16	Ī.90785	26	0.09215	Ī.89071	11	2
59	Ī.79872	15	Ī.90811	26	0.09189	Ī.89060	10	1
60	Ī.79887		Ī.90837		0.09163	Ī.89050		0
	log cos	d	log cot	c. d.	log tan	log sin	d	′

p. p.

	26	25
6	2.6	2.5
7	3.0	2.9
8	3.5	3.3
9	3.9	3.8
10	4.3	4.2
20	8.7	8.3
30	13.0	12.5
40	17.3	16.7
50	21.7	20.8

	17
6	1.7
7	2.0
8	2.3
9	2.6
10	2.8
20	5.7
30	8.5
40	11.3
50	14.2

	16	15
6	1.6	1.5
7	1.9	1.8
8	2.1	2.0
9	2.4	2.3
10	2.7	2.5
20	5.3	5.0
30	8.0	7.5
40	10.7	10.0
50	13.3	12.5

	11
6	1.1
7	1.3
8	1.5
9	1.7
10	1.8
20	3.7
30	5.5
40	7.3
50	9.2

	10	9
6	1.0	0.9
7	1.2	1.1
8	1.3	1.2
9	1.5	1.4
10	1.7	1.5
20	3.3	3.0
30	5.0	4.5
40	6.7	6.0
50	8.3	7.5

p. p.

	26
6	2.6
7	3.0
8	3.5
9	3.9
10	4.3
20	8.7
30	13.0
40	17.3
50	21.7

	25
6	2.5
7	2.9
8	3.3
9	3.8
10	4.2
20	8.3
30	12.5
40	16.7
50	20.8

	16
6	1.6
7	1.9
8	2.1
9	2.4
10	2.7
20	5.3
30	8.0
40	10.7
50	13.3

	15
6	1.5
7	1.8
8	2.0
9	2.3
10	2.5
20	5.0
30	7.5
40	10.0
50	12.5

	11	10
6	1.1	1.0
7	1.3	1.2
8	1.5	1.3
9	1.7	1.5
10	1.8	1.7
20	3.7	3.3
30	5.5	5.0
40	7.3	6.7
50	9.2	8.3

40°

′	log sin	d	log tan	c. d.	log cot	log cos	d	′
0	1̄.80807	15	1̄.92381	26	0.07619	1̄.88425	10	60
1	1̄.80822	15	1̄.92407	26	0.07593	1̄.88415	11	59
2	1̄.80837	15	1̄.92433	25	0.07567	1̄.88404	10	58
3	1̄.80852	15	1̄.92458	26	0.07542	1̄.88394	11	57
4	1̄.80867	15	1̄.92484	26	0.07516	1̄.88383	11	56
5	1̄.80882	15	1̄.92510	25	0.07490	1̄.88372	10	55
6	1̄.80897	15	1̄.92535	26	0.07465	1̄.88362	11	54
7	1̄.80912	15	1̄.92561	26	0.07439	1̄.88351	11	53
8	1̄.80927	15	1̄.92587	25	0.07413	1̄.88340	10	52
9	1̄.80942	15	1̄.92612	26	0.07388	1̄.88330	11	51
10	1̄.80957	15	1̄.92638	25	0.07362	1̄.88319	11	50
11	1̄.80972	15	1̄.92663	26	0.07337	1̄.88308	10	49
12	1̄.80987	15	1̄.92689	26	0.07311	1̄.88298	11	48
13	1̄.81002	15	1̄.92715	25	0.07285	1̄.88287	11	47
14	1̄.81017	15	1̄.92740	26	0.07260	1̄.88276	10	46
15	1̄.81032	15	1̄.92766	26	0.07234	1̄.88266	11	45
16	1̄.81047	14	1̄.92792	25	0.07208	1̄.88255	11	44
17	1̄.81061	15	1̄.92817	26	0.07183	1̄.88244	10	43
18	1̄.81076	15	1̄.92843	25	0.07157	1̄.88234	11	42
19	1̄.81091	15	1̄.92868	26	0.07132	1̄.88223	11	41
20	1̄.81106	15	1̄.92894	26	0.07106	1̄.88212	11	40
21	1̄.81121	15	1̄.92920	25	0.07080	1̄.88201	10	39
22	1̄.81136	15	1̄.92945	26	0.07055	1̄.88191	11	38
23	1̄.81151	15	1̄.92971	25	0.07029	1̄.88180	11	37
24	1̄.81166	14	1̄.92996	26	0.07004	1̄.88169	11	36
25	1̄.81180	15	1̄.93022	26	0.06978	1̄.88158	10	35
26	1̄.81195	15	1̄.93048	25	0.06952	1̄.88148	11	34
27	1̄.81210	15	1̄.93073	26	0.06927	1̄.88137	11	33
28	1̄.81225	15	1̄.93099	25	0.06901	1̄.88126	11	32
29	1̄.81240	14	1̄.93124	26	0.06876	1̄.88115	10	31
30	1̄.81254	15	1̄.93150	25	0.06850	1̄.88105	11	30
31	1̄.81269	15	1̄.93175	26	0.06825	1̄.88094	11	29
32	1̄.81284	15	1̄.93201	26	0.06799	1̄.88083	11	28
33	1̄.81299	15	1̄.93227	25	0.06773	1̄.88072	11	27
34	1̄.81314	14	1̄.93252	26	0.06748	1̄.88061	10	26
35	1̄.81328	15	1̄.93278	25	0.06722	1̄.88051	11	25
36	1̄.81343	15	1̄.93303	26	0.06697	1̄.88040	11	24
37	1̄.81358	14	1̄.93329	25	0.06671	1̄.88029	11	23
38	1̄.81372	15	1̄.93354	26	0.06646	1̄.88018	11	22
39	1̄.81387	15	1̄.93380	26	0.06620	1̄.88007	11	21
40	1̄.81402	15	1̄.93406	25	0.06594	1̄.87996	11	20
41	1̄.81417	14	1̄.93431	26	0.06569	1̄.87985	10	19
42	1̄.81431	15	1̄.93457	25	0.06543	1̄.87975	11	18
43	1̄.81446	15	1̄.93482	26	0.06518	1̄.87964	11	17
44	1̄.81461	14	1̄.93508	25	0.06492	1̄.87953	11	16
45	1̄.81475	15	1̄.93533	26	0.06467	1̄.87942	11	15
46	1̄.81490	15	1̄.93559	25	0.06441	1̄.87931	11	14
47	1̄.81505	14	1̄.93584	26	0.06416	1̄.87920	11	13
48	1̄.81519	15	1̄.93610	26	0.06390	1̄.87909	11	12
49	1̄.81534	15	1̄.93636	25	0.06364	1̄.87898	11	11
50	1̄.81549	14	1̄.93661	26	0.06339	1̄.87887	10	10
51	1̄.81563	15	1̄.93687	25	0.06313	1̄.87877	11	9
52	1̄.81578	14	1̄.93712	26	0.06288	1̄.87866	11	8
53	1̄.81592	15	1̄.93738	25	0.06262	1̄.87855	11	7
54	1̄.81607	15	1̄.93763	26	0.06237	1̄.87844	11	6
55	1̄.81622	14	1̄.93789	25	0.06211	1̄.87833	11	5
56	1̄.81636	15	1̄.93814	26	0.06186	1̄.87822	11	4
57	1̄.81651	14	1̄.93840	25	0.06160	1̄.87811	11	3
58	1̄.81665	15	1̄.93865	26	0.06135	1̄.87800	11	2
59	1̄.81680	14	1̄.93891	25	0.06109	1̄.87789	11	1
60	1̄.81694		1̄.93916		0.06084	1̄.87778		0
	log cos	d	log cot	c. d.	log tan	log sin	d	′

p. p.

26

6	2.6
7	3.0
8	3.5
9	3.9
10	4.3
20	8.7
30	13.0
40	17.3
50	21.7

25

6	2.5
7	2.9
8	3.3
9	3.8
10	4.2
20	8.3
30	12.5
40	16.7
50	20.8

15

6	1.5
7	1.8
8	2.0
9	2.3
10	2.5
20	5.0
30	7.5
40	10.0
50	12.5

14

6	1.4
7	1.6
8	1.9
9	2.1
10	2.3
20	4.7
30	7.0
40	9.3
50	11.7

	11	10
6	1.1	1.0
7	1.3	1.2
8	1.5	1.3
9	1.7	1.5
10	1.8	1.7
20	3.7	3.3
30	5.5	5.0
40	7.3	6.7
50	9.2	8.3

′	log sin	d	log tan	c. d.	log cot	log cos	d	′
0	1̄.81694	15	1̄.93916	26	0.06084	1̄.87778	11	60
1	1̄.81709	14	1̄.93942	25	0.06058	1̄.87767	11	59
2	1̄.81723	15	1̄.93967	26	0.06033	1̄.87756	11	58
3	1̄.81738	14	1̄.93993	25	0.06007	1̄.87745	11	57
4	1̄.81752	15	1̄.94018	26	0.05982	1̄.87734	11	56
5	1̄.81767	14	1̄.94044	25	0.05956	1̄.87723	11	55
6	1̄.81781	15	1̄.94069	26	0.05931	1̄.87712	11	54
7	1̄.81796	14	1̄.94095	25	0.05905	1̄.87701	11	53
8	1̄.81810	15	1̄.94120	26	0.05880	1̄.87690	11	52
9	1̄.81825	14	1̄.94146	25	0.05854	1̄.87679	11	51
10	1̄.81839	15	1̄.94171	26	0.05829	1̄.87668	11	50
11	1̄.81854	14	1̄.94197	25	0.05803	1̄.87657	11	49
12	1̄.81868	14	1̄.94222	26	0.05778	1̄.87646	11	48
13	1̄.81882	15	1̄.94248	25	0.05752	1̄.87635	11	47
14	1̄.81897	14	1̄.94273	26	0.05727	1̄.87624	11	46
15	1̄.81911	15	1̄.94299	25	0.05701	1̄.87613	12	45
16	1̄.81926	14	1̄.94324	26	0.05676	1̄.87601	11	44
17	1̄.81940	15	1̄.94350	25	0.05650	1̄.87590	11	43
18	1̄.81955	14	1̄.94375	26	0.05625	1̄.87579	11	42
19	1̄.81969	14	1̄.94401	25	0.05599	1̄.87568	11	41
20	1̄.81983	15	1̄.94426	26	0.05574	1̄.87557	11	40
21	1̄.81998	14	1̄.94452	25	0.05548	1̄.87546	11	39
22	1̄.82012	14	1̄.94477	26	0.05523	1̄.87535	11	38
23	1̄.82026	15	1̄.94503	25	0.05497	1̄.87524	11	37
24	1̄.82041	14	1̄.94528	26	0.05472	1̄.87513	12	36
25	1̄.82055	14	1̄.94554	25	0.05446	1̄.87501	11	35
26	1̄.82069	15	1̄.94579	25	0.05421	1̄.87490	11	34
27	1̄.82084	14	1̄.94604	26	0.05396	1̄.87479	11	33
28	1̄.82098	14	1̄.94630	25	0.05370	1̄.87468	11	32
29	1̄.82112	14	1̄.94655	26	0.05345	1̄.87457	11	31
30	1̄.82126	15	1̄.94681	25	0.05319	1̄.87446	12	30
31	1̄.82141	14	1̄.94706	26	0.05294	1̄.87434	11	29
32	1̄.82155	14	1̄.94732	25	0.05268	1̄.87423	11	28
33	1̄.82169	15	1̄.94757	26	0.05243	1̄.87412	11	27
34	1̄.82184	14	1̄.94783	25	0.05217	1̄.87401	11	26
35	1̄.82198	14	1̄.94808	26	0.05192	1̄.87390	12	25
36	1̄.82212	14	1̄.94834	25	0.05166	1̄.87378	11	24
37	1̄.82226	14	1̄.94859	25	0.05141	1̄.87367	11	23
38	1̄.82240	15	1̄.94884	26	0.05116	1̄.87356	11	22
39	1̄.82255	14	1̄.94910	25	0.05090	1̄.87345	11	21
40	1̄.82269	14	1̄.94935	26	0.05065	1̄.87334	12	20
41	1̄.82283	14	1̄.94961	25	0.05039	1̄.87322	11	19
42	1̄.82297	14	1̄.94986	26	0.05014	1̄.87311	11	18
43	1̄.82311	15	1̄.95012	25	0.04988	1̄.87300	12	17
44	1̄.82326	14	1̄.95037	25	0.04963	1̄.87288	11	16
45	1̄.82340	14	1̄.95062	26	0.04938	1̄.87277	11	15
46	1̄.82354	14	1̄.95088	25	0.04912	1̄.87266	11	14
47	1̄.82368	14	1̄.95113	26	0.04887	1̄.87255	12	13
48	1̄.82382	14	1̄.95139	25	0.04861	1̄.87243	11	12
49	1̄.82396	14	1̄.95164	26	0.04836	1̄.87232	11	11
50	1̄.82410	14	1̄.95190	25	0.04810	1̄.87221	12	10
51	1̄.82424	15	1̄.95215	25	0.04785	1̄.87209	11	9
52	1̄.82439	14	1̄.95240	26	0.04760	1̄.87198	11	8
53	1̄.82453	14	1̄.95266	25	0.04734	1̄.87187	12	7
54	1̄.82467	14	1̄.95291	26	0.04709	1̄.87175	11	6
55	1̄.82481	14	1̄.95317	25	0.04683	1̄.87164	11	5
56	1̄.82495	14	1̄.95342	26	0.04658	1̄.87153	12	4
57	1̄.82509	14	1̄.95368	25	0.04632	1̄.87141	11	3
58	1̄.82523	14	1̄.95393	25	0.04607	1̄.87130	11	2
59	1̄.82537	14	1̄.95418	26	0.04582	1̄.87119	12	1
60	1̄.82551		1̄.95444		0.04556	1̄.87107		0
	log cos	d	log cot	c. d.	log tan	log sin	d	′

p. p.

26	
6	2.6
7	3.0
8	3.5
9	3.9
10	4.3
20	8.7
30	13.0
40	17.3
50	21.7

25	
6	2.5
7	2.9
8	3.3
9	3.8
10	4.2
20	8.3
30	12.5
40	16.7
50	20.8

15	
6	1.5
7	1.8
8	2.0
9	2.3
10	2.5
20	5.0
30	7.5
40	10.0
50	12.5

14	
6	1.4
7	1.6
8	1.9
9	2.1
10	2.3
20	4.7
30	7.0
40	9.3
50	11.7

	12	11
6	1.2	1.1
7	1.4	1.3
8	1.6	1.5
9	1.8	1.7
10	2.0	1.8
20	4.0	3.7
30	6.0	5.5
40	8.0	7.3
50	10.0	9.2

′	log sin	d	log tan	c. d.	log cot	log cos	d	′
0	$\bar{1}$.82551	14	$\bar{1}$.95444	25	0.04556	$\bar{1}$.87107	11	60
1	$\bar{1}$.82565	14	$\bar{1}$.95469	26	0.04531	$\bar{1}$.87096	11	59
2	$\bar{1}$.82579	14	$\bar{1}$.95495	25	0.04505	$\bar{1}$.87085	12	58
3	$\bar{1}$.82593	14	$\bar{1}$.95520	25	0.04480	$\bar{1}$.87073	11	57
4	$\bar{1}$.82607	14	$\bar{1}$.95545	26	0.04455	$\bar{1}$.87062	12	56
5	$\bar{1}$.82621	14	$\bar{1}$.95571	25	0.04429	$\bar{1}$.87050	11	55
6	$\bar{1}$.82635	14	$\bar{1}$.95596	26	0.04404	$\bar{1}$.87039	11	54
7	$\bar{1}$.82649	14	$\bar{1}$.95622	25	0.04378	$\bar{1}$.87028	12	53
8	$\bar{1}$.82663	14	$\bar{1}$.95647	25	0.04353	$\bar{1}$.87016	11	52
9	$\bar{1}$.82677	14	$\bar{1}$.95672	26	0.04328	$\bar{1}$.87005	12	51
10	$\bar{1}$.82691	14	$\bar{1}$.95698	25	0.04302	$\bar{1}$.86993	11	50
11	$\bar{1}$.82705	14	$\bar{1}$.95723	25	0.04277	$\bar{1}$.86982	12	49
12	$\bar{1}$.82719	14	$\bar{1}$.95748	26	0.04252	$\bar{1}$.86970	11	48
13	$\bar{1}$.82733	14	$\bar{1}$.95774	25	0.04226	$\bar{1}$.86959	12	47
14	$\bar{1}$.82747	14	$\bar{1}$.95799	26	0.04201	$\bar{1}$.86947	11	46
15	$\bar{1}$.82761	14	$\bar{1}$.95825	25	0.04175	$\bar{1}$.86936	12	45
16	$\bar{1}$.82775	13	$\bar{1}$.95850	25	0.04150	$\bar{1}$.86924	11	44
17	$\bar{1}$.82788	14	$\bar{1}$.95875	26	0.04125	$\bar{1}$.86913	11	43
18	$\bar{1}$.82802	14	$\bar{1}$.95901	25	0.04099	$\bar{1}$.86902	12	42
19	$\bar{1}$.82816	14	$\bar{1}$.95926	26	0.04074	$\bar{1}$.86890	11	41
20	$\bar{1}$.82830	14	$\bar{1}$.95952	25	0.04048	$\bar{1}$.86879	12	40
21	$\bar{1}$.82844	14	$\bar{1}$.95977	25	0.04023	$\bar{1}$.86867	12	39
22	$\bar{1}$.82858	14	$\bar{1}$.96002	26	0.03998	$\bar{1}$.86855	11	38
23	$\bar{1}$.82872	13	$\bar{1}$.96028	25	0.03972	$\bar{1}$.86844	12	37
24	$\bar{1}$.82885	14	$\bar{1}$.96053	25	0.03947	$\bar{1}$.86832	11	36
25	$\bar{1}$.82899	14	$\bar{1}$.96078	26	0.03922	$\bar{1}$.86821	12	35
26	$\bar{1}$.82913	14	$\bar{1}$.96104	25	0.03896	$\bar{1}$.86809	11	34
27	$\bar{1}$.82927	14	$\bar{1}$.96129	26	0.03871	$\bar{1}$.86798	12	33
28	$\bar{1}$.82941	14	$\bar{1}$.96155	25	0.03845	$\bar{1}$.86786	11	32
29	$\bar{1}$.82955	14	$\bar{1}$.96180	25	0.03820	$\bar{1}$.86775	12	31
30	$\bar{1}$.82968	13	$\bar{1}$.96205	26	0.03795	$\bar{1}$.86763	11	30
31	$\bar{1}$.82982	14	$\bar{1}$.96231	25	0.03769	$\bar{1}$.86752	12	29
32	$\bar{1}$.82996	14	$\bar{1}$.96256	25	0.03744	$\bar{1}$.86740	12	28
33	$\bar{1}$.83010	14	$\bar{1}$.96281	26	0.03719	$\bar{1}$.86728	11	27
34	$\bar{1}$.83023	13	$\bar{1}$.96307	25	0.03693	$\bar{1}$.86717	12	26
35	$\bar{1}$.83037	14	$\bar{1}$.96332	25	0.03668	$\bar{1}$.86705	11	25
36	$\bar{1}$.83051	14	$\bar{1}$.96357	26	0.03643	$\bar{1}$.86694	12	24
37	$\bar{1}$.83065	14	$\bar{1}$.96383	25	0.03617	$\bar{1}$.86682	12	23
38	$\bar{1}$.83078	13	$\bar{1}$.96408	25	0.03592	$\bar{1}$.86670	11	22
39	$\bar{1}$.83092	14	$\bar{1}$.96433	26	0.03567	$\bar{1}$.86659	12	21
40	$\bar{1}$.83106	14	$\bar{1}$.96459	25	0.03541	$\bar{1}$.86647	12	20
41	$\bar{1}$.83120	13	$\bar{1}$.96484	26	0.03516	$\bar{1}$.86635	11	19
42	$\bar{1}$.83133	14	$\bar{1}$.96510	25	0.03490	$\bar{1}$.86624	12	18
43	$\bar{1}$.83147	14	$\bar{1}$.96535	25	0.03465	$\bar{1}$.86612	12	17
44	$\bar{1}$.83161	13	$\bar{1}$.96560	26	0.03440	$\bar{1}$.86600	11	16
45	$\bar{1}$.83174	14	$\bar{1}$.96586	25	0.03414	$\bar{1}$.86589	12	15
46	$\bar{1}$.83188	14	$\bar{1}$.96611	25	0.03389	$\bar{1}$.86577	12	14
47	$\bar{1}$.83202	13	$\bar{1}$.96636	26	0.03364	$\bar{1}$.86565	11	13
48	$\bar{1}$.83215	14	$\bar{1}$.96662	25	0.03338	$\bar{1}$.86554	12	12
49	$\bar{1}$.83229	13	$\bar{1}$.96687	25	0.03313	$\bar{1}$.86542	12	11
50	$\bar{1}$.83242	14	$\bar{1}$.96712	26	0.03288	$\bar{1}$.86530	12	10
51	$\bar{1}$.83256	14	$\bar{1}$.96738	25	0.03262	$\bar{1}$.86518	11	9
52	$\bar{1}$.83270	13	$\bar{1}$.96763	25	0.03237	$\bar{1}$.86507	12	8
53	$\bar{1}$.83283	14	$\bar{1}$.96788	26	0.03212	$\bar{1}$.86495	12	7
54	$\bar{1}$.83297	13	$\bar{1}$.96814	25	0.03186	$\bar{1}$.86483	11	6
55	$\bar{1}$.83310	14	$\bar{1}$.96839	25	0.03161	$\bar{1}$.86472	12	5
56	$\bar{1}$.83324	14	$\bar{1}$.96864	26	0.03136	$\bar{1}$.86460	12	4
57	$\bar{1}$.83338	13	$\bar{1}$.96890	25	0.03110	$\bar{1}$.86448	12	3
58	$\bar{1}$.83351	14	$\bar{1}$.96915	25	0.03085	$\bar{1}$.86436	11	2
59	$\bar{1}$.83365	13	$\bar{1}$.96940	26	0.03060	$\bar{1}$.86425	12	1
60	$\bar{1}$.83378		$\bar{1}$.96966		0.03034	$\bar{1}$.86413		0
	log cos	d	log cot	c. d.	log tan	log sin	d	′

p. p.

26	
6	2.6
7	3.0
8	3.5
9	3.9
10	4.3
20	8.7
30	13.0
40	17.3
50	21.7

25	
6	2.5
7	2.9
8	3.3
9	3.8
10	4.2
20	8.3
30	12.5
40	16.7
50	20.8

14	
6	1.4
7	1.6
8	1.9
9	2.1
10	2.3
20	4.7
30	7.0
40	9.3
50	11.7

13	
6	1.3
7	1.5
8	1.7
9	2.0
10	2.2
20	4.3
30	6.5
40	8.7
50	10.8

	12	11
6	1.2	1.1
7	1.4	1.3
8	1.6	1.5
9	1.8	1.7
10	2.0	1.8
20	4.0	3.7
30	6.0	5.5
40	8.0	7.3
50	10.0	9.2

′	log sin	d	log tan	c. d.	log cot	log cos	d	
0	1̄.83378	14	1̄.96966	25	0.03034	1̄.86413	12	60
1	1̄.83392	13	1̄.96991	25	0.03009	1̄.86401	12	59
2	1̄.83405	14	1̄.97016	26	0.02984	1̄.86389	12	58
3	1̄.83419	13	1̄.97042	25	0.02958	1̄.86377	11	57
4	1̄.83432	14	1̄.97067	25	0.02933	1̄.86366	12	56
5	1̄.83446	13	1̄.97092	26	0.02908	1̄.86354	12	55
6	1̄.83459	14	1̄.97118	25	0.02882	1̄.86342	12	54
7	1̄.83473	13	1̄.97143	25	0.02857	1̄.86330	12	53
8	1̄.83486	14	1̄.97168	25	0.02832	1̄.86318	12	52
9	1̄.83500	13	1̄.97193	26	0.02807	1̄.86306	11	51
10	1̄.83513	14	1̄.97219	25	0.02781	1̄.86295	12	50
11	1̄.83527	13	1̄.97244	25	0.02756	1̄.86283	12	49
12	1̄.83540	14	1̄.97269	26	0.02731	1̄.86271	12	48
13	1̄.83554	13	1̄.97295	25	0.02705	1̄.86259	12	47
14	1̄.83567	14	1̄.97320	25	0.02680	1̄.86247	12	46
15	1̄.83581	13	1̄.97345	26	0.02655	1̄.86235	12	45
16	1̄.83594	14	1̄.97371	25	0.02629	1̄.86223	12	44
17	1̄.83608	13	1̄.97396	25	0.02604	1̄.86211	11	43
18	1̄.83621	13	1̄.97421	26	0.02579	1̄.86200	12	42
19	1̄.83634	14	1̄.97447	25	0.02553	1̄.86188	12	41
20	1̄.83648	13	1̄.97472	25	0.02528	1̄.86176	12	40
21	1̄.83661	13	1̄.97497	26	0.02503	1̄.86164	12	39
22	1̄.83674	14	1̄.97523	25	0.02477	1̄.86152	12	38
23	1̄.83688	13	1̄.97548	25	0.02452	1̄.86140	12	37
24	1̄.83701	14	1̄.97573	25	0.02427	1̄.86128	12	36
25	1̄.83715	13	1̄.97598	26	0.02402	1̄.86116	12	35
26	1̄.83728	13	1̄.97624	25	0.02376	1̄.86104	12	34
27	1̄.83741	14	1̄.97649	25	0.02351	1̄.86092	12	33
28	1̄.83755	13	1̄.97674	26	0.02326	1̄.86080	12	32
29	1̄.83768	13	1̄.97700	25	0.02300	1̄.86068	12	31
30	1̄.83781	14	1̄.97725	25	0.02275	1̄.86056	12	30
31	1̄.83795	13	1̄.97750	26	0.02250	1̄.86044	12	29
32	1̄.83808	13	1̄.97776	25	0.02224	1̄.86032	12	28
33	1̄.83821	13	1̄.97801	25	0.02199	1̄.86020	12	27
34	1̄.83834	14	1̄.97826	25	0.02174	1̄.86008	12	26
35	1̄.83848	13	1̄.97851	26	0.02149	1̄.85996	12	25
36	1̄.83861	13	1̄.97877	25	0.02123	1̄.85984	12	24
37	1̄.83874	13	1̄.97902	25	0.02098	1̄.85972	12	23
38	1̄.83887	14	1̄.97927	26	0.02073	1̄.85960	12	22
39	1̄.83901	13	1̄.97953	25	0.02047	1̄.85948	12	21
40	1̄.83914	13	1̄.97978	25	0.02022	1̄.85936	12	20
41	1̄.83927	13	1̄.98003	26	0.01997	1̄.85924	12	19
42	1̄.83940	14	1̄.98029	25	0.01971	1̄.85912	12	18
43	1̄.83954	13	1̄.98054	25	0.01946	1̄.85900	12	17
44	1̄.83967	13	1̄.98079	25	0.01921	1̄.85888	12	16
45	1̄.83980	13	1̄.98104	26	0.01896	1̄.85876	12	15
46	1̄.83993	13	1̄.98130	25	0.01870	1̄.85864	13	14
47	1̄.84006	14	1̄.98155	25	0.01845	1̄.85851	12	13
48	1̄.84020	13	1̄.98180	26	0.01820	1̄.85839	12	12
49	1̄.84033	13	1̄.98206	25	0.01794	1̄.85827	12	11
50	1̄.84046	13	1̄.98231	25	0.01769	1̄.85815	12	10
51	1̄.84059	13	1̄.98256	25	0.01744	1̄.85803	12	9
52	1̄.84072	13	1̄.98281	26	0.01719	1̄.85791	12	8
53	1̄.84085	13	1̄.98307	25	0.01693	1̄.85779	13	7
54	1̄.84098	13	1̄.98332	25	0.01668	1̄.85766	12	6
55	1̄.84112	14	1̄.98357	25	0.01643	1̄.85754	12	5
56	1̄.84125	13	1̄.98383	26	0.01617	1̄.85742	12	4
57	1̄.84138	13	1̄.98408	25	0.01592	1̄.85730	12	3
58	1̄.84151	13	1̄.98433	25	0.01567	1̄.85718	12	2
59	1̄.84164	13	1̄.98458	26	0.01542	1̄.85706	13	1
60	1̄.84177		1̄.98484		0.01516	1̄.85693		0

| | log cos | d | log cot | c. d. | log tan | log sin | d | ′ |

p. p.

	26
6	2.6
7	3.0
8	3.5
9	3.9
10	4.3
20	8.7
30	13.0
40	17.3
50	21.7

	25
6	2.5
7	2.9
8	3.3
9	3.8
10	4.2
20	8.3
30	12.5
40	16.7
50	20.8

	14
6	1.4
7	1.6
8	1.9
9	2.1
10	2.3
20	4.7
30	7.0
40	9.3
50	11.7

	13
6	1.3
7	1.5
8	1.7
9	2.0
10	2.2
20	4.3
30	6.5
40	8.7
50	10.8

	12	11
6	1.2	1.1
7	1.4	1.3
8	1.6	1.5
9	1.8	1.7
10	2.0	1.8
20	4.0	3.7
30	6.0	5.5
40	8.0	7.3
50	10.0	9.2

′	log sin	d	log tan	c. d.	log cot	log cos	d		p. p.
0	1̄.84177	13	1̄.98484	25	0.01516	1̄.85693	12	60	
1	1̄.84190	13	1̄.98509	25	0.01491	1̄.85681	12	59	
2	1̄.84203	13	1̄.98534	26	0.01466	1̄.85669	12	58	
3	1̄.84216	13	1̄.98560	25	0.01440	1̄.85657	12	57	
4	1̄.84229	13	1̄.98585	25	0.01415	1̄.85645	13	56	
5	1̄.84242	13	1̄.98610	25	0.01390	1̄.85632	12	55	
6	1̄.84255	14	1̄.98635	26	0.01365	1̄.85620	12	54	
7	1̄.84269	13	1̄.98661	25	0.01339	1̄.85608	12	53	
8	1̄.84282	13	1̄.98686	25	0.01314	1̄.85596	13	52	
9	1̄.84295	13	1̄.98711	26	0.01289	1̄.85583	12	51	
10	1̄.84308	13	1̄.98737	25	0.01263	1̄.85571	12	50	
11	1̄.84321	13	1̄.98762	25	0.01238	1̄.85559	12	49	
12	1̄.84334	13	1̄.98787	25	0.01213	1̄.85547	13	48	
13	1̄.84347	13	1̄.98812	26	0.01188	1̄.85534	12	47	
14	1̄.84360	13	1̄.98838	25	0.01162	1̄.85522	12	46	
15	1̄.84373	12	1̄.98863	25	0.01137	1̄.85510	13	45	
16	1̄.84385	13	1̄.98888	25	0.01112	1̄.85497	12	44	
17	1̄.84398	13	1̄.98913	26	0.01087	1̄.85485	12	43	
18	1̄.84411	13	1̄.98939	25	0.01061	1̄.85473	13	42	
19	1̄.84424	13	1̄.98964	25	0.01036	1̄.85460	12	41	
20	1̄.84437	13	1̄.98989	26	0.01011	1̄.85448	12	40	
21	1̄.84450	13	1̄.99015	25	0.00985	1̄.85436	13	39	
22	1̄.84463	13	1̄.99040	25	0.00960	1̄.85423	12	38	
23	1̄.84476	13	1̄.99065	25	0.00935	1̄.85411	12	37	
24	1̄.84489	13	1̄.99090	26	0.00910	1̄.85399	13	36	
25	1̄.84502	13	1̄.99116	25	0.00884	1̄.85386	12	35	
26	1̄.84515	13	1̄.99141	25	0.00859	1̄.85374	13	34	
27	1̄.84528	12	1̄.99166	25	0.00834	1̄.85361	12	33	
28	1̄.84540	13	1̄.99191	26	0.00809	1̄.85349	12	32	
29	1̄.84553	13	1̄.99217	25	0.00783	1̄.85337	13	31	
30	1̄.84566	13	1̄.99242	25	0.00758	1̄.85324	12	30	
31	1̄.84579	13	1̄.99267	26	0.00733	1̄.85312	13	29	
32	1̄.84592	13	1̄.99293	25	0.00707	1̄.85299	12	28	
33	1̄.84605	13	1̄.99318	25	0.00682	1̄.85287	13	27	
34	1̄.84618	12	1̄.99343	25	0.00657	1̄.85274	12	26	
35	1̄.84630	13	1̄.99368	26	0.00632	1̄.85262	12	25	
36	1̄.84643	13	1̄.99394	25	0.00606	1̄.85250	13	24	
37	1̄.84656	13	1̄.99419	25	0.00581	1̄.85237	12	23	
38	1̄.84669	13	1̄.99444	25	0.00556	1̄.85225	13	22	
39	1̄.84682	12	1̄.99469	26	0.00531	1̄.85212	12	21	
40	1̄.84694	13	1̄.99495	25	0.00505	1̄.85200	13	20	
41	1̄.84707	13	1̄.99520	25	0.00480	1̄.85187	12	19	
42	1̄.84720	13	1̄.99545	25	0.00455	1̄.85175	13	18	
43	1̄.84733	12	1̄.99570	26	0.00430	1̄.85162	12	17	
44	1̄.84745	13	1̄.99596	25	0.00404	1̄.85150	13	16	
45	1̄.84758	13	1̄.99621	25	0.00379	1̄.85137	12	15	
46	1̄.84771	13	1̄.99646	26	0.00354	1̄.85125	13	14	
47	1̄.84784	12	1̄.99672	25	0.00328	1̄.85112	12	13	
48	1̄.84796	13	1̄.99697	25	0.00303	1̄.85100	13	12	
49	1̄.84809	13	1̄.99722	25	0.00278	1̄.85087	13	11	
50	1̄.84822	13	1̄.99747	26	0.00253	1̄.85074	12	10	
51	1̄.84835	12	1̄.99773	25	0.00227	1̄.85062	13	9	
52	1̄.84847	13	1̄.99798	25	0.00202	1̄.85049	12	8	
53	1̄.84860	13	1̄.99823	25	0.00177	1̄.85037	13	7	
54	1̄.84873	12	1̄.99848	26	0.00152	1̄.85024	12	6	
55	1̄.84885	13	1̄.99874	25	0.00126	1̄.85012	13	5	
56	1̄.84898	13	1̄.99899	25	0.00101	1̄.84999	13	4	
57	1̄.84911	12	1̄.99924	25	0.00076	1̄.84986	12	3	
58	1̄.84923	13	1̄.99949	26	0.00051	1̄.84974	13	2	
59	1̄.84936	13	1̄.99975	25	0.00025	1̄.84961	12	1	
60	1̄.84949		0.00000		0.00000	1̄.84949		0	
	log cos	d	log cot	c. d.	log tan	log sin	d	′	p. p.

p. p.

26	
6	2.6
7	3.0
8	3.5
9	3.9
10	4.3
20	8.7
30	13.0
40	17.3
50	21.7

25	
6	2.5
7	2.9
8	3.3
9	3.8
10	4.2
20	8.3
30	12.5
40	16.7
50	20.8

14	
6	1.4
7	1.6
8	1.9
9	2.1
10	2.3
20	4.7
30	7.0
40	9.3
50	11.7

13	
6	1.3
7	1.5
8	1.7
9	2.0
10	2.2
20	4.3
30	6.5
40	8.7
50	10.8

12	
6	1.2
7	1.4
8	1.6
9	1.8
10	2.0
20	4.0
30	6.0
40	8.0
50	10.0

TABLE

OF

COMMON LOGARITHMS

OF NUMBERS

From 1 to 10,000.

N.	Log.	N.	Log.	N.	Log.	N.	Log.	N.	Log.
0	— ∞	**20**	30 103	**40**	60 206	**60**	77 815	**80**	90 309
1	00 000	21	32 222	41	61 278	61	78 533	81	90 849
2	30 103	22	34 242	42	62 325	62	79 239	82	91 381
3	47 712	23	36 173	43	63 347	63	79 934	83	91 908
4	60 206	24	38 021	44	64 345	64	80 618	84	92 428
5	69 897	25	39 794	45	65 321	65	81 291	85	92 942
6	77 815	26	41 497	46	66 276	66	81 954	86	93 450
7	84 510	27	43 136	47	67 210	67	82 607	87	93 952
8	90 309	28	44 716	48	68 124	68	83 251	88	94 448
9	95 424	29	46 240	49	69 020	69	83 885	89	94 939
10	00 000	**30**	47 712	**50**	69 897	**70**	84 510	**90**	95 424
11	04 139	31	49 136	51	70 757	71	85 126	91	95 904
12	07 918	32	50 515	52	71 600	72	85 733	92	96 379
13	11 394	33	51 851	53	72 428	73	86 332	93	96 848
14	14 613	34	53 148	54	73 239	74	86 923	94	97 313
15	17 609	35	54 407	55	74 036	75	87 506	95	97 772
16	20 412	36	55 630	56	74 819	76	88 081	96	98 227
17	23 045	37	56 820	57	75 587	77	88 649	97	98 677
18	25 527	38	57 978	58	76 343	78	89 209	98	99 123
19	27 875	39	59 106	59	77 085	79	89 763	99	99 564
20	30 103	**40**	60 206	**60**	77 815	**80**	90 309	**100**	00 000

LOGARITHMS.

N.	L. 0	1	2	3	4	5	6	7	8	9
100	00 000	043	087	130	173	217	260	303	346	389
101	432	475	518	561	604	647	689	732	775	817
102	860	903	945	988	*030	*072	*115	*157	*199	*242
103	01 284	326	368	410	452	494	536	578	620	662
104	703	745	787	828	870	912	953	995	*036	*078
105	02 119	160	202	243	284	325	366	407	449	490
106	531	572	612	653	694	735	776	816	857	898
107	938	979	*019	*060	*100	*141	*181	*222	*262	*302
108	03 342	383	423	463	503	543	583	623	663	703
109	743	782	822	862	902	941	981	*021	*060	*100
110	04 139	179	218	258	297	336	376	415	454	493
111	532	571	610	650	689	727	766	805	844	883
112	922	961	999	*038	*077	*115	*154	*192	*231	*269
113	05 308	346	385	423	461	500	538	576	614	652
114	690	729	767	805	843	881	918	956	994	*032
115	06 070	108	145	183	221	258	296	333	371	408
116	446	483	521	558	595	633	670	707	744	781
117	819	856	893	930	967	*004	*041	*078	*115	*151
118	07 188	225	262	298	335	372	408	445	482	518
119	555	591	628	664	700	737	773	809	846	882
120	918	954	990	*027	*063	*099	*135	*171	*207	*243
121	08 279	314	350	386	422	458	493	529	565	600
122	636	672	707	743	778	814	849	884	920	955
123	991	*026	*061	*096	*132	*167	*202	*237	*272	*307
124	09 342	377	412	447	482	517	552	587	621	656
125	691	726	760	795	830	864	899	934	968	*003
126	10 037	072	106	140	175	209	243	278	312	346
127	380	415	449	483	517	551	585	619	653	687
128	721	755	789	823	857	890	924	958	992	*025
129	11 059	093	126	160	193	227	261	294	327	361
130	394	428	461	494	528	561	594	628	661	694
131	727	760	793	826	860	893	926	959	992	*024
132	12 057	090	123	156	189	222	254	287	320	352
133	385	418	450	483	516	548	581	613	646	678
134	710	743	775	808	840	872	905	937	969	*001
135	13 033	066	098	130	162	194	226	258	290	322
136	354	386	418	450	481	513	545	577	609	640
137	672	704	735	767	799	830	862	893	925	956
138	988	*019	*051	*082	*114	*145	*176	*208	*239	*270
139	14 301	333	364	395	426	457	489	520	551	582
140	613	644	675	706	737	768	799	829	860	891
141	922	953	983	*014	*045	*076	*106	*137	*168	*198
142	15 229	259	290	320	351	381	412	442	473	503
143	534	564	594	625	655	685	715	746	776	806
144	836	866	897	927	957	987	*017	*047	*077	*107
145	16 137	167	197	227	256	286	316	346	376	406
146	435	465	495	524	554	584	613	643	673	702
147	732	761	791	820	850	879	909	938	967	997
148	17 026	056	085	114	143	173	202	231	260	289
149	319	348	377	406	435	464	493	522	551	580
150	609	638	667	696	725	754	782	811	840	869
N.	L. 0	1	2	3	4	5	6	7	8	9

P. P.

	44	43	42
1	4.4	4.3	4.2
2	8.8	8.6	8.4
3	13.2	12.9	12.6
4	17.6	17.2	16.8
5	22.0	21.5	21.0
6	26.4	25.8	25.2
7	30.8	30.1	29.4
8	35.2	34.4	33.6
9	39.6	38.7	37.8

	41	40	39
1	4.1	4.0	3.9
2	8.2	8.0	7.8
3	12.3	12.0	11.7
4	16.4	16.0	15.6
5	20.5	20.0	19.5
6	24.6	24.0	23.4
7	28.7	28.0	27.3
8	32.8	32.0	31.2
9	36.9	36.0	35.1

	38	37	36
1	3.8	3.7	3.6
2	7.6	7.4	7.2
3	11.4	11.1	10.8
4	15.2	14.8	14.4
5	19.0	18.5	18.0
6	22.8	22.2	21.6
7	26.6	25.9	25.2
8	30.4	29.6	28.8
9	34.2	33.3	32.4

	35	34	33
1	3.5	3.4	3.3
2	7.0	6.8	6.6
3	10.5	10.2	9.9
4	14.0	13.6	13.2
5	17.5	17.0	16.5
6	21.0	20.4	19.8
7	24.5	23.8	23.1
8	28.0	27.2	26.4
9	31.5	30.6	29.7

	32	31	30
1	3.2	3.1	3.0
2	6.4	6.2	6.0
3	9.6	9.3	9.0
4	12.8	12.4	12.0
5	16.0	15.5	15.0
6	19.2	18.6	18.0
7	22.4	21.7	21.0
8	25.6	24.8	24.0
9	28.8	27.9	27.0

N.	L. 0	1	2	3	4	5	6	7	8	9
150	17 609	638	667	696	725	754	782	811	840	869
151	898	926	955	984	*013	*041	*070	*099	*127	*156
152	18 184	213	241	270	298	327	355	384	412	441
153	469	498	526	554	583	611	639	667	696	724
154	752	780	808	837	865	893	921	949	977	*005
155	19 033	061	089	117	145	173	201	229	257	285
156	312	340	368	396	424	451	479	507	535	562
157	590	618	645	673	700	728	756	783	811	838
158	866	893	921	948	976	*003	*030	*058	*085	*112
159	20 140	167	194	222	249	276	303	330	358	385
160	412	439	466	493	520	548	575	602	629	656
161	683	710	737	763	790	817	844	871	898	925
162	952	978	*005	*032	*059	*085	*112	*139	*165	*192
163	21 219	245	272	299	325	352	378	405	431	458
164	484	511	537	564	590	617	643	669	696	722
165	748	775	801	827	854	880	906	932	958	985
166	22 011	037	063	089	115	141	167	194	220	246
167	272	298	324	350	376	401	427	453	479	505
168	531	557	583	608	634	660	686	712	737	763
169	789	814	840	866	891	917	943	968	994	*019
170	23 045	070	096	121	147	172	198	223	249	274
171	300	325	350	376	401	426	452	477	502	528
172	553	578	603	629	654	679	704	729	754	779
173	805	830	855	880	905	930	955	980	*005	*030
174	24 055	080	105	130	155	180	204	229	254	279
175	304	329	353	378	403	428	452	477	502	527
176	551	576	601	625	650	674	699	724	748	773
177	797	822	846	871	895	920	944	969	993	*018
178	25 042	066	091	115	139	164	188	212	237	261
179	285	310	334	358	382	406	431	455	479	503
180	527	551	575	600	624	648	672	696	720	744
181	768	792	816	840	864	888	912	935	959	983
182	26 007	031	055	079	102	126	150	174	198	221
183	245	269	293	316	340	364	387	411	435	458
184	482	505	529	553	576	600	623	647	670	694
185	717	741	764	788	811	834	858	881	905	928
186	951	975	998	*021	*045	*068	*091	*114	*138	*161
187	27 184	207	231	254	277	300	323	346	370	393
188	416	439	462	485	508	531	554	577	600	623
189	646	669	692	715	738	761	784	807	830	852
190	875	898	921	944	967	989	*012	*035	*058	*081
191	28 103	126	149	171	194	217	240	262	285	307
192	330	353	375	398	421	443	466	488	511	533
193	556	578	601	623	646	668	691	713	735	758
194	780	803	825	847	870	892	914	937	959	981
195	29 003	026	048	070	092	115	137	159	181	203
196	226	248	270	292	314	336	358	380	403	425
197	447	469	491	513	535	557	579	601	623	645
198	667	688	710	732	754	776	798	820	842	863
199	885	907	929	951	973	994	*016	*038	*060	*081
200	30 103	125	146	168	190	211	233	255	276	298
N.	L. 0	1	2	3	4	5	6	7	8	9

P. P.

	29	28
1	2.9	2.8
2	5.8	5.6
3	8.7	8.4
4	11.6	11.2
5	14.5	14.0
6	17.4	16.8
7	20.3	19.6
8	23.2	22.4
9	26.1	25.2

	27	26
1	2.7	2.6
2	5.4	5.2
3	8.1	7.8
4	10.8	10.4
5	13.5	13.0
6	16.2	15.6
7	18.9	18.2
8	21.6	20.8
9	24.3	23.4

	25
1	2.5
2	5.0
3	7.5
4	10.0
5	12.5
6	15.0
7	17.5
8	20.0
9	22.5

	24	23
1	2.4	2.3
2	4.8	4.6
3	7.2	6.9
4	9.6	9.2
5	12.0	11.5
6	14.4	13.8
7	16.8	16.1
8	19.2	18.4
9	21.6	20.7

	22	21
1	2.2	2.1
2	4.4	4.2
3	6.6	6.3
4	8.8	8.4
5	11.0	10.5
6	13.2	12.6
7	15.4	14.7
8	17.6	16.8
9	19.8	18.9

P. P.

4	5	6	7	8
190	211	233	255	276
406	428	449	471	492
621	643	664	685	707
835	856	878	899	920
*048	*069	*091	*112	*133
260	281	302	323	345
471	492	513	534	555
681	702	723	744	765
890	911	931	952	973
098	118	139	160	181
305	325	346	366	387
510	531	552	572	593
715	736	756	777	797
919	940	960	980	*001
122	143	163	183	203
325	345	365	385	405
526	546	566	586	606
726	746	766	786	806
925	945	965	985	*005
124	143	163	183	203
321	341		380	400
518	537		577	596
713	733		772	792
908	928		967	986
102	122		160	180
295	315		353	372
488	507		545	564
679	698		736	755
870	889		927	946
*059	*078		*116	*135
248	267		305	324
436	455		493	511
624	642		680	698
810	829		866	884
996	*014		*051	*070
181	199		236	254
365	383		420	438
548	566		603	621
731	749		785	803
912	931		967	985
093	112		148	166
274	292		328	346
453	471		507	525
632	650		686	703
810	828		863	881
987	*005		*041	*058
164	182		217	235
340	358		393	410
515	533		568	585
690	707		742	759
863	881		915	933
4	5			

P. P.

	22	21
1	2.2	2.1
2	4.4	4.2
3	6.6	6.3
4	8.8	8.4
5	11.0	10.5
6	13.2	12.6
7	15.4	14.7
8	17.6	16.8
9	19.8	18.9

	20
1	2.0
2	4.0
3	6.0
4	8.0
5	10.0
6	12.0
7	14.0
8	16.0
9	18.0

	19
1	1.9
2	3.8
3	5.7
4	7.6
5	9.5
6	11.4
7	13.3
8	15.2
9	17.1

	18
1	1.8
2	3.6
3	5.4
4	7.2
5	9.0
6	10.8
7	12.6
8	14.4
9	16.2

	17
1	1.7
2	3.4
3	5.1
4	6.8
5	8.5
6	10.2
7	11.9
8	13.6
9	15.3

LOGARITHMS.

N.	L. 0	1	2	3	4	5	6	7	8	9
250	39 794	811	829	846	863	881	898	915	933	950
251	967	985	*002	*019	*037	*054	*071	*088	*106	*123
252	40 140	157	175	192	209	226	243	261	278	295
253	312	329	346	364	381	398	415	432	449	466
254	483	500	518	535	552	569	586	603	620	637
255	654	671	688	705	722	739	756	773	790	807
256	824	841	858	875	892	909	926	943	960	976
257	993	*010	*027	*044	*061	*078	*095	*111	*128	*145
258	41 162	179	196	212	229	246	263	280	296	313
259	330	347	363	380	397	414	430	447	464	481
260	497	514	531	547	564	581	597	614	631	647
261	664	681	697	714	731	747	764	780	797	814
262	830	847	863	880	896	913	929	946	963	979
263	996	*012	*029	*045	*062	*078	*095	*111	*127	*144
264	42 160	177	193	210	226	243	259	275	292	308
265	325	341	357	374	390	406	423	439	455	472
266	488	504	521	537	553	570	586	602	619	635
267	651	667	684	700	716	732	749	765	781	797
268	813	830	846	862	878	894	911	927	943	959
269	975	991	*008	*024	*040	*056	*072	*088	*104	*120
270	43 136	152	169	185	201	217	233	249	265	281
271	297	313	329	345	361	377	393	409	425	441
272	457	473	489	505	521	537	553	569	584	600
273	616	632	648	664	680	696	712	727	743	759
274	775	791	807	823	838	854	870	886	902	917
275	933	949	965	981	996	*012	*028	*044	*059	*075
276	44 091	107	122	138	154	170	185	201	217	232
277	248	264	279	295	311	326	342	358	373	389
278	404	420	436	451	467	483	498	514	529	545
279	560	576	592	607	623	638	654	669	685	700
280	716	731	747	762	778	793	809	824	840	855
281	871	886	902	917	932	948	963	979	994	*010
282	45 025	040	056	071	086	102	117	133	148	163
283	179	194	209	225	240	255	271	286	301	317
284	332	347	362	378	393	408	423	439	454	469
285	484	500	515	530	545	561	576	591	606	621
286	637	652	667	682	697	712	728	743	758	773
287	788	803	818	834	849	864	879	894	909	924
288	939	954	969	984	*000	*015	*030	*045	*060	*075
289	46 090	105	120	135	150	165	180	195	210	225
290	240	255	270	285	300	315	330	345	359	374
291	389	404	419	434	449	464	479	494	509	523
292	538	553	568	583	598	613	627	642	657	672
293	687	702	716	731	746	761	776	790	805	820
294	835	850	864	879	894	909	923	938	953	967
295	982	997	*012	*026	*041	*056	*070	*085	*100	*114
296	47 129	144	159	173	188	202	217	232	246	261
297	276	290	305	319	334	349	363	378	392	407
298	422	436	451	465	480	494	509	524	538	553
299	567	582	596	611	625	640	654	669	683	698
300	712	727	741	756	770	784	799	813	828	842
N.	L. 0	1	2	3	4	5	6	7	8	9

P. P.

	18		17		16		15		14
1	1.8	1	1.7	1	1.6	1	1.5	1	1.4
2	3.6	2	3.4	2	3.2	2	3.0	2	2.8
3	5.4	3	5.1	3	4.8	3	4.5	3	4.2
4	7.2	4	6.8	4	6.4	4	6.0	4	5.6
5	9.0	5	8.5	5	8.0	5	7.5	5	7.0
6	10.8	6	10.2	6	9.6	6	9.0	6	8.4
7	12.6	7	11.9	7	11.2	7	10.5	7	9.8
8	14.4	8	13.6	8	12.8	8	12.0	8	11.2
9	16.2	9	15.3	9	14.4	9	13.5	9	12.6

N.	L. 0	1	2	3	4	5	6	7	8	9
300	47 712	727	741	756	770	784	799	813	828	842
301	857	871	885	900	914	929	943	958	972	986
302	48 001	015	029	044	058	073	087	101	116	130
303	144	159	173	187	202	216	230	244	259	273
304	287	302	316	330	344	359	373	387	401	416
305	430	444	458	473	487	501	515	530	544	558
306	572	586	601	615	629	643	657	671	686	700
307	714	728	742	756	770	785	799	813	827	841
308	855	869	883	897	911	926	940	954	968	982
309	996	*010	*024	*038	*052	*066	*080	*094	*108	*122
310	49 136	150	164	178	192	206	220	234	248	262
311	276	290	304	318	332	346	360	374	388	402
312	415	429	443	457	471	485	499	513	527	541
313	554	568	582	596	610	624	638	651	665	679
314	693	707	721	734	748	762	776	790	803	817
315	831	845	859	872	886	900	914	927	941	955
316	969	982	996	*010	*024	*037	*051	*065	*079	*092
317	50 106	120	133	147	161	174	188	202	215	229
318	243	256	270	284	297	311	325	338	352	365
319	379	393	406	420	433	447	461	474	488	501
320	515	529	542	556	569	583	596	610	623	637
321	651	664	678	691	705	718	732	745	759	772
322	786	799	813	826	840	853	866	880	893	907
323	920	934	947	961	974	987	*001	*014	*028	*041
324	51 055	068	081	095	108	121	135	148	162	175
325	188	202	215	228	242	255	268	282	295	308
326	322	335	348	362	375	388	402	415	428	441
327	455	468	481	495	508	521	534	548	561	574
328	587	601	614	627	640	654	667	680	693	706
329	720	733	746	759	772	786	799	812	825	838
330	851	865	878	891	904	917	930	943	957	970
331	983	996	*009	*022	*035	*048	*061	*075	*088	*101
332	52 114	127	140	153	166	179	192	205	218	231
333	244	257	270	284	297	310	323	336	349	362
334	375	388	401	414	427	440	453	466	479	492
335	504	517	530	543	556	569	582	595	608	621
336	634	647	660	673	686	699	711	724	737	750
337	763	776	789	802	815	827	840	853	866	879
338	892	905	917	930	943	956	969	982	994	*007
339	53 020	033	046	058	071	084	097	110	122	135
340	148	161	173	186	199	212	224	237	250	263
341	275	288	301	314	326	339	352	364	377	390
342	403	415	428	441	453	466	479	491	504	517
343	529	542	555	567	580	593	605	618	631	643
344	656	668	681	694	706	719	732	744	757	769
345	782	794	807	820	832	845	857	870	882	895
346	908	920	933	945	958	970	983	995	*008	*020
347	54 033	045	058	070	083	095	108	120	133	145
348	158	170	183	195	208	220	233	245	258	270
349	283	295	307	320	332	345	357	370	382	394
350	407	419	432	444	456	469	481	494	506	518
N.	L. 0	1	2	3	4	5	6	7	8	9

P. P.

15

1	1.5
2	3.0
3	4.5
4	6.0
5	7.5
6	9.0
7	10.5
8	12.0
9	13.5

14

1	1.4
2	2.8
3	4.2
4	5.6
5	7.0
6	8.4
7	9 8
8	11.2
9	12.6

13

1	1.3
2	2.6
3	3.9
4	5.2
5	6.5
6	7.8
7	9.1
8	10.4
9	11.7

12

1	1.2
2	2.4
3	3.6
4	4.8
5	6.0
6	7.2
7	8.4
8	9.6
9	10.8

N.		1	2	3	4	5	6	7	8	9
350	54 407	419	432	444	456	469	481	494	506	518
351		543	555	568	580	593	605	617	630	642
352		667	679	691	704	716	728	741	753	765
353		790	802	814	827	839	851	864	876	888
354		913	·925	937	949	962	974	986	998	*011
355	55 023	035	047	060	072	084	096	108	121	133
356	145	157	169	182	194	206	218	230	242	255
357		279	291	303	315	328	340	352	364	376
358		400	413	425	437	449	461	473	485	497
359		522	534	546	558	570	582	594	606	618
360		642	654	666	678	691	703	715	727	739
361		763	775	787	799	811	823	835	847	859
362		883	895	907	919	931	943	955	967	979
363		*003	*015	*027	*038	*050	*062	*074	*086	*098
364	56 110	122	134	146	158	170	182	194	205	217
365		241	253	265	277	289	301	312	324	336
366		360	372	384	396	407	419	431	443	455
367		478	490	502	514	526	538	549	561	573
368		597	608	620	632	644	656	667	679	691
369	703	714	726	738	750	761	773	785	797	808
370		832	844	855	867	879	891	902	914	926
371		949	961	972	984	996	*008	*019	*031	*043
372	57 054	066	078	089	101	113	124	136	148	159
373		183	194	206	217	229	241	252	264	276
374		299	310	322	334	345	·357	368	380	392
375		415	426	438	449	461	473	484	496	507
376		530	542	553	565	576	588	600	611	623
377		646	657	669	680	692	703	715	726	738
378		761	772	784	795	807	818	830	841	852
379		875	887	898	910	921	933	944	955	967
380		990	*001	*013	*024	*035	*047	*058	*070	*081
381	58 092	104	115	127	138	149	161	172	184	195
382		218	229	240	252	263	274	286	297	309
383		331	343	354	365	377	388	399	410	422
384		444	456	467	478	490	501	512	524	535
385		557	569	580	591	602	614	625	636	647
386		670	681	692	704	715	726	737	749	760
387		782	794	805	816	827	838	850	861	872
388		894	906	917	928	939	950	961	973	984
389		*006	*017	*028	*040	*051	*062	*073	*084	*095
390	59 106	118	129	140	151	162	173	184	195	207
391		229	240	251	262	273	284	295	306	318
392		340	351	362	373	384	395	406	417	428
393		450	461	472	483	494	506	517	528	539
394		561	572	583	594	605	616	627	638	649
395		671	682	693	704	715	726	737	748	759
396		780	791	802	813	824	835	846	857	868
397		890	901	912	923	934	945	956	966	977
398		999	*010	*021	*032	*043	*054	*065	*076	*086
399	60 097	108	119	130	141	152	163	173	184	195
400		217	228	239	249	260	271	282	293	304
N.		1	2	3	4	5	6	7	8	9

Proportional parts:

13		**12**		**11**		**10**	
1	1.3	1	1.2	1	1.1	1	1.0
2	2.6	2	2.4	2	2.2	2	2.0
3	3.9	3	3.6	3	3.3	3	3.0
4	5.2	4	4.8	4	4.4	4	4.0
5	6.5	5	6.0	5	5.5	5	5.0
6	7.8	6	7.2	6	6.6	6	6.0
7	9.1	7	8.4	7	7.7	7	7.0
8	10.4	8	9.6	8	8.8	8	8.0
9	11.7	9	10.8	9	9.9	9	9.0

LOGARITHMS.

3	4		3	4
239	249		271	282
347	358		379	390
455	466		487	498
563	574		595	606
670	681		703	713
778	788		810	821
885	895		917	927
991	*002		*023	*034
098	109		130	140
204	215		236	247
310	321		342	352
416	426		448	458
521	532		553	563
627	637		658	669
731	742		763	773
836	847		868	878
941	951		972	982
045	055		076	086
149	159		180	190
252	263		284	294
356	366		387	397
459	469			
562	572			
665	675			
767	778			
870	880			
972	982			
073	083			
175	185			
276	286			
377	387			
478	488			
579	589			
679	689			
779	789			
879	889			
979	988			
078	088			
177	187			
276	286			

	11
1	1.1
2	2.2
3	3.3
4	4.4
5	5.5
6	6.6
7	7.7
8	8.8
9	9.9

	10
1	1.0
2	2.0
3	3.0
4	4.0
5	5.0
6	6.0
7	7.0
8	8.0
9	9.0

	9
1	0.9
2	1.8
3	2.7
4	3.6
5	4.5
6	5.4
7	6.3
8	7.2
9	8.1

L.	0	1	2	3	4	5	6	7	8	9
65	321	331	341	350	360	369	379	389	398	408
	418	427	437	447	456	466	475	485	495	504
	514	523	533	543	552	562	571	581	591	600
	610	619	629	639	648	658	667	677	686	696
	706	715	725	734	744	753	763	772	782	792
	801	811	820	830	839	849	858	868	877	887
	896	906	916	925	935	944	954	963	973	982
	992	*001	*011	*020	*030	*039	*049	*058	*068	*077
66	087	096	106	115	124	134	143	153	162	172
	181	191	200	210	219	229	238	247	257	266
	276	285	295	304	314	323	332	342	351	361
	370	380	389	398	408	417	427	436	445	455
	464	474	483	492	502	511	521	530	539	549
	558	567	577	586	596	605	614	624	633	642
	652	661	671	680	689	699	708	717	727	736
	745	755	764	773	783	792	801	811	820	829
	839	848	857	867	876	885	894	904	913	922
	932	941	950	960	969	978	987	997	*006	*015
67	025	034	043	052	062	071	080	089	099	108
	117	127	136	145	154	164	173	182	191	201
	210	219	228	237	247	256	265	274	284	293
	302	311	321	330	339	348	357	367	376	385
	394	403	413	422	431	440	449	459	468	477
	486	495	504	514	523	532	541	550	560	569
	578	587	596	605	614	624	633	642	651	660
	669	679	688	697	706	715	724	733	742	752
	761	770	779	788	797	806	815	825	834	843
	852	861	870	879	888	897	906	916	925	934
	943	952	961	970	979	988	997	*006	*015	*024
68	034	043	052	061	070	079	088	097	106	115
	124	133	142	151	160	169	178	187	196	205
	215	224	233	242	251	260	269	278	287	296
	305	314	323	332	341	350	359	368	377	386
	395	404	413	422	431	440	449	458	467	476
	485	494	502	511	520	529	538	547	556	565
	574	583	592	601	610	619	628	637	646	655
	664	673	681	690	699	708	717	726	735	744
	753	762	771	780	789	797	806	815	824	833
	842	851	860	869	878	886	895	904	913	922
	931	940	949	958	966	975	984	993	*002	*011
69	020	028	037	046	055	064	073	082	090	099
	108	117	126	135	144	152	161	170	179	188
	197	205	214	223	232	241	249	258	267	276
	285	294	302	311	320	329	338	346	355	364
	373	381	390	399	408	417	425	434	443	452
	461	469	478	487	496	504	513	522	531	539
	548	557	566	574	583	592	601	609	618	627
	636	644	653	662	671	679	688	697	705	714
	723	732	740	749	758	767	775	784	793	801
	810	819	827	836	845	854	862	871	880	888
	897	906	914	923	932			958	966	975
		1	2	3	4			7	8	9

Proportional parts:

10	
1	1.0
2	2.0
3	3.0
4	4.0
5	5.0
6	6.0
7	7.0
8	8.0
9	9.0

9	
1	0.9
2	1.8
3	2.7
4	3.6
5	4.5
6	5.4
7	6.3
8	7.2
9	8.1

8	
1	0.8
2	1.6
3	2.4
4	3.2
5	4.0
6	4.8
7	5.6
8	6.4
9	7.2

N.	L. 0	1	2	3	4	5	6	7	8	9
500	69 897	906	914	923	932	940	949	958	966	975
501	984	992	*001	*010	*018	*027	*036	*044	*053	*062
502	70 070	079	088	096	105	114	122	131	140	148
503	157	165	174	183	191	200	209	217	226	234
504	243	252	260	269	278	286	295	303	312	321
505	329	338	346	355	364	372	381	389	398	406
506	415	424	432	441	449	458	467	475	484	492
507	501	509	518	526	535	544	552	561	569	578
508	586	595	603	612	621	629	638	646	655	663
509	672	680	689	697	706	714	723	731	740	749
510	757	766	774	783	791	800	808	817	825	834
511	842	851	859	868	876	885	893	902	910	919
512	927	935	944	952	961	969	978	986	995	*003
513	71 012	020	029	037	046	054	063	071	079	088
514	096	105	113	122	130	139	147	155	164	172
515	181	189	198	206	214	223	231	240	248	257
516	265	273	282	290	299	307	315	324	332	341
517	349	357	366	374	383	391	399	408	416	425
518	433	441	450	458	466	475	483	492	500	508
519	517	525	533	542	550	559	567	575	584	592
520	600	609	617	625	634	642	650	659	667	675
521	684	692	700	709	717	725	734	742	750	759
522	767	775	784	792	800	809	817	825	834	842
523	850	858	867	875	883	892	900	908	917	925
524	933	941	950	958	966	975	983	991	999	*008
525	72 016	024	032	041	049	057	066	074	082	090
526	099	107	115	123	132	140	148	156	165	173
527	181	189	198	206	214	222	230	239	247	255
528	263	272	280	288	296	304	313	321	329	337
529	346	354	362	370	378	387	395	403	411	419
530	428	436	444	452	460	469	477	485	493	501
531	509	518	526	534	542	550	558	567	575	583
532	591	599	607	616	624	632	640	648	656	665
533	673	681	689	697	705	713	722	730	738	746
534	754	762	770	779	787	795	803	811	819	827
535	835	843	852	860	868	876	884	892	900	908
536	916	925	933	941	949	957	965	973	981	989
537	997	*006	*014	*022	*030	*038	*046	*054	*062	*070
538	73 078	086	094	102	111	119	127	135	143	151
539	159	167	175	183	191	199	207	215	223	231
540	239	247	255	263	272	280	288	296	304	312
541	320	328	336	344	352	360	368	376	384	392
542	400	408	416	424	432	440	448	456	464	472
543	480	488	496	504	512	520	528	536	544	552
544	560	568	576	584	592	600	608	616	624	632
545	640	648	656	664	672	679	687	695	703	711
546	719	727	735	743	751	759	767	775	783	791
547	799	807	815	823	830	838	846	854	862	870
548	878	886	894	902	910	918	926	933	941	949
549	957	965	973	981	989	997	*005	*013	*020	*028
550	74 036	044	052	060	068	076	084	092	099	107
N.	L. 0	1	2	3	4	5	6	7	8	9

Proportional parts:

9	
1	0.9
2	1.8
3	2.7
4	3.6
5	4.5
6	5.4
7	6.3
8	7.2
9	8.1

8	
1	0.8
2	1.6
3	2.4
4	3.2
5	4.0
6	4.8
7	5.6
8	6.4
9	7.2

7	
1	0.7
2	1.4
3	2.1
4	2.8
5	3.5
6	4.2
7	4.9
8	5.6
9	6.3

	0	1	2	3	4	5	6	7	8	9
			052	060	068		084	092	099	107
			131	139	147		162	170	178	186
			210	218	225		241	249	257	265
			288	296	304		320	327	335	343
			367	374	382		398	406	414	421
			445	453	461		476	484	492	500
			523	531	539		554	562	570	578
			601	609	617		632	640	648	656
			679	687	695		710	718	726	733
			757	764	772		788	796	803	811
			834	842	850		865	873	881	889
			912	920	927		943	950	958	966
			989	997	*005		*020	*028	*035	*043
			066	074	082		097	105	113	120
			143	151	159		174	182	189	197
			220	228	236		251	259	266	274
			297	305	312	320	328	335	343	351
	358	366	374	381	389	397	404	412	420	427
	435	442	450	458	465	473	481	488	496	504
	511	519	526	534	542	549	557	565	572	580
	587	595	603	610	618	626	633	641	648	656
	664	671	679	686	694	702	709	717	724	732
	740	747	755	762	770	778	785	793	800	808
	815	823	831	838	846	853	861	868	876	884
	891	899	906	914	921	929	937	944	952	959
	967	974	982	989	997	*005	*012	*020	*027	*035
	76 042	050	057	065	072	080	087	095	103	110
	118	125	133	140	148	155	163	170	178	185
	193	200	208	215	223	230	238	245	253	260
579	268	275	283	290	298	305	313	320	328	335
	343	350	358	365	373	380	388	395	403	410
	418	425	433	440	448	455	462	470	477	485
	492	500	507	515	522	530	537	545	552	559
	567	574	582	589	597	604	612	619	626	634
	641	649	656	664	671	678	686	693	701	708
	716	723	730	738	745	753	760	768	775	782
	790	797	805	812	819	827	834	842	849	856
	864	871	879	886	893	901	908	916	923	930
	938	945	953	960	967	975	982	989	997	*004
	77 012	019	026	034	041	048	056	063	070	078
	085	093	100	107	115	122	129	137	144	151
	159	166	173	181	188	195	203	210	217	225
	232	240	247	254	262	269	276	283	291	298
	305	313	320	327	335	342	349	357	364	371
	379	386	393	401	408	415	422	430	437	444
	452	459	466	474	481	488	495	503	510	517
	525	532	539	546	554	561	568	576	583	590
	597	605	612	619	627	634	641	648	656	663
	670	677	685	692	699	706	714	721	728	735
	743	750	757	764	772	779	786	793	801	808
	815	822	830	837	844	851	859	866	873	880
		1	2	3	4	5	6	7	8	9

P. P.

8

1	0.8
2	1.6
3	2.4
4	3.2
5	4.0
6	4.8
7	5.6
8	6.4
9	7.2

7

1	0.7
2	1.4
3	2.1
4	2.8
5	3.5
6	4.2
7	4.9
8	5.6
9	6.3

N.	L. 0	1	2	3	4	5	6	7	8	9
600	77 815	822	830	837	844	851	859	866	873	880
601	887	895	902	909	916	924	931	938	945	952
602	960	967	974	981	988	996	*003	*010	*017	*025
603	78 032	039	046	053	061	068	075	082	089	097
604	104	111	118	125	132	140	147	154	161	168
605	176	183	190	197	204	211	219	226	233	240
606	247	254	262	269	276	283	290	297	305	312
607	319	326	333	340	347	355	362	369	376	383
608	390	398	405	412	419	426	433	440	447	455
609	462	469	476	483	490	497	504	512	519	526
610	533	540	547	554	561	569	576	583	590	597
611	604	611	618	625	633	640	647	654	661	668
612	675	682	689	696	704	711	718	725	732	739
613	746	753	760	767	774	781	789	796	803	810
614	817	824	831	838	845	852	859	866	873	880
615	888	895	902	909	916	923	930	937	944	951
616	958	965	972	979	986	993	*000	*007	*014	*021
617	79 029	036	043	050	057	064	071	078	085	092
618	099	106	113	120	127	134	141	148	155	162
619	169	176	183	190	197	204	211	218	225	232
620	239	246	253	260	267	274	281	288	295	302
621	309	316	323	330	337	344	351	358	365	372
622	379	386	393	400	407	414	421	428	435	442
623	449	456	463	470	477	484	491	498	505	511
624	518	525	532	539	546	553	560	567	574	581
625	588	595	602	609	616	623	630	637	644	650
626	657	664	671	678	685	692	699	706	713	720
627	727	734	741	748	754	761	768	775	782	789
628	796	803	810	817	824	831	837	844	851	858
629	865	872	879	886	893	900	906	913	920	927
630	934	941	948	955	962	969	975	982	989	996
631	80 003	010	017	024	030	037	044	051	058	065
632	072	079	085	092	099	106	113	120	127	134
633	140	147	154	161	168	175	182	188	195	202
634	209	216	223	229	236	243	250	257	264	271
635	277	284	291	298	305	312	318	325	332	339
636	346	353	359	366	373	380	387	393	400	407
637	414	421	428	434	441	448	455	462	468	475
638	482	489	496	502	509	516	523	530	536	543
639	550	557	564	570	577	584	591	598	604	611
640	618	625	632	638	645	652	659	665	672	679
641	686	693	699	706	713	720	726	733	740	747
642	754	760	767	774	781	787	794	801	808	814
643	821	828	835	841	848	855	862	868	875	882
644	889	895	902	909	916	922	929	936	943	949
645	956	963	969	976	983	990	996	*003	*010	*017
646	81 023	030	037	043	050	057	064	070	077	084
647	090	097	104	111	117	124	131	137	144	151
648	158	164	171	178	184	191	198	204	211	218
649	224	231	238	245	251	258	265	271	278	285
650	291	298	305	311	318	325	331	338	345	351
N.	L. 0	1	2	3	4	5	6	7	8	9

P. P.

	8
1	0.8
2	1.6
3	2.4
4	3.2
5	4.0
6	4.8
7	5.6
8	6.4
9	7.2

	7
1	0.7
2	1.4
3	2.1
4	2.8
5	3.5
6	4.2
7	4.9
8	5.6
9	6.3

	6
1	0.6
2	1.2
3	1.8
4	2.4
5	3.0
6	3.6
7	4.2
8	4.8
9	5.4

	1	2	3	4	5	6	7	8
81 291	298	305	311	318		331	338	345
	365	371	378	385		398	405	411
	431	438	445	451		465	471	478
	498	505	511	518		531	538	544
	564	571	578	584		598	604	611
	631	637	644	651		664	671	677
	697	704	710	717		730	737	743
	763	770	776	783		796	803	809
	829	836	842	849		862	869	875
	895	902	908	915		928	935	941
954	961	968	974	981		994	*000	*007
82 020	027	033	040	046		060	066	073
086	092	099	105	112		125	132	138
151	158	164	171	178		191	197	204
217	223	230	236	243		256	263	269
282	289	295	302	308		321	328	334
347	354	360	367	373		387	393	400
413	419	426	432	439		452	458	465
478	484	491	497	504		517	523	530
543	549	556	562	569		582	588	595
	614	620	627	633		646	653	659
	679	685	692	698		711	718	724
	743	750	756	763		776	782	789
	808	814	821	827	834	840	847	853
	872	879	885	892	898	905	911	918
	937	943	950	956	963	969	975	982
	*001	*008	*014	*020	*027	*033	*040	*046
83 059	065	072	078	085	091	097	104	110
	129	136	142	149	155	161	168	174
	193	200	206	213	219	225	232	238
251	257	264	270	276	283	289	296	302
	321	327	334	340	347	353	359	366
	385	391	398	404	410	417	423	429
	448	455	461	467	474	480	487	493
	512	518	525	531	537	544	550	556
	575	582	588	594	601	607	613	620
	639	645	651	658	664	670	677	683
	702	708	715	721	727	734	740	746
	765	771	778	784	790	797	803	809
	828	835	841	847	853	860	866	872
	891	897	904	910	916	923	929	935
	954	960	967	973	979	985	992	998
84 011	017	023	029	036	042	048	055	061
	080	086	092	098	105	111	117	123
	142	148	155	161	167	173	180	186
	205	211	217	223	230	236	242	248
	267	273	280	286	292	298	305	311
	330	336	342	348	354	361	367	373
	392	398	404	410	417	423	429	435
	454	460	466	473	479	485	491	497
	516	522	528	535	541	547	553	559

P. P.

7

1	0.7
2	1.4
3	2.1
4	2.8
5	3.5
6	4.2
7	4.9
8	5.6
9	6.3

6

1	0.6
2	1.2
3	1.8
4	2.4
5	3.0
6	3.6
7	4.2
8	4.8
9	5.4

7	
1	0.7
2	1.4
3	2.1
4	2.8
5	3.5
6	4.2
7	4.9
8	5.6
9	6.3

6	
1	0.6
2	1.2
3	1.8
4	2.4
5	3.0
6	3.6
7	4.2
8	4.8
9	5.4

5	
1	0.5
2	1.0
3	1.5
4	2.0
5	2.5
6	3.0
7	3.5
8	4.0
9	4.5

N.	L. 0	1	2	3	4	5	6	7	8	9	P. P.
750	87 506	512	518	523	529	535	541	547	552	558	
751	564	570	576	581	587	593	599	604	610	616	
752	622	628	633	639	645	651	656	662	668	674	
753	679	685	691	697	703	708	714	720	726	731	
754	737	743	749	754	760	766	772	777	783	789	
755	795	800	806	812	818	823	829	835	841	846	
756	852	858	864	869	875	881	887	892	898	904	
757	910	915	921	927	933	938	944	950	955	961	
758	967	973	978	984	990	996	*001	*007	*013	*018	
759	88 024	030	036	041	047	053	058	064	070	076	
760	081	087	093	098	104	110	116	121	127	133	
761	138	144	150	156	161	167	173	178	184	190	**6**
762	195	201	207	213	218	224	230	235	241	247	1 0.6
763	252	258	264	270	275	281	287	292	298	304	2 1.2
764	309	315	321	326	332	338	343	349	355	360	3 1.8
765	366	372	377	383	389	395	400	406	412	417	4 2.4
766	423	429	434	440	446	451	457	463	468	474	5 3.0
767	480	485	491	497	502	508	513	519	525	530	6 3.6
768	536	542	547	553	559	564	570	576	581	587	7 4.2
769	593	598	604	610	615	621	627	632	638	643	8 4.8 9 5.4
770	649	655	660	666	672	677	683	689	694	700	
771	705	711	717	722	728	734	739	745	750	756	
772	762	767	773	779	784	790	795	801	807	812	
773	818	824	829	835	840	846	852	857	863	868	
774	874	880	885	891	897	902	908	913	919	925	
775	930	936	941	947	953	958	964	969	975	981	
776	986	992	997	*003	*009	*014	*020	*025	*031	*037	
777	89 042	048	053	059	064	070	076	081	087	092	
778	098	104	109	115	120	126	131	137	143	148	
779	154	159	165	170	176	182	187	193	198	204	
780	209	215	221	226	232	237	243	248	254	260	
781	265	271	276	282	287	293	298	304	310	315	**5**
782	321	326	332	337	343	348	354	360	365	371	1 0.5
783	376	382	387	393	398	404	409	415	421	426	2 1.0
784	432	437	443	448	454	459	465	470	476	481	3 1.5
785	487	492	498	504	509	515	520	526	531	537	4 2.0
786	542	548	553	559	564	570	575	581	586	592	5 2.5
787	597	603	609	614	620	625	631	636	642	647	6 3.0
788	653	658	664	669	675	680	686	691	697	702	7 3.5
789	708	713	719	724	730	735	741	746	752	757	8 4.0 9 4.5
790	763	768	774	779	785	790	796	801	807	812	
791	818	823	829	834	840	845	851	856	862	867	
792	873	878	883	889	894	900	905	911	916	922	
793	927	933	938	944	949	955	960	966	971	977	
794	982	988	993	998	*004	*009	*015	*020	*026	*031	
795	90 037	042	048	053	059	064	069	075	080	086	
796	091	097	102	108	113	119	124	129	135	140	
797	146	151	157	162	168	173	179	184	189	195	
798	200	206	211	217	222	227	233	238	244	249	
799	255	260	266	271	276	282	287	293	298	304	
800	309	314	320	325	331	336	342	347	352	358	
N.	L. 0	1	2	3	4	5	6	7	8	9	P. P.

L.	0	1	2	3	4	5	6	7	8	9
				325	331	336	342	347	352	358
				380	385	390	396	401	407	412
				434	439	445	450	455	461	466
				488	493	499	504	509	515	520
				542	547	553	558	563	569	574
				596	601	607	612	617	623	628
				650	655	660	666	671	677	682
				703	709	714	720	725	730	736
				757	763	768	773	779	784	789
				811	816	822	827	832	838	843
				865	870	875	881	886	891	897
				918	924	929	934	940	945	950
				972	977	982	988	993	998	*004
				025	030	036	041	046	052	057
				078	084	089	094	100	105	110
				132	137	142	148	153	158	164
				185	190	196	201	206	212	217
				238	243	249	254	259	265	270
				291	297	302	307	312	318	323
				344	350	355	360	365	371	376
	381	387	392	397	403	408	413	418	424	429
	434	440	445	450	455	461	466	471	477	482
	487	492	498	503	508	514	519	524	529	535
	540	545	551	556	561	566	572	577	582	587
	593	598	603	609	614	619	624	630	635	640
	645	651	656	661	666	672	677	682	687	693
	698	703	709	714	719	724	730	735	740	745
	751	756	761	766	772	777	782	787	793	798
	803	808	814	819	824	829	834	840	845	850
	855	861	866	871	876	882	887	892	897	903
	908	913	918	924	929	934	939	944	950	955
	960	965	971	976	981	986	991	997	*002	*007
92	012	018	023	028	033	038	044	049	054	059
	065	070	075	080	085	091	096	101	106	111
	117	122	127	132	137	143	148	153	158	163
	169	174	179	184	189	195	200	205	210	215
	221	226	231	236	241	247	252	257	262	267
	273	278	283	288	293	298	304	309	314	319
	324	330	335	340	345	350	355	361	366	371
	376	381	387	392	397	402	407	412	418	423
	428	433	438	443	449	454	459	464	469	474
	480	485	490	495	500	505	511	516	521	526
	531	536	542	547	552	557	562	567	572	578
	583	588	593	598	603	609	614	619	624	629
	634	639	645	650	655	660	665	670	675	681
	686	691	696	701	706	711	716	722	727	732
	737	742	747	752	758	763	768	773	778	783
	788	793	799	804	809	814	819	824	829	834
	840	845	850	855	860	865	870	875	881	886
	891	896	901	906	911	916	921	927	932	937
	942	947	952	957	962	967	973	978	983	988

Proportional parts:

	6
1	0.6
2	1.2
3	1.8
4	2.4
5	3.0
6	3.6
7	4.2
8	4.8
9	5.4

	8
1	0.5
2	1.0
3	1.5
4	2.0
5	2.5
6	3.0
7	3.5
8	4.0
9	4.5

P. P.

N	0	1	2	3	6	7	8	9
	92 942	947	952					
	998	*003						
	93 044	049	054					
	100	105						
	151	156						
	202	207						
	252	258						
	303	308						
	354	359						
	404	409						
	455	460						
	505	510						
	556	561						
	606	611						
	656	661						
	707	712						
	757	762						
	807	812						
	857	862						
	907	912			932	937		
	957	962			982	987	992	997
	007	012			032	037	042	047
	057	062			082	086	091	096
	106	111			131	136	141	146
874	151	156	161		181	186	191	196
875	201	206	211		231	236		
876	250	255	260		280	285		
877	300	305	310		330	335		
878	349	354	359		379	384		
879	399	404	409		429	433		
880	448	453	458	463	478	483		
881	498	503	507	512	527	532		
882	547	552	557	562	576	581		
883	596	601	606	611	626	630		
884	645	650	655	660	675	680		
885	694	699	704	709	724	729		
886	743	748	753	758	773	778		
887	792	797	802	807	822	827		
888	841	846	851	856	871	876		
889	890	895	900	905	919	924		
890	939	944	949	954	968	973		
891	988	993	998	*002	*017	*022		
892	95 036	041	046	051	066	071		
893	085	090	095	100	114	119		
894	134	139	143	148	163	168		
895	182	187	192	197	211	216		
896	231	236	240	245	260	265		
897	279	284	289	294	308	313		
898	328	332	337	342	357	361		
899	376	381	386	390	405	410		
900	424	429	434	439	453	458		
		1	2	3	6	7	8	

P. P.

	6
1	0.6
2	1.2
3	1.8
4	2.4
5	3.0
6	3.6
7	4.2
8	4.8
9	5.4

	5
1	0.5
2	1.0
3	1.5
4	2.0
5	2.5
6	3.0
7	3.5
8	4.0
9	4.5

	4
1	0.4
2	0.8
3	1.2
4	1.6
5	2.0
6	2.4
7	2.8
8	3.2
9	3.6

P. P.

LOGARITHMS.

L.	0	1	2	5	6	7	8	9
95	424	429	434	448	453	458	463	468
	472	477	482	497	501	506	511	516
	521	525	530	545	550	554	559	564
	569	574	578	593	598	602	607	612
	617	622	626	641	646	650	655	660
	665	670	674	689	694	698	703	708
	713	718	722	737	742	746	751	756
	761	766	770	785	789	794	799	804
	809	813	818	832	837	842	847	852
	856	861	866	880	885	890	895	899
	904	909	914	928	933	938	942	947
	952	957	961	976	980	985	990	995
	999	*004	*009	*023	*028	*033	*038	*042
96	047	052	057	071	076	080	085	090
	095	099	104	118	123	128	133	137
	142	147	152	166	171	175	180	185
	190	194	199	213	218	223	227	232
	237	242	246	261	265	270	275	280
	284	289	294	308	313	317	322	327
	332	336	341	355	360	365	369	374
	379	384	388	402	407	412	417	421
	426	431	435	450	454	459	464	468
	473	478	483	497	501	506	511	515
	520	525	530	544	548	553	558	562
	567	572	577	591	595	600	605	609
	614	619	624	638	642	647	652	656
	661	666	670	685	689	694	699	703
	708	713	717	731	736	741	745	750
	755	759	764	778	783	788	792	797
	802	806	811	825	830	834	839	844
	818	853	858	872	876	881	886	890
		900	904	918	923	928	932	937
		946	951	965	970	974	979	984
		993	997	*011	*016	*021	*025	*030
		039	044	058	063	067	072	077
		086	090	104	109	114	118	123
		132	137	151	155	160	165	169
		179	183	197	202	206	211	216
		225	230	243	248	253	257	262
		271	276	290	294	299	304	308
		317	322	336	340	345	350	354
		364	368	382	387	391	396	400
		410	414	428	433	437	442	447
		456	460	474	479	483	488	493
		502	506	520	525	529	534	539
		548	552	566	571	575	580	585
		594	598	612	617	621	626	630
		640	644	658	663	667	672	676
		685	690	704	708	713	717	722
		731	736	749	754	759	763	768
		777	782	795	800	804	809	813

Footer columns: 5 | 6 | 7 | 8 | 9 | P. P.

P. P.

5

1	0.5
2	1.0
3	1.5
4	2.0
5	2.5
6	3.0
7	3.5
8	4.0
9	4.5

4

1	0.4
2	0.8
3	1.2
4	1.6
5	2.0
6	2.4
7	2.8
8	3.2
9	3.6

LOGARITHMS.

804	809
850	855
896	900
941	946
987	991
032	037
078	082
123	127
168	173
214	218
259	263
304	308
349	354
394	399
439	444
484	489
529	534
574	579
619	623
664	668
709	713
753	758
798	802
843	847
887	892
932	936
976	981
*021	*025
065	069
109	114
154	158

967
968
969
970
971
972
973
974
975
976
977
978
979
980
981
982
983
984
985
986

	5
1	0.5
2	1.0
3	1.5
4	2.0
5	2.5
6	3.0
7	3.5
8	4.0
9	4.5

	4
1	0.4
2	0.8
3	1.2
4	1.6
5	2.0
6	2.4
7	2.8
8	3.2
9	3.6

TABLE OF USEFUL NUMBERS.

	Symbol.	Numerical Value.	Logarithm.
Diagonal of square whose side is 1	$\sqrt{2}$	1.4142	0.15052
Side of square whose diagonal is 1	$\frac{1}{2}\sqrt{2}$	0.70710	$\bar{1}$.84948
Diagonal of cube whose edge is 1	$\sqrt{3}$	1.7321	0.23856
Edge of cube whose diagonal is 1	$\frac{1}{3}\sqrt{3}$	0.57735	$\bar{1}$.76144
Circumference of circle whose diameter is 1	π	3.1416	0.49715
Diameter of circle whose circumference is 1	$\frac{1}{\pi}$	0.31831	$\bar{1}$.50285
Area of circle whose diameter is 1	$\frac{1}{4}\pi$	0.7854	$\bar{1}$.89509
Volume of sphere whose diameter is 1 . . .	$\frac{1}{6}\pi$	0.5236	$\bar{1}$.71900
One meter in yards		1.0936	0.03887
One yard in meters		0.91439	$\bar{1}$.96113
One centimeter in inches		0.39370	$\bar{1}$.59517
One inch in centimeters		2.5400	0.40483
One square meter in square feet		10.764	1.03198
One square foot in square meters		0.092901	$\bar{2}$.96802
One cubic centimeter in cubic inches . . .		0.061025	$\bar{2}$.78551
One cubic inch in cubic centimeters		16.387	1.21449
One liter in gallons		0.26418	$\bar{1}$.42190
One gallon in liters		3.7853	0.57810
One kilogram in pounds		2.2046	0.34333
One pound in kilograms		0.45359	$\bar{1}$.65667

Made in the USA
Monee, IL
07 July 2026

56552384R00187